24 0455521 2

ITEP Lectures on

Imperial College &
Science Museum Libraries

London SW7 2AZ Issue Desk Tel. 020-7594 8810

LONG LOAN

To be returned by the last date stamped (normally one term),
or after *three weeks* if the book is reserved by another reader.
A fine may be charged if the book is returned late.

Class Number 53 I 12 SHI

World Scientific Lecture Notes in Physics

Published

Vol. 47: Some Elementary Gauge Theory Concepts
H M Chan and S T Tsou

Vol. 48: Electrodynamics of High Temperature Superconductors
A M Portis

Vol. 49: Field Theory, Disorder and Simulations
G Parisi

Vol. 50: The Quark Structure of Matter
M Jacob

Vol. 51: Selected Topics on the General Properties of Quantum Field Theory
F Strocchi

Vol. 52: Field Theory: A Path Integral Approach
A Das

Vol. 53: Introduction to Nonlinear Dynamics for Physicists
H D I Abarbanel, et al.

Vol. 54: Introduction to the Theory of Spin Glasses and Neural Networks
V Dotsenko

Vol. 55: Lectures in Particle Physics
D Green

Vol. 56: Chaos and Gauge Field Theory
T S Biro, et al.

Vol. 57: Foundations of Quantum Chromodynamics (2nd edn.)
T Muta

Vol. 59: Lattice Gauge Theories: An Introduction (2nd edn.)
H J Rothe

Vol. 60: Massive Neutrinos in Physics and Astrophysics
R N Mohapatra and P B Pal

Vol. 61: Modern Differential Geometry for Physicists (2nd edn.)
C J Isham

World Scientific Lecture Notes in Physics – Vol. 62

ITEP Lectures on

PARTICLE PHYSICS AND FIELD THEORY

Imperial College
London

Published by

World Scientific Publishing Co. Pte. Ltd.

P O Box 128, Farrer Road, Singapore 912805

USA office: Suite 1B, 1060 Main Street, River Edge, NJ 07661

UK office: 57 Shelton Street, Covent Garden, London WC2H 9HE

British Library Cataloguing-in-Publication Data
A catalogue record for this book is available from the British Library.

The cover illustration by Sergey Yurov depicts the Institute of Theoretical and Experimental Physics (ITEP) in Moscow. The concept of the cover illustration by M. A. Shifman.

ITEP LECTURES ON PARTICLE PHYSICS AND FIELD THEORY

Copyright © 1999 by World Scientific Publishing Co. Pte. Ltd.

All rights reserved. This book, or parts thereof, may not be reproduced in any form or by any means, electronic or mechanical, including photocopying, recording or any information storage and retrieval system now known or to be invented, without written permission from the Publisher.

For photocopying of material in this volume, please pay a copying fee through the Copyright Clearance Center, Inc., 222 Rosewood Drive, Danvers, MA 01923, USA. In this case permission to photocopy is not required from the publisher.

ISBN 981-02-2639-X (set)
ISBN 981-02-3947-5 (Vol. I)
ISBN 981-02-3949-1 (Vol. II)

ISBN 981-02-2640-3 (pbk) (set)
ISBN 981-02-3948-3 (pbk) (Vol. I)
ISBN 981-02-3950-5 (pbk) (Vol. II)

Printed in Singapore by Uto-Print.

FOREWORD

My career in theoretical high energy physics began 25 years ago. I was lucky — its beginning coincided with a very exciting time in this field, when a major breakthrough in our understanding of nature took place. The Standard Model of fundamental interactions was born, and Quantum Chromodynamics (QCD) gradually emerged as *the* theory of hadronic matter. Unlike many other theories created later whose relevance to nature is still a big question mark, these two will definitely stay with us forever. And, fortunately, I happened to be in the right place at the right time so that I could appreciate these discoveries early. The infancy of any true theory creates great opportunities for young researchers who suddenly find themselves pioneers in *terra incognita*, with so many interesting, important and challenging problems around.

For about twenty years, I was a member of the ITEP theory group.[1] During those years I gave many lectures at different schools of physics. Some of these are presented below. As a matter of fact, a few of these lectures are recent, given after I left ITEP. Still, I consider them ITEP lectures since I acquired many of the ideas found in them from my formal and informal teachers at ITEP. I am especially grateful to B. Ioffe, A. Vainshtein and V. Zakharov.

Since the ITEP theory group of the seventies and the eighties was quite an endemic phenomenon, it seems worthwhile to introduce it to the reader before I proceed to the scientific issues.

Glimpses of ITEP

ITEP was more than an institute. It was our refuge where the insanity of the surrounding reality was, if not eliminated, reduced to a bearable level. Doing physics there was something which gave a meaning to our lives, making it interesting and even happy. Our theory group was like a large family. As in any big family, of course, this did not mean that everybody loved everybody else, but we knew that we had to stay together and to rely on each other, no matter what, in order to survive and to be able to continue doing physics. This was considered by our teachers to be the most important thing, and this message was always being conveyed, in more than one way, to young people joining the group. We had a wonderful feeling of stability in our small brotherhood. A feeling so rare in the western laboratories where a whirlpool of postdocs,

[1]For those who do not know: ITEP is an abbreviation for Institute of Theoretical and Experimental Physics, in Moscow.

visitors, sabbatical years come and go, there are a lot of new faces, and a lot of people whom you do not care so much about.

The rules of survival were quite strict. First, seminars — what is now known worldwide as the famous Russian-style seminars. The primary goal of the speaker was to *explain* to the audience his or her results, not merely to advertise them. And if the results were nontrivial, or questionable, or just unclear points would surface in the course of the seminar, the standard two hours were not enough to wind up. Then the seminar could last for three or even four hours, until either everything was clear or complete exhaustion, whichever came first. I remember one seminar in Leningrad in 1979, when Gribov was still there, which started at eleven in the morning. A lunch break was announced from two to three, and then it continued from three till seven in the evening.

In ITEP we had three, sometimes more, theoretical seminars a week. The most important were a formal seminar on Mondays, and an informal coffee seminar which at first took place every Friday at 5 o'clock, when the official work day was over, but later was shifted to Thursdays, at the same time. Usually, these were by far the most exciting events of the week. The leaders and the secretaries of the seminars were supposed to find exciting topics, either by recruiting ITEP or other "domestic" authors, or, quite often, by picking up a paper or a preprint from the outside world and asking somebody to learn and report the work to the general audience. This duty was considered to be a moral obligation. The tradition dated back to the time when Pomeranchuk was the head of the theory group, and its isolation had been even more severe than during my time. As a matter of fact, in those days there were no preprints, and getting fresh issues of *Physical Review* or *Nuclear Physics* was not taken for granted at all. When I, as a student, joined the group — this was a few years after Pomeranchuk's death — I was taken, with pride, to the Pomeranchuk memorial library, his former office where a collection of his books and journals was kept. Every paper, in every issue, was marked by Chuk's hand (that's how his students and colleagues would refer to him), either with a minus or a plus sign. If there was a plus, there would also be the name of one of his students who had been asked to "dig into" the paper and give a talk for everyone's benefit. This was not the end of the story, however. Before the scheduled day of the seminar, Pomeranchuk would summon the speaker-to-be to his office to give a pre-talk to him alone, so that he could judge whether the subject had been worked out with sufficient depth and that the speaker was "ripe enough" to face the general audience and their bloodthirsty questions. In my time, the secretaries of the seminars were less inclined to sacrifice themselves to that extent, but, still, it was not uncommon that pre-talks were arranged for unknown, young or inexperienced speakers.

Scientific reports of the few chosen to travel abroad for a conference or just to collaborate for a while with western physicists, were an unquestionable element of the seminar routine. The attendance of an international conference by A or B by no means was considered as a personal matter of A and B alone. Rather, these rare lucky guys were believed to be our ambassadors, and were supposed to represent the whole group. In practical terms, this meant that once you had made your way to a conference, you could be asked to present important results of other members of the group. Moreover, you were supposed to attend as many talks as physically possible, including those which did not exactly belong to your field, make extensive notes and then, after returning home, deliver an exhaustive report of all new developments discussed, all interesting questions raised, rumors, etc.

The scientific rumors, as well as nonscientific impressions, were like an exotic dessert, usually served after nine. I remember that, after his first visit to the Netherlands, Simonov mentioned that he was very surprised to see a lot of people on the streets just smiling. He said he could not understand why they looked so relaxed. Then he added that he finally figured out why: "... because they were not concerned with building communism..." This remark almost immediately became known to "Big Brother" who was obviously watching us this evening, as usual, and it cost Simonov a few years of sudden "unexplainable allergy" to any western exposure. His "health condition", of course, would not allow him to accept any invitation to travel there. I cannot help mentioning another curious episode with Big Brother. Coffee, which we used to have during the coffee seminars, was prepared in turn, by all members of the group. Once, when it was Ioffe's turn, he brought a small bottle of cognac and added a droplet or two in every cup. I do not remember why, perhaps, it was his birthday or something like that. That was Friday evening. Very early on the next Monday morning, he was summoned to the corresponding ITEP branch office to give explanations concerning his "obviously subversive activities"!

The coffee seminars typically lasted until nine, sometimes much later, for instance, in the stormy days of the November revolution in 1974. The few months following the discovery of J/ψ were the star days of QCD and, probably, the highest emotional peak of the ITEP theory group. Never were the mysteries of physics taken so close to our hearts as then. There was a spontaneously arranged team of enthusiasts working practically nonstop. A limit to our discussions was set only by the schedule of the Moscow metro — those who needed to catch the last train had to be leaving before 1 a.m.

The ITEP seminars were certainly one of the key elements in shaping the principles and ideals of our small community, but not the only one. The process of selecting students who could eventually grow up into particle theorists played a crucial role and was, probably, as elaborate as the process of becoming a knight of the British crown. Every year we had about 20 new students, at the level roughly corresponding to that of graduate students in American universities. They came mostly from the Moscow Institute for Physics and Technology, a small elite institution near the city, a counterpart of MIT in the States. Some students were from the Moscow Engineering and Physics Institute, and a few from the Moscow State University. They were offered (actually, obliged to take) such a spectrum of courses in special disciplines which I have never heard of anywhere else in the world: everything from radiophysics and accelerator physics; several levels of topics in quantum mechanics, including intricacies of theory of scattering; radiation theory and nuclear physics; mathematical physics (consisting of several separate parts); not less than three courses in particle phenomenology (weak, electromagnetic and strong interactions); quantum electrodynamics, numerous problem-solving sessions, etc. And yet, only those who successfully passed additional examinations, covering the famous course of theoretical physics by Landau and Lifshitz, were allowed, after showing broad erudition and ingenuity in solving all sorts of tricky problems, to join the theory group. Others were supposed to end up as experimentalists or engineers. Needless to say, the process of passing these examinations could take months, even years, and was notoriously exhausting, but there was never a lack of volunteers trying their luck. They were always seen around Ter-Martirosian and Okun who were sort of responsible for the program. It should be added that the set of values to be passed from the elders to the young generations included the idea that high energy physics is an experimental science that *must* be very closely related to phenomena taking place in nature. Only those theoretical ideas which, at the end of the day, could produce a number which could be confronted with phenomenology were cherished. Too abstract and speculative constructions, and theoretical phantoms, were not encouraged, to put it mildly. The atmosphere was strongly polarized against what is now sometimes called "theoretical theory". Even extremely bright students, who were too mathematically oriented, like, say, Vadim Knizhnik, were having problems in passing these examinations. Vadim, by the way, never made it to the end, got upset and left ITEP. Well, nothing is perfect in this world, and I do not want to make an impression that the examination routine in the ITEP theory group was without flaws.

The ITEP theory group was large — about 50 theorists — and diverse. Moreover, it was a natural center of attraction for the whole Moscow particle physics community. Living in the capital of the last world empire had its advantages. There is no question, it was the evil empire, but what was good, as it usually happens with any empire, all intellectual forces tended to cluster in the capital. So, we had a very dynamic group where virtually every direction was represented by at least several theorists, experts in the given field. If you needed to learn something new, there was an easy way to do it, much faster and more efficient than through reading journals or textbooks. You just needed to talk to the right person. Educating others, sharing your knowledge and expertise with everybody who might be interested, was another rule of survival in our isolated community. In such an environment, different discussion groups and large collaborations were naturally emerging all the time, creating a strong and positive coherent effect. The brain-storming sessions used to produce, among other results, a lot of noise, so once you were inside the old mansion occupied by the theorists, it was very easy to figure out which task force was where — just step out in the corridor and listen. And, certainly, all these sessions were open to everybody.

Now I would like to mention one more aspect which concerns me at present, a very strong pressure existing in our community, to stay in the "mainstream", to work only on fashionable directions and problems which, currently, are under investigation in dozens of other laboratories. This pressure is especially damaging for young people who have little alternative. Of course, a certain amount of cohesion is needed, but the scale of the phenomenon we are witnessing now is unhealthy, beyond any doubt. The isolation of the ITEP theory group had a positive side effect. Everybody, including the youngest members, could afford to work on problems not belonging to the fashion of the day, without publishing a single line for a year or two. Who cared about what we were doing there anyway? This was okay. On the other hand, it was considered indecent to publish results of dubious novelty, incomplete results (of the status report type) or just papers with too many words per given number of formulae. Producing dense papers was a norm. This style, which was probably perceived by the outside readers as a chain of riddles, is partly explained by tradition, presumably dating back to the Landau times. It was also due to specific Soviet conditions, where everything was regulated, including the maximal number of pages any given paper could have. Compressing derivations and arguments to the level considered acceptable, was an art which had its grandmasters. Arkady Vainshtein was especially good at inventing all sorts of tricks which allowed him to squeeze in extra formulae with very few explanatory remarks. I remember that in 1976, when we were working on the large JETP paper on penguins in weak decays,[2] we had to make 30 pages out of the original 60-page preprint version, and he managed to do that without losing any equations and even inserting a few extra ones! This left a strong impression on me.

By the way, about penguins. From time to time students ask how this word could possibly penetrate high energy physics. This is a funny story indeed. The first paper where the graphs that are now called penguins were considered in the weak decays appeared[3] in JETP Letters in 1975, and there they did not look like penguins at all. Later on they were made to look like penguins:

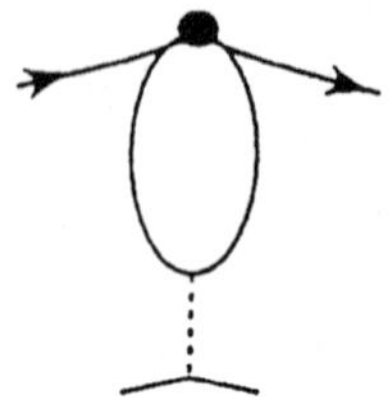

and called penguins by John Ellis. Here is his story as he recollects it himself.

"Mary K. [Gaillard], Dimitri [Nanopoulos] and I first got interested in what are now called penguin diagrams while we were studying CP violation in the Standard Model in 1976... The penguin name came in 1977, as follows.

In the spring of 1977, Mike Chanowitz, Mary K and I wrote a paper on GUTs predicting the b quark mass before it was found. When it was found a few weeks later, Mary K, Dimitri, Serge Rudaz and I immediately started working on its phenomenology. That summer, there was a student at CERN, Melissa Franklin who is now an experimentalist at Harvard. One evening, she, I and Serge went to a pub, and she and I started a game of darts. We made a bet that if I lost I had to put the word penguin into my next paper. She

[2]By "we" I mean Zakharov, Vainshtein and myself. Arkady Vainshtein had a permanent position at the Budker Institute of Nuclear Physics in Novosibirk. He commuted between Moscow and Novosibirsk for many years, and was considered, essentially, as a member of the ITEP theory group.

[3]A. Vainshtein, V. Zakharov and M. Shifman, *Pis'ma ZhETF* **22** (1975) 123 [*JETP Lett.* **22** (1975) 55].

actually left the darts game before the end, and was replaced by Serge, who beat me. Nevertheless, I felt obligated to carry out the conditions of the bet.

For some time, it was not clear to me how to get the word into this b quark paper that we were writing at the time. Then, one evening, after working at CERN, I stopped on my way back to my apartment to visit some friends living in Meyrin where I smoked some illegal substance. Later, when I got back to my apartment and continued working on our paper, I had a sudden flash that the famous diagrams look like penguins. So we put the name into our paper, and the rest, as they say, is history."

About These Lectures

Below the reader will find lectures devoted to different topics in theoretical high energy physics which occupied me during the last 15 years. The choice of topics might seem somewhat chaotic, at first sight. The selection criteria were simple: I tried to select only those topics which are of interest today. Besides, I was limited by the fact that not all the lecture notes are available; some have never been published and have disappeared with time.

These are *lectures*, not reviews; they were intended for beginners — mostly graduate students — who were just about to submerge into the subject and needed some initial impetus and general idea and guidance. Therefore, the pedagogical element was most important. I made no attempt at complete coverage, the lists of references are usually quite fragmentary and so on. These lectures can be used for the initial exposure. Those who would like to master the corresponding topics in full, will need to proceed to more detailed reviews and the original literature. At the end of each lecture I recommend a few sources for further studies.

The lectures written in the eighties were revised and updated specifically for this volume. I significantly expanded them. At the same time, I had to resist the temptation to completely rewrite them. This process would take too much time, and this volume would never appear had I not settled on a compromise.

October 29, 1995

GENERAL ACKNOWLEDGMENTS

I started working on this book in November 1994. If I had known that the task would be so labor-consuming, I would probably have never ventured in this endeavor. And I would have never finished if it were not for the invaluable contributions, encouragement and assistance of many people. First and foremost, I must thank my co-authors, Victor Novikov, Valya Zakharov and, especially, Arkady Vainshtein. My collaboration with Arkady has lasted for almost 30 years. He was and still is one of my teachers; over the years we have discussed all aspects of high energy physics an infinite number of times. Chapter 6 (Instantons Versus Supersymmetry: Fifteen Years Later) which Arkady and I wrote specifically for this book was the hardest.

I am grateful to my colleagues, the members of the Theoretical Physics Institute of the University of Minnesota. The creative atmosphere and excellent conditions which I have enjoyed at the Institute ever since I joined it in 1990, were absolutely instrumental. Special thanks go to the secretaries, Sally Menefee and Jenny Curtis.

Finally, it is my pleasure to thank the editors and the employees of the World Scientific Publishing Company, for their technical assistance and, most importantly, patience. I would like to thank Ms. Lim Feng Nee, and in the later stages, Ms. Lakshmi Narayan. This work was supported in part by DOE under the grant No. DE-FG02-94ER40823.

Minneapolis, February 4, 1999

CONTENTS

Volume I

Volume II

Chapter I

Lectures on Heavy Quarks in Quantum Chromodynamics

M. A. Shifman

Theoretical Physics Institute, University of Minnesota, Minneapolis, MN 55455, USA

An extended version of the lectures given at Theoretical Advanced Study Institute, *QCD and Beyond*, University of Colorado, Boulder, Colorado, June 1995.

Abstract

A pedagogical introduction to the heavy quark theory is given. It is explained that various expansions in the inverse heavy quark mass $1/m_Q$ present a version of the Wilson operator product expansion in QCD. A systematic approach is developed and many practically interesting problems are considered. I show how the $1/m_Q$ expansions can be built using the background field technique and how they work in particular applications. The interplay between perturbative and nonperturbative aspects of the heavy quark theory is discussed.

Contents

SECTION 1
Heavy Quark Symmetry

The statement that Quantum Chromodynamics (QCD) is *the* theory of hadrons has become commonplace. It is a very strange theory, since many questions concerning dynamics of the quarks and gluons at large distances — however simple they might seem — remain unanswered or, at best, understood only at a qualitative level. Progress in the direction of the quantitative description of the hadronic properties is slow — every step bringing us closer to such a description is painfully difficult. At the same time, new results, albeit modest, have a special weight for obvious reasons — QCD, unlike many other trendy theories in modern high energy physics, definitely has a direct relation to nature and will stay with us forever.

Every hadron in a sense is built from quarks and/or gluons. I say "in a sense" because these are no ordinary building blocks. The number of degrees of freedom fluctuates and is not fixed; this we know for sure. At large distances we have to deal with a genuine strongly coupled field theory, and, as usual, the strong coupling creates complicated structures which cannot be treated by perturbative methods. Then we feel helpless and are ready to use every opportunity, no matter where it comes from, as long as it gives the slightest hope of yielding a solid quantitative approach based on QCD.

QCD has two faces, two components — hard and soft. The hard component is the realm of perturbative QCD. Not much will be said in these lectures about this aspect. Instead, we will concentrate on the soft component. Many years ago, at the dawn of the QCD era, it was noted [1] that heavy quarks are, probably, the best probe of the soft component of the gluon fields out of all probes we have at our disposal. The developments we witnessed in recent years confirm this conclusion.

The dynamics of soft degrees of freedom in QCD is the realm of nonperturbative phenomena. Having said this I hasten to add that there is an element of luck — transition from the perturbative regime to the nonperturbative one is very abrupt in QCD. In a sense the gauge coupling constant is abnormally small. I do not mean here the conventional logarithmic suppression of the running constant but, rather, the fact that b, the first coefficient in the Gell-Mann-Low function, is numerically large. This fact allows us to forget, in the first approximation, about perturbative effects and focus on nonperturbative ones in a wide range of problems. It is more precise to say that we will concentrate on studying the soft degrees of freedom, but due to the fortunate circumstance of "abnormal" smallness of $\alpha_s(\mu)/\pi$ for as low normalization point as $\mu \sim 1$ GeV, all effects due to the soft degrees of freedom are essentially nonperturbative. I will elucidate the precise meaning of this statement later.

It would be great if we could just switch off — by adjusting some parameter — all hard processes in QCD without changing its soft component. Then we would be left with the confining dynamics in a clean and uncontaminated form; formulation of the theory would be much easier. The only parameter which might do the job is b. If we could tend $b \to \infty$ with $\Lambda_{\rm QCD}$ fixed, the hard gluons would be suppressed by powers of $1/b$ while the soft component would presumably remain unaltered or almost unaltered. Unfortunately, nobody knows how to make the enhancement of b parametric. (The limit of the large number of colors, $N_c \to \infty$, does not work since, although b is definitely proportional to N_c in this case, the perturbative expansion for all planar graphs goes in N_c/b, not in $1/b$ [2].) Therefore, we will have to rely on the numerical enhancement of b. In the first lectures I will merely assume that the hard gluon exchanges are nonexistent. Later on, at the very end, we will return to this issue and will briefly discuss the impact of hard gluons.

The purpose of these lectures is mainly pedagogical — the coverage of the topic is neither chronological nor comprehensive. Technically sophisticated issues and calculations are avoided whenever possible; instead I discuss particularly illuminating problems, in a simplified setting. Readers interested in specific advanced applications (e.g. combining the $1/m_Q$ expansions with the chiral perturbation theory [3]) are referred to the original publications and the review papers [4] summarizing a wealth of results obtained in the heavy quark theory after 1990. The presentation of the heavy quark theory below as a rule does not follow the standard pattern and is, rather, complementary with respect to the more traditional reviews [4]. We try to emphasize that the heavy quark theory and the heavy quark expansion is nothing else than a version of the Wilson operator product expansion (OPE) [5], an aspect which usually remains fogged.

1.1. *Why Heavy Quarks?*

The quark–gluon dynamics is governed by the QCD Lagrangian

$$\begin{aligned}\mathcal{L} &= -\frac{1}{4}G^a_{\mu\nu}G^a_{\mu\nu} + \sum_q \bar{q}i\,\not{D}q + \sum_Q \bar{Q}(i\,\not{D} - m_Q)Q \\ &= \mathcal{L}_{\rm light} + \sum_Q \bar{Q}(i\,\not{D} - m_Q)Q \qquad (1.1)\end{aligned}$$

where $G^a_{\mu\nu}$ is the gluon field strength tensor, the light quark fields (u, d and s) are generically denoted by q and are assumed, for simplicity, to be massless while the heavy quark fields are generically denoted by Q. To qualify as a heavy quark Q the corresponding mass term m_Q must be much larger than $\Lambda_{\rm QCD}$. The charmed quark c can be called heavy only with some reservations and, in discussing the heavy quark theory, it is more appropriate to keep in mind b quarks. The hadrons to be considered are composed from one heavy quark Q, a light antiquark $\bar{q}$, or diquark qq, and a gluon cloud which can also contain light quark–antiquark pairs. The role of the cloud is, of course, to keep all these objects together, in a colorless bound state which will be generically denoted by H_Q.

Quite naturally in the heavy quark theory, the gamma matrices used are those of the standard representation,

$$\gamma^0 = \begin{pmatrix} 1 & 0 \\ 0 & -1 \end{pmatrix}, \quad \vec{\gamma} = \begin{pmatrix} 0 & \vec{\sigma} \\ -\vec{\sigma} & 0 \end{pmatrix}, \quad \gamma^5 = \begin{pmatrix} 0 & -1 \\ -1 & 0 \end{pmatrix}. \tag{1.2}$$

With these definitions of the gamma matrices the left-handed spinor has the form $\psi_L = (1+\gamma_5)\psi$.

The light component of H_Q, its light cloud,[1] has a complicated structure — the soft modes of the light fields are strongly coupled and strongly fluctuate. Basically, the only fact which we know for sure is that the light cloud is indeed light; typical frequencies are of the order $\Lambda_{\rm QCD}$. One can try to visualize the light cloud as a soft medium. The heavy quark Q is then submerged in this medium. If the hard gluon exchanges are discarded the momentum which the heavy quark can borrow from the light cloud is of the order $\Lambda_{\rm QCD}$, and the corresponding uncertainty in the energy of the heavy quark is of the order $\Lambda^2_{\rm QCD}/m_Q$. Since these quantities are much smaller than m_Q, this means, in particular, that the heavy quark–antiquark pairs cannot play a role. In other words, the field-theoretic (second-quantized) description of the heavy quark becomes redundant, and under the circumstances, it is perfectly sufficient to treat one single heavy quark Q within quantum mechanics, which is infinitely simpler, of course, than any field theory. Moreover, one can systematically expand in $1/m_Q$. Thus, in the limit $m_Q/\Lambda_{\rm QCD} \to \infty$, the heavy quark component of H_Q becomes easily manageable, allowing one to use the heavy quark

[1] In some papers devoted to the subject the light cloud is referred to as 'brown muck'. I think it is absolutely unfair with respect to the soft components of the quark and gluon fields to call them 'brown muck' only because we are not smart enough to fully understand the corresponding dynamics.

as a probe of the light cloud dynamics. The special advantages of this limit in QCD were first emphasized by Shuryak [6].

1.2. *Descending Down*

In field theory one has to specify the normalization point μ where all operators are defined; in particular, the gauge coupling constant g and the quark mass m_Q are functions of μ. The original QCD Lagrangian (1.1) is formulated at very short distances, or, which is the same, at a high normalization point $\mu = M_0$ where M_0 is the mass of an ultraviolet regulator. In other words, the normalization point is assumed to be much higher than all mass scales in the theory, $\mu \gg m_Q$. Constructing an effective theory intended for the description of the low energy properties of the heavy flavor hadrons, we must evolve the Lagrangian from the original high scale M_0 down to a normalization point μ lying below the heavy quark masses m_Q. By evolving down I mean that we integrate out, step by step, all high frequency modes in the theory, thus calculating the Lagrangian $\mathcal{L}(\mu)$ describing dynamics of the soft modes, with characteristic frequencies less than μ. The hard (high frequency) modes determine the coefficient functions in $\mathcal{L}(\mu)$ while the contribution of the soft modes is hidden in the matrix elements of (an infinite set of) operators appearing in $\mathcal{L}(\mu)$. This approach, which in the context of QCD, was put forward by K. Wilson long ago, has become common. It is widely recognized and exploited in countless applications — from the ancient problem of the K meson decays to fresh trends in the lattice calculations [7]. The peculiarity of the heavy quark theory is due to the fact that the *in* and *out* states we deal with contain heavy quarks. Therefore, although we do integrate out the field fluctuations with the frequencies down to μ, the heavy quark fields themselves are not integrated out, since we will be interested in physics in the sector with the Q charge $\neq 0$. The effective Lagrangian $\mathcal{L}(\mu)$ acts in this sector.

If QCD was solved we could include in our explicit calculation of the effective Lagrangian all modes, descending down to $\mu = 0$. The Lagrangian obtained in this way would be built in terms of the fields of physical mesons and baryons, not in terms of quarks and gluons, since the latter become irrelevant degrees of freedom in the infrared limit $\mu \to 0$. This Lagrangian would give us the full set of all conceivable amplitudes and would, thus, represent the final answer for the theory. There would be no need for any further calculations — one would just pick up the amplitude of interest and compare it with experimental data.

This picture is quite utopian, of course. The real QCD is not solved in the closed form, and in doing explicit calculations of the coefficients in the effective Lagrangian, one cannot put $\mu = 0$. The lower the value of μ the larger part of dynamics is accounted for in the explicit calculation. Therefore, we would like to have μ as low as possible; definitely $\mu \ll m_Q$. The heavy quark can be treated as a nonrelativistic object moving in the soft background field only provided the latter condition is met. On the other hand, to keep theoretical control over the explicit calculations of the coefficient functions, we must stop at some $\mu \gg \Lambda_{\rm QCD}$, so that $\alpha_s(\mu)/\pi$ is still a sufficiently small expansion parameter. In practice this means that the best choice (which we will always stick to) is $\mu \sim$ several units times $\Lambda_{\rm QCD}$. All coefficients in the effective Lagrangian obtained in this way will be functions of μ.

Since μ is an auxiliary parameter predictions for physical quantities must of course be μ independent. The μ dependence of the coefficients must be canceled by that coming from the physical matrix elements of the operators in $\mathcal{L}(\mu)$. However, in calculating in the hard and soft domains (i.e. above μ and below μ), we make different approximations, so that the exact μ independence of the physical quantities can be lost. Since the transition from the hard to soft physics is very steep, one may hope that our predictions will be very insensitive to the precise choice of μ, provided that $\mu \sim$ several units times $\Lambda_{\rm QCD}$. Below, if not stated to the contrary, we will assume that the normalization point μ is chosen in this way.

In descending from M_0 down to μ, the form of the Lagrangian (1.1) changes, and a series of operators of higher dimension appears. It is important that all these operators are Lorentz scalars. For instance, the heavy quark part of the Lagrangian takes the form

$$\mathcal{L}_{\rm heavy} = \sum_Q \left\{ \bar{Q}(i\not{D} - m_Q)Q + \frac{c_G}{2m_Q}\bar{Q}(i/2)\sigma_{\mu\nu}G_{\mu\nu}Q + \sum_{\Gamma,\, q} \frac{d_{Qq}^{(\Gamma)}}{m_Q^2}\bar{Q}\Gamma Q\bar{q}\Gamma q \right\} + \mathcal{O}\left(\frac{1}{m_Q^3}\right) \tag{1.3}$$

where c_G and $d_{Qq}^{(\Gamma)}$ are coefficient functions, $G_{\mu\nu} \equiv gG^a_{\mu\nu}t^a$ and t^a is the color generator, $(\mathrm{Tr}\, t^a t^b = \delta^{ab}/2)$; below we will often use the short-hand notation $i\sigma G = i\sigma_{\mu\nu}G_{\mu\nu} = i\gamma_\mu\gamma_\nu G_{\mu\nu}$. The sum over the light quark flavors is shown explicitly as well as the sum over possible structures Γ of the four-fermion operators. All masses and couplings, as well as the coefficient functions c_G and $d^{(\Gamma)}$, depend on the normalization point. For example, the coefficient c_G in the leading logarithmic approximation can be written as

$$c_G(\mu) = \left(\frac{\alpha_s(\mu)}{\alpha_s(m_Q)}\right)^{-\frac{3}{b}} - 1\,, \quad b = 11 - \frac{2}{3}n_f\,, \tag{1.4}$$

where n_f is the number of the light flavors. The power $-3/b$ was first calculated in Ref. [8]. In Sec. 5.3, I will explain the derivation of Eq. (1.4).

The operators of dimension five and higher in Eq. (1.3) are due to the contribution of hard gluons, with off-shellness from μ up to M_0. Since we agreed that in this section we will ignore the existence of such gluons, we will forget about these operators for the time being. Does this mean that what remains from the Lagrangian (1.3) contains no $1/m_Q$ terms?

The answer to this question is negative. The $1/m_Q$ expansion is generated by the first ("tree level") term in the Lagrangian (1.3),

$$\mathcal{L}^0_{\text{heavy}} = \bar{Q}(\not{\mathcal{P}} - m_Q)Q\,. \tag{1.5}$$

Although the field Q in this Lagrangian is normalized at a low point μ, the field Q carries a hidden large parameter, m_Q; isolating this parameter opens the way to the $1/m_Q$ expansion. Indeed, the interaction of the heavy quark with the light degrees of freedom enters through $\mathcal{P}_\mu = iD_\mu$, where

$$D_\mu = \partial_\mu - igA^a_\mu t^a\,.$$

The background gluon field A_μ is weak if measured in the scale m_Q, which means, of course, that there is a large "mechanical" part in the x dependence of $Q(x)$, known from the very beginning [9],

$$Q(x) = e^{-im_Q t}\tilde{Q}(x) \tag{1.6}$$

where $\tilde{Q}(x)$ is a "rescaled" bispinor field which, in the leading approximation, carries no information about the heavy quark mass. It describes a residual motion of the heavy quark inside the heavy hadron [10] with typical momenta of order Λ_{QCD}. Remnants of the heavy quark mass appear in $\tilde{Q}$ only at the level of $1/m_Q$ corrections.

Equation (1.6) is written in the rest frame of H_Q. In the arbitrary frame one singles out, instead, the factor $\exp(-im_Q v_\mu x_\mu)$ where v_μ the four-velocity of the heavy hadron,

$$v_\mu = p_\mu / M_{H_Q}\,.$$

The covariant momentum operator $\mathcal{P}_\mu$ acting on the original field Q, when acting on the rescaled field $\tilde{Q}$, is substituted by the operator $m_Q v_\mu + \pi_\mu$,

$$
\begin{aligned}
iD_\mu Q(x) &= e^{-im_Q v_\mu x_\mu} \left(m_Q v_\mu + iD_\mu\right) \tilde{Q}(x) \\
&\equiv e^{-im_Q v_\mu x_\mu} \left(m_Q v_\mu + \pi_\mu\right) \tilde{Q}(x) \,.
\end{aligned} \tag{1.7}
$$

Below we will consistently use different letters, $\mathcal{P}_\mu$ and π_μ, for the momentum operators iD_μ acting on Q and $\tilde{Q}$, respectively. If not stated to the contrary, we will use the rescaled field $\tilde{Q}$, *omitting* the tilde in all expressions where there is no risk of confusion.[2] Using these distinct notations for the momentum operator is convenient since all expressions written in terms of π_μ and $\tilde{Q}$ do not contain implicitly the large parameter m_Q.

I pause here to make a reservation. The rescaled field $\tilde{Q}$ is a *four-component* Dirac bispinor, not a two-component nonrelativistic spinor which is usually introduced in the heavy quark effective theory (HQET) [10]. HQET is a formalism invented at the very beginning of the 90's [10], which is very often used in connection with heavy quark physics [4]. It is convenient in a range of problems but can be quite misleading in some other problems. I prefer to discuss the heavy quark expansions directly and systematically in *full QCD* in the framework of the Wilson OPE. In many instances the careful reader will certainly recognize a significant overlap, but the Wilson language, being more general, seems to give a better understanding and command over the $1/m_Q$ expansions. Moreover, some issues cannot be addressed in the framework of HQET at all.

The Dirac equation $(\not{\mathcal{P}} - m_Q)Q = 0$, in terms of the rescaled field, can be written as follows:

$$
\frac{1-\gamma_0}{2} Q = \frac{\not{\pi}}{2m_Q} Q \tag{1.8}
$$

and

$$
\pi_0 Q = -\frac{\pi^2 + (i/2)\sigma G}{2m_Q} Q \,. \tag{1.9}
$$

The last equation is actually the squared Dirac equation,

$$
\frac{1}{2m_Q} \left(\not{\mathcal{P}} + m_Q\right) \left(\not{\mathcal{P}} - m_Q\right) Q = \frac{1}{2m_Q} \left(\mathcal{P}^2 + \frac{i}{2}\sigma G - m_Q^2\right) Q = 0 \,.
$$

[2]Whenever one sees an expression containing π's one may be sure that it refers to the rescaled fields $\tilde{Q}$ even if the tildes are not written out explicitly.

In deriving Eq. (1.9) we used the fact that

$$[\mathcal{P}_\mu, \mathcal{P}_\nu] = [\pi_\mu, \pi_\nu] = igG^a_{\mu\nu}t^a \,. \tag{1.10}$$

Armed with this knowledge one can easily obtain the $1/m_Q$ expansion of $\mathcal{L}^0_{\rm heavy}$, up to terms $1/m_Q^2$,

$$\begin{aligned}\mathcal{L}^0_{\rm heavy} = \bar{Q}(i\not{\mathcal{D}} - m_Q)Q \;=\; & \bar{Q}\frac{1+\gamma_0}{2}\left(1+\frac{(\vec{\sigma}\vec{\pi})^2}{8m_Q^2}\right)\left[\pi_0 - \frac{1}{2m_Q}(\vec{\pi}\vec{\sigma})^2\right.\\ & \left.-\frac{1}{8m_Q^2}\left(-(\vec{D}\vec{E}) + 2\vec{\sigma}\cdot\vec{E}\times\vec{\pi}\right)\right]\left(1+\frac{(\vec{\sigma}\vec{\pi})^2}{8m_Q^2}\right)\frac{1+\gamma_0}{2}Q\\ & + \mathcal{O}\left(\frac{1}{m_Q^3}\right),\end{aligned} \tag{1.11}$$

where $\vec{\sigma}$ denote the Pauli matrices and

$$(\vec{\pi}\vec{\sigma})^2 = \vec{\pi}^2 + \vec{\sigma}\vec{B}\,,$$

$\vec{E}$ and $\vec{B}$ denote the background chromoelectric and chromomagnetic fields, respectively. The coupling constant g and the color matrix t^a are included in the definition of these fields. The derivation of this Lagrangian is a good home exercise. I encourage everyone to obtain Eq. (1.11) by using the commutation relation (1.10) and the properties of the gamma matrices. Those who have problems obtaining Eq. (1.11) should consult Chapter 4 of Bjorken and Drell [11] or Sec. 33 of the Landau-Lifshitz course [12] from where this Lagrangian follows immediately. It is worth noting that

$$\mathcal{L}^0_{\rm heavy} \equiv \varphi^+(\pi_0 - \mathcal{H}_Q)\varphi\,, \tag{1.12}$$

where

$$\varphi = \left(1+\frac{(\vec{\sigma}\vec{\pi})^2}{8m_Q^2}\right)\frac{1+\gamma_0}{2}Q \tag{1.13}$$

and $\mathcal{H}_Q$ is a nonrelativistic Hamiltonian, through second order in $1/m_Q$,

$$\mathcal{H}_Q = \frac{1}{2m_Q}(\vec{\pi}^2 + \vec{\sigma}\vec{B}) + \frac{1}{8m_Q^2}\left(-(\vec{D}\vec{E}) + 2\vec{\sigma}\cdot\vec{E}\times\vec{\pi}\right), \tag{1.14}$$

well-known (in the Abelian case) from the textbook expressions [11, 12]. Equation (1.13) is merely the Foldy-Wouthuysen transformation which is necessary to keep the term linear in π_0 in its canonic form.

1.3. $m_Q \to \infty$; *The Heavy Quark Symmetry*

Let us first neglect all $1/m_Q$ corrections altogether. In this limit m_Q drops out from $\mathcal{L}^0_{\text{heavy}}$,

$$\mathcal{L}^0_{\text{heavy}} = \bar{Q}\frac{1+\gamma_0}{2}\pi_0 Q\,. \tag{1.15}$$

This expression takes place in the rest frame of H_Q; in the arbitrary frame [10]

$$\mathcal{L}^0_{\text{heavy}} = \bar{Q}\frac{1+\not{v}}{2}\pi_\mu v_\mu Q\,. \tag{1.16}$$

In the limit $m_Q \to \infty$ the masses of all Q-containing hadrons become equal to that of the heavy quark Q,

$$M_{H_Q} = m_Q + \mathcal{O}(\Lambda_{\text{QCD}})\,.$$

The mass splittings between different hadrons are generically of order $\Lambda_{\text{QCD}} \ll m_Q$. Soon, we will relate these mass splittings to the expectation values of certain operators.

The assertion that all Q-containing hadrons are degenerate to the zeroth order in m_Q is trivial. This "degeneracy" by no means implies that the internal structure of all Q-containing hadrons is the same. A little less trivial is the fact that there exist hadrons whose masses are degenerate to much better accuracy, $\mathcal{O}(m_Q^{-1})$, and whose internal structure is, indeed, identical in the limit $m_Q \to \infty$.

Since all effects due to the heavy quark spin are, obviously, proportional to $1/m_Q$, in this limit the heavy quark spin becomes irrelevant, see Eqs. (1.11) and (1.16). Correspondingly, there emerges a symmetry between the states which differ only by the spin orientation of the heavy quark. The pseudoscalar and vector mesons of the type B and B^* (both are the ground state S wave mesons) present an example of such a spin family. In the limit $m_Q \to \infty$ their masses must be degenerate up to terms $\mathcal{O}(m_Q^{-1})$, and the light clouds of B and B^* coincide. If there is more than one heavy quark, say Q_1 and Q_2, the theory is symmetric with respect to the interchange $Q_1 \leftrightarrow Q_2$, even if their masses are not close to each other (in physical applications we, of course, keep in mind b and c). Indeed, the heavy quark Q_i plays the role of the static force center inside H_{Q_i}; the light cloud is flavor-blind and does not notice the substitution of Q_1 by Q_2 provided that the four-velocities of both quarks are the same. Notice that at this level the four-velocity of the heavy quark coincides with that of the heavy hadron. (Only when higher order corrections in $1/m_Q$ are taken into account does the difference between the four-velocities become important and the symmetry $Q_1 \leftrightarrow Q_2$ is violated. At the level of $1/m_Q$ the spin symmetry is not valid any more.) If the hard gluon effects are neglected the interaction with the light cloud cannot change the heavy quark four-velocity; therefore, this quantity is conserved in strong interactions [10]. (This conservation is, of course, destroyed by the hard gluons which can easily carry away a finite fraction of the heavy quark momentum.)

The symmetry connecting Q_1 and Q_2 emerges in the limit $m_{Q_{1,2}} \to \infty$ even if the masses of the heavy quarks are not close to each other. What is important is that both must be much larger than $\Lambda_{\rm QCD}$. We encounter here a situation which is conceptually close to the problem of the isotopic symmetry of strong interactions. Everybody knows that the strong amplitudes are isotopically invariant with the accuracy up to a few percent, and, at the same time, the masses of the d and u quarks are not too close to each other, $m_d/m_u \sim 2$. It is not the proximity of these masses which counts, but the fact that both masses are much less than the QCD scale $\Lambda_{\rm QCD}$.

Usually, the existence of an internal symmetry implies a degeneracy of the spectrum. For instance, the isotopic symmetry mentioned above, apart from certain relations between the scattering amplitudes, predicts that the proton and neutron masses are the same, up to small corrections, due to the symmetry-breaking effects. The heavy quark symmetry does not manifest itself as a degeneracy in the spectrum — the D and B masses are very far from each other. One has to subtract the mechanical part of the heavy quark mass in order to see that all dynamical parameters are insensitive to the substitution $Q_1 \leftrightarrow Q_2$ in the limit $m_{Q_{1,2}} \to \infty$ [13]. Perhaps, this is the reason why it was discovered so late.

To elucidate the issue of the heavy quark symmetry let us consider a practical problem, semileptonic decay of the B meson induced by the weak $b \to c$ transition. The initial B meson decays into an electron-neutrino pair plus the D meson. Since we do not now discuss the $1/m_Q$ corrections, we may make no distinction between the four-velocities of the quark Q and the hadron H_Q, and between their masses. Assume that the B meson is at rest. Furthermore, let us assume that the four-momentum q carried away by the lepton pair is maximal, $q^2 = (M_B - M_D)^2$. This means that the D meson produced is also at rest — the hadronic system experiences no recoil. The corresponding regime is sometimes called the point of zero recoil.

In this regime the $B \to D$ transition form factor is exactly unity! More exactly,

$$\langle D|\bar{c}\gamma_0 b|B\rangle = (2M_B 2M_D)^{1/2} \times \text{unity (at zero recoil)} \tag{1.17}$$

where the square root factors are due to the relativistic normalization of our amplitudes. By the same token,

$$\langle D^*|\bar{c}\gamma_i\gamma_5 b|B\rangle = i(2M_B 2M_D)^{1/2} D_i^* \times \text{unity (at zero recoil)} \tag{1.18}$$

where D_i^* is the polarization vector of D^*. As is well-known, the exact relations of this type always reflect an underlying symmetry. They can never emerge accidentally because only a symmetry can protect the form factors from renormalizations.

It is very easy to understand why Eqs. (1.17) and (1.18) take place. Indeed, the space-time picture is very transparent. The b quark at rest is surrounded by its light cloud, the latter being the eigenstate of the problem of color interaction with a static force center. At time zero the weak current instantaneously substitutes the b quark by c; the charmed quark is also at rest, and since the color interactions are flavor-blind the same light cloud, continues to be the eigenstate, this time with the c quark as the static center. If, instead of the field-theoretic light cloud we had a quantum-mechanical problem, one could say that the overlap integral for these identical wave functions is 1. The light cloud will feel the substitution $b \to c$ only to the extent to which the heavy quark momentum inside the heavy meson does not vanish exactly — this effect is, of course, suppressed by powers of $1/m_Q$. As we will see later corrections in the right-hand side of Eqs. (1.17) and (1.18) are actually of order $1/m_Q^2$; there are no linear corrections in $1/m_Q$. In the $B \to D^*$ transition generated by the axial-vector current, the current, additionally, changes the orientation of the heavy quark spin. As was already mentioned, all effects related to the heavy quark spin are suppressed by $1/m_Q$; D and D^* are in the same multiplet, and the $B \to D^*$ transition is governed by the same symmetry. This symmetry allows one to rotate arbitrarily four states,

$$b \text{ spin up}, \;\; b \text{ spin down}, \;\; c \text{ spin up}, \;\; c \text{ spin down};$$

therefore, we are obviously dealing with an SU(4) invariance here.

The symmetry relations (1.17) and (1.18) were first derived in Refs. [14, 15]. Then it was realized [16] that the actual symmetry is much stronger — the SU(4) invariance takes place for any given value of v_μ, the four-velocity of the recoiling c quark, not necessarily at the point of zero recoil or close to it. Thus, many different form factors connecting (B, B^*) and (D, D^*) can be expressed in terms of one function depending only on the velocity of the recoiling hadron (in the rest frame of the decaying hadron). The universal form factor is called the *Isgur-Wise function.*

1.4. *The Isgur-Wise Function*

Now we are finally ready to discuss a very elegant observation due to Isgur and Wise [16]. Let us now consider the amplitudes induced by the transition $\bar{c}\Gamma b$ off the zero recoil point. Here Γ is any Lorentz matrix; of special interest are, of course, the vector and the axial-vector cases,

$$\Gamma = \gamma_\mu \text{ or } \gamma_\mu\gamma_5;$$

the weak decays of the B meson are induced by the V–A currents. The physically measurable amplitudes are $\langle D|\bar{c}\Gamma b|B\rangle$ and $\langle D^*|\bar{c}\Gamma b|B\rangle$; for completeness one can also consider the amplitudes of the type $\langle D|\bar{c}\Gamma b|B^*\rangle$, or $\langle D^*|\bar{c}\Gamma b|B^*\rangle$, or $\langle B|\bar{b}\Gamma b|B\rangle$ — this adds nothing new. The four-velocity of the particle H_Q is defined as

$$v_\mu = \frac{(p_{H_Q})_\mu}{M_{H_Q}}; \tag{1.19}$$

the four-velocities of the initial particles will be denoted by v while those of the final particles by v'. It is obvious that $v^2 = 1$, and, additionally, in the rest frame $v = \{1, 0, 0, 0\}$. In the most general case the amplitude $\langle D|\bar{c}\gamma_\mu b|B\rangle$ can be expressed in terms of two form factors, the amplitude $\langle D^*|\bar{c}\gamma_\mu\gamma_5 b|B\rangle$ in terms of three form factors and the amplitude $\langle D^*|\bar{c}\gamma_\mu b|B\rangle$ in terms of one form factor. The heavy quark symmetry tells us that, in the limit $m_Q \to \infty$, these six functions, *a priori* independent, reduce to one and the same function which depends only on the scalar product vv'. Specifically,

$$\langle D|\bar{c}\gamma_\mu b|B\rangle = \sqrt{M_B M_D}\left[(v+v')_\mu\right]\xi(y)\,; \tag{1.20}$$

$$\langle D^*|\bar{c}\gamma_\mu\gamma_5 b|B\rangle = \sqrt{M_B M_D}\, i\left[D^*_\mu(1+vv') - (D^*_\alpha v_\alpha)\, v'_\mu\right]\xi(y)\,; \tag{1.21}$$

$$\langle D^*|\bar{c}\gamma_\mu b|B\rangle = \sqrt{M_B M_D}\left[-\epsilon_{\mu\nu\lambda\sigma} D^*_\mu v'_\lambda v_\sigma\right]\xi(y) \tag{1.22}$$

where

$$y = vv'$$

and $\xi(y)$ is the Isgur-Wise function, D^*_μ is the polarization vector of D^*. The Isgur-Wise function is independent of the heavy quark masses. The square root $\sqrt{M_B M_D}$ reflects the relativistic normalization of the states. The symmetry relations (1.17) and (1.18) imply that the normalization of the Isgur-Wise function at zero recoil is fixed,

$$\xi(y = 1) = 1\,. \tag{1.23}$$

It is perhaps worth noting that the phases in Eqs. (1.21) and (1.22) differ from what you might see in the literature. They are, of course, a matter of convention and reflect the definition of the states. The definition I follow is in accord with the standard relativistic convention, see Eq. (1.25) and Sec. 3.5.

The fact that a large set of form factors degenerate into a single function depending only on y might seem a miracle; but after the assertion is made, with the knowledge you already have, it should not be difficult to understand why it happens. Indeed, let us turn again to the space-time picture described above. A b quark at rest, surrounded by the light cloud, instantaneously converts into a c quark. This time the four-momentum carried away by the lepton pair is not maximal; therefore, the c quark is not at rest. This force center flies away with the velocity $\vec{v}'$. But the light cloud stays intact. So, the question is: "what is the amplitude for the flying c quark and the cloud at rest to form a D or D^* meson?" We can look at this process in another way. After the $b \to c$ transition happened let us proceed to the rest frame of c. In this reference frame the c quark produced is at rest, but the cloud, as a whole, moves away with the velocity $-\vec{v}'$. It is clear that this system — the static charmed force center plus a moving light cloud — has a projection on D or D^*. The amplitude *per se*, with the kinematic structures *excluded*, can depend only on $|\vec{v}'|$ — there is no preferred orientation in the space, and the direction of $\vec{v}'$ is irrelevant. Using covariant notations one can say that the amplitude depends only on vv' since in the B rest frame $vv' = \sqrt{(1+\vec{v}'^2)}$. There is simply no place for the dependence on the heavy quark masses, apart from the overall normalization factors appearing because we stick to the relativistic normalization of states.[3]

Since the heavy quark spin is irrelevant in the limit $m_Q \to \infty$, to warm up, let us consider a toy model where the heavy quarks are deprived of their spins

[3]Warning: an additional dependence on the heavy quark masses may emerge if we include the hard gluon exchanges neglected so far. For details see Ref. [17].

from the very beginning. In other words, I replace the genuine spin-1/2 heavy quarks of QCD by spin-0 color triplets with the same mass. We will turn to this toy model more than once below.

In QCD, B and B^* form a multiplet which includes four states: the total angular momentum of the light cloud (1/2) combines with the heavy quark spin (1/2) to produce either spin-0 state (B) or three spin-1 states (B^*). In our toy model the analog of this ground state multiplet is obviously a baryon of spin 1/2; let us denote the corresponding field by N_Q^α, where α is the spinorial index. The current generating the transition $N_b \to N_c$ has the form $c^\dagger b$, where the fields b and c are assumed to be scalar now. At first sight the amplitude $\langle N_c|c^\dagger b|N_b\rangle$ might contain four different kinematic structures,

$$\bar{N}_c N_b \,, \quad \bar{N}_c \not{v} N_b \,, \quad \bar{N}_c \not{v}' N_b \,, \quad \bar{N}_c \sigma_{\mu\nu} N_b v_\mu v'_\nu \,,$$

where I list only P-even structures, of course. On mass shell they all reduce to the first one, however; for instance, $\not{v} N_b = N_b$. Hence

$$\langle N_c|c^\dagger b|N_b\rangle = \bar{N}_c N_b \xi(y) \,.$$

Returning to real QCD what remains to be done is to work out the consequences of spin. The most concise general formula can be written in terms of the matrices

$$\mathcal{M} = B(i\gamma_5) + B_\mu \gamma_\mu \text{ and } \mathcal{M}' = D^*(i\gamma_5) + D^*_\mu \gamma_\mu \,, \tag{1.24}$$

where B_μ and D_μ are the polarization vectors of B^* and D^*, respectively. The (heavy quark) spin independence of the strong interactions at $m_Q \to \infty$ manifests itself in the fact that the couplings of the ground state pseudoscalar to $i\gamma_5$ and the ground state vector to γ_μ are the same, see Sec. 3.4. Now, the whole set of the transition amplitudes can be expressed by one compact cute formula,[4]

$$\frac{1}{\sqrt{2M_D 2M_B}}\langle H_c(v')|\bar{c}\Gamma b|H_b(v)\rangle = \frac{1}{2}\mathrm{Tr}\left\{\mathcal{M}'\frac{1+\not{v}'}{2}\Gamma\frac{1+\not{v}}{2}\mathcal{M}\right\}\xi(y=vv') \,. \tag{1.25}$$

Completing the trace we recover Eqs. (1.20)–(1.22).

[4]Strictly speaking, for the outgoing particles, one must use $\overline{\mathcal{M}}' = \gamma_0(\mathcal{M}')^\dagger\gamma_0$. With our conventions, however, $\overline{\mathcal{M}}' = \mathcal{M}'$.

Equation (1.25) can be derived in many different ways. Originally it was obtained in Ref. [17] (see also Ref. [18]). In Sec. 3.5 we will discuss one of the possible derivations — perhaps, not the simplest, but very instructive. Before we will be able to do that it is necessary to make a digression and study some elements of the background field technique.

Equations (1.20)–(1.22) are valid not only in the space-like domain (the form factor kinematics) but also in the time-like domain. The latter assertion calls for an immediate reservation, though. The heavy quark symmetry implies that $M_B = M_{B^*}$. In the real world this equality is not exact: the heavy quark symmetry is violated by small $1/m_Q$ terms. This small violation can be strongly enhanced in the near-threshold domain, $E \approx 2M_B$, where the symmetry-breaking parameter turns out to be of order one [19]. Indeed, let us consider a kinematical point above the threshold of the $B\bar{B}$ production but below $B\bar{B}^*$. In this domain all form factors describing three amplitudes

$$\langle B\bar{B}|J_\mu|0\rangle\,, \quad \langle B\bar{B}^*|J_\mu|0\rangle\,, \quad \langle B^*\bar{B}^*|J_\mu|0\rangle\,,$$

(here J_μ is some heavy quark current, say, $J_\mu = \bar{b}\gamma_\mu b$) have imaginary parts associated with the normal thresholds due to the intermediate state $B\bar{B}$. On the other hand, there is no contribution to the imaginary part from the intermediate state $B\bar{B}^*$ and $B^*\bar{B}^*$. In the pseudoscalar meson B the spin of the heavy quark Q is rigidly correlated with that of the light cloud. Hence, the spin independence of the heavy quark interaction is totally lost in the imaginary part in this point. In particular, in the amplitude $\langle B^*\bar{B}^*|J_\mu|0\rangle$ a kinematic structure forbidden by the Isgur-Wise formula appears. An even more pronounced effect of the heavy quark symmetry violation takes place in the anomalous thresholds generated by the pion exchange, which can start parametrically much below the normal thresholds, depending on the interplay between $M_{B^*}^2 - M_B^2$ and the pion mass [20].

1.5. *The Mass Formula*

To complete our first encounter with the basics of the heavy quark theory we will now derive a $1/m_Q$ expansion for the masses of the Q-containing hadrons.

It is intuitively clear that the heavy hadron mass can be expanded in terms of that of the heavy quark as follows:

$$M_{H_Q} = m_Q + \bar{\Lambda} + \mathcal{O}(m_Q^{-1}) \tag{1.26}$$

where $\bar{\Lambda}$ is a constant, of order of Λ_{QCD}, which depends on the light quark content and the quantum numbers of H_Q but is independent of m_Q. (It first appeared in Ref. [21].) Later on we will see that this expression is not as trivial as it might naively seem and requires thoughtful definitions of all parameters involved. In particular, since the quarks are never observed as isolated objects, one may ask what the quark mass m_Q actually means. In due time we will return to this question, of course. For the time being we agreed to disregard hard gluon exchanges; then m_Q is just the mass parameter in the Lagrangian (1.1).

Formally Eq. (1.26) can be most easily derived by analyzing the trace of the energy-momentum tensor in QCD,

$$\theta_{\mu\mu} = m_Q \bar{Q} Q + \frac{\beta(\alpha_s)}{4\alpha_s} G^a_{\mu\nu} G^a_{\mu\nu} \tag{1.27}$$

where $\beta(\alpha_s)$ is the Gell-Mann-Low function. For simplicity we assume the light quarks to be massless; the introduction of the light quark masses changes only technical details at intermediate stages of our analysis. If the mass term of the light quarks is set equal to zero, the light quark fields do not appear explicitly in the trace of the energy-momentum tensor. The expression (1.27) contains two terms: the first one is a mechanical part while the second term is the famous trace anomaly of QCD [22] (for a review see e.g. Ref. [23]).

Furthermore, as is well-known, for any given one-particle state, the expectation value of the trace of the energy-momentum tensor reduces to the mass of the state. Then, the hadron mass can be expressed in terms of two expectation values,

$$M_{H_Q} = \frac{1}{2M_{H_Q}} \langle H_Q | m_Q \bar{Q} Q | H_Q \rangle + \frac{1}{2M_{H_Q}} \left\langle H_Q \left| \frac{\beta(\alpha_s)}{4\alpha_s} G^2 \right| H_Q \right\rangle , \tag{1.28}$$

where the relativistic normalization of the states is implied,

$$\langle H_Q | H_Q \rangle = 2 M_{H_Q} V$$

in the rest frame; V is the normalization volume. We will always use only the relativistic normalization of states which will routinely result in the factors $(2M_{H_Q})^{-1}$ in all expressions.

Let us discuss the expectation values of the operators in Eq. (1.28) in turn. The first one is explicitly proportional to m_Q. To be more quantitative we must determine the matrix element of the heavy quark density $\bar{Q}Q$. To this end it is convenient to use an argument suggested in Ref. [24] which will show us that the expectation value of $\bar{Q}Q$ is very close to unity; as a matter of fact, with our present accuracy it is just equal to unity. The second expectation value reduces to $\bar{\Lambda}$.

Indeed, in the rest frame of H_Q a typical momentum of Q is of order $\Lambda_{\rm QCD}$, i.e. the heavy quark is very slow. This means that the lower components of the bispinor field Q are small compared to the upper ones and, hence, the scalar density of the heavy quark is close to its vector charge, $\bar{Q}Q \approx \bar{Q}\gamma_0 Q$. The difference is only due to the lower components. The vector charge, however, just measures the number of the heavy quarks inside H_Q; therefore, its matrix element is exactly unity.

It is instructive to do the simple derivation outlined above in some detail. Combining the equations of motion, (1.8) and (1.9), it is easy to get that

$$\frac{1-\gamma_0}{2}Q = -\frac{1}{2m_Q}\vec{\pi}\vec{\gamma}\frac{1+\gamma_0}{2}Q + \mathcal{O}(m_Q^{-2})\,, \tag{1.29}$$

which implies, in turn,

$$\bar{Q}Q = \bar{Q}\gamma_0 Q - \frac{1}{2m_Q^2}\bar{Q}\left(\vec{\pi}^2 + \vec{\sigma}\vec{B}\right)Q + \text{higher orders} \tag{1.30}$$

where $\vec{B}$ is the chromomagnetic field, $B_i = \epsilon_{ijk}G_{jk}$. Equation (1.30) is the desired result demonstrating that

$$\frac{1}{2M_{H_Q}}\langle H_Q|\bar{Q}Q|H_Q\rangle = \frac{1}{2M_{H_Q}}\langle H_Q|\bar{Q}\gamma_0 Q|H_Q\rangle + \mathcal{O}(m_Q^{-2}) = 1 + \mathcal{O}(m_Q^{-2})\,. \tag{1.31}$$

The matrix element of the vector charge (appropriately normalized) is set equal to unity, as was discussed above.

This digression has been undertaken merely to familiarize the reader with the basics of the $1/m_Q$ expansion in QCD. As our understanding progresses the level of the explanatory remarks will be reduced so that in the subsequent lectures many derivations of a more technical nature will be suggested as an exercise. Thus, we have established that the first expectation value in Eq. (1.28) produces m_Q in the expansion for the heavy hadron mass. The second expectation value which also has the dimension of mass obviously does not scale with m_Q in the limit $m_Q \to \infty$, so one can define[5]

$$\bar{\Lambda} = \frac{1}{2M_{H_Q}} \left\langle H_Q \left| \frac{\beta(\alpha_s)}{4\alpha_s} G^2 \right| H_Q \right\rangle_{m_Q \to \infty} . \tag{1.32}$$

Thus, the parameter $\bar{\Lambda}$ of the heavy quark theory is, in a sense, similar to the gluon condensate [27]. The latter is the expectation value of the same gluon operator over the vacuum state. In the case of $\bar{\Lambda}$ the gluon operator is averaged over the lowest state of the system with the given (unit) value of the heavy quark charge. The lowest state is, of course, the ground state pseudoscalar meson, B. Generally speaking, H_Q can be any Q-containing hadron. B mesons are most interesting from the point of view of applications; of practical interest also are Q-containing baryons which are the lowest lying states in the given channel with the baryon quantum numbers. Therefore, strictly speaking, unlike the gluon condensate, there exist many different parameters $\bar{\Lambda}$, one for every channel considered. Usually we will tacitly assume that $\bar{\Lambda}$ is defined with respect to the B mesons.

Both expectation values,

$$\frac{1}{2M_{H_Q}} \langle H_Q | m_Q \bar{Q} Q | H_Q \rangle \text{ and } \frac{1}{2M_{H_Q}} \left\langle H_Q \left| \frac{\beta(\alpha_s)}{4\alpha_s} G^2 \right| H_Q \right\rangle ,$$

have $1/m_Q$ corrections which show up at the level $\mathcal{O}(m_Q^{-1})$ in Eq. (1.26). Later on we will derive the expansion for M_{H_Q} which takes into account these $\mathcal{O}(m_Q^{-1})$ terms.

The $1/m_Q$ corrections in the expectation value of the gluon anomaly are due to the fact that in our approach the states $|H_Q\rangle$ are physical heavy flavor states, rather than the asymptotic states corresponding to $m_Q = \infty$ which are usually considered within HQET. Instead of working with these fictitious states I prefer to explicitly keep track of all $1/m_Q$ corrections, both in the operators and in the definition of the states, appealing directly to the Wilsonean operator product expansion.

---

[5]This expression relating $\bar{\Lambda}$ to the expectation value of the gluon anomaly operator was obtained in Ref. [25]. Some subtleties left aside in the derivation presented here are discussed in detail in this paper. It is instructive to compare Eq. (1.32) with a similar expression for the nucleon mass,

$$M_N = \frac{1}{2M_N} \left\langle N \left| \frac{\beta(\alpha_s)}{4\alpha_s} G^2 \right| N \right\rangle ,$$

known from ancient times [26].

SECTION 2
Basics of the Background Field Technique

The essence of our approach is the separation of all momenta into two classes — hard and soft. For the time being we will continue to pretend that the role of the gluon degrees of freedom reduces to a soft gluon medium. This is an ideal situation where the gluons can be treated as a background field. A powerful method allowing one to put calculations in the background fields on an industrial basis was developed by Schwinger in electrodynamics many years ago. In the eighties it was adapted to QCD. We will be unable to go into all details of this technique, and will, rather, present some basic elements in particular examples. The review paper [28] is recommended for further education. This section will be rather technical — its primary goal is to teach how the heavy quark mass expansions can be constructed in a systematic way in different problems.

The starting point of the method is the decomposition of fields into two parts — the quantum part and the background one. The propagation of quanta is described by the correlation functions of the quantum part of the fields considered in the external field. Later on the external field is to be considered as a fluctuating field of the light cloud, but this stage need not concern us at the moment.

Let us start with a brief review of the Schwinger method, as it can be applied in QCD. We introduce the coordinate and momentum operators, X_μ and p_μ, respectively, $[p_\mu, X_\nu] = ig_{\mu\nu}$, $[p_\mu, p_\nu] = [X_\mu, X_\nu] = 0$. Moreover, we introduce a formal set of states $|x)$ which are the eigenstates of the coordinate operator X_μ,

$$X_\mu|x) = x_\mu|x). \tag{2.1}$$

Please note that $|x)$ has nothing to do with the field-theoretic eigenstates, e.g. $|H_Q\rangle$. To emphasize this fact we use the regular bracket) in the notation, instead of the angle one, which is reserved for the field-theoretic eigenstates.

Then, define the covariant momentum operator $\mathcal{P}_\mu$ satisfying the following commutation relations

$$[\mathcal{P}_\mu, X_\nu] = ig_{\mu\nu}, \quad [\mathcal{P}_\mu, \mathcal{P}_\nu] = igt^a G^a_{\mu\nu}, \tag{2.2}$$

where t^a are the generators of the color group, $G^a_{\mu\nu}$ is the external field.

The algebra (2.2) is the basic tool of the Schwinger formalism. We will expand the Green functions in the background field, and in each order of the expansion we will need to use only this algebra.

In the coordinate basis $\mathcal{P}_\mu$ acts as a covariant derivative, namely

$$(y|\mathcal{P}_\mu|x) = \left(i\frac{\partial}{\partial x_\mu} + gt^a A^a_\mu(x)\right)\delta(x-y)\,, \tag{2.3}$$

if

$$(y|x) = \delta(x-y). \tag{2.4}$$

Now we can write formal expressions for the Green functions. For instance, for the quark Green function (mass m_q) describing propagation from the point 0 to the point x, we have

$$S(x,0) = \left(x\left|\frac{1}{\not{\mathcal{P}} - m_q}\right|0\right). \tag{2.5}$$

Equation (2.5), rather obvious by itself, is readily verified by applying the Dirac operator to both sides of Eq. (2.5). Furthermore, it can be identically rewritten as follows:

$$S(x,0) = \left(x\left|(\not{\mathcal{P}} + m_q)\frac{1}{\mathcal{P}^2 - m_q^2 + (i/2)G_{\mu\nu}\sigma_{\mu\nu}}\right|0\right) \tag{2.6}$$

where

$$G_{\mu\nu} \equiv gt^a G^a_{\mu\nu}\,.$$

Please note that the ordering is important here since $\mathcal{P}_\mu$ does not commute with $\mathcal{P}_\nu$ of $G_{\mu\nu}$.

If we aim at calculating the coefficient functions in the Born approximation, then we need nothing else — Eq. (2.6) is just systematically expanded in powers of the background field by using the commutation relations (2.2).

Observe that one can always shift $\mathcal{P}_\mu$ by a c-number vector due to the fact that

$$e^{iqX}\mathcal{P}_\mu e^{-iqX} = \mathcal{P}_\mu + q_\mu. \tag{2.7}$$

Hence, the Fourier-transformed propagator reduces to

$$\begin{aligned}\int d^4x e^{iqx} S(x,0) &= \int d^4x \left(x\left|e^{iqX}\frac{1}{\not{\mathcal{P}} - m_q}e^{-iqX}\right|0\right)\\ &= \int d^4x \left(x\left|\frac{1}{\not{\mathcal{P}} + \not{q} - m_q}\right|0\right).\end{aligned}$$

This simple trick allows one to readily develop the expansion sought for. Indeed, assume that q is large (hard momentum) and $\mathcal{P}$ represents soft modes and is small in this sense. Then we can expand in $\mathcal{P}$,

$$\int d^4x e^{iqx} S(x,0) \to \frac{1}{\not{q} - m_q} - \frac{1}{\not{q} - m_q} \not{\mathcal{P}} \frac{1}{\not{q} - m_q} + \ldots \tag{2.8}$$

Next, we transpose $\mathcal{P}$ to the right-most (left-most) position and act on the states using the equations of motion.

It may seem that, so far, we got almost nothing compared to the standard Feynman graph calculations. Let us demonstrate the efficiency of the background field technique in a few examples.

2.1. *Inclusive Decay of the Heavy Quark — Toy Model*

One of the most important practical problems in the heavy quark theory is the description of the inclusive decays of heavy flavors. The semileptonic and radiative decays of the B mesons $B \to X_c l\nu$ and $B \to X_s\gamma$ are particular examples. Both are two key elements of the ongoing experimental efforts, in quest of new physics. Needless to say that a reliable QCD-based theory of such decays is badly needed. In this chapter we start discussing the basics of such a theory.

Since this is our first exercise, for pedagogical reasons, it seems reasonable to "peel off" all inessential technicalities, like the quark spins, and resort to a simplified model. In this toy model we will consider the inclusive decay of a spinless heavy quark into a spinless lighter quark plus a photon. Of course, our photon is also a toy photon. We will assume it to be scalar and the corresponding field will be denoted by ϕ.

The Lagrangian describing the transition of a heavy quark Q into a lighter quark q and a "photon" has the form

$$\mathcal{L}_\phi = h\bar{Q}\phi q \; + \; \text{h.c.}\,, \tag{2.9}$$

where h is the coupling constant and $\bar{Q} = Q^\dagger$. The masses of the quarks Q and q are both *large* (and I remind that they are both spinless). Moreover, to further simplify the problem, we will analyze a special limit (the so-called small velocity or Shifman-Voloshin (SV) limit suggested in Ref. [15]) in which

$$\Lambda_{\rm QCD} \ll m_Q - m_q \ll m_Q\,. \tag{2.10}$$

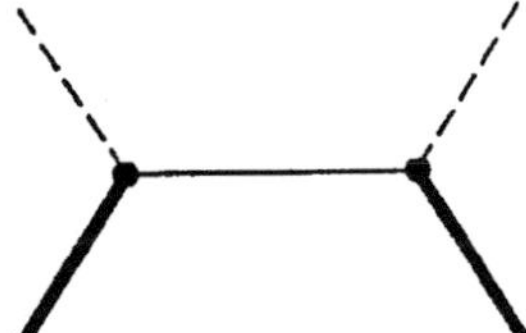

Fig. 1. The forward scattering amplitude $Q \to Q$. The dashed line denotes the ϕ quantum. The solid line connecting two vertices is the q quark Green function in the background gluon field. The thick solid lines describe the Q quarks in the background field.

The field ϕ carries color charge zero; the reaction $Q \to q+\phi$ could be considered a toy model for the radiative decays of the type $B \to X_s\gamma$, where X_s is an arbitrary inclusive hadronic state containing the s quark produced in the b quark decay.

It is very easy to calculate the total width for the *free* quark decay $Q \to q + \phi$,

$$\Gamma_{\text{free quark}}(Q \to q\phi) = \frac{h^2 E_0}{8\pi m_Q^2} \equiv \Gamma_0 \tag{2.11}$$

where

$$E_0 = \frac{m_Q^2 - m_q^2}{2m_Q}\,. \tag{2.12}$$

This free quark expression is valid for the total inclusive probability in the asymptotic limit when $m_Q \to \infty$. We are interested, however, in the preasymptotic corrections proportional to powers of $1/m_Q$.

First of all we must formulate what object we must deal with in order to be able to calculate these corrections systematically. Upon reflection one concludes that it cannot be the decay amplitude $Q \to q\phi$ itself. Instead we must consider the $Q \to Q$ forward "scattering" amplitude depicted in Fig. 1. By scattering I mean that Q scatters off the ϕ quantum and off the background gluon field which is not shown in Fig. 1 explicitly but is implied. It is implied that all quark lines, Q and q, are submerged into this soft gluon background field. Through the optical theorem the imaginary part of the amplitude of Fig. 1 is related to the inclusive probability of the $Q \to q\phi$ transition. More specifically, if we introduce the *transition operator*

$$\hat{T} = i \int d^4x \, \mathrm{e}^{-iqx} T\{\bar{Q}(x)q(x)\,,\, \bar{q}(0)Q(0)\}\,, \tag{2.13}$$

then the energy spectrum of the ϕ particle in the inclusive decay is obtained from $\hat{T}$ in the following way:

$$\frac{d\Gamma}{dE} = \frac{h^2 E}{4\pi^2 M_{H_Q}} \mathrm{Im}\, \langle H_Q|\hat{T}|H_Q\rangle \,. \tag{2.14}$$

Here as usual H_Q denotes a hadron built from the heavy quark Q and the light cloud (including the light antiquark), q in the exponent is the four-momentum carried away by ϕ and E is the energy of the ϕ quantum.

Equation (2.14) immediately translates the $1/m_Q$ expansion for the transition operator in the $1/m_Q$ expansion for the inclusive decay rate. The fact that the transition operator must be the primary object of the analysis in all problems of this type was realized in Refs. [9, 29, 30]. Now we use what we have already learnt about the background field technique to write the transition operator in the form

$$\begin{aligned}\hat{T} &= -\int d^4x\, \mathrm{e}^{-iqx} \left(x\left|\bar{Q}\frac{1}{\mathcal{P}^2 - m_q^2}Q\right|0\right) \\ &= -\int d^4x \left(x\left|\bar{\tilde{Q}}\frac{1}{(P_0 - q + \pi)^2 - m_q^2}\tilde{Q}\right|0\right) \,, \end{aligned} \tag{2.15}$$

where

$$(P_0)_\mu = m_Q v_\mu \,.$$

The Green function of the quark q differs from the Green function given in Eq. (2.6) in an obvious way, since we assume for the time being that our quarks Q and q have spin zero, and, correspondingly, instead of Eq. (2.6) referring to the spinor quarks, we have

$$S(x,0) = \left(x\left|\frac{1}{\mathcal{P}^2 - m_q^2}\right|0\right) \,. \tag{2.16}$$

In the second line of Eq. (2.15) we proceeded to the rescaled fields $\tilde{Q}$ which singles out the mechanical part of the momentum operator.

One more thing which will be needed is the equation of motion for the scalar field Q substituting the Dirac equation. Starting from $(\mathcal{P}^2 - m_Q^2)Q = 0$ we obviously get

$$\pi_0 \tilde{Q} = -\frac{\pi^2}{2m_Q}\tilde{Q} \,. \tag{2.17}$$

Finally we are ready to begin constructing the $1/m_Q$ expansion. Since π is of the order Λ_{QCD} while $P_0 - q$ scales as m_Q, in the leading approximation, π in the denominator of Eq. (2.15) can be neglected altogether. Then, obviously,

$$\hat{T}^{(0)} = \bar{Q}Q\frac{1}{m_q^2 - k^2} \tag{2.18}$$

where

$$k = P_0 - q\,.$$

We see that the leading operator appearing in the expansion is $\bar{Q}Q$ and it has dimension 2 (let us recall that the scalar Q field has dimension 1 in contrast to the real quark fields of dimension 3/2, which leads in particular to different normalization factors in the matrix elements). Taking the imaginary part we conclude that in the leading approximation

$$\frac{1}{\pi}\mathrm{Im}\,\langle H_Q|\hat{T}^{(0)}|H_Q\rangle = \frac{\langle\bar{Q}Q\rangle}{2m_Q}\delta(E - E_0)\,. \tag{2.19}$$

Here and below I will use a very convenient short-hand notation

$$\langle\bar{Q}Q\rangle \equiv \langle H_Q|\bar{Q}Q|H_Q\rangle\,.$$

The delta function in the imaginary part is characteristic of a two-body decay. As a matter of fact, combining Eq. (2.19) with the general expression (2.14) and approximating $\langle\bar{Q}Q\rangle$ by unity — which can and must be done in the leading order in $1/m_Q$ — we get the delta function spectrum of the the free quark decay. Integrating over the energy we then arrive at the free quark decay width (2.11).

Although this little achievement is quite gratifying and shows that we are on the right track the real $1/m_Q$ expansion begins when the preasymptotic terms switch on. To this end the terms with π in the denominator of Eq. (2.15) must be kept, and then the expansion in $(k\pi + \pi^2)$ must be carried out. The general term of this expansion is

$$\hat{T} = \frac{1}{m_q^2 - k^2}\sum_{n=0}^{\infty}\bar{Q}\left(\frac{2m_Q\pi_0 + \pi^2 - 2q\pi}{m_q^2 - k^2}\right)^n Q\,. \tag{2.20}$$

In Sec. 4 where the theory of the end point spectrum will be presented we will need the whole sum. At the moment our purpose is more limited — we aim at getting the first correction in the total decay width. This task does not require the infinite sum; only two terms, with $n = 1$ and $n = 2$, are relevant. Both terms are especially simple.

Indeed, if $n = 1$, the combination $2m_Q\pi_0 + \pi^2$ in the numerator acting on Q is nothing else than the equation of motion, and can be dropped, see Eq. (2.17). We can further discard the $\vec{q}\vec{\pi}$ part — since the H_Q spin is assumed to be zero, there is no preferred orientation and, hence, $\langle \bar{Q}\vec{\pi}Q\rangle = 0$. In this way we arrive at

$$\langle H_Q|\hat{T}^{(1)}|H_Q\rangle = -\frac{q_0}{m_Q}\frac{\langle\vec{\pi}^2\rangle}{(m_q^2 - k^2)^2} \tag{2.21}$$

plus terms of higher order in $1/m_Q$. Here the same equation of motion (2.17) was applied to eliminate π_0 in favor of π^2 which, in the given order in $1/m_Q$, coincides with $-\vec{\pi}^2$. The notation is even more concise than it was previously, namely $\langle\vec{\pi}^2\rangle$ stands for $\langle H_Q|\bar{Q}\vec{\pi}^2Q|H_Q\rangle$. You will often see similar short-hand notation below.

I pause here to make a side remark. The physical meaning of the matrix element $\langle\vec{\pi}^2\rangle$ is quite transparent — it merely represents the average value of the square of the momentum of the heavy quark Q inside the heavy hadron H_Q. This quantity is of the order $\Lambda^2_{\rm QCD}$. This is one of the most important parameters of the heavy quark theory, along with $\bar{\Lambda}$. Note the gap in dimensions of the operators appearing in the expansion. The dimension 2 operator $\bar{Q}Q$ is followed by dimension 4 operator $\bar{Q}\vec{\pi}^2Q$. No relevant operator of dimension 3 exists. Due to this reason the contribution of $\hat{T}^{(1)}$ in the total width is "unnaturally" suppressed by two powers of the inverse heavy quark mass, not one power as one would expect *a priori*. The observation of the dimension gap was first made in Ref. [30] in the context of HQET; it is crucial in phenomenological applications.

Let us return now to the construction of the $1/m_Q$ expansion, and consider the term with $n = 2$ in the sum (2.20). One of two factors $(2m_Q\pi_0 + \pi^2)$ can be applied to the right, the other one to the left. The difference between π applied to the right and to the left is a total derivative which vanishes anyway in the forward matrix element $\langle H_Q|\ldots|H_Q\rangle$. This simple observation implies that the combination $(2m_Q\pi_0 + \pi^2)$ in the numerator again vanishes by virtue of the equation of motion and we are left with

$$\langle H_Q|\hat{T}^{(2)}|H_Q\rangle = \frac{4}{3}\vec{q}^2\frac{1}{(m_q^2 - k^2)^3}\langle\vec{\pi}^2\rangle\,, \tag{2.22}$$

where I have singled out and retained only the spin-0 part of the operator $\bar{Q}\pi_i\pi_jQ \to (1/3)\delta_{ij}\bar{Q}\vec{\pi}^2Q$ for the reasons explained above.

We are almost done. The imaginary parts of $\hat{T}^{(1)}$ and $\hat{T}^{(2)}$ are expressible in terms of the first and second derivatives of the delta function, and after some simple algebra it is not difficult to get

$$\frac{1}{\pi}\mathrm{Im}\,\langle\hat{T}\rangle = \left(\frac{\langle\bar{Q}Q\rangle}{2m_Q} - \frac{\langle\bar{Q}\vec{\pi}^2 Q\rangle}{12m_Q^3}\right)\delta(E-E_0)$$

$$-\frac{E_0\langle\bar{Q}\vec{\pi}^2 Q\rangle}{12m_Q^3}\delta'(E-E_0) + \frac{E_0^2\langle\bar{Q}\vec{\pi}^2 Q\rangle}{12m_Q^3}\delta''(E-E_0) + \cdots \qquad (2.23)$$

where operators of higher dimension are ignored; I have taken into account that $q_0 = E$ and $\vec{q}^2 = E^2$ and played a little with the delta functions.

The expansion of $\mathrm{Im}\,\hat{T}$ into local operators generates more and more singular terms at the point where the ϕ spectrum would be concentrated in the free quark approximation. You should not be surprised by this circumstance which will be elucidated in every detail in due time. What is important is that the physical spectrum is a smooth function of E. One could derive a smooth spectrum by summing up the infinite set of operators in Eq. (2.20) — this will be the subject of Sec. 4. There is no need to carry out this summation now, however, since we are interested only in the integral characteristics of the type of the total probability. As far as such integral characteristics are concerned, the expansion in Eq. (2.23) is perfectly legitimate.

At first, we calculate the total width by substituting Eq. (2.23) into Eq. (2.14) and integrating over E,

$$\Gamma = \int dE\frac{d\Gamma}{dE} = \Gamma_0\frac{m_Q}{M_{H_Q}}\langle\bar{Q}Q\rangle \qquad (2.24)$$

where the integration runs from 0 to the physical boundary E_0^{phys}, expressed in terms of the hadron masses

$$E_0^{\mathrm{phys}} = \frac{M_{H_Q}^2 - M_{H_q}^2}{2M_{H_Q}}\,. \qquad (2.25)$$

The power correction proportional to $\langle\bar{Q}\vec{\pi}^2 Q\rangle/m_Q^2$ which might have appeared cancels in the total width! Is this cancellation unexpected? No, we could have anticipated it on general grounds. Indeed, the total width Γ is a Lorentz scalar, and, quite naturally, the $1/m_Q$ expansion for this quantity must run over the Lorentz scalar operators; $\bar{Q}Q$ is Lorentz scalar while $\bar{Q}\vec{\pi}^2 Q$ is not. The fact that there are no explicit $1/m_Q$ corrections in Eq. (2.24) does not mean that they are absent in Γ at all. They could appear through $\langle\bar{Q}Q\rangle m_Q/M_{H_Q}$. Hence, our next task is to find the expansion for $\langle\bar{Q}Q\rangle$ in the toy model at hand. To solve the problem we will use the very same idea as in Sec. 1.5; the only difference is the form of the heavy quark current. For the scalar quarks the current whose diagonal matrix element counts the number of quarks is $\bar{Q}\, i\overleftrightarrow{D}_\mu Q$. Hence, in the rest frame of H_Q, we have

$$\frac{1}{2M_{H_Q}}\langle H_Q|\bar{Q}\, i\overleftrightarrow{D}_0 Q|H_Q\rangle = 1 \ . \tag{2.26}$$

Passing to the rescaled fields we arrive at the relation

$$\begin{aligned} 1 &= \frac{1}{M_{H_Q}}\langle H_Q|m_Q\bar{Q}Q + \bar{Q}\pi_0 Q|H_Q\rangle \\ &= \frac{m_Q}{M_{H_Q}}\langle H_Q|\bar{Q}Q|H_Q\rangle + \frac{1}{2M_{H_Q}m_Q}\langle H_Q|\bar{Q}\vec{\pi}^2 Q|H_Q\rangle\,, \end{aligned} \tag{2.27}$$

where the second line is due to the equation of motion. Equation (2.27) leads us to the result sought for,

$$\frac{m_Q}{M_{H_Q}}\langle H_Q|\bar{Q}Q|H_Q\rangle = \left(1 - \frac{\mu_\pi^2}{2m_Q^2} + \dots\right) . \tag{2.28}$$

I have introduced here the standard notation for the expectation value of $\vec{\pi}^2$,

$$\left(\mu_\pi^2\right)_{\text{toy model}} = \frac{1}{2M_{H_Q}}\langle H_Q|2m_Q\bar{Q}\vec{\pi}^2 Q|H_Q\rangle \ . \tag{2.29}$$

As mentioned, μ_π^2 is a crucial parameter of the heavy quark theory. Its definition in QCD will be slightly different from that in our toy model, but the physical meaning will be the same.

Plugging Eq. (2.28) in (2.24) we finally arrive at the desired expression,

$$\Gamma = \Gamma_0\left(1 - \frac{\mu_\pi^2}{2m_Q^2}\right) . \tag{2.30}$$

The inclusive width coincides with the parton model result up to the terms of order $1/m_Q^2$. There is no correction of order $1/m_Q$! Moreover, the term $1/m_Q^2$ is calculable and its physical meaning is quite transparent: it reflects the time dilation for the moving quark inside the heavy hadron at rest (the Doppler effect). The coefficient $(-1/2)$ in front of $\langle\vec{\pi}^2\rangle/m_Q^2$ could, therefore, have been guessed from the very beginning, without explicit calculations, were we a little bit smarter.

This situation is quite general, it takes place not only in the toy model at hand but in real QCD as well. The absence of the correction of order $1/m_Q$ in the total inclusive widths (say, the semileptonic width of the B mesons, or $\Gamma(B \to X_s\gamma)$, and so on) is called[6] the CGG/BUV theorem [30, 24].

I hasten to add, though, that the absence of the $1/m_Q$ correction is *not* merely a consequence of the dimension gap in the set of the relevant operators, as it is sometimes stated in the literature. Indeed, let me give a counterexample. Let us calculate the average energy of the ϕ particle, or, more exactly, the first moment of the spectrum,

$$I_1 = \int_0^{E_0^{\rm phys}} dE\,(E_0^{\rm phys} - E)\,\frac{1}{\Gamma_0}\frac{d\Gamma}{dE}\,. \tag{2.31}$$

In the parton model the ϕ spectrum is a pure delta function and, consequently, I_1 vanishes. The heavy quark expansion does generate a nonvanishing result, a $1/m_Q$ effect. To see that this is indeed the case we integrate the theoretical spectrum (2.23) which yields

$$I_1 = \Delta - \frac{\mu_\pi^2 E_0^{\rm phys}}{2m_Q^2} \tag{2.32}$$

where Δ is defined as

$$\Delta = E_0^{\rm phys} - E_0, \tag{2.33}$$

and the parameters E_0 and $E_0^{\rm phys}$ are given in Eqs. (2.12) and (2.25). Now, invoking what we have already learnt in Sec. 1.5 about the heavy hadron masses we find that

$$\Delta = \frac{1}{2}v_0^2\bar{\Lambda} + \mathcal{O}(\Lambda_{\rm QCD}^2) \tag{2.34}$$

where

$$v_0 = \frac{M_{H_Q} - M_{H_q}}{M_{H_Q}}\,.$$

In the SV limit v_0 is small and coincides with the velocity of the final hadron produced in the transition $Q \to q\phi$.

With all these definitions

$$I_1 = \frac{1}{2}v_0^2\bar{\Lambda} + \mathcal{O}(\Lambda_{\rm QCD}^2)\,, \tag{2.35}$$

---

[6]CGG/BUV stands for Chay-Georgi-Grinstein-Bigi-Uraltsev-Vainshtein.

i.e. the preasymptotic correction in the first moment is of the order of $\Lambda_{\rm QCD}$, not $\Lambda^2_{\rm QCD}$. The reason for the occurrence of a "wrong" power of the QCD parameter is that the leading correction term in the $1/m_Q$ expansion in this particular quantity is unrelated to any local operator. As we will see later on, such a situation is not rare in the heavy quark theory. The sum rule (2.35) is just a version of Voloshin's optical sum rule [31], while that of Eq. (2.30) can be interpreted in terms of the Bjorken sum rule [32]. We will dwell on both sum rules in real QCD in Sec. 3.6.

I apologize for this little waterfall of new letters and definitions and hope that a simple picture behind our results is not overshadowed. Notice that in the SV limit $E_0^{\rm phys}-E$ reduces to the excitation energy of the final hadron produced in the decay. The factor $E_0^{\rm phys}-E$ in the integrand eliminates the "elastic" peak, so that the integral is saturated only by the inelastic contributions. Say, in the $b\to c\phi$ transition, the contribution of $B\to D\phi$ is eliminated, only the excited D mesons survive in the first moment. Since the excitation energies are of the order of $\Lambda_{\rm QCD}$, or $\bar\Lambda$, the prediction (2.35) means that the probabilities of the inelastic transitions $B\to$ excited D's are all proportional to v^2. This is in full agreement with the theorem [15] discussed in Sec. 1.3 — that in the point of zero recoil the only transition that can occur is the elastic $B\to D$ transition, with the unit probability. Away from the point of the zero recoil (but in the SV limit) the inelastic transitions are generated. However, Eq. (2.30) shows, that up to small corrections $\mathcal{O}(\Lambda^2_{\rm QCD}/m_Q^2)$ which can be neglected if we are interested only in the linear in $\Lambda_{\rm QCD}$ effects, the total probability remains unity. In other words, the total probability is just reshuffled: a small v^2 part is taken away from the elastic transition and is given to the inelastic transitions. The QCD analog of this assertion is the essence of the Bjorken sum rule [32].

It is quite evident that the series of such sum rules can readily be continued further. For the next moment, for instance, we get

$$I_2=\int_0^{E_0^{\rm phys}} dE\,(E_0^{\rm phys}-E)^2\,\frac{1}{\Gamma_0}\frac{d\Gamma}{dE}\;=\;\Delta^2+\frac{\mu_\pi^2E_0^2}{3m_Q^2}\;. \tag{2.36}$$

Analyzing this sum rule in the SV limit, one obtains, in principle, additional information, not included in the results of Refs. [15, 31, 32]. It is worth emphasizing that in Eqs. (2.24), (2.32) and (2.36) we have collected all terms through the order $\Lambda^2_{\rm QCD}$, whereas those of the order $\Lambda^3_{\rm QCD}$ are systematically omitted. Predictions for higher moments would require calculating the terms $\mathcal{O}(\Lambda^3_{\rm QCD})$ and higher.

Concluding this part, let me suggest to you an exercise which will show whether the technology introduced above is well understood by you. Try to repeat in real QCD, with the quark spins switched on, everything we have done in the toy model. Of particular interest to us will be the transition operator

$$\hat{T} = i \int d^4x \, \mathrm{e}^{-iqx} T\{\bar{Q}(x)\Gamma_\mu q(x) \, , \, \bar{q}(0)\Gamma_\nu Q(0)\} \tag{2.37}$$

where Γ_μ is either γ_μ or $\gamma_\mu\gamma_5$. This transition operator is relevant for the semileptonic b to c decays. To facilitate the task consider special kinematics: (i) zero recoil (the vanishing spatial momentum of the lepton pair, $\vec{q} = 0$; (ii) small velocity limit $\vec{q}^2 \ll m_Q^2$. To further facilitate the task limit yourself to the spatial components of Γ_μ. If you still have problems go over this section again and consult the original works [33, 34, 35]. The full answer for the transition operator (2.37) is given, for instance, in Appendix of Ref. [33].

2.2. *The Fock-Schwinger Gauge*

In some situations (especially when one deals with massless quarks) a variant of the background field technique based on the so-called Fock-Schwinger gauge for the external field turns out to be very efficient (for a review and extensive list of references see Ref. [28]). The gauge condition on the background gluon field has the form

$$x_\mu A_\mu(x) = 0 \, . \tag{2.38}$$

What is remarkable in this condition is that in this gauge the gauge four-potential can be represented as an expansion which runs only over the gauge-covariant quantities, the gluon field strength tensor and its covariant derivatives,

$$A_\mu(x) = \frac{1}{2} x_\rho G_{\rho\mu}(0) + \frac{1}{3}\frac{1}{1!} x_\alpha x_\rho D_\alpha G_{\rho\mu}(0) + \ldots \, . \tag{2.39}$$

This expression implies, in particular, that $A(0) = 0$. It is worth noting that the gauge condition (2.38) singles out the origin and, hence, breaks the translational invariance. The latter is restored only in the final answer for the gauge-invariant amplitudes.

It is rather easy to show (see Ref. [28]) that the massless quark Green function in the coordinate space is

$$S(x,0) = \frac{1}{2\pi^2}\frac{\not{x}}{x^4} - \frac{1}{8\pi^2}\frac{x_\alpha}{x^2}\tilde{G}_{\alpha\phi}\gamma_\phi\gamma_5 + \ldots \, , \quad \tilde{G}_{\alpha\phi} = \frac{1}{2}\epsilon_{\alpha\phi\mu\nu}G_{\mu\nu} . \tag{2.40}$$

One can also construct a similar expansion for the Green function in the momentum space $S(q)$.

If the quark is not massless, $m_q \neq 0$, the expansion of the Green function in the background field becomes much more cumbersome. Although we will hardly need it in full I quote it here for the sake of completeness,

$$S(x,0) = \frac{1}{2\pi^2}\frac{\not{x}}{x^4}\left\{-\frac{1}{2}m^2x^2K_2(m\sqrt{-x^2})\right\}$$
$$-\frac{1}{8\pi^2}\frac{x_\alpha}{x^2}\tilde{G}_{\alpha\phi}\gamma_\phi\gamma_5\left\{\frac{-x^2mK_1(m\sqrt{-x^2})}{\sqrt{-x^2}}\right\}$$
$$-\frac{im^2}{4\pi^2}\frac{K_1(m\sqrt{-x^2})}{\sqrt{-x^2}} - \frac{im}{16\pi^2}G_{\rho\lambda}\sigma^{\rho\lambda}K_0(m\sqrt{-x^2}) + \dots \quad (2.41)$$

Here K is the McDonald function. This result was obtained in Ref. [36]. Further education on the Fock-Schwinger gauge technique can be obtained from Ref. [28]. The best way to master those aspects which are most common in the heavy quark theory is merely to play with this tool kit. Let us see how it works in the calculation of the $1/m_Q^2$ correction in the total probability of the semileptonic decay of the heavy quark in real QCD. Unlike Sec. 2.1 we will address directly the total width, bypassing the stage of the spectrum. The first calculation of the power correction in $\Gamma(B \to X_c l\nu)$ along these lines was carried out in Ref. [24] (see also Ref. [37]).

2.3. *The $1/m_Q$ Corrections to the Semileptonic Inclusive Width in QCD*

In this Chapter I will describe probably the most elegant application of the ideas developed above — the calculation of the leading correction in the total semileptonic widths. In the toy model considered in the previous chapter, it was established that the $1/m_Q$ correction was absent, and the first nontrivial correction $1/m_Q^2$ was associated with the matrix element of $\langle\bar{Q}Q\rangle$. As a matter of fact, through the heavy quark expansion, we managed to express it in terms of the matrix element $\langle\bar{Q}\vec{\pi}^2Q\rangle$.

When we take the quark spins into account a new dimension 5 Lorentz scalar operator appears, $\langle\bar{Q}(i/2)\sigma GQ\rangle$. On general grounds one may expect that the main lesson abstracted from the toy model — the absence of the $1/m_Q$ term — persists but the $1/m_Q^2$ correction will receive a contribution from the operator $\bar{Q}(i/2)\sigma GQ$. The conclusion will be confirmed by the analysis presented below. You will see how efficiently the Fock-Schwinger technique is in this case.

Thus, let us proceed to the calculation of semileptonic widths. The final quark mass m_q is arbitrary — we do not assume the SV limit now, nor is any other constraint imposed on m_q. The weak Lagrangian responsible for the semileptonic decays has the generic form

$$\mathcal{L} = \frac{G_F}{\sqrt{2}} V_{Qq} (\bar{q}\Gamma_\mu Q)(\bar{l}\Gamma_\mu \nu)\,, \quad \Gamma_\mu = \gamma_\mu(1+\gamma_5)\,, \tag{2.42}$$

where l is a charged lepton, electron for definiteness. The mass of the charged lepton will be neglected. Moreover, G_F and V_{Qq} are constants irrelevant for our purposes. As usual, at the first stage we construct the transition operator $\hat{T}(Q \to X \to Q)$,

$$\hat{T} = i \int d^4x T\{\mathcal{L}(x)\mathcal{L}(0)\} = \sum_i C_i \mathcal{O}_i \tag{2.43}$$

describing a diagonal amplitude with the heavy quark Q in the initial and final state (with identical momenta). The lowest dimension operator in the expansion of $\hat{T}(Q \to X \to Q)$ is $\bar{Q}Q$, and the complete perturbative prediction — the spectator model — corresponds to the perturbative calculation of the coefficient of this operator. For the time being we are not interested in perturbative calculations. Our task is the analysis of the influence of the soft modes in the gluon field manifesting themselves as a series of higher dimension operators in $\hat{T}$.

At the second stage we average $\hat{T}$ over the hadronic state of interest, say, B mesons. At this stage the nonperturbative large distance dynamics enters through matrix elements of the operators of dimension 5 and higher.

Finally, the imaginary part of $\langle H_Q|\hat{T}|H_Q\rangle$ presents the H_Q semileptonic width sought for,

$$\Gamma = \frac{1}{M_{H_Q}} \mathrm{Im}\langle H_Q|\hat{T}|H_Q\rangle\,. \tag{2.44}$$

The diagram determining the transition operator is depicted in Fig. 2. The lepton propagators are, of course, free — they do not feel the background gluon field. Thick lines refer to the initial quark Q. Although the gluon field is not shown one should understand that the lines corresponding to Q and q are submerged into a soft gluon background.

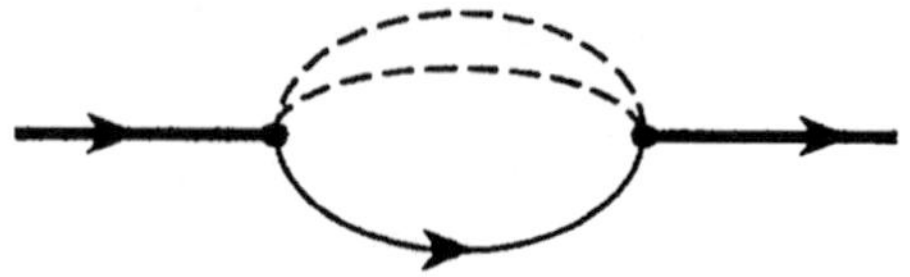

Fig. 2. The transition operator relevant to the total semileptonic width of the heavy mesons. The dashed lines denote the leptons, l and ν. Other notations are the same as in Fig. 1.

In the Fock-Schwinger gauge the line corresponding to the final quark q (Fig. 2) remains free, and the only source of the dimension 5 operators is the external line corresponding to Q (or $\bar{Q}$). Let us elaborate this point in more detail.

If we do not target corrections higher than $1/m_Q^2$ it is sufficient to use the expression for the quark Green function given in Eq. (2.40) or (2.41). The particular form is absolutely inessential; the only important point is the chiral structure of the vertices in the weak Lagrangian and the fact that the leptons are massless.

The currents in the weak Lagrangian (2.42) are left-handed. Therefore, the Green function of the quark q is sandwiched between Γ_μ and Γ_ν. This means that $1+\gamma_5$ projectors annihilate the part of the Green function with the even number of the γ matrices. Then the only potential contribution is associated with the first and the second lines in Eq. (2.41).

The nonperturbative term is the one containing $\tilde{G}_{\alpha\phi}$. This term vanishes, however, after convoluting it with the lepton part. Indeed, the lepton loop (with the massless leptons) has the form

$$L_{\mu\nu} = -\frac{2}{\pi^4}\frac{1}{x^8}(2x_\mu x_\nu - x^2 g_{\mu\nu}). \tag{2.45}$$

Here I take the product of two massless fermion propagators in the coordinate space, with the appropriate γ matrices inserted, and do the trace. Actually we need to know only the expression in the brackets. Now, convoluting it with the $\tilde{G}$ term from the quark Green function we get

$$(\Gamma_\mu x_\alpha \tilde{G}_{\alpha\phi}\gamma_\phi\Gamma_\nu)(2x_\mu x_\nu - x^2 g_{\mu\nu}) \equiv 0,$$

q.e.d. [24, 38].

Thus, if one uses the Fock-Schwinger gauge the only source for the $1/m_Q$ correction in the total semileptonic widths (at the level up to $1/m_Q^2$) is through the equations of motion for Q. Here is how it works.

The expression for the amplitude corresponding to the diagram of Fig. 2 can be generically written as follows:

$$\mathcal{A} = \int d^4x \bar{Q}(0) F(x) Q(x), \tag{2.46}$$

where a function $F(x)$ incorporates the lepton loop and the q quark Green function. It may include Lorentz and color matrices, etc. Now, let us single out the large, mechanical part of the motion of the heavy quark,

$$Q(x) = \mathrm{e}^{-iP_0 x} \tilde{Q}(x)\,.$$

Then

$$\begin{aligned}
\mathcal{A} &= \int d^4x \left\{ \bar{\tilde{Q}}(0) F(x) \mathrm{e}^{-iP_0 x} \left[\tilde{Q}(0) + x_\mu \partial_\mu \tilde{Q}(0) + \frac{1}{2} x_\mu x_\nu \partial_\mu \partial_\nu \tilde{Q}(0) + \ldots \right] \right\} \\
&= \left\{ \bar{\tilde{Q}}(0) \left[\tilde{F}(P_0) \tilde{Q}(0) + i \frac{\partial}{\partial P_{0\mu}} \tilde{F}(P_0) \partial_\mu \tilde{Q}(0) \right.\right. \\
&\quad \left.\left. + i^2 \frac{1}{2} \frac{\partial^2}{\partial P_{0\mu} \partial P_{0\nu}} \tilde{F}(P_0) \partial_\mu \partial_\nu \tilde{Q}(0) + \ldots \right] \right\} \\
&= \bar{\tilde{Q}}(0) \tilde{F}(P_0 + i\partial) \tilde{Q}(0) = \bar{Q}(0) \tilde{F}(i\partial) Q(0),
\end{aligned} \tag{2.47}$$

where $\tilde{F}$ is the Fourier transform of $F(x)$.

Our next goal is to convert $i\partial$ in the covariant derivative and then use the equation of motion, $i \not{D} Q = m_Q Q$. More exactly, we start from the expressions of the form

$$\bar{Q}(0) \not{p} (p^2)^k Q(0), \;\; p_\mu = i\partial_\mu, \tag{2.48}$$

rewrite p_μ in terms of $\mathcal{P}_\mu = iD_\mu$ plus terms with the gluon field strength tensor (in the Fock-Schwinger gauge) and then substitute $\not{\mathcal{P}}$ acting on Q by m_Q. Expressions (2.48) appear in $\mathrm{Im}\hat{T}$. If the final q quark is massless, $m_q = 0$, the only relevant power is $k = 2$. Switching on the quark mass, $m_q \neq 0$, brings in other values of k as well. (Warning: in the procedure sketched above all operators p in Eq. (2.48) should be considered as acting either only to the right or only to the left. I will assume they act to the right. We cannot make some of them act to the right and others to the left and neglect full derivatives. Question: do you understand why?)

Since we focus now on $\bar{Q}\sigma G Q$ it is sufficient to keep only the terms linear in the gluon field strength tensor; the terms with derivatives of $G_{\mu\nu}$ are to be neglected as well. In this approximation in the Fock-Schwinger gauge $A_\mu = (1/2)x_\rho G_{\rho\alpha}$.

Furthermore,

$$p^2 = \mathcal{P}^2 - 2Ap \tag{2.49}$$

where we neglected the terms quadratic in A and used the fact that $[p_\mu, A_\mu] = 0$. Equation (2.49) should be substituted in Eq. (2.48), and then we start transposing Ap trying to put it in the left-most position, next to $\bar{Q}(0)$. If Ap appears in this position the result is zero since $A(0) = 0$. Notice that

$$[p^2, Ap] \propto G_{\alpha\beta} p_\alpha p_\beta = 0,$$

so, one can freely transpose Ap through p^2. In this way we arrive at

$$\bar{Q}(0)\, \not{p}(p^2)^k Q(0) = \bar{Q}(0)\left[\not{\mathcal{P}}(\mathcal{P}^2)^k - 2k\, \not{\mathcal{P}}(A\mathcal{P})(\mathcal{P}^2)^{k-1}\right] Q(0)\,. \tag{2.50}$$

Moreover, with our accuracy the last term reduces to

$$\begin{aligned}\gamma_\alpha[\mathcal{P}_\alpha, A_\mu]\mathcal{P}_\mu(\mathcal{P}^2)^{k-1} &= \gamma_\alpha \frac{i}{2} G_{\alpha\mu}\mathcal{P}_\mu(\mathcal{P}^2)^{k-1} \\ &= -\frac{i}{8}(\not{\mathcal{P}}\sigma G - \sigma G\, \not{\mathcal{P}})\not{\mathcal{P}}^{2k-2}\end{aligned} \tag{2.51}$$

and, hence, using the equations of motion we conclude that

$$\bar{Q}(0)\, \not{\mathcal{P}}(A\mathcal{P})(\mathcal{P}^2)^{k-1} Q(0) = 0. \tag{2.52}$$

As a result, the σG terms emerge only due to the fact that

$$\mathcal{P}^2 = \not{\mathcal{P}}^2 - \frac{i}{2}\sigma G,$$

and the final expression is as follows:

$$\bar{Q}(0)\, \not{p}(p^2)^k Q(0) \to m_Q^{2k+1}\bar{Q}(0)Q(0) - \frac{ik}{2}\bar{Q}(0)\sigma G Q(0) m_Q^{2k-1}. \tag{2.53}$$

After these explanatory remarks the procedure of calculating the leading $1/m_Q^2$ correction in the total semileptonic width should be perfectly clear. Let us summarize it in the form of a prescription.

(i) Calculate the semileptonic width in the parton model. The result has the form

$$\frac{G_F^2|V_{Qq}|^2}{192\pi^3} m_Q^5 F_0\left(\frac{m_q^2}{m_Q^2}\right) \tag{2.54}$$

where $F_0(m_q^2/m_Q^2)$ is the phase space factor, a function of the ratio m_q^2/m_Q^2 well-known in the literature (see Eq. (2.57) below).

(ii) Then construct the expression for Γ including the $\mathcal{O}(m_Q^{-2})$ corrections in the following way [24]:

$$\Gamma = \frac{G_F^2|V_{Qq}|^2}{192\pi^3} m_Q^5 \left\{ \frac{1}{2M_{H_Q}} \langle H_Q|\bar{Q}Q|H_Q\rangle F_0(\rho) + \frac{\mu_G^2}{m_Q^2} \left(\rho \frac{d}{d\rho} - 2 \right) F_0(\rho) \right\} \tag{2.55}$$

where

$$\mu_G^2 = \frac{1}{2M_{H_Q}} \langle H_Q|\bar{Q}(i/2)\sigma G Q|H_Q\rangle \tag{2.56}$$

and

$$\rho = \frac{m_q^2}{m_Q^2}\,.$$

A few comments are in order concerning this beautiful expression for the total semileptonic width. The expansion contains two Lorentz scalar operators, $\bar{Q}Q$ and $\bar{Q}(i/2)\sigma G Q$, of dimensions 3 and 5, respectively. The fact that only the Lorentz scalars contribute is obvious since Γ is a Lorentz-invariant quantity. We observe here the very same gap in dimensions mentioned previously in the context of the toy model — there is no operator of dimension 4 [30]. The only element still needed to complete the derivation is the matrix element $\langle \bar{Q}Q \rangle$. Fortunately, the corresponding heavy quark expansion has already been built, see Eq. (1.30).

Borrowing the explicit expression for $F_0(\rho)$ from textbooks (it is singled out below in the braces) and assembling all other pieces together we finally get

$$\Gamma = \frac{G_F^2|V_{Qq}|^2}{192\,\pi^3} m_Q^5 \times \left[\left(1 + \frac{\mu_G^2 - \mu_\pi^2}{2m_Q^2} \right) \left\{ 1 - 8\rho + 8\rho^3 - \rho^4 - 12\rho^2 \ln\rho \right\} - 2\frac{\mu_G^2}{m_Q^2}(1-\rho)^4 \right], \tag{2.57}$$

where μ_π^2 is defined in QCD as

$$\mu_\pi^2 = \frac{1}{2M_{H_Q}} \langle H_Q|\bar{Q}\vec{\pi}^2 Q|H_Q\rangle\,. \tag{2.58}$$

This result is due to Bigi *et al.* [24]. The absence of the $1/m_Q$ correction is a manifestation of the CGG/BUV theorem.

If the mass of the final charged lepton is nonnegligible, the property of no soft gluon emission from the q quark line (in the Fock-Schwinger gauge) is lost. The expansion for $\Gamma(B \to X_c \tau \nu_\tau)$ in this case is much more cumbersome; it was constructed in Ref. [39].

Dimension 5 operators are responsible for the *leading* nonperturbative corrections in the total semileptonic widths. To assess the convergence of the expansion it may be instructive to have an idea of the higher order terms in the expansion. The $1/m_Q^3$ terms were estimated in Ref. [40]. This is a rather messy and time-consuming analysis, and it is hardly in order to comment on it in this section. Surprising though it is, from what we already know, it is practically trivial to find the coefficient of the dimension 7 operator

$$\bar{Q} G_{\mu\nu} G_{\mu\nu} Q, \tag{2.59}$$

with two gluon field strength tensors fully contacted over the Lorentz indices for massless final quarks (say, $b \to u$ transition). Of course, this is a purely academic exercise, for many reasons. In particular, because it is only one of a rather large number of dimension 7 operators. Since we do not know their matrix elements anyhow, it seems to be meaningless to carry out full classification and calculate all coefficients. The operator (2.59) is chosen since one can at least use factorization for a rough estimate of the corresponding matrix element and since we get its coefficient essentially for free.

The point is that the massless quark propagator (in the Fock-Schwinger gauge) does not contain $G_{\mu\nu} G_{\mu\nu}$ term at all (see Ref. [41]). This fact implies that the only source of the operator (2.59) is the same as in the case of $\bar{Q} \sigma G Q$. It is not difficult to get that

$$\begin{aligned} \bar{Q}(0)\, \not{p} p^4 Q(0) &= \bar{Q}(0)\, \not{\mathcal{P}} \mathcal{P}^4 Q(0) + \frac{1}{2} \bar{Q}(0) G_{\mu\nu} G_{\mu\nu} Q(0) \\ &= \bar{Q}(0) \not{\mathcal{P}}^5 Q(0) + \bar{Q}(0) G_{\mu\nu} G_{\mu\nu} Q(0)\,, \end{aligned} \tag{2.60}$$

where we omitted structures of the type $[G_{\mu\nu} G_{\mu\alpha} - (1/4) g_{\nu\alpha} G_{\mu\rho} G_{\mu\rho}]$.

2.4. *Digression*

This section is intended for curious readers — those who are anxious to find out where and how else, beyond the theory of the H_Q states, the background field technique can be used to obtain interesting predictions. Here I will discuss an estimate of the mass splittings between the levels of the highly excited

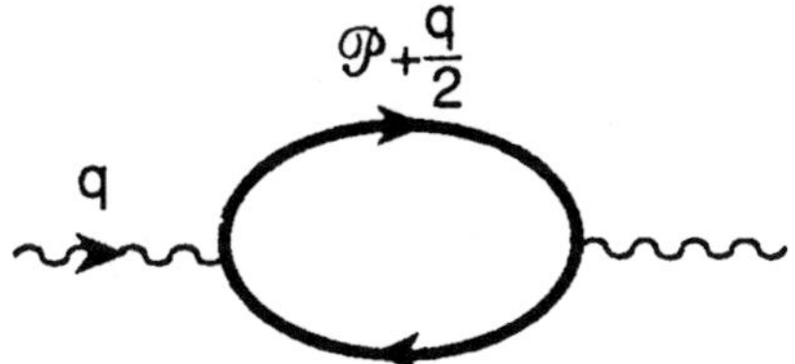

Fig. 3. The two-point function (2.62). The wavy line is the external scalar or pseudoscalar current; the quark propagator is in the background field.

quarkonium states. This part can be safely omitted in the first reading since it is unrelated to the remainder of these lectures.

The quarkonium states to be considered below consist of one quark Q, one antiquark $\bar{Q}$ and the soft gluon cloud connecting them together. To begin we will assume that m_Q is large, $m_Q \gg \Lambda_{\rm QCD}$ (later on this assumption will be relaxed). The $Q\bar{Q}$ mesons can have different quantum numbers. We will analyze the excited S and P wave states with the quantum numbers 0^- and 0^+, respectively. The naive quark model language is used to name the states; this does not mean, of course, that we accept any of the dynamical assumptions of the naive quark model. It would be more accurate to say that the mesons of interest are produced from the vacuum by the currents

$$J_P = \bar{Q} i\gamma_5 Q \quad \text{and} \quad J_S = \bar{Q}Q\,. \tag{2.61}$$

The central object of our analysis is the difference between the two-point functions in the pseudoscalar and scalar channels. In terms of the Green functions in the background field this difference takes the form (see Fig. 3)

$$\begin{aligned}
\Pi_P - \Pi_S &= i\int dx e^{iqx}\left(\langle \text{vac}|T\{J_P(x)J_P(0)\}|\text{vac}\rangle - \langle \text{vac}|T\{J_S(x)J_S(0)\}|\text{vac}\rangle\right) \\
&= i\text{Tr}\left\{ i\gamma_5 \frac{1}{\not{\mathcal{P}} + (\not{q}/2) - m_Q} i\gamma_5 \frac{1}{\not{\mathcal{P}} - (\not{q}/2) - m_Q} \right. \\
&\quad \left. - \frac{1}{\not{\mathcal{P}} + (\not{q}/2) - m_Q}\, \frac{1}{\not{\mathcal{P}} - (\not{q}/2) - m_Q} \right\}.
\end{aligned} \tag{2.62}$$

A comment is in order here concerning the trace operation in this expression. It implies not only the trace over the Lorentz and color indices, as usual, but also the trace in the momentum space substituting $\int d^4p/(2\pi)^4$ in the conventional

Feynman integral. With the help of Eq. (2.6) the difference $\Pi_P - \Pi_S$ can be identically rewritten as

$$\Pi_P - \Pi_S = -2m_Q^2 i\mathrm{Tr}\left\{\frac{1}{D_+}\frac{1}{D_-}\right\} \tag{2.63}$$

where

$$D_\pm = [\mathcal{P} \pm (q/2)]^2 - m_Q^2 + (i/2)\sigma G\,.$$

We continue to ignore hard gluons assuming that the only role of the gluon field is to provide a soft cementing background. This is certainly an idealization, but let us see how far one can go within the framework of this simplified picture. Neglecting hard gluons means, in particular, that we will be unable to analyze the low-lying levels of heavy quarkonium where an essential role is played by the short distance Coulomb interaction. Each σG insertion in the denominator is of the order of $\Lambda^2_{\rm QCD}$, i.e. it does not scale with the external momentum q when q is large. Let us expand Eq. (2.63) in σG and take the trace over the Lorentz indices. Then the first order term drops out; the first surviving term is bilinear in G,

$$\begin{aligned}\Pi_P - \Pi_S = -8m_Q^2 i\mathrm{Tr}&\left\{\frac{1}{D_+^0}\frac{1}{D_-^0}\right.\\ &+\frac{1}{2}\frac{1}{D_+^0}G_{\alpha\beta}\frac{1}{D_+^0}G_{\alpha\beta}\frac{1}{D_+^0}\,\frac{1}{D_-^0}+\frac{1}{2}\frac{1}{D_-^0}G_{\alpha\beta}\frac{1}{D_-^0}G_{\alpha\beta}\frac{1}{D_-^0}\,\frac{1}{D_+^0}\\ &\left.+\frac{1}{2}\,\frac{1}{D_+^0}G_{\alpha\beta}\frac{1}{D_+^0}\,\frac{1}{D_-^0}G_{\alpha\beta}\frac{1}{D_-^0}\right\}+\dots\,,\end{aligned} \tag{2.64}$$

where $D_\pm^0 = [\mathcal{P} \pm (q/2)]^2 - m_Q^2$.

A closer look at this expression reveals some peculiar features. First of all, one can interpret each term as a certain correlation function in the theory where the quark Q is scalar, not spinor. Take, for instance, this first line. It is nothing else than the two-point function of the $L = 0$ quarkonium in scalar QCD (i.e. QCD with the scalar quarks; L is the total angular momentum of the meson). The current producing the scalar quarkonium from the vacuum in scalar QCD is $Q^\dagger Q$ (Fig. 4). The second and the third lines, together, represent the four-point function of the type depicted in Fig. 5. The current denoted by the dashed line in this figure is $Q^\dagger G_{\alpha\beta} Q$; the momentum flowing through this line vanishes. Two insertions of G imply that this four-point function is explicitly proportional to $\Lambda^4_{\rm QCD}$.

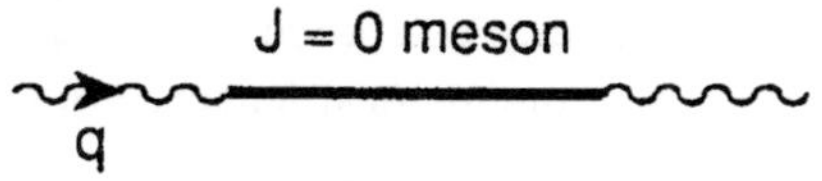

J = 0 meson

q

Fig. 4. The two-point function of two scalar currents in scalar QCD.

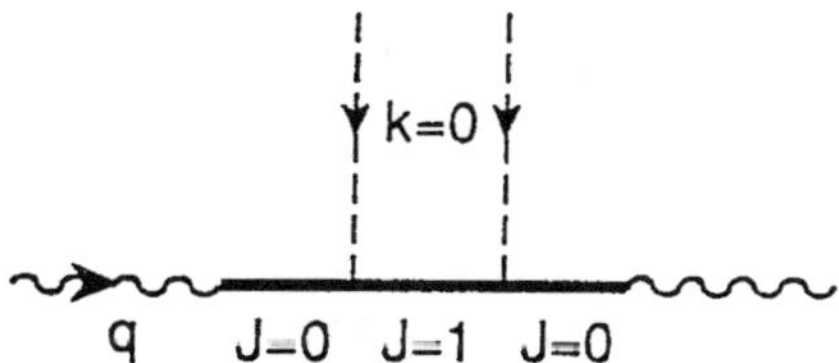

Fig. 5. The four-point function appearing in Eq. (2.64). The dashed line denotes the current $Q^{\dagger}G_{\alpha\beta}Q$.

Now let us examine Eq. (2.64) in the complex q^2 plane. At some positive values of q^2 the two-point function of Fig. 4 has simple poles corresponding to positions of the $L = 0$ quarkonium levels in scalar QCD. The four-point function of Fig. 5 has double and single poles at the very same values of q^2 and single poles at some other values of q^2 corresponding to the production of $L = 1$ states in scalar QCD. The latter are produced from the $L = 0$ states by applying to them the current $Q^{\dagger}G_{\alpha\beta}Q$.

On the other hand, the original difference $\Pi_P - \Pi_S$ in real QCD has only single poles at the positions of the S and P wave states. These positions are shifted compared to the levels in scalar QCD. Expanding in the shift one generates double poles. The $L = 1$ pole — the partner to the P wave meson states in $\Pi_P - \Pi_S$ appears only in the four-point function of Fig. 5. From this figure it is quite clear that the residue of the $L = 1$ pole scales as $\Lambda^4_{\mathrm{QCD}}/(\Delta(M^2))^2$ where $\Delta(M^2)$ is a characteristic splitting between the $L = 0$ and $L = 1$ states. On the other hand, the residue of the P wave meson in $\Pi_P - \Pi_S$, on general grounds, scales as $\vec{p}^{\,2}/M^2$ where $\vec{p}$ is a characteristic quark momentum, and I assume that $\Lambda_{\mathrm{QCD}} \ll |\vec{p}| \ll M$. Equating these two estimates we find that

$$\Delta M \sim \Lambda^2_{\mathrm{QCD}}/|\vec{p}|\,.$$

One may observe, with satisfaction, that this is exactly the characteristic level splitting (between radial or orbital excitations) for two heavy quarks interacting through a string ("linear potential"). What is remarkable is that in no place does our estimate invoke any reference to the linear potential or other models. It was based only on some general features of QCD.

What changes if the quarks are light or even massless, $m_Q \to 0$? The only difference is that now the residues of the P wave mesons in $\Pi_P - \Pi_S$ are of the same order as those of the S wave mesons for highly excited states, which implies that

$$\Lambda_{\rm QCD}^4/(\Delta(M^2))^2 \sim 1\,,$$

or

$$\Delta(M^2) \sim \Lambda_{\rm QCD}^2\,.$$

In other words, we got the linear Regge trajectories, at least for highly excited states. Moreover, this analysis makes clear a potentially important point — the empirical observation that even the lowest states in every channel lie on the linear Regge trajectories looks like a numerical coincidence and cannot be exact.

SECTION 3

Classic Problems with Heavy Quarks

The number of problems successfully solved within the heavy quark expansion is quite large. Even a brief review of the main applications is beyond the scope of these lectures. Some issues, however, are quite general and are important in a variety of applications. Everybody, not only the heavy quark practitioners, should know them. In this section we will discuss several such topics — the scaling of the heavy meson coupling constants, some properties of the Isgur-Wise function, and, finally, the analysis of corrections violating the heavy quark symmetry at the point of zero recoil. We begin, however, from a systematic classification of all *local* operators which appear in the heavy quark expansion up to the level $\mathcal{O}(m_Q^{-3})$. Some terms of the order $1/m_Q^3$, in particular heavy quark expansions, are actually not expressible in terms of the local operators and are, rather, related to nonlocal correlation functions. A full classification of such corrections also exists [42, 25], but we will not go into details, only marginally mentioning them here and there. The interested reader is referred to the original publications [42, 25].

3.1. *Catalog of Relevant Operators*

The local operators in the heavy quark expansion are bilinear in the heavy quark field. They are certainly gauge-invariant, and in many instances, when the expansion is built for scalar quantities, the operators must be Lorentz scalars. As with any operator product expansion in QCD they can be ordered according to their dimension. We will limit ourselves here to dimension 6 and lower. This leaves us with quite a few possibilities listed below. We start with the Lorentz scalar operators. The only appropriate operators are

$$\bar{Q}Q\,,\ \ \frac{i}{2}\bar{Q}\sigma_{\mu\nu}G_{\mu\nu}Q\,,\ \ \text{and}\ \ \bar{Q}\Gamma Q\bar{q}\Gamma q \tag{3.1}$$

where Γ stands here for a combination of γ and color matrices. All other structures that might come to one's mind reduce to those listed above and full derivatives by virtue of the equations of motion.

(Exercise: prove that this is the case, for instance, for the operators $\bar{Q}D^2Q$ and $\bar{Q}G_{\mu\nu}\gamma_\mu D_\nu Q$.)

(i) The only operator of dimension 3 is $\bar{Q}Q$. This operator is related to the heavy quark current $\bar{Q}\gamma_0 Q$ plus terms suppressed by powers of $1/m_Q$. The leading term of this expansion has already been discussed, see Eq. (1.30). Actually it is not difficult to continue this expansion one step further. The following relation is *exact*:

$$\begin{aligned}\bar{Q}\,Q &= \bar{Q}\gamma_0 Q \ + \ 2\bar{Q}\left(\frac{1-\gamma_0}{2}\right)^2 Q \\ &= \bar{Q}\gamma_0 Q \ - \ 2\bar{Q}\,\frac{\overleftarrow{\not{\pi}}}{2m_Q}\cdot\frac{\overrightarrow{\not{\pi}}}{2m_Q}\,Q \\ &= \bar{Q}\gamma_0 Q \ + \ \bar{Q}\frac{\not{\pi}^2}{2m_Q^2}Q \ + \ \text{a total derivative}\ ;\end{aligned} \tag{3.2}$$

in the first relation above the operators $\overleftarrow{\pi}_\mu$ act on the $\bar{Q}$ field. Keeping in mind that we always consider only the forward matrix elements, with the zero momentum transfer, we can drop all terms with total derivatives. Applying now the equations of motion (1.8) and (1.9) generates the $1/m_Q$ expansion for the scalar density,

$$\bar{Q}\,Q = \bar{Q}\,\gamma_0\,Q + \frac{1}{2m_Q^2}\bar{Q}\left(\pi^2 + \frac{i}{2}\sigma G\right)Q$$
$$= \bar{Q}\,\gamma_0\,Q - \frac{1}{2m_Q^2}\bar{Q}\,(\vec{\pi}\vec{\sigma})^2 Q - \frac{1}{4m_Q^3}\,\bar{Q}(-(\vec{D}\vec{E}) + 2\vec{\sigma}\cdot\vec{E}\times\vec{\pi})Q$$
$$+\mathcal{O}\left(\frac{1}{m_Q^4}\right). \tag{3.3}$$

Here $E_i = G_{i0}$ is the chromoelectric field, and its covariant derivative is defined as[7] $D_j E_k = -i[\pi_j, E_k]$; we have omitted the term $\bar{Q}([\pi_k, [\pi_0, \pi_i]] - [\pi_i, [\pi_0, \pi_k]])Q$ using the Jacobi identity. Moreover,

$$\vec{D}\vec{E} = g^2 t^a J_0^a$$

by virtue of the QCD equation of motion (here $J_\mu^a = \sum_q \bar{q}\gamma_\mu t^a q$ is the color quark current). Therefore the first of the $1/m_Q^3$ terms can be rewritten as a four-fermion operator.

(ii) As has already been mentioned, no operators of dimension 4 exist.

(iii) Dimension five. There is only one such operator,

$$\mathcal{O}_G = \bar{Q}\frac{i}{2}\sigma_{\mu\nu}G_{\mu\nu}Q\,, \tag{3.4}$$

where $\sigma_{\mu\nu} = (1/2)[\gamma_\mu, \gamma_\nu]$. Again, it can be expanded in the powers of $1/m_Q$,

$$\mathcal{O}_G = -\bar{Q}\vec{\sigma}\vec{B}Q - \frac{1}{2m_Q}\bar{Q}(-(\vec{D}\vec{E}) + 2\vec{\sigma}\cdot\vec{E}\times\vec{\pi})Q + \mathcal{O}\left(\frac{1}{m_Q^2}\right), \tag{3.5}$$

where $\vec{B}$ is the chromomagnetic field, $\vec{B} = \vec{\nabla}\times\vec{A} = -(1/2)\epsilon_{ijk}G_{jk}$.

(iv) Dimension 6 four-quark operators $\sum_i \bar{Q}\Gamma_i Q\bar{q}\Gamma_i q$. Generally speaking, the matrix Γ_i can be any Lorentz matrix $(1, \gamma_5, \gamma_\mu, \gamma_\mu\gamma_5$ or $\sigma_{\mu\nu})$ or any of the above multiplied by t^a. Of course, in specific problems, only a subset of these matrices may appear. The four-quark operators differ by the chiral properties of the light quark field q. Some of them carry nonzero chirality (they are nonsinglet with respect to $\mathrm{SU}(N_f)_L \times \mathrm{SU}(N_f)_R$). Hence, they do not show up in the transitions associated with the weak currents of the $V - A$ type.

[7]Note that in our notations $\vec{D} = -\partial/\partial\vec{x} - i\vec{A}$, therefore $(\vec{D}\vec{E}) = -\mathrm{div}\,\mathbf{E}$ in the Abelian case.

Further remarks will concern operators that are spatial scalars but not Lorentz scalars. They appear in the low energy effective Lagrangian (1.11) and in the expansions of the type (3.3) and (3.5). The most important operator from the class is

$$\mathcal{O}_\pi = \bar{Q}\vec{\pi}^2 Q\,, \tag{3.6}$$

which we have already encountered more than once.

Dimension 4 operators are all reducible to those of dimension 5 and higher. For instance,

$$\bar{Q}\vec{\gamma}\vec{\pi}Q = \frac{1}{m_Q}\bar{Q}\left(\vec{\pi}^2 - \frac{i}{2}\sigma G\right)Q + \mathcal{O}(m_Q^{-2})\,,$$

$$\bar{Q}\vec{\gamma}\vec{\pi}\gamma_0 Q = \mathcal{O}(m_Q^{-2}),$$

$$\bar{Q}\pi_0 Q = \frac{1}{2m_Q}\bar{Q}\left(\vec{\pi}^2 - \frac{i}{2}\sigma G\right)Q + \mathcal{O}(m_Q^{-2})\,.$$

At the level of dimension 6 only one additional operator emerges (apart from the four-fermion operators), namely,

$$\bar{Q}\vec{\sigma}\cdot\vec{E}\times\vec{\pi}Q\,. \tag{3.7}$$

At first sight it might seem that one could build extra operators of dimension 6, from the gluon fields, e.g.

$$\bar{Q}\pi_i E_i Q \quad \text{or} \quad \bar{Q}\sigma_i\epsilon_{ijk}(D_j E_k)Q.$$

Actually they are reducible to operators of higher dimension via the equations of motion. Indeed, using the fact that

$$E_i = -i[\pi_0\pi_i]\,,$$

one can rewrite

$$\begin{aligned}
\bar{Q}\pi_i E_i Q &= -i\bar{Q}\pi_i[\pi_0\pi_i]Q = -i\bar{Q}(\pi_i\pi_0\pi_i - \pi_i^2\pi_0)Q \\
&= -i\bar{Q}([\pi_i\pi_0]\pi_i + \pi_0\pi_i^2 - \pi_i^2\pi_0)Q \\
&= -i\bar{Q}\left([\pi_i\pi_0]\pi_i + \frac{1}{2m_Q}(\vec{\sigma}\vec{\pi})^2\pi_i^2 - \frac{1}{2m_Q}\pi_i^2(\vec{\sigma}\vec{\pi})^2\right)Q \\
&\quad -\frac{i}{2}\bar{Q}[\pi_i[\pi_0\pi_i]]Q + \text{ dimension 7} \\
&= -i\bar{Q}(\mathrm{div}\vec{E})Q + \text{ dimension 7}\,.
\end{aligned} \tag{3.8}$$

This is a four-fermion operator. By the same token,

$$\begin{aligned}\bar{Q}\sigma_i\epsilon_{ijk}(D_jE_k)Q &= \bar{Q}\sigma_i D_0 B_i Q \\ &= -i\bar{Q}\sigma_i[\pi_0 B_i]Q \\ &= -\frac{i}{2m_Q}\bar{Q}[(\vec{\sigma}\vec{\pi})^2, \vec{\sigma}\vec{B}]Q \end{aligned} \tag{3.9}$$

which is obviously of the next order in $1/m_Q$ (a dimension 7 operator).

3.2. *Extracting/Determining the Matrix Elements*

The construction of the operator product expansion is only the first step in any theoretical analysis. The heavy quark expansion must be converted into predictions for the physical quantities. To this end it is necessary to take the matrix elements of the operators involved in the expansion. The latter carry all information about the large distance dynamics responsible for the hadronic structure, in all its peculiarity. These matrix elements in our QCD-based approach play the same role as the wave functions in the nonrelativistic quark models.

In this chapter we will summarize what is known about the matrix elements of the operators from the list presented above.

(i) The most favorable situation takes place at the level of dimension 3. Indeed, the only Lorentz scalar operator of dimension 3 is $\bar{Q}Q$ which has a nice expansion (3.3). The operator $\bar{Q}\gamma_0 Q$ is the time component of the conserved current, measuring the number of the quarks Q in H_Q. Therefore, both for mesons and baryons,

$$\frac{1}{2M_H}\langle H_Q|\bar{Q}\gamma_0 Q|H_Q\rangle = 1\,. \tag{3.10}$$

(As usual, we stick to the rest frame of H_Q; in the case of baryons averaging over the baryon spin is implied.)

(ii) The status of two operators of dimension 5 is different. Let us consider first $\mathcal{O}_G$ whose matrix elements are expressible in terms of experimentally measurable quantities. To leading order in $1/m_Q$ the parameter μ_G^2 defined in Eq. (2.56) reduces to[8]

$$\mu_G^2 = \frac{1}{2M_{H_Q}}\langle H_Q|\mathcal{O}_G|H_Q\rangle = -\frac{1}{2M_{H_Q}}\langle H_Q|\bar{Q}\vec{\sigma}\vec{B}Q|H_Q\rangle\,. \tag{3.11}$$

[8]Some authors prefer a different nomenclature [43]. The expectation values of the chromomagnetic and kinetic energy operators, Eqs. (3.11) and (3.6), respectively, are sometimes called λ_2 and λ_1.

For pseudoscalar mesons this quantity can be related to the measured hyperfine mass splittings. Indeed, $\bar{Q}\vec{\sigma}\vec{B}Q$ is the leading spin-dependent operator in the heavy quark Hamiltonian (1.14). Hence

$$\mu_G^2(B_Q) = \frac{3}{4}(M_{B^*}^2 - M_B^2) \tag{3.12}$$

where B^* and B are generic notations for the vector and pseudoscalar mesons, respectively, and the limit $m_Q \to \infty$ is implied. Assuming that the b quark already belongs to this asymptotic limit one estimates μ_G^2 from the measured B meson masses,

$$\mu_G^2 \approx 0.35 \ \text{GeV}^2\,.$$

Furthermore, in the baryon family, four baryons are expected to decay weakly and are, thus, long-living states: Λ_Q, Σ_Q, Ξ_Q and Ω_Q. In the first three of them the total angular momentum of the light cloud is zero; hence the chromomagnetic field has no vector to be aligned with, and

$$\mu_G^2(\Lambda_Q) = \mu_G^2(\Sigma_Q) = \mu_G^2(\Xi_Q) = 0. \tag{3.13}$$

In the case of Ω_Q the total angular momentum of the light cloud is 1. Hence,

$$\mu_G^2(\Omega_Q) = \frac{2}{3}\left(M^2_{\Omega_Q^{3/2}} - M^2_{\Omega_Q^{1/2}}\right) \tag{3.14}$$

where the superscripts 3/2 and 1/2 mark the spin of the baryon. Although the mass splitting on the right-hand side of Eq. (3.14) is in principle measurable, it is not known at present, and if one wants to get an estimate, one has to resort to quark models or lattice calculations. Both approaches are not mature enough at the moment to give reliable predictions for this quantity and I suggest we wait until experimental measurements appear.

Let us proceed now to the discussion of the matrix element of the operator $\mathcal{O}_\pi$. The physical meaning of this matrix element is the average kinetic energy (more exactly, the spatial momentum squared) of the heavy quark Q inside H_Q. This operator is spin-independent, and it is much harder to extract μ_π^2 (the parameter μ_π^2 is defined in Eq. (2.58)) from phenomenology, although such an extraction is possible, in principle (see Ref. [25] and Secs. 4 and 5 for details). Since the phenomenological analysis has not been carried out yet one has to rely on theoretical estimates. Several calculations of μ_π^2 within the QCD sum rules yield [44, 45]

$$\mu_\pi^2(B) = \frac{1}{2M_B}\langle B|\mathcal{O}_\pi|B\rangle = 0.5 \pm 0.15 \ \text{GeV}^2\,.$$

A remarkable model-independent lower bound on $\mu_\pi^2(B)$ exists in the literature,

$$\mu_\pi^2(B) > \mu_G^2(B) \approx 0.35 \text{ GeV}^2 . \tag{3.15}$$

The quantum-mechanical derivation of this inequality due to Voloshin (see Ref. [4]) is straightforward. Indeed, start from the square of the Hermitean operator $(\vec{\sigma}\vec{\pi})^2$ and average it over the B meson state. It is obvious then that $\langle(\vec{\sigma}\vec{\pi})^2\rangle > 0$. Using the fact that $(\vec{\sigma}\vec{\pi})^2 = \vec{\pi}^2 + \vec{\sigma}\vec{B}$ we immediately arrive at Eq. (3.15). A field-theoretic derivation of the same result can be found in Ref. [25]. It is remarkable that the inequality (3.15) almost saturates the QCD sum rule estimate quoted above. Another lower bound on $\mu_\pi^2(B)$, obtained from a totally different line of reasoning, is discussed in Sec. 3.6. It turns out to be close to Eq. (3.15) numerically.

It is plausible that μ_π^2 for mesons and baryons is different — there is no reason why they should coincide. The task of estimating μ_π^2 for baryons remains open.

These parameters, μ_π^2 and μ_G^2, along with $\bar{\Lambda}$, are most important in applications. In most applications one deals with the expectation values over the B meson state. Therefore, let us agree that μ_π^2, μ_G^2, and $\bar{\Lambda}$, with no subscripts or arguments, are defined with respect to the B mesons. This is a standard convention. In a few cases when these quantities are defined with respect to other heavy flavor hadrons we will mark them by the corresponding subscripts or indicate with parentheses.

(iii) Operators of dimension 6 are studied to a much lesser extent than those of dimension 5. Perhaps, the least favorable is the situation with the operator $\mathcal{O}_E$ given in Eq. (3.7). Let us parametrize its matrix element as follows:

$$\frac{1}{2M_H}\langle H_Q|\bar{Q}\vec{\sigma} \times \vec{E}\vec{\pi}Q|H_Q\rangle = \mu_E^3. \tag{3.16}$$

This operator in the heavy quark Hamiltonian is responsible for the spin-orbit interactions and consequently generates the spin-orbit splittings between the masses of the ground states and the orbital excitations. Hence, in the nonrelativistic limit (nonrelativistic with respect to the spectator light quark) μ_E^3 vanishes for the S wave states. Of course, the nonrelativistic approximation with respect to the light quark is very bad. The estimate of μ_E^3 existing in the literature [40] is so rough that it, probably, does not deserve to be discussed here.

As for the four-quark operators the only method of estimating their matrix elements, which does not rely heavily on the most primitive (and hence totally unreliable) quark models, is the old idea of factorization applicable only in mesons but — alas — not in baryons.

First of all let us observe that each of the four-quark operators exists in two variants differing by the color flow. One can always rearrange the operators, using the Fierz identities, in the form

$$\mathcal{O}_{4q} = \bar{Q}\Gamma q\bar{q}\Gamma Q \tag{3.17}$$

and

$$\tilde{\mathcal{O}}_{4q} = \bar{Q}t^a\Gamma q\bar{q}t^a\Gamma Q\,. \tag{3.18}$$

Take for definiteness $\Gamma = \gamma_\mu\gamma_5$. (Other γ matrices can also appear, of course.) In the first operator color is transferred from the initial heavy to the final light quark and from the initial light to the final heavy quark. The second operator is essentially color-exchange. Now, if we are interested in the matrix elements over the meson states we can simply factorize the currents appearing in Eqs. (3.17) and (3.18) (i.e. saturate by the vacuum intermediate state),

$$\frac{1}{2M_B}\langle B_Q|\mathcal{O}_{4q}|B_Q\rangle = \frac{1}{2}M_B f_B^2, \quad \frac{1}{2M_B}\langle B_Q|\tilde{\mathcal{O}}_{4q}|B_Q\rangle = 0 \tag{3.19}$$

where f_B is the pseudoscalar decay constant,

$$\langle 0|\bar{Q}\gamma_\mu\gamma_5 q|B_Q\rangle = if_B p_\mu\,. \tag{3.20}$$

As we will discuss shortly, in the limit $m_Q \to \infty$, the combination $M_B f_B^2$ scales as m_Q^0 (modulo logarithmic corrections) so that the right-hand side of Eq. (3.19) is the cube of a typical hadronic mass, as it should be.

Factorization in Eq. (3.19) is justified by $1/N_c$ arguments. Indeed, corrections to Eq. (3.19) are of the order of N_c^{-1}. Thus, from the whole set of the four-quark operators we can say something about the meson expectation values of those operators which are reducible to

$$(\bar{Q}\Gamma q)(\bar{q}\Gamma Q) \tag{3.21}$$

where Γ stands for a Lorentz matrix but not for the color one, and the color indices are contracted within each of the two brackets separately. Up to the terms $\mathcal{O}(N_c^{-1})$ two brackets can be factorized.

To get an idea about the numerical value of $(1/2)f_B^2 M_B$ we should substitute a numerical value for f_B which is not measured so far. Theoretical ideas about this fundamental constant will be discussed later. Now let me say only that $(1/2)f_B^2 M_B \sim 0.1$ GeV^3, with a significant uncertainty. Those matrix elements which are due to nonfactorizable contributions (see Eq. (3.18)) are essentially undetermined, although they are expected to be suppressed compared to the factorizable matrix elements $(1/2)f_B^2 M_B$.

As for the baryon matrix elements of the four-quark operators, next-to-nothing is known about them at the moment. Some very crude estimates within the naive quark model are available [46] but they are very unreliable.

In conclusion of this chapter a remark is in order concerning numerical estimates of the key parameter of the heavy quark theory, $\bar{\Lambda}$. I postponed discussing the issue because its value continues to be controversial. QCD sum rules indicate [45, 47–49] that $\bar{\Lambda} \sim 0.5$ GeV. This number is in full agreement with the lower bound stemming from Voloshin's sum rule, see Sec. 3.6. However, some lattice calculations yield a factor of 2 lower estimate. I am inclined to think that there is something wrong in the lattice results. Perhaps, the lattice definition of $\bar{\Lambda}$ does not fully correspond to that of the continuum theory. It is inconceivable that such a low value of $\bar{\Lambda}$ as 0.2 or even 0.3 GeV could be reconciled with the lower bound implied by Voloshin's sum rule.

In the discussion above we have totally disregarded logarithmic dependence of the operators and their matrix elements due to anomalous dimensions — i.e. the issue of the normalization point (including the normalization point of $\bar{\Lambda}$). This is in line with our approach since so far we pretend that hard gluons do not exist. A brief excursion into this topic will be undertaken later; here it is only worth mentioning that all numerical estimates presented above refer to a low normalization point, of the order of several units times $\Lambda_{\rm QCD}$.

We are now ready to review classic problems of the heavy quark theory. We will gradually move from simpler to more sophisticated problems.

3.3. *Mass Formula Revisited*

In Sec. 1.5 we have found the first subleading term in the mass formula for the heavy flavor hadrons. The parameter $\bar{\Lambda}$ was related to the expectation value of the gluon anomaly, see Eq. (1.32). It is very easy to continue the expansion one step further and find the next subleading term, of the order $1/m_Q$. One could have extended the derivation along the lines suggested in Sec. 1.5. This was done in Ref. [25]. This is not the fastest route, however. Instead, let us observe that the $1/m_Q$ term in the Hamiltonian (1.14) can be considered as the first order perturbation; the corresponding correction to the mass is merely the expectation value of this perturbation,

$$M_{H_Q} = m_Q + \bar{\Lambda} + \frac{1}{2m_Q}(2M_{H_Q})^{-1}\langle H_Q|\vec{\pi}^2 + \vec{\sigma}\vec{B}|H_Q\rangle + \dots$$
$$= m_Q + \bar{\Lambda} + \frac{(\mu_\pi^2 - \mu_G^2)_{H_Q}}{2m_Q} + \dots . \tag{3.22}$$

The terms of the order $1/m_Q^2$ are neglected. If we keep only the terms up to $1/m_Q$ it does not matter whether the state H_Q we average over is an asymptotic state (corresponding to $m_Q = \infty$) or the real physical heavy flavor state. I remind that, unlike HQET, we work with the physical states. The difference becomes noticeable only at the level $1/m_Q^2$. In this order the mass formula does not reduce any more to the expectation values of local operators. A part of the $1/m_Q^2$ correction is due to nonlocal correlation functions, see Ref. [25] for further details. Equation (3.22) was first presented in Ref. [43].

3.4. *The Scaling Law of the Pseudoscalar and Vector Coupling Constants*

The pseudoscalar and vector meson constants f_P and f_V are defined as

$$\langle 0|\bar{Q}\gamma_\mu\gamma_5 q|B_Q\rangle = if_P p_\mu, \quad \langle 0|\bar{Q}\gamma_\mu q|B_Q^*\rangle = if_V M_V \epsilon_\mu . \tag{3.23}$$

An alternative definition of the pseudoscalar constant can be given in terms of the pseudoscalar current,

$$\langle 0|\bar{Q}i\gamma_5 q|B_Q\rangle = f_P' M_B . \tag{3.24}$$

The constant f_B is one of the key parameters of the heavy quark physics, just in the same way the constant f_π is a key parameter of the soft pion physics. Below we will show that in the limit $m_Q \to \infty$ all three constants, f_P, f_V and f_P', coincide with each other and scale as $m_Q^{-1/2}$ modulo a weak logarithmic dependence on m_Q. (Needless to say that both masses, M_P and M_V, also coincide in this limit.) For definiteness let us consider f_P'. Two other constants can be treated in a similar manner. The subscripts will be omitted in the remainder of this section to avoid overloaded expressions.

We start from the two-point function

$$\mathcal{A}(k) = i\int e^{ikx} d^4x \langle \mathrm{T}\{\bar{Q}(x)i\gamma_5 q(x)\ \bar{q}(0)i\gamma_5 Q(0)\}\rangle , \tag{3.25}$$

where k is the external momentum. The two-point function $\mathcal{A}(k)$ develops a pole at $k^2 = M^2$, the position of the ground state pseudoscalar,

$$\mathcal{A}(k) = \frac{f^2 M^2}{k^2 - M^2} + \text{excitations}\,. \tag{3.26}$$

Of course, the currents produce from the vacuum not only the ground state mesons but also all excitations in the given channel. It is clear that to isolate the lowest lying pole we should keep k^2 close to M^2. Keeping in mind Eq. (1.26) it is natural to represent k as

$$k_\mu = \{m_Q + \epsilon, 0, 0, 0\}$$

where ϵ scales like $\Lambda_{\rm QCD}$ while $m_Q \to \infty$. With this parametrization of k_μ we merely separate the mechanical (uninteresting) part of the momentum. The pole is achieved at $\epsilon = \bar\Lambda$; near the pole

$$\mathcal{A}(\epsilon) \approx \frac{f^2 M}{2(\epsilon - \bar\Lambda)}\,. \tag{3.27}$$

The value of the coupling constant is obtained by amputating the pole,

$$f^2 = \lim_{\epsilon \to \bar\Lambda} \left\{ \frac{2(\epsilon - \bar\Lambda)}{M} \mathcal{A}(\epsilon) \right\}\,. \tag{3.28}$$

Let us now examine the theoretical expression for the same two-point function. In the background field technique (which is already pretty familiar, right?) we write

$$\mathcal{A}(k) = i\mathrm{Tr} \left\{ i\gamma_5 \frac{1}{\not{\mathcal{P}} - m_q} i\gamma_5 \frac{1}{\not{k} + \not{\mathcal{P}} - m_Q} \right\}\,. \tag{3.29}$$

Superficially this expression looks the same as if the quarks were treated as free; they are not, however; the coupling to the background field is reflected in the fact that $\mathcal{P}_\mu$ is the momentum operator, not just a c-number four-vector.

Now we will take advantage of the fact that $m_Q \to \infty$. As usual, we close our eyes on any possible hard contributions, assuming that $\mathcal{P}$, the momentum operator of the light quark, is soft, i.e. does not scale with m_Q in the large mass limit but rather, $\mathcal{P} \sim \Lambda_{\rm QCD}$. (This is the reason, by the way, why the large external momentum k was directed through the heavy quark line in Eq. (3.29).) Intuitively it is clear that the hard components of $\mathcal{P}$ should be irrelevant for

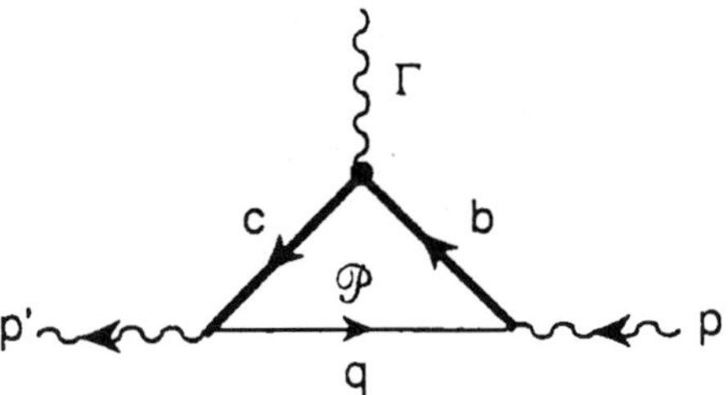

Fig. 6. The three-point function relevant to the proof of the Isgur-Wise formula.

the lowest lying state whose "excitation energy" measured from m_Q is of the order Λ_{QCD}.

If $\mathcal{P}$ is soft and $m_Q \to \infty$ the heavy quark Green function in the leading approximation takes the form

$$\frac{1}{\not{k} + \not{\mathcal{P}} - m_Q} = (\not{k} + \not{\mathcal{P}} + m_Q)\,\frac{1}{(k+\mathcal{P})^2 - m_Q^2 + (i/2)\sigma G}$$
$$= \frac{\gamma_0 + 1}{2}\,\frac{1}{\epsilon + \mathcal{P}_0}\,, \tag{3.30}$$

where in the second line all $1/m_Q$ terms are omitted. No explicit m_Q dependence is left! This means that $\mathcal{A}(\epsilon)$ scales as m_Q^0. Equation (3.28) immediately implies then that f scales as

$$f \sim m_Q^{-1/2}\,. \tag{3.31}$$

Equation (3.30) for the heavy quark Green function in the limit $m_Q \to \infty$ is in one-to-one correspondence with the leading term $\bar{Q}\pi_0[(1+\gamma_0)/2]Q$ in the low energy Lagrangian (1.11). The analysis of the scaling law of the coupling constants presented above is a simplified version of that carried out many years ago by Shuryak [6]. Later it was established that the power dependence on m_Q in Eq. (3.31) is supplemented by a logarithmic dependence appearing due to the hard gluon exchanges [50].

A few words about the numerical value of the coupling constants. Although in principle f_D and f_B are measurable experimentally, practically it is a very hard measurement, especially for B. No experimental number for f_B exists so far. Its value was estimated in the QCD sum rules and on lattices more than once. Leaving aside a dramatic evolution of the issue, I will only say that the recent and most reliable results cluster around 160 MeV, both in the sum rules [47, 48, 51] and in the lattice calculations [52]. It is curious to note that in $(m_b)^{1/2} f_B$ the preasymptotic $1/m_Q$ correction turned out to be unexpectedly large and negative [45, 47, 52, 53]; at the same time in $(m_b)^{1/2} f'_B$, the preasymptotic $1/m_Q$ correction is much more modest [47, 48].

3.5. *Proof of the Isgur-Wise Formula*

I return to my promise to prove the Isgur-Wise formula (1.25). Consider the three-point function depicted in Fig. 6. The sides of the triangle are the Green functions of the quarks in the background gluon field. The reduction theorems tell us that in order to get the transition amplitudes $\langle H_c|\bar{c}\Gamma b|H_b\rangle$ from this three-point function we must "amputate" it: multiply by $(p^2 - M_B^2)$ and $(p'^2 - M_D^2)$, tending p^2 to M_B^2 and p'^2 to M_D^2. This singles out the meson states we want to pick out. For the vector mesons we must also multiply the three-point functions by its polarization vector ϵ_μ. The last step necessary for amputation is dividing by the coupling constants (residues) connecting the currents $\bar{b}i\gamma_5 c$ and $\bar{b}\gamma_\mu c$ to the respective mesons. If the currents are normalized appropriately — and we will always do that — the corresponding coupling constants in the pseudoscalar and vector channels are the same, fM (see Sec. 3.4). It is convenient to combine the pseudoscalar and vector channels together by introducing the currents

$$J = B\bar{b}i\gamma_5 q + B_\mu \bar{b}\gamma_\mu q \text{ and } J' = D\bar{q}i\gamma_5 c + D^*_\mu \bar{q}\gamma_\mu c \tag{3.32}$$

where B and D are external constants marking the annihilation and creation of the initial and final B's and D's (B_μ and D_μ denote the polarization vectors of B^* and D^*, respectively).

The expression for the three-point function of Fig. 6 takes the form

$$i\mathrm{Tr}\left\{\mathcal{M}'\frac{1}{\not{p}' + \not{P} - m_c}\Gamma\frac{1}{\not{p} + \not{P} - m_b}\mathcal{M}\frac{1}{\not{P} - m_q}\right\} \tag{3.33}$$

where the matrices $\mathcal{M}'$ and $\mathcal{M}$ are introduced in Eq. (1.24). In Sec. 3.4 the $m_Q \to \infty$ limit of the quark propagator was obtained in the rest frame. Here we have two heavy flavor states, initial and final, and both cannot be at rest simultaneously. Therefore, we need the very same propagator in the arbitrary frame. Let $p_\mu = m_Q v_\mu + \epsilon_\mu$. Then a trivial generalization of Eq. (3.30) is

$$\frac{1}{\not p + \not{\mathcal{P}} - m_Q} = (\not p + \not{\mathcal{P}} + m_Q)\,\frac{1}{(p+\mathcal{P})^2 - m_Q^2 + (i/2)\sigma G}$$
$$\to \frac{\not v + 1}{2}\,\frac{1}{(\epsilon + \mathcal{P})v}\,. \tag{3.34}$$

Using this propagator in Eq. (3.33) we rewrite the three-point function of Fig. 6 as follows:

$$\left(\mathcal{M}'\frac{\not v' + 1}{2}\Gamma\frac{\not v + 1}{2}\mathcal{M}\right)_{\alpha\beta} \times i\mathrm{Tr}\left\{\frac{1}{(\epsilon' + \mathcal{P})v'}\,\frac{1}{(\epsilon + \mathcal{P})v}\left(\frac{1}{\not{\mathcal{P}} - m_q}\right)_{\beta\alpha}\right\}. \tag{3.35}$$

The expression in the braces is independent of the heavy quark masses; moreover, it is proportional to the three-point function in the theory with the scalar heavy quarks considered in Sec. 1.4. As was explained there, in this theory, in the limit $m_Q \to \infty$, only one form factor survives.

One subtle point deserves to be discussed here. When I speak about the heavy flavor mesons I keep in mind particles built from the heavy quark and a light antiquark, which is not always in line with the accepted nomenclature. Say, they call B meson a particle with the b antiquark, not quark. Since this distinction plays no role in my lectures, I will continue to ignore this linguistic nuance referring to the $b\bar{q}$ states as B mesons. All equations presented above assume that the b quark in the initial state annihilates to produce the c quark in the final state. Simultaneously, a light antiquark in the initial meson is annihilated and the same light antiquark reappears in the final meson.

We return now to the model with spinless heavy quarks. The heavy flavor hadrons we now deal with are spin-1/2 baryons. More exactly, we have *anti*baryon in the initial state and *anti*baryon in the final state. This means that near the mass shell the expression in the braces in Eq. (3.35) takes the form

$$\left(\frac{-\not v + 1}{2}\,\frac{-\not v' + 1}{2}\right)_{\beta\alpha}\sqrt{M_B M_D}\,f_B f_D \frac{1}{v'\epsilon' - \bar\Lambda}\,\frac{1}{v\epsilon - \bar\Lambda}\,\xi(y)\,; \tag{3.36}$$

the minus sign between the unit term and the v term in the density matrices is due to the fact that we deal with the antibaryons. Amputating the legs and combining Eqs. (3.36) and (3.35) we get the Isgur-Wise formula (1.25) since $\mathcal{M}(-\not v + 1) = (\not v + 1)\mathcal{M}$, and so on.

3.6. *The Bjorken Sum Rule and All That*

The Isgur-Wise function $\xi(y)$ carries information about the structure of the light cloud. Needless to say, the heavy quark expansion *per se* does not help to calculate this function. One has to rely on methods applicable in the strong coupling regime which are outside the scope of my lectures (QCD sum rules, lattices, ...). Still, some interesting and important relations emerge. Here we will discuss a sum rule for the slope of the Isgur-Wise function and related topics.

We are already familiar with the sum rule technology in the heavy quark theory. In Sect 2.1 we dwelled on a simplified problem: inclusive decays of a spinless heavy quark Q into a lighter spinless quark q and a fictitious spin-0 photon ϕ. The "photon" was assumed to be on mass shell, $q^2 = 0$. The predictions obtained referred to the moments of the "photon" energy. Now you are mature enough to face actual problems from real life. We will concentrate on the decays of a b-containing hadron into a c-containing hadron plus the lepton pair $l\nu$. The four-momentum of the lepton pair is a free parameter, in particular, $q^2 \neq 0$. We can and will choose the value of q to our advantage.

Consider a transition $H_b \to H_c$ induced by some particular current, say, axial-vector. At zero recoil $\xi = 1$. In the SV limit where the velocity of the recoiling hadron is small,

$$\xi(y) = 1 - \rho^2(y-1) + \dots = 1 - \rho^2\frac{\vec{v}^2}{2} + \dots \tag{3.37}$$

where $y = vv'$, $\vec{v}$ is the spatial velocity of H_c in the H_b rest frame, and the slope parameter ρ^2 was introduced in Ref. [32]. It plays the same role as, say, the charge radius of pions.

To get relations involving ρ^2 we start from the consideration of the transition operator similar to that of Eq. (2.13). The expectation value of the transition operator over the B meson state yields the hadronic amplitude whose imaginary part is proportional to the probability of the inclusive decay $B \to X_c l\nu$ with the fixed value of q, the momentum carried away by the lepton pair $l\nu$. (Here X_c denotes an inclusive hadronic state containing one c quark.) A new element compared to the toy model of Sec. 2.1 is the heavy quark spin. Another distinction is the fact that, to achieve the SV limit, we do not need now to assume that m_c is close to m_b. In the semileptonic decay $B \to X_c l\nu$ one can fine-tune the lepton pair momentum in such a way that q^2 is close to its maximal value, $q^2_{\rm max} = (M_B - M_D)^2$; then the c-containing hadronic state produced is almost at rest, and we are in the SV limit even though the charmed quark is significantly lighter than the b quark. In other words, for such values of q^2 the c quark is always slow.

This transition operator describes the forward scattering of B to B via intermediate states D^* and "excitations". (We focus for definiteness on the axial-vector current, $\bar{c}\gamma_\mu\gamma_5 b$. The vector current can be treated in a similar way.) The excitations can include, for instance, $D^*\pi\pi$, and so on. In general, all intermediate states except the lowest lying D^* will be referred to as excitations. The transition operator

$$\hat{T}_{\mu\nu} = i\int d^4x e^{-iqx} T\{\bar{b}(x)\gamma_\mu\gamma_5 c(x)\,,\,\bar{c}(0)\gamma_\mu\gamma_5 b(0)\} \tag{3.38}$$

in the Born approximation is given by the diagram of Fig. 1. The hadronic amplitude obtained by averaging $\hat{T}_{\mu\nu}$ over the B meson state,

$$h_{\mu\nu} = \frac{1}{2M_B}\langle B|\hat{T}_{\mu\nu}|B\rangle \tag{3.39}$$

contains various kinematical factors. In the general case the hadronic tensor $h_{\mu\nu}$ consists of five different covariant structures [30, 33]:

$$h_{\mu\nu} = -h_1 g_{\mu\nu} + h_2 v_\mu v_\nu - ih_3\epsilon_{\mu\nu\alpha\beta}v_\alpha q_\beta + h_4 q_\mu q_\nu + h_5(q_\mu v_\nu + q_\nu v_\mu). \tag{3.40}$$

Moreover, the invariant hadronic functions h_1 to h_5 depend on two variables, q_0 and q^2, or q_0 and $|\vec{q}|$. For $\vec{q} = 0$ only one variable survives, and only two of five tensor structures in $h_{\mu\nu}$ are independent.

Each of these hadronic invariant functions satisfies a dispersion relation in q_0,

$$h_i(q_0) = \frac{1}{2\pi}\int \frac{w_i(\tilde{q}_0)d\tilde{q}_0}{\tilde{q}_0 - q_0} + \text{polynomial} \tag{3.41}$$

where w_i are observable structure functions,

$$w_i = 2\,\mathrm{Im}\,h_i\,.$$

For our purposes it is quite sufficient to analyze only one function, namely, h_1. Moreover, for the time being, we will disregard all nonperturbative corrections $\mathcal{O}(\Lambda^2_{\rm QCD})$ which means that operators in the expansion of $\hat{T}_{\mu\nu}$ other than $\bar{b}b$ can be neglected, and the B meson expectation value of $\bar{b}b$ can be replaced by unity. Calculating h_1 in this approximation is a trivial problem (it was a part of the exercise suggested in Sec. 2.1). Specifically,

$$-h_1 = (m_b + m_c - q_0)\frac{1}{z} + \mathcal{O}(\Lambda^2_{\rm QCD}) \tag{3.42}$$

where

$$z = (m_b - q_0 - E_c)(m_b - q_0 + E_c)\,, \;\; E_c^2 = m_c^2 + \vec{q}^{\,2}\,. \tag{3.43}$$

I remind that $\vec{q}$ is assumed to be fixed, and $\Lambda_{\rm QCD} \ll |\vec{q}| \ll M_D$, so that actually, I will expand in $\vec{q}$, keeping only the terms up to second order. It is convenient to shift q_0 by introducing a new variable

$$\epsilon = q_{0\,\rm max} - q_0 \tag{3.44}$$

where

$$q_{0max} = M_B - E_{D^*}, \;\; E_{D^*} = M_{D^*} + \frac{\vec{q}^{\,2}}{2M_{D^*}}\,. \tag{3.45}$$

When ϵ is real and positive we are on the physical cut where the actual intermediate states (e.g. D^*) are produced. Here the imaginary part of h_1 is given by the "elastic" contribution of D^* plus inelastic excitations. For negative ϵ we are below the cut. The result for h_1 above can be trusted if $-\epsilon \gg \Lambda_{\rm QCD}$ since the expansion actually runs in $\Lambda_{\rm QCD}/\epsilon$. The expansion in the inverse heavy quark mass also requires, of course, that $|\epsilon| \ll m_{c,b}$. A bridge between the physical domain of positive ϵ and the Euclidean domain of negative ϵ where the calculation is done is provided by the dispersion relations.

At the next stage the amplitude h_1 is expanded in powers of $\Lambda_{\rm QCD}/\epsilon$ and $\epsilon/m_{b,c}$. Polynomials in ϵ can be discarded since they have no imaginary part. We are interested only in negative powers of ϵ. The coefficients in front of $1/\epsilon^n$ are related, through dispersion relations, to the integrals over the imaginary part of h_1 with the weight functions proportional to the excitation energy to the power $n-1$. Indeed,

$$\begin{aligned}-h_1(\epsilon, \vec{q}^{\,2}) &= \frac{1}{2\pi}\int d\tilde{\epsilon}\,\frac{w_1(\tilde{\epsilon},\vec{q}^{\,2})}{\epsilon - \tilde{\epsilon}} \\ &= \frac{1}{\epsilon}\cdot\frac{1}{2\pi}\int d\tilde{\epsilon}\,w_1(\tilde{\epsilon},\vec{q}^{\,2}) + \frac{1}{\epsilon^2}\cdot\frac{1}{2\pi}\int d\tilde{\epsilon}\,\tilde{\epsilon}\,w_1(\tilde{\epsilon},\vec{q}^{\,2}) \\ &\quad + \frac{1}{\epsilon^3}\cdot\frac{1}{2\pi}\int d\tilde{\epsilon}\,\tilde{\epsilon}^2 w_1(\tilde{\epsilon},\vec{q}^{\,2}) + \ldots\end{aligned} \tag{3.46}$$

Thus, our immediate task is to build the $1/\epsilon$ expansion from the amplitude (3.42).

The theoretical expression for the amplitude h_1 above knows nothing, of course, about the meson masses; it contains only the quark masses. Correspondingly, it is very convenient to build first the expansion of h_1 in an auxiliary quantity,

$$\epsilon_q = m_b - E_c - q_0 , \quad E_c = m_c + \frac{\vec{q}^2}{2m_c} . \tag{3.47}$$

Then, if necessary, we re-express the expansion obtained in this way in terms of ϵ,

$$\frac{1}{\epsilon_q} = \frac{1}{\epsilon} + \frac{(\epsilon - \epsilon_q)}{\epsilon^2} + \dots \tag{3.48}$$

The difference between ϵ_q and ϵ is $\mathcal{O}(\Lambda_{\rm QCD} \cdot \vec{q}^2/m_{b,c}^2)$ and $\mathcal{O}(\Lambda_{\rm QCD}^2/m_{b,c})$.

After these introductory remarks, assembling all information in our disposal, we get

$$-h_1 = \left(1 - \frac{\vec{v}^2}{4}\right) \frac{1}{\epsilon} + \bar{\Lambda} \frac{\vec{v}^2}{2} \frac{1}{\epsilon^2} + \dots \tag{3.49}$$

plus terms of higher order in $\vec{q}^2$ or $\Lambda_{\rm QCD}$. In deriving Eq. (3.49) I used the fact that $\epsilon - \epsilon_q = \bar{\Lambda}\vec{v}^2/2$ plus terms of higher order in $\vec{q}^2$ or $\Lambda_{\rm QCD}$. This completes the theoretical aspect of the calculation. The coefficients in front of $1/\epsilon$ and $1/\epsilon^2$ in h_1 are known; Eq. (3.46) tells us that these coefficients are equal to integrals over w_1, the spectral density. So, what remains to be done is to express the spectral density in terms of the contribution coming from the physical intermediate states. Let us assume for simplicity that the spectrum of the intermediate states is discrete. Denote the mass of the ith state by M_i and the energy by E_i; the lowest lying meson, D^*, corresponds to $i = 0$. Then the propagator of the ith meson

$$\frac{1}{(M_B - q_0 - E_i)(M_B - q_0 + E_i)}$$

at positive ϵ has the imaginary part

$$(2E_i)^{-1} \pi \delta(\epsilon - \delta_i)$$

where δ_i is the excitation energy (including the corresponding kinetic energy),

$$\delta_i = E_i - E_{D^*} .$$

For the "elastic" $B \to D^*$ transition δ_0 vanishes, of course.

Now it is rather obvious that the structure function w_1 reduces to

$$w_1(\epsilon) = \sum_{i=0}^{\infty} \frac{|f_{B\to i}|^2}{2E_i} \, 2\pi\delta(\epsilon - \delta_i), \tag{3.50}$$

where the sum runs over all possible final hadronic states, the term with $i = 0$ corresponds to the "elastic" transition $B \to D^*$ while $i = 1, 2, \ldots$ represent excited states with the energies $E_i = M_i + \vec{q}^2/(2M_i)$; furthermore, $|f_{B\to i}|^2$ looks like the square of a form factor. Strictly speaking $|f_{B\to i}|^2$ is not exactly the square of a form factor; rather this is the (appropriately normalized) contribution to the given structure function coming from the multiplet of the degenerate states which includes summation over spin states as well. By appropriate normalization I mean that we routinely insert the normalization factor $(2M_B)^{-1}$. In the particular example considered (the axial-vector current) D is not produced in the elastic transition, so that in the elastic part one needs to sum only over polarizations of D^*. Say, at zero recoil $f_{B\to D^*} = \sqrt{2M_{D^*}} F_{B\to D^*}$ where $F_{B\to D^*}$ is the $B \to D^*$ form factor at zero recoil, see Eq. (3.58) below.

Let us examine in more detail the elastic contribution, $i = 0$. The form factor of the $B \to D^*$ transition generated by the axial-vector current is given in Eq. (1.21). Using this expression we readily obtain

$$(2E_{D^*})^{-1}|f_{B\to D^*}|^2 = \frac{M_{D^*}}{E_{D^*}} \left(\frac{1 + vv'}{2}\right)^2 |\xi(vv')|^2 \approx 1 - \rho^2\vec{v}^2 \,. \tag{3.51}$$

Comparing the $1/\epsilon$ coefficient in the dispersion representation (3.46) with that of Eq. (3.49) we conclude that

$$\frac{1}{2\pi}\int d\epsilon \; w_1(\epsilon) = 1 - \rho^2\vec{v}^2 + \sum_{i=1}^{\infty} \frac{|f_{B\to i}|^2}{2E_i} = 1 - \frac{\vec{v}^2}{4}\,, \tag{3.52}$$

which implies, in turn, that

$$\rho^2 = \frac{1}{4} + \sum_{i=1}^{\infty} \frac{|f_{B\to i}|^2}{2M_i\vec{v}^2}\,. \tag{3.53}$$

In Sec. 1.3 we learnt that at zero recoil (i.e. $\vec{v} = 0$) only the elastic transition survives. As a consequence of the heavy quark symmetry for all nonelastic transitions $|F_{B\to i}|^2 \sim \vec{v}^2$. The ratio $|f_{B\to i}|^2/\vec{v}^2$ stays finite in the limit of small $\vec{v}$. Equation (3.53) is the Bjorken sum rule proper. Since the contribution of the excited states on the right-hand side is obviously positive it tells us, in particular, that $\rho^2 > 1/4$. This inequality is not very informative, though, since both, the QCD sum rule [48, 54] and lattice calculations indicate that ρ^2 is only slightly less than unity, perhaps, close to 0.8.

Leaving technicalities aside let me summarize the physical meaning of the result obtained. The coefficient in front of $1/\epsilon$ in Eq. (3.49) does not contain $\Lambda_{\rm QCD}$. This means that we calculate the probability of the decay $b \to c$"$l\nu$" with the given value of $\vec{v}^2$ merely in the parton model; this probability is equal to that of the physical decay $B \to X_c$"$l\nu$"; the latter is comprised of the elastic transition $B \to D^*$"$l\nu$" and the transition of B into excitations. (The quotation marks are used to emphasize the fact that the decays that are measured are induced by both, the axial-vector and vector, currents, while we focus now only on the transitions induced by the axial-vector current.) All probabilities of production of the excited states are proportional to $\vec{v}^2$ (at small $\vec{v}^2$), and so is the part of the elastic transition containing ρ^2. The sum of these two contributions must coincide with the $\vec{v}^2$ term obtained in the parton model. The very same analysis, by the way, presents a proof of the fact that $\xi(y=1) = 1$. (Do you see this?)

Now, we make the next step, proceeding to the $1/\epsilon^2$ terms. The $1/\epsilon^2$ term in Eq. (3.49) is proportional to $\bar{\Lambda}$, hence the result we are about to get evidently goes beyond the parton model. Combining Eq. (3.49) with the dispersion representation (3.46) we find

$$\bar{\Lambda}\frac{\vec{v}^2}{2} = \sum_{i=1}^{\infty} \frac{|f_{B\to i}|^2}{2E_i}\,\delta_i = \sum_{i=1}^{\infty} \frac{|f_{B\to i}|^2}{2M_i}\,(M_i - M_{D^*})\,. \tag{3.54}$$

The sum runs not from zero to infinity but from 1 to infinity since $\delta_0 = 0$. Moreover, since all $|F_{B\to i}|^2$ for $i = 1, 2, \ldots$ are proportional to $\vec{v}^2$, and we are interested only in the $\vec{v}^2$ term, it is legitimate to substitute, as I did, E_i by M_i and δ_i by $M_i - M_{D^*}$.

Equation (3.54) is the optical (or Voloshin's) sum rule; superficially it looks the same as in the toy model of Sec. 2.1. Please, remember this sum rule — it gives a unique opportunity to *measure* $\bar{\Lambda}$, one of the key parameters of the heavy quark theory. To this end one has to measure the inelastic transition probabilities in the semileptonic decays $B \to X_c l\nu$ in the SV limit. This is a difficult measurement, but not impossible, at least in principle. Before venturing into this noble task — the extraction of $\bar{\Lambda}$ from experimental data — I must warn you that acceptable accuracy can be achieved only provided that the perturbative corrections (hard gluons) as well as nonperturbative ones, of the next order in $\Lambda_{\rm QCD}$, are included in the sum rules. We will briefly discuss the impact of the perturbative corrections in Sec. 5.

Those who are anxious to get something practical from the optical sum rule in the absence of the necessary measurements should not be discouraged. We can still get a lower bound on $\bar{\Lambda}$. Indeed, let us rewrite Eq. (3.54) as follows:

$$\bar{\Lambda}\frac{\vec{v}^2}{2} = \sum_{i=1}^{\infty}\frac{|f_{B\to i}|^2}{2M_i}\left(M_1 - M_{D^*}\right) + \sum_{i=2}^{\infty}\frac{|f_{B\to i}|^2}{2M_i}\left(M_i - M_1\right)$$

$$= \left(M_1 - M_{D^*}\right)\left(\rho^2 - \frac{1}{4}\right)\vec{v}^2 + \sum_{i=2}^{\infty}\frac{|f_{B\to i}|^2}{2M_i}\left(M_i - M_1\right), \qquad (3.55)$$

where Eq. (3.53) is substituted. Since the second term on the right-hand side is obviously positive we conclude that

$$\bar{\Lambda} > 2(M_1 - M_{D^*})\left(\rho^2 - \frac{1}{4}\right) \sim 500 \text{ MeV} \qquad (3.56)$$

where M_1 is the mass of the first excited resonance with the quantum numbers of D^* ($M_1 - M_{D^*} \sim 0.5$ GeV).

Following the same line of reasoning one can derive the "third" sum rule relating μ_π^2 to an appropriately weighted sum over excitations [55]. The corresponding inequality analogous to Eq. (3.56) takes the form

$$\mu_\pi^2 > 3(M_1 - M_{D^*})^2\left(\rho^2 - \frac{1}{4}\right) \sim 0.45 \text{ GeV}^2 . \qquad (3.57)$$

3.7. *Deviations of the $B \to D^*$ Form Factor from Unity at Zero Recoil*

The heavy quark theory began from the observation that the $B \to D^*$ axial-vector form factor at zero recoil is exactly unity in the limit $m_b \to \infty$, $m_c \to \infty$, m_b/m_c arbitrary, see Sec. 1.3. This is purely a symmetry statement, as usual, dynamics resides in the corrections. In this section we will discuss deviations from unity.

When the heavy quark mass is infinite it is nailed at the origin, both in the initial B meson and in the final D^*. The light cloud then does not notice the replacement of one quark by another, the overlap is unity. If we make the quark masses finite they start jiggling inside the mesons, and this motion is different in B and D^* since the heavy quark velocities are different. On top of this the difference in the relative spin orientations of the heavy quarks and light clouds shows up. These two effects lead to deviations from unity. At a heuristic level there is no doubt that the deviations are of the order of (i) the square of the characteristic heavy quark momentum ($\vec{p}$ itself cannot enter since there is no preferred orientation) or (ii) chromomagnetic correlation $\vec{\sigma}\vec{B}$. In both cases dimensional arguments prompt us that the deviation from unity at zero recoil is proportional to $1/m_{c,b}^2$; linear effects in $1/m_{c,b}$ are absent. The assertion was first formulated in Ref. [15] and was cast in the form of a theorem (Luke's theorem) in Ref. [21]. The proof presented below is abstracted from the recent work [56].

Let us define the $B \to D^*$ form factor at zero recoil as follows:

$$\langle D^*|\bar{c}\gamma_\mu\gamma_5 b|B\rangle = i\sqrt{4M_B M_D^*}F_{B\to D^*}D_\mu^*\,, \tag{3.58}$$

to be compared with Eq. (1.18). Conceptually our present derivation is very close to that leading to the Bjorken and Voloshin sum rules (Sec. 3.6). We will again consider the transition operator induced by the axial-vector current limiting ourselves to the spatial components of the current. Technically it is simultaneously simpler and more involved. Simpler — because at the point of zero recoil, one must put $\vec{q} = 0$, so that kinematics is trivial. In particular, from the very beginning, only one structure (h_1) survives in the general decomposition (3.40). The calculation is more complicated on the other hand since now one has to keep track of the terms of the order $\Lambda_{\rm QCD}^2$. Those of the order $\Lambda_{\rm QCD}$ are simply absent!

The quantity ϵ is defined now as

$$\epsilon = M_B - M_{D^*} - q_0 \tag{3.59}$$

and we continue to assume that $\Lambda_{\rm QCD} \ll |\epsilon| \ll m_{c,b}$ and continue to examine our old acquaintance, h_1, expanded in powers of $1/\epsilon$ and $\epsilon/m_{c,b}$. The result of a relatively simple calculation (which the reader is encouraged to do) is

$$-h_1 = \{1-\Delta\}\frac{1}{\epsilon} + \mathcal{O}(\Lambda_{\rm QCD}^3)\frac{1}{\epsilon^2} + \dots\,, \tag{3.60}$$

where

$$\Delta = \frac{1}{3}\frac{\mu_G^2}{m_c^2} + \frac{\mu_\pi^2 - \mu_G^2}{4}\left(\frac{1}{m_c^2} + \frac{1}{m_b^2} + \frac{2}{3m_c m_b}\right).$$

An explanatory remark is in order here concerning the $1/\epsilon^2$ term in h_1. The theoretical expression for h_1, as it naturally emerges from the computations, depends on ϵ_q, not on ϵ where ϵ_q is the energy measured from the "quark" threshold, see Eq. (3.47). In the kinematics at hand, when we are at the point of zero recoil,

$$\begin{aligned}\epsilon - \epsilon_q &= M_B - M_{D^*} - (m_b - m_c) \\ &= -(\mu_\pi^2 - \mu_G^2)\left(\frac{1}{2m_c} - \frac{1}{2m_b}\right) - \frac{2}{3m_c}\mu_G^2 + \ldots\end{aligned} \tag{3.61}$$

where I invoked Eq. (3.22). (Can you figure out why the coefficients in this expression and in Eq. (3.61) are different? Hint: The parameters μ_G^2 and μ_π^2 in Eq. (3.61) are defined as the expectation values of the corresponding operators over the *pseudoscalar* meson. This is not the case in Eq. (3.22).)

We first expand $-h_1$ in $1/\epsilon_q$ and then pass to the physical variable ϵ and rearrange the expansion. In the $1/\epsilon_q$ expansion the corrections of order $\mathcal{O}(\Lambda_{\rm QCD})$ are absent from the very beginning — an obvious fact hardly requiring further comments — while the term $\mathcal{O}(\Lambda_{\rm QCD}^2/\epsilon_q^2)$ does appear explicitly. This term, however, is killed in passing from $1/\epsilon_q$ to $1/\epsilon$, see Eq. (3.61).

That is why I assert that the coefficient in front of $1/\epsilon^2$ is $\mathcal{O}(\Lambda_{\rm QCD}^3)$. Moreover, as we will see shortly this coefficient in Eq. (3.60) is positive. In principle, it is calculable (more exactly, expressible in terms of several new phenomenological parameters) but this will be of no concern to us in this section.

Repeating, step by step, the derivation of Sec. 3.6 we conclude that

$$|F_{B\to D^*}|^2 + \sum_{i=1,2,\ldots} |F_{B\to \rm excit}|^2 = 1 - \Delta \tag{3.62}$$

and

$$\sum_{i=1,2,\ldots} |F_{B\to \rm excit}|^2 (M_i - M_{D^*}) = \mathcal{O}(\Lambda_{\rm QCD}^3)\,. \tag{3.63}$$

The latter sum rule, by the way, is the reason why we know that the coefficient $\mathcal{O}(\Lambda_{\rm QCD}^3)$ in front of $1/\epsilon^2$ is positive — the left-hand side of the sum rule is obviously positive-definite. These two relations, taken together, plus positivity of $|F_{B\to i}|^2$, imply that $|F_{B\to D^*}|^2$ is limited from below and from above,

$$1 - \Delta - \frac{\mathcal{O}(\Lambda_{\rm QCD}^3)}{M_1 - M_{D^*}} < |F_{B\to D^*}|^2 < 1 - \Delta\,, \tag{3.64}$$

where M_1 is the mass of the first excited state produced by the axial-vector current, $M_1 - M_{D^*} = \mathcal{O}(\Lambda_{\rm QCD})$.

Not only is it seen that the deviation of $|F_{B\to D^*}|^2$ from unity starts from terms scaling like $1/m_{c,b}^2$, with no $1/m_{c,b}$ corrections, but we understand now the reasons lying behind this remarkable fact. Moreover, we have an idea of how large the actual deviations are since Eq. (3.64) establishes a lower limit for these deviations in terms of the parameter Δ which is determined numerically rather well. In this aspect the derivation I present here goes beyond the more conventional analysis of Refs. [42, 43]. The reader is nevertheless advised to consult the latter works to get a broader perspective of the heavy quark theory — the more approaches you master the better for you.

Qualitatively it is quite clear why the deviation of $|F_{B\to D^*}|^2$ from unity starts from $\Lambda_{\rm QCD}^2/m_{c,b}^2$. Indeed, let us return to the Bjorken formula (3.37). In this formula, it is assumed $\vec{v} \gg \Lambda_{\rm QCD}/m_Q$, so that, actually, we do not distinguish between the velocity of the recoiling final heavy hadron and that of the final quark. At zero recoil the heavy hadron is nailed, but not the heavy quark. The latter experiences a primordial motion inside the nailed hadron, with the velocity $\vec{v}^2 \sim \Lambda_{\rm QCD}^2/m_Q^2$. So, a reasonable guess would be to extrapolate Eq. (3.37) down to $\vec{v}^2 \sim \Lambda_{\rm QCD}^2/m_Q^2$. As we see, this guess works.

It is worth emphasizing that our analysis need not be confined to the transitions induced by the spatial components of the axial-vector current. We could consider the temporal components, or vector currents, or something else. Each time we get additional information, for instance, from the transition operator induced by the vector currents, we get a sum rule proving the inequality $\mu_\pi^2 > \mu_G^2$ obtained in Sec. 3.2 from a quantum-mechanical argument.

SECTION 4
Theory of the Line Shape

In this section I will discuss one of the most interesting and practically important applications of the heavy quark theory, the spectra in the end point domain in the inclusive decays. Inclusive weak decays of heavy flavors, in particular, semileptonic decays, are close relatives of famous deep inelastic scattering — the processes where a highly virtual photon scatters off nucleons to produce an inclusive multiparticle state. The latter are related to the former via channel crossing. Deep inelastic lepton–nucleon scattering was in the focus of theoretical activity in the late sixties and the beginning of seventies and was instrumental in discovering and developing QCD [57]. It is thus quite surprising that for a long time there were hardly any attempts to treat the beauty decays in QCD proper along essentially the same lines as it was done in deep inelastic scattering. The realization of the idea that the $1/m_Q$ expansion in the theory of the line shape can play the same role as the twist $1/Q^2$ expansion in DIS came with the 20 years delay [58, 59, 60] — I see absolutely no reasons why the corresponding theory was worked out only recently and not 20 years ago.

The theory of the line shape in QCD resembles that of the Mössbauer effect. To explain what I mean it is convenient to consider, for definiteness, the transition $B \to X_s\gamma$ where X_s denotes the inclusive hadronic state with the s quark. This decay has been recently observed experimentally. (The description of $B \to X_q l\nu$ is conceptually similar but is more technically involved.)

Again, to avoid inessential technicalities I will neglect the quark and photon spins. So we will consider the transition $Q \to q\phi$ where all fields Q, q and ϕ are spinless. Thus, to begin with, we will limit ourselves to the toy model described in Sec. 2.1, see Eq. (2.9). The mass of the final quark m_q will be treated as a free parameter which can vary from zero almost up to m_Q. For our approach to be valid we still need that $\Delta m \equiv m_Q - m_q \gg \Lambda_{\rm QCD}$ although the mass difference may be small compared to the quark masses.

To warm up we will put the final quark mass to zero. At the level of the free quark decay the photon energy is then fixed by the two-body kinematics of the decay $Q \to q\phi$, namely, $E_\phi = m_Q/2$. In other words, in the rest frame of the decaying Q quark, the photon energy spectrum is a monochromatic line at $E_\phi = m_Q/2$ (Fig. 7). On the other hand, in the actual hadronic decays $H_Q \to X_q\phi$, the kinematical boundary of the spectrum lies at $M_{H_Q}/2$. Moreover, due to multiparticle final states (which are, of course, present at the level of the hadronic decays) the "photon" line will be smeared. In particular, the *window* — a gap between $m_Q/2$ and $M_{H_Q}/2$ — will be closed (Fig. 7). There are two mechanisms smearing the monochromatic line of the free quark decay. The first is purely perturbative: the final quark q can shake off a hard gluon, thus leading to the three-body kinematics. This mechanism tends to diminish the photon energy and may be important at $E < m_Q/2$. We will defer its discussion till later. The second mechanism is due to the "primordial" motion of the heavy quark Q inside H_Q and is nonperturbative. Even if the decaying B is nailed at the origin so that its velocity vanishes, the b quark moves inside the light cloud, its momentum being of the order $\Lambda_{\rm QCD}$. This is

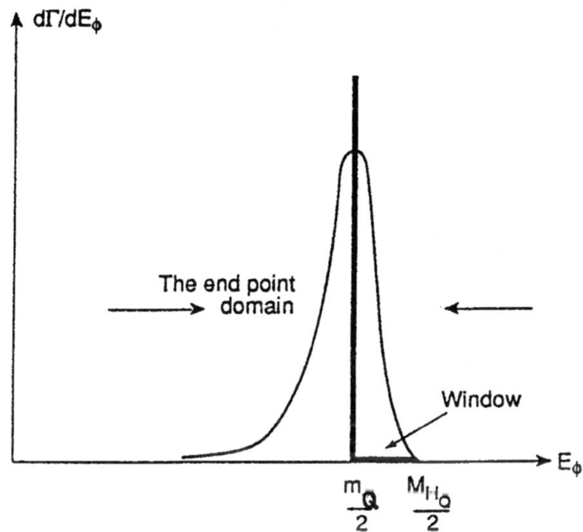

Fig. 7. The "photon" spectrum in the decay $H_Q \to X_q\phi$. The final quark is assumed to be massless. The thick line represents the delta function spectrum of the free quark approximation. The solid line is a sketch of the actual hadronic spectrum in the end point domain (possible radiation of hard gluons is neglected).

the QCD analog of the Fermi motion of the nucleons in the nuclei. It is quite clear that this motion affects the decay spectra. Say, if the "primordial" heavy quark momentum is parallel to that of the photon, the photon produced gets more energy, and vice versa, for the antiparallel momenta, it gets less. It is quite clear that this effect is preasymptotic (suppressed by inverse powers of m_b): while typical energies of the decay products are of the order m_b, a shift due to the heavy quark motion is of the order $\Lambda_{\rm QCD}$.

Only the second mechanism will be of interest to us in this section.

The window (i.e. the domain kinematically inaccessible for free quarks) plus the adjacent domain below the window, of width several units $\times\Lambda_{\rm QCD}$, taken together, form what is called the end point domain. Below I will outline the main elements of the theory allowing one to translate an intuitive picture of the Q quark primordial motion inside H_Q in QCD-based predictions for the spectrum in the end point domain.[9] The spectrum below the end point domain is the realm of the perturbative physics (hard gluon emission).

---

[9]The basics of the theory of the line shape were worked out in Refs. [58–60]. Further crucial steps were undertaken in Refs. [61–64]. In my presentation I follow mainly Bigi *et al.* [60].

4.1. *Formalism*

Let us return to Sec. 2.1 and consider the transition operator defined there. Since we are interested in the energy spectrum the "photon" momentum q must be fixed. Let us assemble Eqs. (2.14) and (2.23) together. For convenience I will reproduce the result here again, taking into account the fact that now m_q is assumed to vanish,

$$\frac{d\Gamma}{dE} = \frac{m_Q}{M_{H_Q}}\Gamma_0 \left\{ \langle\bar{Q}Q\rangle\delta\left(E - \frac{m_Q}{2}\right) - \frac{\langle\vec{\pi}^2\rangle}{4m_Q}\delta'\left(E - \frac{m_Q}{2}\right) \right.$$
$$\left. + \frac{\langle\vec{\pi}^2\rangle}{24}\delta''\left(E - \frac{m_Q}{2}\right) + \ldots \right\}. \tag{4.1}$$

If, in the leading approximation the spectrum is just a delta function, the corrections are more and more singular! The higher the correction the stronger the singularity. Nonsense? No, this was to be expected: the width of the ϕ line in the transition $H_Q \to X_q\phi$ is of the order $\Lambda_{\rm QCD}$. We expand in the powers of $\Lambda_{\rm QCD}/m_Q$; hence we must expect the enhancement of the singularities in each successive order. Equation (4.1) gives all terms up to $\Lambda^2_{\rm QCD}$. It is clear that to describe the shape of the line one needs to sum up the infinite number of terms in this expansion.

Then in the approximation of Fig. 1 (no hard gluon exchanges) the transition operator is given by Eq. (2.15) with m_q set equal to zero. To construct the operator product expansion to all orders we observe that the momentum operator π corresponding to the residual motion of the heavy quark is $\sim \Lambda_{\rm QCD}$ and the expansion in π/k is possible. Unlike the problem of the total widths, however, in the end point domain, k^2 is anomalously small, the expansion parameter is of the order unity, and there exists an infinite set of terms where all terms are of the same order of magnitude.

To elucidate this statement let us examine different terms in the denominator of the propagator,

$$k^2 + 2\pi k + \pi^2 \,.$$

In the end point domain

$$|E - (1/2)m_Q| \sim \bar{\Lambda}\,. \tag{4.2}$$

It is quite trivial to find that in this domain

$$k_0 \sim |\vec{k}| \sim m_Q/2,\ k^2 \sim m_Q\bar{\Lambda}\,;$$

in particular, at the kinematical boundary (for the maximal value of the "photon" energy) $k_0 = M_Q/2$ and $k^2 = -m_Q\bar{\Lambda}$. Hence, in the end point domain

$$k^2 \sim k\pi \gg \pi^2.$$

In other words, when one expands the propagator of the final quark in the transition operator,

$$\bar{Q}\frac{1}{k^2 + 2\pi k + \pi^2}Q\,, \tag{4.3}$$

in π, in the leading approximation, all terms $(2k\pi/k^2)^n$ must be taken into account while terms containing π^2 can be omitted. The first subleading correction would contain one π^2 and arbitrary number of $2k\pi$'s, etc.

Thus, in this problem it is the twist of the operators ($\equiv$ dimension — Lorentz spin) in the operator product expansion, not their dimension, that counts. For connoisseurs I will add that this aspect makes the theory of the line shape in the end point domain akin to that of deep inelastic scattering (DIS). Keeping only those terms in the expansion that do not vanish in the limit $m_Q \to \infty$ (analogs of the twist-2 operators in DIS) we get the following series for the transition operator:

$$\hat{T} = -\frac{1}{k^2}\sum_{n=0}^{\infty}\left(-\frac{2}{k^2}\right)^n k^{\mu_1}\ldots k^{\mu_n}\left(\bar{Q}\pi_{\mu_1}\ldots\pi_{\mu_n}Q - \text{traces}\right). \tag{4.4}$$

Traces are subtracted by hand since they are irrelevant anyway; their contribution is suppressed as $k^2\pi^2/(k\pi)^2 \sim \Lambda_{\text{QCD}}/m_Q$ to a positive power. Another way to make the same statement is to say that in Eq. (4.4) the four-vector k can be considered as *light-like*, $k^2 = 0$.

4.2. *The Light-Cone Distribution Function*

After the transition operator is built the next step is averaging $\hat{T}$ over the hadronic state H_Q. Using only the general arguments of the Lorentz covariance one can write

$$\langle H_Q|\bar{Q}\pi_{\mu_1}\ldots\pi_{\mu_n}Q - \text{traces}|H_Q\rangle = a_n\bar{\Lambda}^n(v_{\mu_1}\ldots v_{\mu_n} - \text{traces}) \tag{4.5}$$

where a_n are constants parametrizing the matrix elements. Their physical meaning will become clear momentarily. Right now it is worth noting that the term with $n = 1$ drops out ($a_1 = 0$). Indeed, $\langle H_Q|\bar{Q}\vec{\pi}Q|H_Q\rangle$ is obviously zero for spinless H_Q while π_0 through the equation of motion reduces to $\vec{\pi}^2/(2m_Q)$

and is of the next order in $1/m_Q$. The disappearance of $\bar{Q}\pi_\mu Q$ means that there is a gap in the dimensions of the relevant operators.

Let us write a_n's as moments of some function $F(x)$,

$$a_n = \int dx x^n F(x) \,. \tag{4.6}$$

Then, $F(x)$ is nothing else than the primordial line-shape function! (That is to say, $F(x)$ determines the shape of the line before it is deformed by hard gluon radiation; this latter deformation is controllable by perturbative QCD.) The variable x is related to the photon energy,

$$x = \frac{2}{\bar{\Lambda}}\left(E - \frac{m_Q}{2}\right) .$$

If this interpretation is accepted — and I will prove that it is correct — it immediately implies that (i) $F(x) > 0$; (ii) the upper limit of integration in Eq. (4.6) is 1; (iii) $F(x)$ exponentially falls off at negative values of x so that practically the integration domain in Eq. (4.6) is limited from below at $-x_0$ where x_0 is a positive number of the order unity.

To see that the above statement is indeed valid we substitute Eqs. (4.5), (4.6) in $\hat{T}$,

$$\langle H_Q|\hat{T}|H_Q\rangle = -\frac{1}{k^2}\sum_n \int dy F(y)\, y^n \left(-\frac{2\bar{\Lambda}kv}{k^2+i0}\right)^n , \tag{4.7}$$

and sum up the series. The $i0$ regularization will prompt us how to take the imaginary part at the very end. In this way we arrive at

$$\frac{d\Gamma}{dE} = -\frac{4}{\pi}\Gamma_0 \frac{m_Q E}{M_{H_Q}} \mathrm{Im} \int dy F(y) \frac{1}{k^2 + 2y\bar{\Lambda}kv + i0} = (2/\bar{\Lambda})\Gamma_0 F(x) \,, \tag{4.8}$$

where the variable $x = k^2/(2\bar{\Lambda}kv)$ was written out above in terms of the "photon" energy, and Γ_0 is the total decay width in the parton approximation. Corrections to Eq. (4.8) are of the order $\bar{\Lambda}/m_Q$.

Thus, we succeeded in getting the desired smearing: the monochromatic line of the parton approximation is replaced by a finite size line whose width is of the order $\bar{\Lambda}$. The pre-asymptotic effect we deal with is linear in Λ_{QCD}/m_Q.

At this point you might ask me how this could possibly happen. We have already learnt that there is a gap in dimensions of the operators in the expansion — no operators of dimension 4 exist — and the correction to $\bar{Q}Q$ is also quadratic in $1/m_Q$ (the CGG/BUV theorem). There are no miracles — the occurrence of the effect linear in $\Lambda_{\rm QCD}/m_Q$ became possible due to the summation of the infinite series in Eq. (4.4); no individual term in this series gives rise to $\Lambda_{\rm QCD}/m_Q$.

To avoid misunderstanding it is worth explicitly stating that the primordial distribution function $F(x)$ is *not* calculated; rather, $F(x)$ is related to the light-cone distribution function of the heavy quark inside H_Q, namely $\langle \bar{Q}(n\pi)^n Q\rangle$, $n^2 = 0$, or more explicitly

$$F(x) \propto \int dt e^{ixt\bar{\Lambda}} \langle H_Q|\bar{Q}(x=0) e^{-i\int_0^t nA(n\tau)d\tau} Q(x_\mu = n_\mu t)|H_Q\rangle, \qquad (4.9)$$

where n is a light-like vector[10]

$$n_\mu = (1, 0, 0, 1).$$

Unfortunately, this primordial function is not the one that will be eventually measured from $d\Gamma/dE$; the actual measured line shape will be essentially deformed by the radiation of hard gluons. I will say a few words about this in Sec. 5.

The primordial distribution function $F(x)$ which we defined here can be called the *light-cone* distribution function. This is clear from the expression (4.9) which has a very transparent physical meaning. The quark q produced is massless and, therefore, propagates along the light-cone from the point of emission to the point of absorption in the transition operator defining the distribution function.

If we looked at the physical line shape sketched in Fig. 7 more attentively, through a microscope, we would notice that a smooth curve is obtained as a result of adding up many channels, specific decay modes. A typical interval in E that contains already enough channels to yield a smooth curve after summation is $\sim \Lambda^2_{\rm QCD}/m_Q$. Roughly one can say that the spectrum of Fig. 7 covers altogether $m_Q/\Lambda_{\rm QCD}$ resonance states produced in the H_Q decays and composed of q plus the spectator (I keep in mind here that the final hadronic state is produced through decays of highly excited resonances, as in the multicolor QCD). These states span the window between $m_Q/2$ and $M_{H_Q}/2$ and the adjacent domain to the left of the maximum at $E = m_Q/2$.

[10] Similar light-cone distributions for light quarks are well-known [65] in the theory of deep inelastic scattering, see also [66].

4.3. Varying the Mass of the Final Quark

So far I was discussing the transition into a massless final quark. It is very interesting to trace what happens with the line shape and the primordial distribution function as the final quark mass m_q increases.

The inspection of the transition operator shows that as long as $m_q^2 \ll m_Q\bar{\Lambda}$ nothing changes in our formulae at all in the leading twist approximation. Since the characteristic values of k^2 in the end point domain are of order $m_Q\bar{\Lambda}$ and $m_q^2 \ll m_Q\bar{\Lambda}$ one can simply neglect the final quark mass altogether.

A more interesting regime is $m_q \sim (m_Q\bar{\Lambda})^{1/2}$. In this regime one cannot neglect m_q^2 in the denominator. It is not difficult to see, however [63], that, as in the massless case, all traces can be neglected since $k^2\pi^2/(k\pi)^2 \sim \Lambda_{\rm QCD}/m_Q$. This means that the same light-cone distribution function $F(x)$ that emerged in the massless case describes the line shape if $m_q \sim m_Q\bar{\Lambda}$ as well. The only change that occurs is a shift of the end point spectrum, as a whole, to the left. Indeed, if previously the variable x was defined as $(2/\bar{\Lambda})[E - (1/2)m_Q]$, now when $m_q \neq 0$,

$$x = \frac{2}{\bar{\Lambda}}(E - E_0)$$

where $E_0 = (2m_Q)^{-1}(m_Q^2 - m_q^2)$. The maximum of the distribution, in particular, shifts from $m_Q/2$ to $E_0 = m_Q/2 - \mathcal{O}(\bar{\Lambda})$.

What happens if one continues to increase m_q? Increasing the quark mass further results in more drastic changes. The trace terms cannot be omitted any more, and the light-cone function gives place to other distribution functions. This is obvious already from a simple kinematical argument. Indeed, with m_q increasing, the window shrinks. When we eventually come to the SV limit

$$\Lambda_{\rm QCD} \ll \Delta m \equiv m_Q - m_q \ll m_{Q,q}$$

it shrinks to zero. In this limit the photon energy in the two-body quark decay, $\Delta m(1 + \Delta m(2m_Q)^{-1})$ differs from the maximal photon energy in the hadronic decay, $\Delta M(1 + \Delta M(2M_Q)^{-1})$, only by a tiny amount inversely proportional to m_Q (Δm and ΔM stand for the quark and meson mass differences, respectively).

Thus, the kinematical consideration prompts us that the line shape must essentially change. Anticipating the results of the calculation let me describe the situation pictorially. Simultaneously with the shrinkage of the window the peak becomes more asymmetric and develops a two-component structure (Fig. 8). The dominant component of the peak, on its right-hand side, becomes

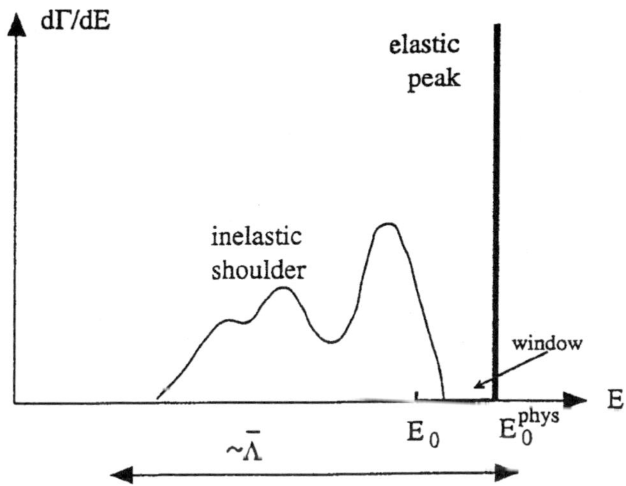

Fig. 8. Evolution of the spectrum of Fig. 7 as the mass of the final quark increases (the schematic plot refers to $m_q \sim m_Q/2$). The effects of the hard gluon bremstrahhlung are not included.

narrower and eventually collapses into a delta function when m_q becomes a finite fraction of m_Q. A shoulder develops on the left-hand side; the number of the hadronic states populating the end point domain becomes smaller — instead of $m_Q/\Lambda_{\rm QCD}$ states at $m_q = 0$ we are speaking of just several states at m_q = a finite fraction of m_Q. When we approach the SV limit the height of the shoulder corresponding to the production of the excited states becomes very small, proportional to $\vec{v}^2 \ll 1$ (Fig. 9). This is the end of the evolution — starting from the light-cone distribution function at $m_q = 0$ we continuously pass to the *temporal* distribution function in the SV limit. It is the temporal distribution function that shapes the inelastic shoulder in Fig. 9.

This rather sophisticated picture, hardly reproducible in naive quark models, emerges from the operator product expansion (in the leading approximation) if one follows along the same lines as previously. The transition operator $\hat{T}$ for $m_q \neq 0$ is given in Eq. (2.20); I reproduce it here again for convenience,

$$\hat{T} = \frac{1}{m_q^2 - k^2} \sum_{n=0}^{\infty} \bar{Q} \left(\frac{2m_Q\pi_0 + \pi^2 - 2q\pi}{m_q^2 - k^2} \right)^n Q\,. \tag{4.10}$$

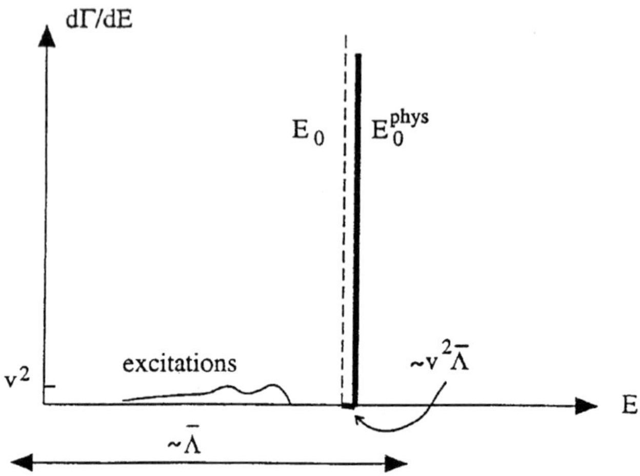

Fig. 9. The photon spectrum in the SV limit, $\vec{v}^2 \ll 1$. The dashed line shows the would-be spectrum of the free quark decay.

Notice that $2m_Q\pi_0 + \pi^2$ acting on Q yields zero (the equation of motion) and in the SV limit q must be treated as a small parameter,

$$q_0/m_Q \equiv E/m_Q = v \ll 1;$$

v is the spatial velocity of the heavy quark produced. Although v is small the inclusive description is still valid provided that $\Delta m \gg \Lambda_{\rm QCD}$.

In the zeroth order in q the only term surviving in the sum (4.10) is that with $n = 0$, and we are left with the single pole, the elastic contribution depicted in Fig. 9. This is the extreme realization of the quark–hadron duality. The inclusive width is fully saturated by a single elastic peak. We have already discussed this phenomenon in Sec. 2. What might seem to be a miracle at first sight has a symmetry explanation — the phenomenon is explained by the heavy quark symmetry. The fact that the parton model monochromatic line is a survivor of hadronization is akin to the Mössbauer effect.

If terms $\mathcal{O}(v^2)$ are switched on, the transition operator acquires an additional part,

$$\hat{T}_{v^2} = \frac{4}{3}\vec{q}^2 \frac{1}{(m_q^2 - k^2)^3} \sum_{n=0}^{\infty} \left(\frac{2m_Q}{m_q^2 - k^2}\right)^n \bar{Q}\pi_i\pi_0^n\pi_i Q\,. \tag{4.11}$$

From this expression it is obvious that the shape of the v^2 shoulder is given by the *temporal* distribution function $G(x)$ whose moments are introduced through the matrix elements

$$\langle H_Q|\bar{Q}\pi_i\pi_0^n\pi_i Q|H_Q\rangle = \bar{\Lambda}^{n+2}\int dx x^n G(x)\,. \tag{4.12}$$

Alternatively, $G(x)$ can be written as a Wilson line along the time direction,

$$G(x) \propto \int dt e^{ixt\bar{\Lambda}}\langle H_Q|\bar{Q}(t=0,\vec{x}=0)\pi_i e^{-i\int_0^t A_0(\tau)d\tau}\pi_i Q(t,\vec{x}=0)|H_Q\rangle\,. \tag{4.13}$$

Intuitively it is quite clear why the light-cone distribution function gives place to the temporal one in the SV limit. Indeed, if the massless final quark propagates along the light-cone, for $\Delta m \ll m$, the quark q is at rest in the rest frame of Q, i.e. propagates only in time.

In terms of $G(x)$ our prediction for the line shape following from Eq. (4.11) takes the form

$$\frac{d\Gamma}{dE} \propto \left[1-\frac{v^2}{3}\int\left(\frac{1}{y^2}+\frac{\bar{\Lambda}/E_{\max}}{y}\right)G(y)dy\right]\delta(x)+\frac{v^2}{3}\left(\frac{1}{x^2}+\frac{\bar{\Lambda}/E_{\max}}{x}\right)G(x)\,, \tag{4.14}$$

where $x = (E-E_{\max})/\bar{\Lambda}$. The v^2 corrections affect both, the elastic peak (they reduce the height of the peak) and the shoulder (they create the shoulder). The total decay rate stays intact, however: the suppression of the elastic peak is compensated by the integral over the inelastic contributions in the shoulder. This is the Bjorken sum rule thoroughly considered in Sec. 3. It is important that we do not have to guess or make *ad hoc* assumptions — a situation typical of model-building — QCD itself tells us what distribution function enters in this or that case and in what particular way.

4.4. *Real QCD: Inclusive Semileptonic Decays*

From the analysis presented above the following remarkable fact should be clear. The very same primordial distribution functions that determine the line shape in the radiative transitions appear in the problem of the spectra in the semileptonic decays. In particular, in $b \to ul\nu$ we deal with $F(x)$.

Of course, kinematical conditions are different. Now the hadronic part of the process, $B \to X_u l\nu$ inclusive decay, depends on two variables, for instance, q_0 and q^2, or q_0 and $|\vec{q}|$. The probability of the decay, in the free quark approximation, is proportional to $\delta(m_Q - q_0 - |\vec{q}|)$ [67]. In other words, in this

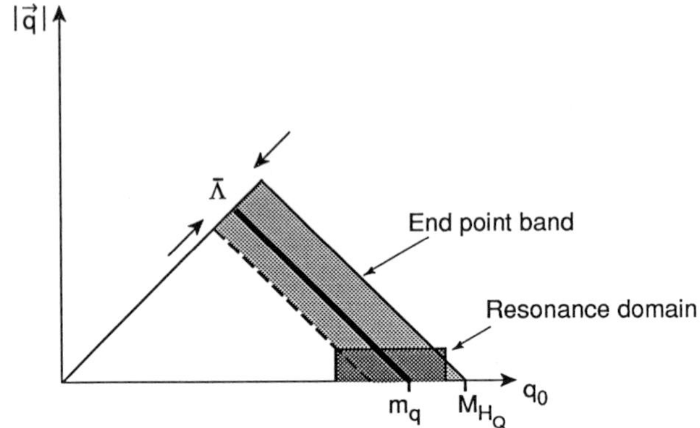

Fig. 10. Kinematically allowed domain in the transition $B \to X_u l\nu$. The thick line indicates the populated phase space in the free quark decay. The shaded area of width $\sim \bar{\Lambda}$ is the end point domain populated due to the primordial motion of the b quark inside B. The shaded square in the lower right corner is the exclusive resonance domain where the inclusive approach developed here is inapplicable.

approximation only a line on the $q_0, |\vec{q}|$ plane is populated (Fig. 10). (I assume that we are not interested in the individual momenta of l and ν and measure just the total momentum of the lepton pair. This is quite a fantastic formulation of the problem since experimentally the neutrino energy and momentum are not measured, of course; only the electron energy is usually measured. Nevermind, let us keep in mind a *gedanken* experiment.) The end point domain is defined now as a band whose width is several units $\times\ \Lambda_{\rm QCD}$ adjacent to the above quark line (Fig. 10). Needless to say that in the physical decay the whole large triangle is populated; the inner part of the triangle, to the left of the end point band, is due to the hard gluon emission. The smearing of the delta-like spectrum in the band is due to the primordial motion of b inside B, and is described by the light-cone distribution.

A trivial modification compared to Sec. 4.3 is the occurrence of several structure functions. All five structure functions are expressible, however, in terms of the same light-cone primordial distribution function $F(x)$ where, as previously, $x = -\bar{\Lambda}^{-1}k^2/2k_0$. Since q_0 and $\vec{q}$ are independent variables, in the case at hand,

$$x = -\bar{\Lambda}^{-1}k^2/2k_0 = -\bar{\Lambda}^{-1}k^2/(k_0 + |\vec{k}|) = -\bar{\Lambda}^{-1}(m_Q - q_0 - |\vec{q}|) \qquad (4.15)$$

where in the denominator the difference between k_0 and $|\vec{k}|$ is neglected which is perfectly legitimate in the end point band. In this band $k_0 - |\vec{k}| = \mathcal{O}(\Lambda_{\rm QCD})$, and the difference between k_0 and $|\vec{k}|$ becomes important only at the level of the subleading twists which are not included anyway.

Thus, we observe a scaling behavior: the structure functions that generally speaking could depend on two variables, q_0 and $|\vec{q}|$, actually depend only on the single light-cone combination (4.15). This is the analog of the Bjorken scaling in deep inelastic scattering! In the rest of the phase space, outside the end point band, the approximate equality $k_0 \approx |\vec{k}|$ is not valid, of course, and the above scaling is not going to take place. The primordial distribution falls off — presumably exponentially — outside the end point band. The hard gluon emissions will populate the phase space outside this domain creating long logarithmic tails. The primordial part is buried under these tails. Therefore, outside the band one cannot expect that the structure functions depend on the single combination $q_0 + |\vec{q}|$ anyway.

Guesses about a scaling behavior in the inclusive semileptonic decays are known in the literature [67]. Now we are finally able to say for sure what sort of scaling takes place, where it is expected to hold and where and how it will be violated.

I will not go into further details which are certainly important if one addresses the problem of the extraction of V_{ub} from experimental data. Some of them are discussed in the literature, others still have to be worked out. Applications of the theory to data analysis is a separate topic going beyond the scope of this section.

What can be said about the light-cone distribution function $F(x)$? This function depends on the structure of the light cloud of the B meson and, thus, belongs to the realm of the soft physics. The moments of this function are related to the expectation values of the operators $\bar{Q}\pi_{\mu_1}\ldots\pi_{\mu_n}Q$ [see Eq. (4.5)]; in real QCD the properly normalized matrix elements on the left-hand side include the factor $(2M_B)^{-1}$). The knowledge of the infinite set of these expectation values would be equivalent to the knowledge of the structure of the light cloud. Needless to say that this is beyond our abilities at present. Still, we know a few first moments of $F(x)$ and have a general idea of the shape of this function. It must be positive everywhere in the physical domain, vanish at $x = 1$ and have exponential fall-off at large negative x. The latter property ensures the existence of all moments. Moreover,

$$a_0 = \int dx F(x) = 1\,,$$

$$a_1 = \int dx x F(x) = 0\,,$$

and

$$a_2 = \int dx\, x^2 F(x) = \frac{\mu_\pi^2}{3\bar{\Lambda}^2}\,.$$

Estimates of the third moment also exist in the literature [60, 42]. I cannot dwell on this issue now and will only mention that a_3 is constrained by exact inequalities, i.e. [64]

$$a_2 < \frac{1}{4} + \sqrt{\frac{1}{4} - a_3}\,, \quad a_3 < \frac{1}{4} - \left(a_2 - \frac{1}{2}\right)^2\,. \tag{4.16}$$

To derive these inequalities one merely observes that for any t the integral from $-\infty$ to 1 over x over the function $(1-x)(x-t)^2F(x)$ is positive; on the other hand, this integral is a second-order polynomial in t and, hence, its discriminant must be negative.

A sketch of a function satisfying all these requirements is given in Fig. 11.

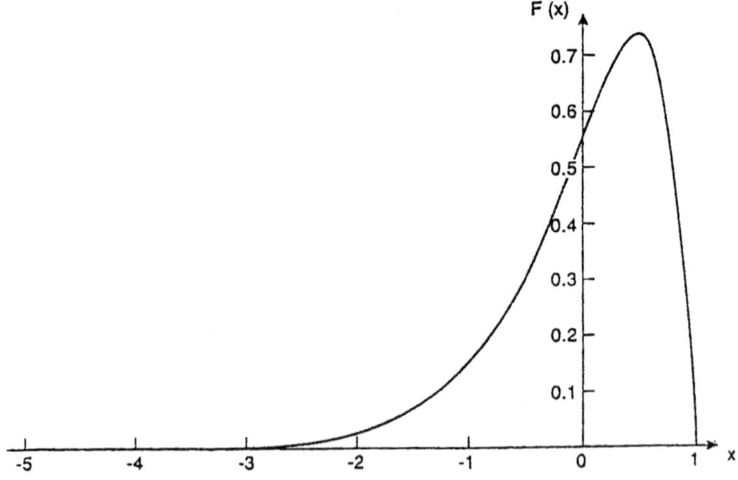

Fig. 11. More or less realistic light-cone primordial distribution function versus x (borrowed from Ref. [64]).

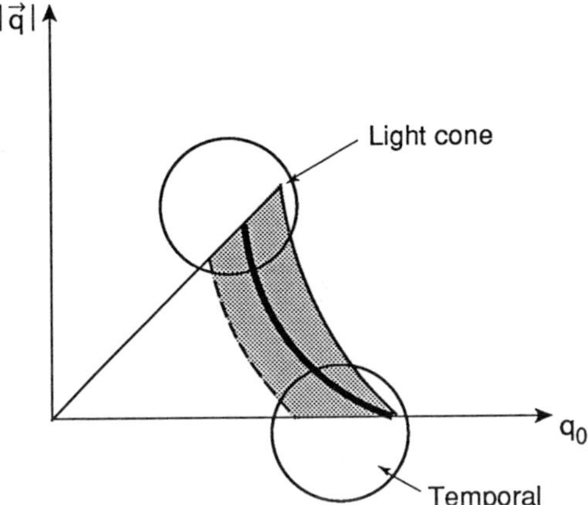

Fig. 12. Kinematically allowed domain in the decay $B \to X_c l\nu$. Two large circles show the domains where the description should be based on the light-cone and temporal distribution functions, respectively.

A natural desire to extend the formalism described above to the semileptonic inclusive transitions $b \to cl\nu$ encounters serious technical difficulties. The essence of the problem is as follows. The final quark c can be treated as heavy, although at the same time, $m_c^2 \ll m_b^2$. The ratio $m_c^2/m_b^2 \approx 0.07$ is a small parameter while $m_c^2/(\bar{\Lambda} m_b) \sim 1$. Under the circumstances the type of the distribution function describing the primordial motion of b inside B and determining the measurable structure functions w_1 to w_5 will depend on the value of $|\vec{q}|$, and the scaling property — dependence on one particular combination of variables — is lost.

The $q_0, |\vec{q}|$ plane is shown in Fig. 12. In the free quark approximation the transition probability is proportional to $\delta(m_b - q_0 - E_c)$ where $E_q = (m_c^2 + \vec{q}^2)^{1/2}$, and all events are concentrated along the line indicated in Fig. 12. At the hadronic level the phase space consists of the full triangle, with one side curved. The end point band is also curved.

The fact that one side of the triangle is distorted compared to $b \to ul\nu$ is not crucial. What is important is the change of dynamics as we move from the upper left corner to the lower right one. In the case of $b \to ul\nu$, moving along the end point band in this direction does not affect the measured structure functions (apart from the extreme domain of soft u — the exclusive resonance domain — where our description fails altogether). The situation is different in the $b \to cl\nu$ transition.

If $|\vec{q}|^2 \gg m_c^2$ one recovers [63] the same light-cone function $F(x)$ as in the transition $b \to ul\nu$ or $b \to s\gamma$. Modifications are marginal. First, some extra terms explicitly proportional to m_c/m_b are generated in the structure functions due to the fact that $\not{P} + m_c$ replaces $\not{P}$ in the numerator of the quark Green function. Moreover, if in the $b \to ul\nu$ transitions the scaling variable in the end point domain is

$$x = \bar{\Lambda}^{-1}(q_0 + |\vec{q}| - m_b), \tag{4.17}$$

in the $b \to cl\nu$ transition, it is shifted by a constant term of the order one,

$$x = \bar{\Lambda}^{-1}(q_0 + |\vec{q}| - m_b) + \frac{m_c^2}{\bar{\Lambda} m_b} . \tag{4.18}$$

To see how this shift occurs [63] and to reveal limitations of the approximation let us start from the parton model variable $m_b - q_0 - E_c$. In the limit $|\vec{q}|^2/m_c^2 \to \infty$ inside the end point band formally one may substitute E_c by

$$E_c \to |\vec{q}| + \frac{m_c^2}{2|\vec{q}|} \to |\vec{q}| + \frac{m_c^2}{m_b}\left(1 + \mathcal{O}(m_c^2/m_b^2)\right) .$$

If $m_c^2 = \mathcal{O}(\bar{\Lambda} m_b)$ formally one may discard the $\mathcal{O}(m_c^2/m_b^2)$ correction. In this way we arrive at the scaling variable (4.18).

It is worth emphasizing that the occurrence of the light-cone distribution function in this regime, the same as in the $b \to u$ transition, is a remarkable fact. Indeed, if we could examine the measured structure functions "in the microscope" in these two cases, we would see that their microstructure is quite different. As was already mentioned, in the $b \to u$ transition, the end point band is saturated by the production of $\sim m_b/\bar{\Lambda}$ states, with the spacing between the individual states of the order $\bar{\Lambda}^2/m_b$. In the $b \to c$ transition, even if we are in the upper left corner of the phase space where the light-cone distribution is relevant, the number of the states produced is $\sim m_b/m_c \sim m_c/\bar{\Lambda}$ and the spacing is of the order $\bar{\Lambda}^2/m_c$. We deal with a much coarser structure in the latter case, and still all resonance contributions being summed up must add up to produce the light-cone distribution, formally the same one that is created by a much larger number of resonances in the $b \to u$ transition. (Purely theoretically we cannot predict fine grain versus coarse grain composition of the structure functions if we limit ourselves to the leading twist. Only an analysis of all twists could resolve these details, would this analysis be possible.)

The word "formally" is used above three times, not accidentally. Practically in the $b \to cl\nu$ transition $|\vec{q}|$ can never be much larger than m_c. Indeed, the *maximal* value of $|\vec{q}|$, corresponding to $q^2 = 0$ is $(m_b^2 - m_c^2)/2m_b \sim 2$ GeV, that is only $\sim 1.5m_c$. Therefore, only by stretching a point and only in a narrow domain near $q^2 = 0$, one can expect that the light-cone function of the variable (4.18) is, perhaps, more or less relevant.

As $|\vec{q}|$ decreases and becomes less than m_c (this regime takes place in a large part of the phase space) the light-cone distribution function becomes irrelevant. The measurable structure functions are determined by a different distribution — the light-like vector n_μ in Eq. (4.9) is replaced by

$$w_\mu = \left(1, \frac{\vec{q}}{E_c}\right),$$

and it becomes clear that the would-be scaling variable $x = \bar{\Lambda}^{-1}(q_0 + E_c - m_b)$ fails to represent all dependence of the structure functions on q_0 and $|\vec{q}|$. When q^2 approaches its maximal value,

$$q_{\max}^2 = (M_B - M_D)^2,$$

$|\vec{q}|$ tends to zero and we eventually approach the SV regime which I have already discussed, with fascination, in the toy example above. In the SV limit the velocity of H_c produced is small, and the structure functions probe the primordial motion described by the temporal distribution function $G(x)$ where now

$$x \approx \bar{\Lambda}^{-1}(q_0 - \Delta m),$$

Δm is the quark mass difference coinciding, to the leading order with $M_B - M_D$.

Thus, changing q^2 from zero to $q_{\max}^2$ results in an evolution of the distribution function appearing in theoretical formulae for $d\Gamma(B \to X_c l\nu)$, from light-cone to temporal, through a series of intermediate distributions. The physical reason for this evolution is quite clear — what distribution function is actually measured depends on the parton model velocity of the quark produced in the b decay. In the limiting cases of very large recoil and very small recoil the problem is solved in the sense that the structure functions are expressed in terms of the light-cone and temporal distribution functions, respectively. The intermediate case $|\vec{q}| \sim m_c$ is not worked out in detail so far. It is beyond any doubt, however, that the parton model type scaling will not take place.

SECTION 5

Including Hard Gluons. Generalities of the Operator Product Expansion

Finally the time comes when I cannot ignore the existence of hard gluons any more. Hard gluons are mere nuisance from the point of view of the theory of hadrons since they play no, or very little, role in the structure of the low lying hadronic states. Yet, if we want to go beyond purely academic exercises, however beautiful they might look, and descend down into a messy world of real hadronic physics, hard gluons cannot be forgotten since they "contaminate" nearly every experimentally measurable quantity. To make contact with the real world we have to consider the interplay between the soft and hard physics.

The hard gluons manifest themselves in many ways. They contribute to the coefficient functions in the effective Lagrangian (1.3) obtained by integrating out all degrees of freedom with the characteristic frequencies down to μ. They show up in the calculations of the total decay rates and spectra discussed in Sections 3 and 4 resulting in perturbative corrections which, in some instances, change the answer quite drastically. They result in the fact that all basic parameters of the heavy quark physics — the heavy quark mass, $\bar{\Lambda}$, μ_π^2 and so on — generally speaking, become μ-dependent and cannot be treated as universal constants. Here we will address some of these issues in brief.

5.1. *Calculation of the Effective Lagrangian*

I have already started discussing this topic in Sec. 1.2. The original QCD Lagrangian (1.1) is formulated at very short distances. In principle, it codes all information necessary for the calculation of all observable amplitudes. We just do the functional integral and ... alas, there are very few functional integrals that can be calculated analytically; numerical evaluation on lattices may take years, and I even dare to assert that some amplitudes will never be calculated that way. So, we take the original Lagrangian and start evolving it down, integrating out all fluctuations with the frequencies $\mu < \omega < M_0$ where M_0 is the original normalization point, and μ will be treated, for the time being, as a current parameter. In this way we get the Lagrangian which has the form

$$\mathcal{L} = \sum_n C_n(M_0; \mu) \mathcal{O}_n(\mu) \,. \tag{5.1}$$

The coefficient functions C_n represent the contribution of virtual momenta from μ to M_0. The operators $\mathcal{O}_n$ enjoy the full rights of the Heisenberg operators with respect to all field fluctuations with frequencies less than μ. The sum in Eq. (5.1) is infinite — it runs over all possible Lorentz singlet gauge-invariant operators with the appropriate quantum numbers; for instance, if CP is conserved, only CP-even operators will appear in (5.1). If, say, the electromagnetic processes are included, the operators in the Lagrangian (5.1) may contain the photon and electron fields, and so on. All operators can be ordered according to their dimension; moreover, we can use the equations of motion stemming from the original QCD Lagrangian to get rid of some of the operators in the sum. Those operators that are reducible to full derivatives give vanishing contributions to the physical (on mass shell) matrix elements and can thus be discarded as well.

If one just abstractly writes the expression (5.1) one is free to take any value of μ; in particular, $\mu = 0$ would mean that *everything* is calculated and we have the full S matrix, all conceivable amplitudes, at our disposal. Nothing is left to be done. In this case Eq. (5.1) is just a sum of all possible amplitudes. This sum then must be written in terms of the physical hadronic states, of course, not in terms of the quark and gluon operators since the latter degrees of freedom are simply nonexistent at large distances.

This is day-dreaming, of course. Needless to say that in our explicit calculation of the coefficient functions we have to stop somewhere, at such virtualities that the quark and gluon degrees of freedom are still relevant, and the coefficient functions $C_n(M_0, \mu)$ are still explicitly calculable. On the other hand, for obvious reasons it is highly desirable to have μ as low as possible. In the heavy quark theory there is an additional requirement that μ must be much less than m_Q. The process of calculating the coefficients $C_n(M_0, \mu)$ is called *matching* in the more standard presentation of HQET. Actually we see that this procedure is nothing else than a generalization of Wilson's idea of the renormalization group and the (Wilsonean) operator product expansion. Using the standard OPE language has an evident advantage: all well-studied elements of the latter approach can be immediately adapted in the environment of the heavy quark expansions. In particular all parameters that one can read off from the Lagrangian (5.1) depend on μ (including, say, the heavy quark mass). Let us assume that μ is large enough so that $\alpha_s(\mu)/\pi \ll 1$, on the one hand, and small enough so that there is no large gap between $\Lambda_{\rm QCD}$ and μ. The possibility to make such a choice of μ could not be anticipated *a priori* and is an extremely fortunate feature of QCD, a gift from the Gods. Quarks and gluons with the offshellness larger than μ chosen that way are called hard.

Needless to say that the parameter μ is in our minds, not in nature. All observable amplitudes must be μ independent. The μ dependence of the coefficient functions C_n must conspire with that of the matrix elements of the operators $\mathcal{O}_n$ in such a way as to ensure this μ independence of the physical amplitudes.

What can be said about the calculation of the coefficients C_n? Since μ is sufficiently large, as seen above, the main contribution comes from perturbation theory. We just draw all relevant Feynman graphs and calculate them, generating an expansion in $\alpha_s(\mu)$ which for brevity I will denote by α_s, with the argument omitted,

$$C_n = \sum_l a_l \alpha_s^l \,.$$

Sometimes some graphs will contain not only powers of $\alpha_s(\mu)$ but powers of $\alpha_s \ln(m_Q/\mu)$ too. This happens if the anomalous dimension of the operator $\mathcal{O}_n$ is nonvanishing — quite a typical situation — or if a part of a contribution to C_n comes from characteristic momenta of order m_Q and is, thus, expressible in terms of $\alpha_s(m_Q)$, and we rewrite it in terms of $\alpha_s(\mu)$. Nevermind, this is a trivial technicality. You are supposed to know how to sum up these logarithms.

As a matter of fact the expression for C_n above is not quite accurate theoretically. One should not forget that, in doing the loop integrations, in C_n we *must* discard the domain of virtual momenta below μ, by definition of $C_n(\mu)$. Subtracting this domain from the perturbative loop integrals we introduce in C_n power corrections of the type $(\mu/m_Q)^n$ by hand. In principle, one should recognize the existence of such corrections and try to learn how to deal with them. The fact that they are there was realized long ago (see e.g. V. Novikov *et al.*, Ref. [5]) and then largely ignored. If it is possible to choose μ sufficiently small these corrections may be insignificant numerically and can be omitted. This is what is actually done in practice. This is one of the elements of a simplification of the Wilsonean operator product expansion. The simplified version is called the *practical version of OPE*, see below. Certainly, at the modern stage of the theoretical development it is desirable to return to the issue to engineer a better procedure than just discarding these μ/m_Q terms in the coefficient functions. Attempts in this direction are under way [68].

Even if perturbation theory dominates in the coefficient functions they still contain also nonperturbative terms coming from short distances. Sometimes they are referred to as noncondensate nonperturbative terms. An example is provided by the so-called direct instantons with the sizes of order m_Q^{-1}. These contributions fall off as high powers of $\Lambda_{\rm QCD}/m_Q$ and are very poorly controllable theoretically. Since the fall-off of the noncondensate nonperturbative corrections is extremely steep, basically the only thing we need to know is a critical value of m_Q. For lower values of m_Q no reliable theoretical predictions are possible at present. For higher values of m_Q one can ignore the noncondensate nonperturbative contributions. There are good reasons to believe that the b quark, fortunately, lies above the critical point. Again, I must add that the noncondensate nonperturbative contributions are neglected in the practical version of OPE.

(Do we see seeds of the nonperturbative contribution in the expansion $C_n = \Sigma a_l \alpha_s^l$? Yes, we do. At any finite order the perturbative contribution is well defined. At the same time, if the coefficients a_l in the series grow factorially with l — and this is actually the case — the tail of the series, $l > 1/\alpha_s$, must be regularized, which may bring in terms of the order

$$\exp\left(-C/\alpha_s(m_Q)\right) \sim \left(\frac{\Lambda_{\rm QCD}}{m_Q}\right)^{\gamma} \tag{5.2}$$

where C is some positive constant and the exponent γ need not be an integer. In a sense, one may say that contributions to C_n of this type are vaguely related to diagrams with $1/\alpha_s$ hard gluon loops.)

Thus, two sources of nonperturbative corrections in the physical amplitudes are indicated. Those due to nonperturbative terms in the coefficient functions are systematically ignored (and, perhaps, rightly so, as I tried to convince you) in these lectures and in all works based on the practical version of OPE which constitute the overwhelming majority of all works devoted to the $1/m_Q$ expansions. The second source is operators of higher dimensions in the Lagrangian (5.1), the so-called condensate corrections. The latter were in the center of our attention; they generate the $1/m_Q$ expansions discussed above. One new element which I would like to add here is that the series of $1/m_Q$ terms generated by higher dimensional operators is also asymptotic and divergent in high-orders [69]. Of course, we always calculate only one, at best two, first $1/m_Q$ corrections, truncating the series. If, however, one would ask what the impact of the high-order tail of the power series is, the answer would be: this tail is reflected in exponentially small terms $\sim \exp(-m_Q)$. This type of contribution is certainly not seen in OPE truncated at any finite order. A transparent example is again provided by instantons. This time one has to fix the size of the instanton ρ by hand, $\rho_0 \sim \Lambda^{-1}$. Then their contribution to physical amplitudes is $\mathcal{O}(\exp(-m_Q\rho_0))$. The relation between $\exp(-m_Q\rho_0)$ piece and the high-order terms of the power series is conceptually akin to the connection between $\exp(-1/\alpha_s)$ terms and $l \sim 1/\alpha_s$ orders in the perturbative expansion.

Summarizing, Wilsonean OPE (5.1) leads to expansions in different parameters. Purely logarithmic terms $(\ln m_Q)^{-l}$ are due to ordinary perturbation theory. Terms of the type $(m_Q^2)^{-k}(\ln m_Q)^{-\gamma}$ reflect higher dimension operators and direct instantons. In the former case the values of k are integer, the latter case may produce noninteger values of k. In the practical version of OPE we calculate the coefficient functions perturbatively. All nonperturbative terms come from condensates within this approximation. The condensate power series is truncated: only those operators whose dimension is smaller than some number are retained.

The practical version of OPE was heavily used in connection with the QCD sum rule method (Chapter II). It was checked [27] that in the majority of channels this is a valid approximation allowing one to calculate in the Euclidean domain down to μ as low as 0.6 or 0.7 GeV. The validity of this approximation is an element of luck; it relies, among other things, on the fact that $\Lambda_{\rm QCD}$ is significantly smaller than 1 GeV, and $\alpha_s(1{\rm GeV})/\pi$ is already a small parameter.

I hasten to add that some exceptional channels where the practical version of OPE fails at much larger values of μ were detected in the analysis of glueballs [70]. It would be interesting to explore the issue in the context of the heavy quark theory. The existing theory gives no clues for establishing the domain of validity of the practical version of OPE from first principles, neither does it tell us about when the exponential terms, not visible by standard methods, become negligibly small. At this point we have to rely on indirect methods and phenomenological information.

5.2. *Untangling Hard Gluons from Soft Ones*

The coefficient functions C_n in Eq. (5.1) contain, generally speaking, an infinite number of perturbative terms, and nonperturbative contributions of different types. In practice, we often calculate them to the first nontrivial order. For instance, in Section 3 we treated the transition operator in the Born approximation; thus, all coefficients in OPE were found to order α_s^0. For a number of purposes (although not always, of course) such a calculation, ignoring the hard

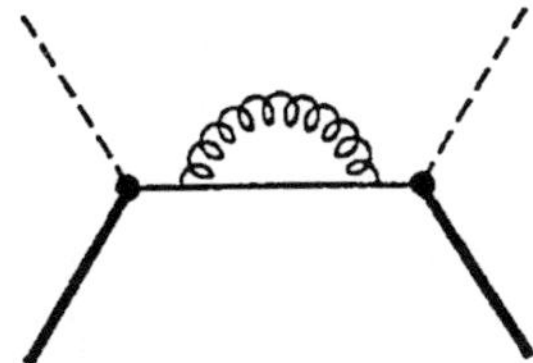

Fig. 13. Correction of the first order in α_s to the transition operator of Fig. 1. Shown is the only graph contributing to the imaginary part in the Coulomb gauge. The gluon is denoted by the curly line.

gluon exchanges altogether, is quite sufficient. Let me remind that, by hard gluons, I mean those with offshellness from μ up to m_Q. Let us ask a question — can one find a theoretical parameter which would justify the approximation of no hard gluon exchanges? In other words, does a parameter exist that would allow one to switch the hard gluons on/off?

Each extra hard loop contains the running gauge coupling $\alpha_s(\mu)$,

$$\frac{\alpha_s(\mu)}{\pi} = \frac{2}{b}\left[\ln\left(\frac{\mu}{\Lambda_{\rm QCD}}\right)\right]^{-1} \tag{5.3}$$

where b is the first coefficient of the Gell-Mann-Low function,

$$b = \frac{11}{3}N_c - \frac{2}{3}n_f\,.$$

If we could make b very large the running law of α_s would be very steep, effectively switching off all hard gluons. Indeed, once μ is bigger than, say, $2\Lambda_{\rm QCD}$ and $b \to \infty$, the gauge coupling constant $\alpha_s(\mu) \to 0$. The first idea which immediately comes to one's mind is to make b large by tending the number of colors N_c to infinity. Alas, this idea does not work. It is known from the early days of QCD that the expansion parameter in all planar diagrams is $N_c\alpha_s$, not α_s itself [2]. Thus, the diagram of Fig. 13 is of the same order in N_c as the Born graph of Fig. 1. So, we have to rely on the numerical smallness of $1/b$. For instance, in the theory with three light flavors and three colors $b = 9$, quite a large number. This is not the first time in physics that we have to deal with numerical enhancements. It is true that it is always better to have an adjustable parameter, which could be sent to infinity at will, than to deal with just a large fixed number. It is quite unfortunate that we do not have such a parameter at our disposal in the real world QCD. If one still wants to have b as an adjustable parameter one could try a trick. Let us assume that, apart from quarks and gluons, our theory contains quark *ghost* fields. These ghost fields are perfectly the same as the quark fields, with a single exception — each ghost loop has an extra minus sign. The quark ghost fields may or may not have a mass term. Let us say that they do have a mass term $m_{\rm gh}$ equal to $\Lambda_{\rm QCD}$. Then they would automatically decouple in the soft contributions. The action of such a crazy theory has the form

$$
\begin{aligned}
iS &= iS_{\rm QCD} + i\sum_q \bar{q}_{\rm gh}(i\not{D} - m_{\rm gh})q_{\rm gh} \\
&= iS_{\rm QCD} - N_{\rm gh}\ln\det\{i\not{D} - m_{\rm gh}\}
\end{aligned}
\tag{5.4}
$$

where $S_{\rm QCD}$ is the action of quantum chromodynamics, see Eq. (1.1), and $N_{\rm gh}$ is the number of the quark ghosts, a free parameter assumed to be large. Notice the ghostly minus sign in front of the logarithm of the determinant. After some thinking, one may conclude that, perhaps, this theory is not so crazy. Let us postulate that the initial particles we consider belong to our world — B, D and so on — i.e. they do not carry these quark ghosts. Of course, if B decays the quark ghosts do appear in the final state, and the probability of their emission is negative. This does not mean, however, that the total amplitude is not unitary, as one could suspect from the fact that we introduced the fields with a wrong metric. Indeed, it is obvious that the only role of the quark ghosts[11] is to switch off all hard gluons in the limit $N_{\rm gh}\to\infty$, since in this limit

$$
b = 11 - \frac{2}{3}N_f + \frac{2}{3}N_{\rm gh} \to \infty
$$

and $\alpha_s(\mu)\to 0$, according to Eq. (5.3). In particular, the diagram of Fig. 13 where the gluon line is dressed with the bubble insertions vanishes. All soft contributions with $\mu \lesssim \Lambda_{\rm QCD}$ remain intact, however, and the positivity of the forward scattering amplitudes is not violated.

If there exists a stringy representation of QCD it should refer to the fake "QCD", Eq. (5.4), rather than to the real one, since in the string amplitudes, there is no place for hard gluons.

The idea of treating b as a numerically large parameter is not new in QCD. In purely perturbative calculations it constitutes the basis of the so-called Brodsky-Lepage-Mackenzie (BLM) approach [71]. Originally the BLM approach was engineered as a scale-setting procedure intended as a substitute

---

[11] Instead of introducing the "quark ghosts" one could have formally assumed that the number of the quark flavors is large and *negative*.

for full computations of $\mathcal{O}(\alpha_s^2)$ corrections. Assume that $\mathcal{O}(\alpha_s)$ corrections in some amplitude are known exactly. In order α_s^2, typically, one has to deal with a large number of graphs. The idea is to pick up only those which contain a "large parameter", $b\alpha_s^2$, presuming that the graphs without b are numerically suppressed. Typically there are very few graphs producing $b\alpha_s^2$. By doing so we can approximately determine the scale μ in the $\mathcal{O}(\alpha_s)$ term without labor and a time-consuming calculation of a large number of all α_s^2 contributions. Later, it was suggested [72, 73] to extend the prescription of the "b graph dominance" to even higher orders, a more extremist approach. In both cases the limit of large b is used to get some information about perturbation theory. I use this limit in order to switch off the perturbative hard gluons in the first place, pushing the theory to the mode where only the soft gluons survive, hopefully providing a more transparent picture of the infrared dynamics determining the regularities of the hadronic world.

5.3. *Impact of Hard Gluons*

Having said all that let us return to the real world where b is fixed, not infinity, and examine several examples of corrections due to hard gluons. An instructive example to begin with is the calculation of the coefficient in front of the chromomagnetic operator $\mathcal{O}_G$ in the effective Lagrangian $\mathcal{L}_{\rm heavy}(\mu)$, see Eq. (1.3).[12] This coefficient takes into account virtual gluons with offshellness from μ to m_Q.

The line of reasoning is as follows. Our starting point is $\mu = m_Q$. At this normalization point the Lagrangian we deal with is the QCD Lagrangian (1.1) with the coupling constant and heavy quark mass normalized at m_Q. We then descend a little further, down to μ equal to a finite fraction of m_Q, say, $m_Q/5$. This is sufficient to make the Q quark nonrelativistic and make all nonrelativistic expansions work. Being interested only in logarithms of m_Q we ignore any nonlogarithmic α_s corrections that may appear at this stage. The nonrelativistic expansion of the Lagrangian $\bar{Q}(i\not{D} - m_Q)Q$ implies that the operator $\bar{Q}(i/2)\sigma G Q$ appears with the coefficient $C_0 = 1/(2m_Q)$. Further evolution down to $\mu =$ several units $\times\ \Lambda_{\rm QCD}$ will change C; in particular, at one-loop

$$C_0 \to C(\mu) = C_0\left(1 + \gamma\frac{\alpha_s}{4\pi}\ln\frac{m_Q^2}{\mu^2} + \text{nonlog terms}\right) \tag{5.5}$$

[12] In the limit $b \to \infty$ the coefficient given in this expression does indeed vanish, in full accord with the argument of the previous section.

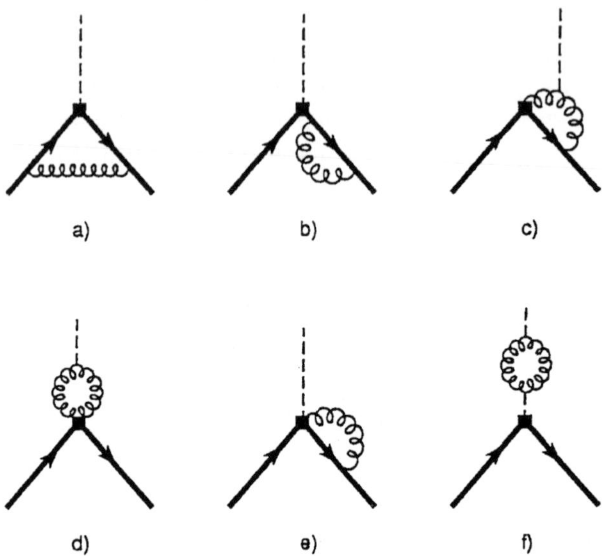

Fig. 14. One-loop diagrams determining the coefficient c_G. The wavy line denotes the gluon quanta, dashed line background gluon field.

where γ is a number. Our goal is to find γ and then sum up all leading logarithms. This is not the end of the story, however, if one wants to represent the result in the form (1.3), where the sum over the operators includes only the Lorentz-invariant ones. The leading operator is $\mathcal{L}^0_{\text{heavy}}(\mu)$. The coefficient $C(\mu)$ should be represented as

$$C(\mu) = C_0 + (C(\mu) - C_0) ;$$

then C_0 is swallowed back in the definition of $\mathcal{L}^0_{\text{heavy}}(\mu)$ while the expression in the brackets represents c_G in Eq. (1.3).

The relevant one-loop graphs are depicted in Fig. 14. At first sight the number of diagrams is rather large, and the computation might seem rather cumbersome. My task is to reduce it to a back-of-the-envelope calculation by using several smart observations and the background field technique.

First of all, as was already mentioned, we will be hunting only for the terms containing $\alpha_s \ln m_Q/\mu$ omitting all α_s terms without logarithms. The logarithms $\ln m_Q/\mu$ have a dual nature — they appear from the loop integrations where the integrands present an infrared limit with respect to heavy quarks Q while presenting simultaneously the ultraviolet limit with respect to gluons. That is why they were called hybrid in Ref. [50], the paper where these logarithms were discovered. In the language of HQET they are referred to as matching logarithms.

Secondly, in this perturbative calculation we will naturally discard all $1/m_Q$ corrections.

The closed box in the diagrams of Fig. 14 denotes the vertex $(i/2)\sigma^{ij}G_{ij} = -\vec{\sigma}\vec{B}$. Let us consider for definiteness only one term with $i,j = 1,2$, i.e. $-\sigma_z B_z$, keeping in mind that other terms will give the same.

It is absolutely obvious that the graph of Fig. 14(e) gives no contribution in our approximation. Indeed, the very existence of this graph is due to the nonlinear term $[A_1 A_2]$ in the definition of G_{12}. However, neither A_1 nor A_2 interact with the heavy quark in the leading in $1/m_Q$ approximation, as it is clear from Eq. (1.15), only A_0. (We work in the rest frame of the heavy quark Q.)

Next, let us analyze the diagrams *c* and *d*. To this end it is convenient to write the gluon Green function in the background field. For a detailed exposition of the technique the reader is referred to the review paper [28]. For our purposes we need so little that it is quite in order to carry out all necessary derivations here. Let us split the four-potential A_μ in two parts — the external field $(A_\mu)_{\rm ext}$ and the quantum part a_μ which will propagate in loops,

$$A_\mu = (A_\mu)_{\rm ext} + a_\mu\,. \tag{5.6}$$

As explained in Ref. [28] the gauge conditions on $(A_\mu)_{\rm ext}$ and a_μ may be different, for instance, the Fock-Schwinger gauge with respect to the background field and the Feynman gauge with respect to the quantum field. Here we do not need to discuss the gauge condition on $(A_\mu)_{\rm ext}$. The quantum field a_μ will be treated in the Feynman gauge. The definition of the gluon propagator in the background field is standard:

$$D^{ab}_{\mu\nu} = \langle T\{a^a_\mu(x), a^b_\nu(0)\}\rangle\,. \tag{5.7}$$

The Lagrangian of the quantum gluon field in the Feynman gauge has the form

$$\mathcal{L} = -\frac{1}{2}(D^{\rm ext}_\mu a^a_\nu)^2 + g a^a_\mu (G^b_{\mu\nu})_{\rm ext} a^c_\nu f^{abc} \tag{5.8}$$

plus cubic and higher order terms in a_μ plus the ghost terms — all irrelevant for the calculation at hand. Here

$$D^{\rm ext}_\mu a^a_\nu = \partial_\mu a^a_\nu + g f^{abc}(A^b_\mu)_{\rm ext} a^c_\nu\,.$$

The second term in the Lagrangian (5.8) describes the interaction of the magnetic moment of the gluon quantum with the background field. If we switch off this magnetic term for a short while we immediately observe that both graphs, Figs. 14(c) and 14(d), vanish. Indeed, the Lorentz structure of the first term in Eq. (5.8) is such that the Green function generated by it is obviously proportional to $g_{\mu\nu}$. Hence the loops displayed in Figs. 14(c) and 14(d) cannot be formed. Say, the diagram 14(c) requires converting the a_i quantum leaving the vertex into the a_0 quantum coupled to the heavy quark. Let us now switch on the magnetic term and take into account the fact that the background field is chromomagnetic, not chromoelectric (I remind that we are interested in the vertex $-\vec{\sigma}\vec{B}$). This means that the graph 14(c) still vanishes since the conversion of a_i into a_0 can only take place in the chromoelectric background ($G_{0i}^{\rm ext}$). The diagram 14(d) is not vanishing, however, and is readily calculable. We start from the vertex

$$(i/2)\bar{Q}\sigma^{12}G_{12}^{c}t^{c}Q \to (i/2)\bar{Q}\sigma^{12}ga_1^a a_2^b f^{abc}t^c Q\,,$$

make one insertion of the magnetic term in the Lagrangian (5.8),

$$iga^{\rho\bar{a}}(G^{\bar{b}}_{\rho\phi})_{\rm ext}a^{\phi\bar{c}}f^{\bar{a}\bar{b}\bar{c}}\,,$$

and after that one can take the free gluon propagators, which yields

$$(i/2)2g^2\bar{Q}\sigma^{12}f^{abc}t^cQf^{a\bar{b}b}G_{12}^{\bar{b}}(-i)^2 i\int\frac{1}{k^4}\frac{d^4k}{(2\pi)^4}\,, \tag{5.9}$$

where the factor 2 comes from two different ways of pairings. The integral over dk is evidently logarithmically divergent both at the upper and lower ends and should be cut off at m_Q from above and at μ from below. This logarithmic divergence should be welcome since in this way we are going to get the desired hybrid logarithm. Equation (5.9) immediately leads to

$$-2\frac{N_c g^2}{16\pi^2}\,(i/2)\,\bar{Q}\sigma^{12}G_{12}^{c}t^{c}Q\ln\left(\frac{m_Q^2}{\mu^2}\right)$$

where N_c is the number of colors ($N_c = 3$). In other words the factor produced by the one-loop graph of Fig. 14(d) is

$$\gamma\,|_{\text{Fig. 14(d)}} = -2N_c\,. \tag{5.10}$$

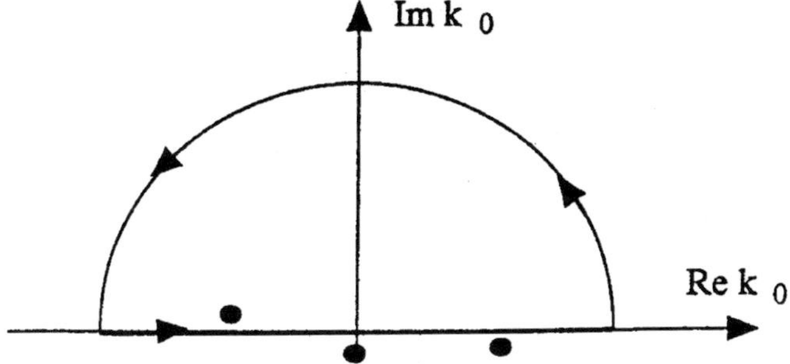

Fig. 15. Integration contour in the k_0 plane.

The last step in our exercise is the calculation of the diagrams of Figs. 14(a) and 14(b). The Feynman integral for the diagram a is quite trivial,

$$g^2 \int \frac{d^4k}{k^4} \bar{Q} \frac{1}{k_0} t^a (-\sigma_z B_z^b t^b) t^a \frac{1}{k_0} Q(-i) \frac{1}{k^2} \tag{5.11}$$

where k is the virtual gluon momentum. Now, a minute reflection shows that in the Abelian theory (i.e. if the gluons were photons and the diagram of Fig. 14(a) was considered in QED) this contribution must be exactly canceled by that coming from diagrams 14(b). This assertion can be traced back to the non-renormalization of the $\bar{Q}\gamma_\mu Q$ vertex in QED. (We should take into account the fact that the hybrid logarithms do not depend on the Lorentz structure of the vertex at all [50] and are the same for γ_μ and $\sigma_{\mu\nu}$.) This observation implies that in QCD the net effect of the two diagrams 14(b) reduces to replacing Eq. (5.11) by

$$\begin{aligned} g^2 \int \frac{d^4k}{k^4} \frac{1}{2} \bar{Q} \left(t^a [t^b t^a] + [t^a t^b] t^a \right) (-\sigma_z B_z^b) Q \frac{-i}{k_0^2} \frac{1}{k^2} \\ = \frac{N_c g^2}{2} \int \frac{d^4k}{k^4} \bar{Q}(-\sigma_z B_z^b t^b) Q \frac{i}{(k_0 + i\epsilon)^2} \frac{1}{k^2 + i\epsilon} \,. \end{aligned} \tag{5.12}$$

The $i\epsilon$ prescription indicated explicitly defines the integration contour (Fig. 15). We first do the k_0 integration using the residue theorem, then the remaining d^3k integration and arrive at

$$\frac{N_c g^2}{16\pi^2} (i/2) \bar{Q}(-\sigma_z B_z^c t^c) Q \ln \left(\frac{m_Q^2}{\mu^2} \right) \tag{5.13}$$

leading to the following "dressing" factor due to diagrams 14(b) and a:

$$\gamma\,|_{\rm Figs.\ 14(a)+(b)} = N_c\,. \tag{5.14}$$

The diagram 14(f) must be discarded in the background field calculation — it merely renormalizes the gauge coupling constant included in the definition of $\mathcal{O}_G$.

The overall one-loop dressing factor is obtained by adding up Eqs. (5.10) and (5.14),

$$\gamma = -N_c = -3\,. \tag{5.15}$$

Now the renormalization group allows us to sum up all leading log terms, in a standard manner; the summation leads to Eq. (1.4). The same result can be rephrased as follows: in the $1/m_Q$ expanded effective Lagrangian $\mathcal{L}_{\rm heavy}(\mu)$ the overall coefficient in front of $\mathcal{O}_G$ is

$$C(\mu) = \left(\frac{\alpha_s(\mu)}{\alpha_s(m_Q)}\right)^{-\frac{3}{b}}\,. \tag{5.16}$$

This is nothing else other than the reflection of the hybrid anomalous dimension of the operator $\mathcal{O}_G$ found in Ref. [8].

It is curious to note that $\mathcal{O}_\pi$, the second operator of dimension 5 (see Eq. (3.6)), has vanishing anomalous dimension which can be proven with no calculations in no time.

To see that this is indeed the case we merely repeat the argument preceding and following Eq. (5.5). Let us assume for a short while that the hybrid anomalous dimension of the operator $\mathcal{O}_\pi$ is nonvanishing. Then, after evolving to a low normalization point, its coefficient gets renormalized, and there is no way one could absorb $\mathcal{O}_\pi$ back into a Lorentz invariant expression $\bar{Q}(\not{\mathcal{P}} - m_Q)$ in the effective Lagrangian. Needless to say that $\mathcal{L}_{\rm heavy}(\mu)$ (before the $1/m_Q$ expansion) must be expressible in terms of the Lorentz-invariant structures.

A close line of reasoning leading to the same conclusion takes advantage of the expansion (3.3),

$$\bar{Q}Q - \frac{1}{2m_Q^2}\bar{Q}\frac{i}{2}\sigma G Q = \bar{Q}\gamma_0 Q - \frac{1}{2m_Q^2}\bar{Q}\vec{\pi}^2 Q + \ldots \tag{5.17}$$

where the dots denote terms of the higher order in $1/m_Q$. The left-hand side is Lorentz scalar while the right-hand side is written as a sum of terms that are *not* Lorentz scalars individually. The matrix element of $\bar{Q}\gamma_0 Q$ has the meaning of energy E (which at small velocities reduces to $m + \vec{p}^2/2m$) while that of the second term on the right-hand side has the meaning of $-\vec{p}^2/2m$. The first term is not renormalized by the gluon dressings, of course. If the coefficient of the second term was distorted by the anomalous dimension, the cancellation of the Lorentz noninvariant part would be ruined, and the right-hand side could not be equal to the left-hand side.

Concluding this chapter let us discuss the impact of the hard gluons on the scaling law of, say, pseudoscalar coupling f_P defined in Sec. 3.4. In this section it was shown that $f_P \sim m_Q^{-1/2}$ modulo logarithmic corrections. Now we address the issue of the logarithmic corrections due to the hybrid anomalous dimension of the current $\bar{Q}\gamma_\mu\gamma_5 q$. Let us add to the original QCD Lagrangian the term $\Delta\mathcal{L} = \mathcal{A}_\mu\bar{Q}\gamma_\mu\gamma_5 q$ where $\mathcal{A}_\mu$ is an auxiliary c-number field and evolve $\Delta\mathcal{L}$ down to μ. The result of this evolution is the anomalous dimension

$$(\bar{Q}\gamma_\mu\gamma_5 q)_{m_Q} = \left(\frac{\alpha_s(\mu)}{\alpha_s(m_Q)}\right)^{\frac{2}{b}} (\bar{Q}\gamma_\mu\gamma_5 q)_\mu\,;$$

the subscript here indicates the normalization point. The corresponding calculation is even simpler than that of the anomalous dimension of $\mathcal{O}_G$ and will not be discussed here. The interested reader is referred to Ref. [50] or to review papers [4].[13] Correspondingly the complete asymptotic scaling law of f_P is $f_P \sim m_Q^{-1/2}(\alpha_s(m_Q))^{-2/b}$.

5.4. *μ Dependence of the Basic Parameters of the Heavy Quark Theory. Measuring $\bar{\Lambda}(\mu)$*

The Lagrangian (1.3) summarizes the evolution from a high normalization point down to μ. Since all operators in this Lagrangian are normalized at μ it is perfectly natural that their matrix elements are also μ-dependent. In particular, the matrix elements of $\mathcal{O}_G$ and $\mathcal{O}_\pi$ denoted by μ_G^2 and μ_π^2 in Sec. 3 depend on μ. Actually, μ_G^2 depends on μ rather strongly, through logarithms of μ — this is explicitly demonstrated by the fact that $(\mathcal{O}_G)_{m_Q} = (\alpha_s(\mu)/\alpha_s(m_Q))^{-3/b}(\mathcal{O}_G)_\mu$. In this case hardly anybody would even think about tending $\mu \to 0$. As for the operator $\mathcal{O}_\pi$, the situation here is trickier. As we have seen, it has no diagonal anomalous dimension; still, some μ dependence appears through mixing with $\bar{Q}Q$, see Ref. [25] for details.

[13] The wording in these reviews is somewhat different. You will read about the matching logarithms of HQET for the axial current. Technically speaking this is perfectly the same as the anomalous dimension within our approach.

Now let us discuss $\bar{\Lambda}$, another basic parameter of the heavy quark theory. The issue of its μ dependence was at the epicenter of a heated debate recently. By itself $\bar{\Lambda}$ never appears in $\mathcal{L}_{\rm heavy}$; moreover the quark mass m_Q appears in the $1/m_Q$ expanded effective Lagrangian (1.11) only through $1/m_Q$ corrections. Therefore, in the limit $m_Q \to \infty$ (which is often identified with HQET) it is quite tempting to say that $\bar{\Lambda} = M_{H_Q} - m_Q$ is a universal constant. For a few years it was taken for granted that such a constant exists. Within the framework of our approach based on the Wilsonean treatment of full QCD it is perfectly clear that this is not the case. The quark mass in Eq. (1.3) explicitly depends on μ resulting in a μ dependence of $\bar{\Lambda}$.

Since the issue is of importance let us rephrase this statement as follows. Since quarks are permanently confined the notion of the heavy quark mass becomes ambiguous. To eliminate this ambiguity one must explicitly specify the procedure of measuring "the heavy quark mass". The definition through the effective Lagrangian (1.3) is consistent. Other definitions are certainly conceivable; *any* consistent procedure will necessarily involve a cut-off parameter μ, and then $\bar{\Lambda}(\mu) = M_Q - m_Q(\mu)$.[14]

The effective Lagrangian (1.3) is not something you directly measure, neither is m_Q. Defining $\bar{\Lambda}$ or m_Q is equivalent to saying how they are measured, how the parameters in the effective Lagrangian are related to measurable quantities. To this end one can use any suitable prediction of the heavy quark theory, in particular, Voloshin's sum rule (2.35). To avoid inessential technicalities I will discuss the issue in the framework of the toy model of Sec. 2.1. All results can be immediately extended to the real QCD. Equation (2.35) gives a nice definition of $\bar{\Lambda}$ in terms of a measurable quantity, the average value of $E_0^{\rm phys} - E$ where E is the energy of the ϕ quantum. The problem is that in Sec. 2.1 we discussed the issue switching off all hard gluons, so that the above average value looked like a μ independent number. To see where the μ dependence comes from we must include hard gluon corrections.

---

[14] In the literature you can find assertions that an "absolute" heavy quark mass, or the so-called pole mass, can be defined and can be shown to be a universal number independent of any cut-offs. These assertions are false. The notion of the pole mass exists only to a given *finite* order of perturbation theory. No consistent definition of the pole mass can be given already at the level of the leading nonperturbative corrections $\mathcal{O}(1/m_Q)$. The notion of the pole mass is absolutely foreign to the approach I present here, therefore, I do not want to go into details, see [74]. I will only say that it assumes it is possible to separate perturbative contributions from nonperturbative (?!) in contradiction with our approach which separates soft contributions from hard.

If the gluon field is treated only as a soft medium the spectrum of the decay $H_Q \to X_q + \phi$ looks roughly as in Fig. 9. The shoulder to the left of the elastic peak arises due to the production of the excited states. It is important that in this approximation the spectrum rapidly (exponentially) decreases outside the end point domain, so that the entire region of E from zero up to $E_0^{\rm phys}-$ several units $\times\Lambda_{\rm QCD}$ remains unpopulated. The average

$$\int_0^{E_0^{\rm phys}} dE\,(E_0^{\rm phys}-E)\,\frac{1}{\Gamma_0}\frac{d\Gamma}{dE}$$

is then a well-defined number independent of m_Q or any cut-offs.

The situation drastically changes once we include hard gluon emission. In calculating radiative gluon correction we can disregard, in the leading approximation, nonperturbative effects, like the difference between m_Q and M_{H_Q} or the motion of the initial quark inside H_Q. Thus we deal with the decay of the free quark Q at rest into $q+\phi+$ gluon. The virtual gluon contribution merely renormalizes the constant h in the analysis presented above and is irrelevant.

The effects from real gluon emission are most simply calculated in the Coulomb gauge, where only the graph shown in Fig. 13 contributes. A straightforward computation yields [25, 31] to leading order in $\vec{v}^{\,2}$

$$\frac{d\Gamma^{(1)}}{dE}=\Gamma_0\frac{8\alpha_s}{9\pi}\frac{E^3}{E_0 m_Q^2}\frac{1}{E_0-E}\,. \tag{5.18}$$

The hard gluon emission obviously contributes to the spectrum in the entire interval $0<E<E_0^{\rm phys}$ creating a long "radiative" tail to the left of the end point domain. (Note that in this calculation one can put $E_0^{\rm phys}$ to E_0.) In the first order calculation α_s does not run, of course. Its scale dependence shows up only in the two-loop calculation; it is quite evident, however, that it is $\alpha_s(E-E_0)$ that enters. Therefore, strictly speaking, one cannot apply Eq. (5.18) too close to E_0. Even leaving aside the blowing up of $\alpha_s(E-E_0)$, there exists another reason not to use Eq. (5.18) in the vicinity of E_0: if E is close to E_0, the emitted gluon is soft; such gluons are to be treated as belonging to the soft gluon medium in order to avoid double counting.

The separation between soft and hard gluons is achieved by explicitly introducing a normalization point μ. The value of μ should be large enough to justify a small value for $\alpha_s(\mu)$. On the other hand we would like to choose μ as small as possible. We then draw a line: to the left of $E_0-\mu$ the gluon is considered to be hard, to the right soft. At $E<E_0-\mu$ the experimentally measured spectrum must follow the one-loop formula (5.18), see Fig. 16.

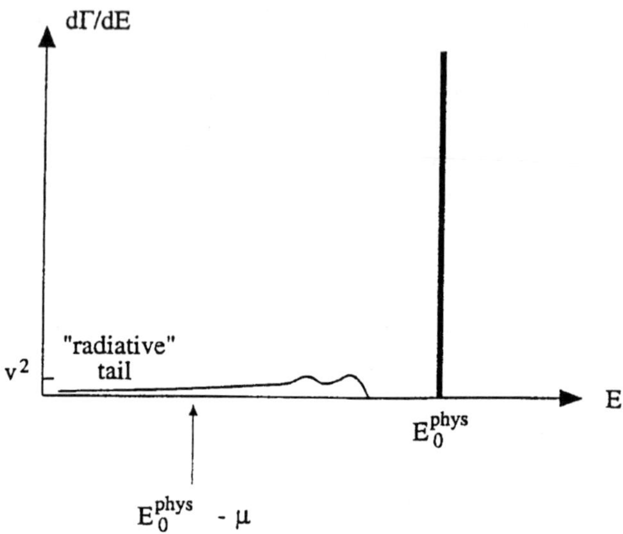

Fig. 16. A sketch of the photon spectrum in the SV limit with the hard gluon radiation included.

Let us return now to Voloshin's sum rule, i.e. the first moment of $E_0^{\rm phys} - E$, with radiative corrections included. A qualitative sketch of how $d\Gamma/dE$ looks now is presented in Fig. 16. Because of the tail to the left of the end point domain we cannot define $\bar{\Lambda}$ as the value of $E_0^{\rm phys} - E$ averaged over the entire range of the ϕ energy, $0 < E < E_0^{\rm phys}$. The integral would be proportional to $\alpha_s m_Q$ because of the domain of small E. Besides, this would contradict the physical meaning of what we want to define. By evolving the effective Lagrangian down to μ we include all gluons harder than μ in m_Q, thus excluding them from $\bar{\Lambda}$. Thus, we must accept that

$$\bar{\Lambda}(\mu) = \int_{E_0^{\rm phys}-\mu}^{E_0^{\rm phys}} \frac{2}{v_0^2}\frac{1}{\Gamma_0}\frac{d\Gamma}{dE}(E_0^{\rm phys} - E)\, dE. \tag{5.19}$$

Since the explicit form of the tail to the left of the end point domain is known (for small $\alpha_s(\mu)/\pi$ the physical spectrum is supposed to tend to the perturbative result) the μ dependence of $\bar{\Lambda}$ becomes obvious,

$$\delta\overline{\Lambda} = \delta\mu \frac{16}{9}\frac{\alpha_s(\mu)}{\pi}\,. \tag{5.20}$$

Equation (5.19) provides us with one possible physical definition of $\bar{\Lambda}(\mu)$ (among others) relating this quantity to an integral over a physically measurable spectral density. The pole-mass-based definition, being applied to our example, would involve three steps: (i) Take the radiative perturbative tail to the left of the shoulder and extrapolate it all the way to the point $E = E_0$; (ii) subtract the result from the measured spectrum; (iii) integrate the difference over dE with the weight function $(E - E_0^{\rm phys})$. The elastic peak drops out and the remaining integral is equal to $\Gamma_0(\vec{v}^{\,2}/2)\bar{\Lambda}$. It is quite clear that this procedure cannot be carried out consistently — there exists no unambiguous way to extrapolate the perturbative tail too close to $E_0^{\rm phys}$, the end point of the spectrum. Our procedure, with the normalization point μ introduced explicitly, is free from this ambiguity.

In practice, the μ dependence of $\bar{\Lambda}(\mu)$ may turn out to be rather weak. This is the case if the spectral density is such as shown in Fig. 16, where the contribution of the first excitations (lying within $\sim \Lambda_{\rm QCD}$ from $E_0^{\rm phys}$) is numerically much larger than the radiative tail representing high excitations. It is quite clear that if the physical spectral density resembles that of Fig. 16 and μ = several units $\times\ \Lambda_{\rm QCD}$, the running $\bar{\Lambda}(\mu)$ is rather insensitive to the particular choice of μ.

It remains to be added that a similar definition of $\bar{\Lambda}(\mu)$ works in real QCD. Here it may be defined through an integral over the spectrum in the decay $B \to X_c l\nu$ measured in the domain where the recoil of the hadronic system is small, $|\vec{q}| \ll m_c$, i.e. in the SV limit.

5.5. *Hard Gluons and the Line Shape*

Actually we have already started considering this question in the previous section where the radiative correction to the spectrum in the decay $H_Q \to X_q + \phi$ was found in the SV limit. In this case the impact of the hard gluons is mild — they provide a long but squeezed tail outside the end point domain which could be evaluated in the leading (one-loop) approximation. The reason why there are no violent distortions of the spectrum is simple: the q quark produced is slow, and slow quarks do not like to emit hard gluons. If the final quark was fast it would produce gluons like crazy through bremsstrahlung, and the impact of such bremsstrahlung on the line shape would be much more drastic. As a matter of fact, if $m_q \to 0$, one cannot limit oneself to any finite number of gluons — an infinite sequence of the so-called Sudakov (or double-log) corrections must be summed over.

By definition the Sudakov corrections are those in which each power of α_s is accompanied by two powers of logarithm $\ln m_Q/(m_Q - 2E)$. When one approaches the end point domain the logarithm inevitably becomes large, and overcompensates the smallness of the gauge coupling constant α_s. So, the more gluons emitted, the higher the probability. The phenomenon is classical in nature and has a transparent physical interpretation. Indeed, in the initial state H_Q, the color field in the light cloud corresponds to a static source. The final quark produced is very fast. The stationary state of the color field corresponding to a fast-moving color charge is strongly different from that of the stationary charge. Therefore, the excess of the color field is just shaken off in the form of the multiple emission of gluons. If you forbid to emit a large number of gluons and insist that the final state is just "one quark" (this would correspond to the two-body decay kinematics and the delta-function-like narrow spectrum) then the probability of such an improbable event is terribly suppressed. This explains why for the massless final quarks the narrow peak in the end point domain obtained in Sec. 4.2 will be drastically distorted, and a well-developed tail to the left of the end point domain will appear.

The theory of the Sudakov corrections constitutes a noticeable part of the perturbative QCD, and here, of course, I have no possibility even to scratch the surface. I will give just a few hints referring the interested reader to the original papers and textbooks [75].

The first order probability of emission of the massless gluon in the $b \to s\gamma$ decay is

$$\frac{1}{\Gamma}\frac{d^2\Gamma_g}{d\omega d\vartheta^2} = \frac{2\alpha_s}{3\pi\omega(1-\cos\vartheta)} \tag{5.21}$$

where ω is the gluon momentum and ϑ is its angle relative to the momentum of the q quark. In this expression it is assumed that $\omega \ll m_Q$. As a matter of fact it is perfectly legitimate to make this assumption since the double logarithm comes only from this domain of integration. The ϕ energy in the presence of a gluon in the final state is given by

$$E = \frac{m_Q^2 - 2\omega(m_Q-\omega)(1-\cos\vartheta)}{2m_Q - 2\omega(1-\cos\vartheta)} \simeq \frac{m_Q}{2} - \frac{k_\perp^2}{4\omega}\,, \tag{5.22}$$

$$k_\perp \approx \omega\vartheta\,,$$

where it is assumed that $\vartheta \ll 1$. As we will see shortly, this assumption is justified in the approximation of double logarithms.

One starts from computing the (first-order) probability $w(E)$ for the gluon to be emitted with such momentum that the ϕ quantum gets energy below given E. This probability is obtained by integrating the distribution (5.21) with the constraint that $(m_Q/2) - (k_\perp^2/4\omega)$ is less than the given E,

$$w(E) = \int d\omega\, d\vartheta^2\, \frac{1}{\Gamma}\frac{d^2\Gamma_g}{d\omega\, d\vartheta^2}\,\theta\left(\frac{k_\perp^2}{4\omega} - \left(\frac{m_Q}{2} - E\right)\right) = \frac{2\alpha_s}{3\pi}\ln^2\frac{m_Q}{m_Q - 2E}\,. \tag{5.23}$$

I integrate over ϑ^2 first; the upper limit of integration is of the order one, the lower limit is determined from the θ function in Eq. (5.23). Then we can carry out the ω integration. The upper limit is $m_Q/2$ while the lower limit is seen from the same θ function,

$$\omega \lesssim \frac{m_Q}{2} - E\,.$$

The function $w(E)$ has the meaning of the probability of emission of a sufficiently hard gluon lowering the ϕ energy below E. The all-order summation of double logs amounts then to merely exponentiating this probability [75],

$$S(E) = \mathrm{e}^{-w(E)}\,. \tag{5.24}$$

The spectrum then takes the form

$$\frac{1}{\Gamma}\frac{d\Gamma}{dE} = -\frac{dS}{dE}\,. \tag{5.25}$$

We see that as E approaches the end point, E close to $m_Q/2$, the spectrum gets suppressed, in full accord with our expectations, since the presence of the very hard ϕ does not allow, purely kinematically, the gluon shower to develop and the color field to restructure itself. Notice that the Sudakov corrections merely redistribute the probability, since the full integral over the spectrum remains unchanged. They pump events out from the end point domain to lower values of E.

The double log approximation *per se* does not allow us to determine the scale of α_s in the Sudakov exponent $S(E)$. For practical purposes the scale setting is of course very important since $\exp(-w(E))$ is a steep function. The question goes far beyond the scope of this section. Some partial answers can be found in Refs. [76, 77], see also Ref. [64]; suffice it to mention here that α_s in Eq. (5.23) turns out to be [77] $\alpha_s(\sqrt{(m_Q - 2E)m_Q})$. Another point that deserves to be stressed is that, with the classical Sudakov formula, one cannot travel over the energy axis too close to the end point $E = m_Q/2$ (even after the scale setting). Indeed, if $E > (m_Q/2) - \mu$, then the gluons emitted become too soft; such gluons constitute the soft gluon medium and have nothing to do with the perturbative calculation; they have to be referred to the primordial distribution function. Equation (5.23) is applicable provided that

$$\bar\Lambda \ll m_Q - 2E \ll m_Q\,.$$

If we come closer to the end point domain the classical Sudakov factor must be modified by cutting off and discarding the contribution of the soft gluons. This idea gained recognition only recently; it is obviously premature to further immerse into this topic for the time being.

If the effect of the hard (perturbative) gluon emission is known the full physical spectrum is obtained by convoluting the perturbative one with the primordial distribution function,

$$\frac{d\Gamma(E)}{dE} = \theta(E)\int dy F(y)\frac{d\Gamma_Q^{\rm pert}(E-(\bar\Lambda/2)y)}{dE}\ . \tag{5.26}$$

The integration over y runs from $-\infty$ to 1 (more exactly, the lower limit of integration is $y_0 = -m_Q/\bar\Lambda$, but this difference can be ignored). One should keep in mind that $d\Gamma_Q^{\rm pert}/dE$ is nonvanishing only in the interval $(0, m_Q/2)$. The convolution formula above is legitimate only for as long as one does not apply it to the very low energy part, $E \sim \bar\Lambda$; further details are presented in Ref. [64].

Acknowledgments

I am grateful to Prof. D. Soper and Prof. K. Mahanthappa for their kind invitation to deliver these lectures at TASI-95. The first two lectures were prepared during my visit to Nagoya University in November and December 1994. The visit became possible due to the Research Fellowship provided by JSPS. It is my pleasure to thank Prof. S. Sawada, Prof. T. Sanda, Prof. K. Yamawaki and other members of the Theory Group of the Nagoya University for their kind hospitality. Countless fruitful discussions on all topics touched upon in these lectures with N. Uraltsev, A. Vainshtein and M. Voloshin are gratefully acknowledged. I would like to thank V. Braun for useful comments concerning the preliminary version of the text. This work was supported in part by DOE under the grant number DE-FG02-94ER40823.

References

[1] V. Novikov, L. Okun, M. Shifman, A. Vainshtein, M. Voloshin and V. Zakharov, *Phys. Rep.* **41** (1978) 1.

[2] G. 't Hooft, *Nucl. Phys.* **B72** (1974) 461; E. Witten, *Nucl. Phys.* **B149** (1979) 285; for a review see e.g. S. Coleman, *Aspects of Symmetry* (Cambridge University Press, 1985), p. 351.

[3] M. B. Wise, *Phys. Rev.* **D45** (1992) 2188; Tung-Mow Yan, Hai-Yang Cheng, Chi-Yee Cheung, Guey-Lin Lin, Y.C. Lin and Hoi-Lai Yu, *Phys. Rev.* **D46** (1992) 1148; G. Burdman and J. F. Donoghue, *Phys. Lett.* **B280** (1992) 287-291; E. Jenkins, *Recent Results in Chiral Perturbation Theory for Mesons Containing a Single Heavy Quark*, in "QCD and High-Energy Hadronic Interactions", Proc. 28 Rencontres de Moriond, Ed. J. Tran Thanh Van (Editions Frontieres, Gif-sur-Yvette, 1993), p. 201; M. B. Wise, *Combining Chiral and Heavy Quark Symmetry*, in "Particle Physics at the Fermi Scale", Ed. Yang Pang *et al.* (Gordon and Breach, 1995), p. 71.

[4] For recent reviews see e.g. N. Isgur and M. B. Wise, in *B Decays*, Ed. B. Stone (World Scientific, Singapore, 1992); H. Georgi, in *Perspectives in the Standard Model*, Proc. of the Theoretical Advanced Study Institute, Boulder, Colorado, 1991, Eds. R. K. Ellis, C. T. Hill and J. D. Lykken (World Scientific, Singapore, 1992); T. Mannel, *Heavy Quark Mass Expansions in QCD*, in Proc. of the Workshop *QCD — 20 Years Later*, Eds. P. M. Zerwas and H. A. Kastrup (World Scientific, Singapore, 1993), vol. 2, p. 634; B. Grinstein, *Ann. Rev. Nucl. Part. Sci.* **42** (1992) 101; M. Neubert, *Phys. Rep.* **245** (1994) 259; M. Voloshin, *Surv. High En. Phys.* **8** (1995) 27; B. Grinstein, *An Introduction to Heavy Mesons*, in Proc. 6th Mexican School of Particles and Fields, Villahermosa, Mexico, October 1994, Ed. J. C. D'Olivo *et al.* (World Scientific, Singapore, 1995), p. 122; I. Bigi, M. Shifman and N. Uraltsev, *Ann. Rev. Nucl. Part. Sci.* **47** (1997) 591 [hep-ph/9703290].

[5] K. Wilson, *Phys. Rev.* **179** (1969) 1499; K. Wilson and J. Kogut, *Phys. Rep.* **12** (1974) 75; see also V. Novikov, M. Shifman, A. Vainshtein and V. Zakharov, *Nucl. Phys.* **B249** (1985) 445.

[6] E. Shuryak, *Nucl. Phys.* **B198** (1982) 83.

[7] G. P. Lepage, *Lattice QCD for Small Computers*, in *The Building Blocks of Creation*, Proc. of the 1993 Theoretical Advanced Study Institute, Eds. S. Raby and T. Walker (World Scientific, Singapore, 1994) [hep-lat/9403018]; M. Alford, W. Dimm, G. P. Lepage, G. Hockney and P. B. Mackenzie, *Phys. Lett.* **B361** (1995) 87.

[8] A. Falk, B. Grinstein and M. Luke, *Nucl. Phys.* **B357** (1991) 185.

[9] M. Shifman and M. Voloshin, 1982, unpublished, see in V. Khoze and M. Shifman, *Uspekhi Fiz. Nauk* **140** (1983) 3 [*Sov. Phys. Uspekhi* **26** (1983) 387].

[10] E. Eichten and B. Hill, *Phys. Lett.* **B234** (1990) 511; H. Georgi, *Phys. Lett.* **B240** (1990) 447.

[11] J. D. Bjorken and S. D. Drell, *Relativistic Quantum Mechanics* (McGraw-Hill, 1964).
[12] V. B. Berestetskii, E. M. Lifshitz and L. P. Pitaevskii, *Quantum Electrodynamics*, 2nd Edition (Pergamon Press, Oxford, 1982).
[13] E. Shuryak, *Phys. Lett.* **B93** (1980) 134.
[14] S. Nussinov and W. Wetzel, *Phys. Rev.* **D36** (1987) 130.
[15] M. Voloshin and M. Shifman, *Yad. Fiz.* **47** (1988) 801 [*Sov. J. Nucl. Phys.* **47** (1988) 511].
[16] N. Isgur and M. Wise, *Phys. Lett.* **B232** (1989) 113; *Phys. Lett.* **B237** (1990) 527.
[17] A. F. Falk, H. Georgi, B. Grinstein and M. B. Wise, *Nucl. Phys.* **B343** (1990) 1.
[18] J. D. Bjorken, in *Proc. of the 4th Rencontres de Physique de la Vallèe d'Aoste*, La Thuille, Italy, 1990, Ed. M. Greco (Editions Frontières, Gif-Sur-Yvette, France, 1990), p. 583.
[19] T. Mannel and Z. Ryzak, *Phys. Lett.* **B247** (1990) 412; A. Falk and B. Grinstein, *Phys. Lett.* **B249** (1990) 314; J. Körner, D. Pirjol and C. Dominguez, *Phys. Lett.* **B301** (1993) 257; B. Grinstein and P. Mende, *Phys. Lett.* **B299** (1993) 127.
[20] P. Ball, H. Dosch and M. Shifman, *Phys. Rev.* **D47** (1993) 4077.
[21] M. E. Luke, *Phys. Lett* **B252** (1990) 447.
[22] R. Crewther, *Phys. Rev. Lett.* **28** (1972) 1421; M. Chanowitz and J. Ellis, *Phys. Lett.* **40B** (1972) 397; J. Collins, L. Duncan and S. Joglekar, *Phys. Rev.* **D16** (1977) 438.
[23] M. Shifman, *Phys. Rep.* **209** (1991) 341.
[24] I. Bigi, N. Uraltsev and A. Vainshtein, *Phys. Lett.* **B293** (1992) 430; (E) **B297** (1993) 477.
[25] I. Bigi, M. Shifman, N. G. Uraltsev and A. Vainshtein, *Phys. Rev.* **D52** (1995) 196.
[26] M. Shifman, A. Vainshtein and V. Zakharov, *Phys. Lett.* **78B** (1978) 443 [reprinted in *The Standard Model Higgs Boson*, Ed. M. Einhorn (North-Holland, Amsterdam, 1991), p. 84.]
[27] M. Shifman, A. Vainshtein and V. Zakharov, *Nucl. Phys.* **B147** (1979) 385; 448 [for a review see *Vacuum Structure and QCD Sum Rules*, Ed. M. Shifman (North-Holland, Amsterdam, 1992).]
[28] V. Novikov, M. Shifman, A. Vainshtein and V. Zakharov, *Fortsch. Phys.* **32** (1984) 585 [a part of this review is reprinted in *Vacuum Structure and QCD Sum Rules*, Ed. M. Shifman (North-Holland, Amsterdam, 1992), p. 240.]
[29] M. Voloshin and M. Shifman, *Yad. Fiz.* **41** (1985) 187 [*Sov. J. Nucl. Phys.* **41** (1985) 120]; *ZhETF* **91** (1986) 1180 [*Sov. Phys. — JETP* **64** (1986) 698.]
[30] J. Chay, H. Georgi and B. Grinstein, *Phys. Lett.* **B247** (1990) 399.
[31] M. Voloshin, *Phys. Rev.* **D46** (1992) 3062.
[32] J. D. Bjorken, in *Proc. of the 4th Rencontres de Physique de la Vallèe d'Aoste*, La Thuille, Italy, 1990, Ed. M. Greco (Editions Frontières, Gif-Sur-Yvette, France, 1990), p. 583; J. D. Bjorken, I. Dunietz and J. Taron, *Nucl. Phys.* **B371** (1992) 111; see also N. Isgur and M. Wise, *Phys. Rev.* **D43** (1991) 819.

[33] B. Blok, L. Koyrakh, M. Shifman and A. Vainshtein, *Phys. Rev.* **D49** (1994) 3356; (E) **D50** (1994) 3572.

[34] A. Manohar and M. Wise, *Phys. Rev.* **D49** (1994) 1310.

[35] T. Mannel, *Nucl. Phys.* **B413** (1994) 396.

[36] V. Belyaev and B. Blok, *Z. Phys.* **C30** (1986) 151.

[37] I. Bigi, M. Shifman, N. G. Uraltsev and A. Vainshtein, *Phys. Rev. Lett.* **71** (1993) 496.

[38] B. Blok and M. Shifman, *Nucl. Phys.* **B399** (1993) 441.

[39] L. Koyrakh, *Phys. Rev.* **D49** (1994) 3379.

[40] B. Blok, R. Dikeman and M. Shifman, *Phys. Rev.* **D51** (1995) 6167.

[41] E. Shuryak and A. Vainshtein, *Nucl. Phys.* **B199** (1982) 451; **B201** (1982) 143.

[42] T. Mannel, *Phys. Rev.* **D50** (1994) 428.

[43] A. Falk and M. Neubert, *Phys. Rev.* **D47** (1993) 2965; 2982.

[44] P. Ball and V. Braun, *Phys. Rev.* **D49** (1994) 2472.

[45] For a recent update see E. Bagan, P. Ball, V. Braun and P. Gosdzinsky, *Phys. Lett.* **B342** (1995) 362.

[46] B. Blok and M. Shifman, in Proc. of the *Workshop on the Tau-Charm Factory*, Marbella, Spain, 1993, Eds. J. Kirkby and R. Kirkby (Editions Frontieres, Gif-sur-Yvette, 1994), p. 247. For earlier estimates see e.g. B. Guberina, R. Rückl and J. Trampetić, *Z. Phys.* **C33** (1986) 297.

[47] V. Eletsky and V. Shuryak, *Phys. Lett.* **B276** (1992) 191.

[48] M. Neubert, *Phys. Rev.* **D45** (1992) 2451.

[49] S. Narison, *Phys. Lett.* **B308** (1993) 365.

[50] M. Voloshin and M. Shifman, *Yad. Fiz* **45** (1987) 463 [*Sov. J. Nucl. Phys.*, **45** (1987) 292]; H. D. Politzer and M. Wise, *Phys. Lett.* **206B** (1988) 681.

[51] E. Bagan, P. Ball, V. Braun and H. G. Dosch, *Phys. Lett.* **B278** (1992) 457. In this paper $1/m_b$ corrections to f_B are discussed in great detail. It also gives a thorough analysis of the hybrid logarithms in the sum rule for f_B.

[52] H. Wittig, *Lattice Results for Heavy Quark Physics*, in *Proc. VI International Conference on Hadron Spectroscopy HADRON'95*, eds. M. C. Birse, G. D. Lafferty, J. A. McGovern (World Scientific, Singapore, 1996) p. 248 [hep-ph/9509292]; C. Bernard *et al.*, *Nucl. Phys. B* (*Proc. Suppl.*) **47** (1996) 459.

[53] P. Ball, *Nucl. Phys.* **B421** (1994) 593.

[54] P. Ball, *Phys. Lett.* **B281** (1992) 133; B. Blok and M. Shifman, *Phys. Rev.* **D47** (1993) 2949; see also Ref. [31].

[55] I. Bigi, A. G. Grozin, M. Shifman, N. Uraltsev and A. Vainshtein, *Phys. Lett.* **B339** (1994) 160.

[56] M. Shifman, N. G. Uraltsev and A. Vainshtein, *Phys. Rev.* **D51** (1995) 2217.

[57] For reviews see e.g. B. L. Ioffe, V. A. Khoze and L. N. Lipatov, *Hard Processes* (North-Holland, Amsterdam, 1984); R. G. Roberts, *The Structure of the Proton*, (Cambridge University Press, 1995).

[58] R. Jaffe and L. Randall, *Nucl. Phys.* **B412** (1994) 79.

[59] M. Neubert, *Phys. Rev.* **D49** (1994) 3392.

[60] I. Bigi, M. Shifman, N. G. Uraltsev and A. Vainshtein, *Int. J. Mod. Phys.* **A9** (1994) 2467.
[61] A. Falk, E. Jenkins, A. Manohar and M. Wise, *Phys. Rev.* **D49** (1994) 4553.
[62] M. Neubert, *Phys. Rev.* **D49** (1994) 4623.
[63] T. Mannel and M. Neubert, *Phys. Rev.* **D50** (1994) 2037.
[64] R. Dikeman, M. Shifman and N. Uraltsev, *Int. J. Mod. Phys.* **A11** (1996) 571.
[65] J. Collins and D. Soper, *Nucl. Phys.* **B194** (1982) 445.
[66] I. Balitsky and V. Braun, *Nucl. Phys.* **B311** (1988) 541; **B361** (1991) 93.
[67] A. Bareiss and E. Paschos, *Nucl. Phys.* **B327** (1989) 353; C. Jin, W. Palmer and E. Paschos, *Phys. Lett.* **B329** (1994) 364.
[68] X. Ji, *Wilson's Expansion with Power Accuracy*, Preprint MIT-CTP-2437/1995 [hep-ph/9506216]; *Matching Perturbative and NonPerturbative Physics With Power Accuracy in Heavy-Quark Effective Theory*, Preprint MIT-CTP-2453/1995 [hep-ph/9507322].
[69] M. Shifman, *Theory of Preasymptotic Effects in Weak Inclusive Decays*, in Proc. of the Workshop *Continuous Advances in QCD*, Ed. A. Smilga (World Scientific, Singapore, 1994), p. 249 [hep-ph/9405246]; *Recent Progress in the Heavy Quark Theory*, in Proc. V Int. Symp. on Particles, Strings and Cosmology — PASCOS — March 1995, Ed. J. Bagger *et al.* (World Scientific, Singapore, 1996), p. 69.
[70] V. Novikov, M. Shifman, A. Vainshtein and V. Zakharov, *Nucl. Phys.* **B191** (1981) 301.
[71] S. J. Brodsky, G. P. Lepage and P. B. Mackenzie, *Phys. Rev.* **D28** (1983) 228; G. P. Lepage and P. B. Mackenzie, *Phys. Rev.* **D48** (1993) 2250.
[72] M. Beneke and V. M. Braun, *Phys. Lett.* **B348** (1995) 513.
[73] M. Neubert, *Phys. Rev.* **D51** (1995) 5924; attempts of applying these ideas are presented in M. Neubert, *Nucl. Phys.* **B463** (1996) 511.
[74] I. Bigi, M. Shifman, N. Uraltsev and A. Vainshtein, *Phys. Rev.* **D50** (1994) 2234; M. Beneke and V. Braun, *Nucl. Phys.* **B426** (1994) 301.
[75] For initial exposure see Yu. Dokshitzer, V. Khoze, A. Mueller and S. Troyan, *Basics of Perturbative QCD* (Editions Frontieres, Gif-sur-Yvette, 1991); J. Collins, *Sudakov Form Factors*, in *Perturbative Quantum Chromodynamics*, Ed. A. Mueller (World Scientific, Singapore, 1989), p. 573.
[76] Yu. L. Dokshitzer, D. I. Dyakonov and S. I. Troyan, *Phys. Reps.* **58** (1980) 270; D. Amati, A. Bassetto, M. Ciafaloni, G. Marchesini and G. Veneziano, *Nucl. Phys.* **B173** (1980) 429.
[77] A. Smilga, *Nucl. Phys.* **B161** (1979) 449; G. Korchemsky, *Phys. Lett.* **B217** (1989) 330.

Recommended Literature

J. D. Bjorken and S. D. Drell, *Relativistic Quantum Mechanics* (McGraw-Hill, 1964), Chap. 4.

V. B. Berestetskii, E. M. Lifshitz and L. P. Pitaevskii, *Quantum Electrodynamics*, 2nd Edition (Pergamon Press, Oxford, 1982), Sec. 33.

A. V. Manohar and M. B. Wise, *Heavy Quark Physics* (Cambridge University Press, 1999).

V. Novikov, L. Okun, M. Shifman, A. Vainshtein, M. Voloshin and V. Zakharov, *Charmonium and Gluons*, *Phys. Rep.* **41** (1978) 1.

A general introduction to the background field technique in QCD is given in V. Novikov, M. Shifman, A. Vainshtein and V. Zakharov, *Fortsch. Phys.* **32** (1984) 585.

N. Isgur and M. B. Wise, *Heavy Quark Symmetry*, in *B Decays*, 2nd Edition, Ed. B. Stone (World Scientific, Singapore, 1992), p. 231.

H. Georgi, *Heavy Quark Effective Field Theory* in *Perspectives in the Standard Model*, Proc. of the Theoretical Advanced Study Institute, Boulder, Colorado, 1991, Eds. R. K. Ellis, C. T. Hill and J. D. Lykken (World Scientific, Singapore, 1992).

I. Bigi, M. Shifman and N. Uraltsev, *Aspects of Heavy-Quark Theory, Ann. Rev. Nucl. Part. Sci.* **47** (1997) 591.

Chapter II

Snapshots of Hadrons or the Story of How the Vacuum Medium Determines the Properties of the Classical Mesons Which Are Produced, Live and Die in the QCD Vaccum

M. A. Shifman

Theoretical Physics Institute, University of Minnesota, Minneapolis, MN 55455, USA

A combined version of lectures given at Ettore Majorana International School on Subnuclear Physics *Vacuum and Vacua: The Physics of Nothing*, Erice, Sicily, July 8, 1995, the ALEPH τ Group Workshop, CERN, November 14, 1996, and at the 1997 Yukawa International Seminar *Nonperturbative QCD — Structure of the QCD Vacuum*, Kyoto, December 2–12, 1997.

Abstract

These lectures give a modern perspective on the QCD sum rule method of Shifman, Vainshtein and Zakharov. I describe basic elements of the method, its strong and weak sides, as well as sample applications and recent developments. One of the most important practical questions still unresolved in full is the violation of the quark–hadron duality. I explain what this notion means in the context of the Wilsonean operator product expansion. Novel ideas and attempts at solving the problem are discussed.

Contents

1. QCD Sum Rules: Twenty Years Later

I will discuss a method of treating the nonperturbative dynamics of QCD which was created almost twenty years ago [1] in an attempt to understand a variety of properties and behavior patterns in the hadronic family in terms of several basic parameters of the vacuum state. The method goes under the name *QCD Sum Rules* — rather awkward, for many reasons. First and foremost, it does not emphasize the essence of the method. Second, in quantum chromodynamics there exist many other sum rules, having nothing to do with those suggested in Ref. [1]. Finally, some authors add further confusion by using *ad hoc* names, e.g. the Laplace sum rules, spectral sum rules, and so on, which are even foggier and are not generally accepted.

It would be more accurate to say "the method of expansion of the correlation functions in the vacuum condensates with the subsequent matching via the dispersion relations." This is evidently far too long a string to put into circulation. Therefore, for clarity I will refer to the Shifman-Vainshtein-Zakharov (SVZ) sum rules. Sometimes, I will resort to abbreviations such as "the condensate expansion."

Twenty years ago, next to nothing was known about the nonperturbative aspects of QCD. The condensate expansion was the first quantitative approach which proved to be successful in dozens of problems. Since then, many things changed. Various new ideas and models were suggested concerning the peculiar infrared behavior in quantum chromodynamics. Lattice QCD grew into a powerful computational scheme which promises, with time, to produce the most accurate results, if not for the whole set of the hadronic parameters, at least, for a significant part.

It seems timely to survey the ideas and technology constituting the core of the SVZ sum rules from the modern perspective, when the method became just one among several theoretical components in a modern highly competitive environment. An exhaustive review of a wealth of "classical", old elements of the method and applications was given in Ref. [2]. There is hardly any need for an abbreviated version of such a report. New applications which were worked out in the last decade or so definitely do deserve a detailed discussion. As far as I know, no comprehensive coverage of the topic exists in the literature. Unfortunately, in these lectures I will not be able to provide such a coverage, which thus remains a task for the future.[1] Instead, I will focus on those qualitative aspects where understanding became deeper. This is the first goal. Secondly, selected new applications will be considered to the extent that they illustrate the theoretical ideas of the last decade. And last but not least, I will try to outline an ecological niche which belongs to the SVZ method today. As a matter of fact, over the years, slow but steady advances have taken place in our knowledge of the hadronic world. Some old and largely forgotten predictions of the SVZ sum rules were confirmed recently by other investigations based on totally different principles. These predictions are extremely nontrivial. For instance, about 15 years ago, it was discovered [3] that not all hadrons are alike; there are remarkable distinctions between them, especially in the glueball sector. The fact that not all hadrons are alike is now becoming more and more evident from lattice results as well. Other examples are known too. By confronting them with alternative sources of information, such as lattices, we get a much fuller picture of the QCD vacuum. This process might be very beneficial to both sides. Unfortunately, at present the lattice and analytic QCD communities are largely disconnected, and seem to rarely talk to each other. My task is to show that borrowing from each other can make everybody richer!

2. QCD Vacuum and Basics of the SVZ Method

2.1. *General Ideas*

The color dynamics described by QCD is very peculiar. If we have two probe color charges, their interaction approaches the Coulomb law at short distances, with a weak coupling constant. The Coulomb interaction is due to the one gluon exchange (Fig. 1). At larger distances, the gluon starts branching (Fig. 2), which leads to a remarkable phenomenon known as antiscreening, or asymptotic freedom [4]. In normal theories, like QED, the virtual cloud screens the bare charge making the charge visible at larger distances smaller than the bare one. This situation is perfectly transparent intuitively.

[1] Work on systematically reviewing a variety of developments that took place since the mid-1980s and numerous new applications is under way (a private communication from B. L. Ioffe). A survey devoted to the relation between the sum rule and lattice results is being written by A. Khodjamirian.

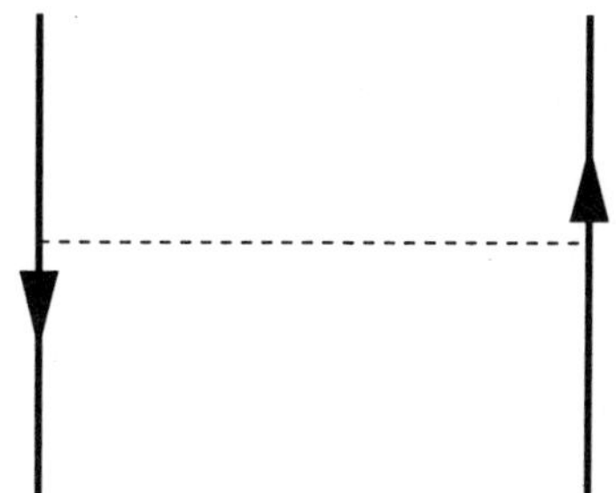

Fig. 1. Heavy color charges (solid lines) interacting through the one-gluon exchange (dashed line).

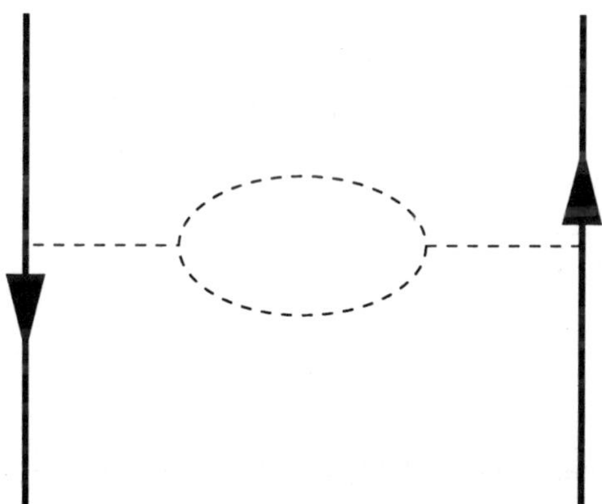

Fig. 2. Gluon branching responsible for the logarithmic running of the effective gauge coupling constant in perturbation theory (shown is a sample one-loop graph).

In QCD, instead of screening, the branching processes result in a totally counter-intuitive behavior — the antiscreening. Those of you who would like to know the physical origin of antiscreening are referred to the very pedagogical review [Ref. [5], Sec. 1.3]. If the distances are not too large one can apply perturbation theory to quantitatively describe the gluon branchings, and derive the famous formula of asymptotic freedom

$$\alpha_s = \text{Const.}/\ln\frac{r_0}{r} \qquad (1)$$

where r_0 is a dynamically generated scale parameter of QCD, $r_0 \sim 1$ fm. At small separations the effective coupling constant dies off logarithmically.

Usually one considers the running coupling constant in the momentum space. Then in the leading (one-loop) approximation corresponding to Eq. (1),

$$\alpha_s(\mu^2) = \frac{4\pi}{b\ln(\mu^2/\Lambda^2)}\,, \qquad b = \frac{11}{3}N_c - \frac{2}{3}N_f\,, \tag{2}$$

where N_c is the number of colors and N_f is the number of flavors and b is the first coefficient in the Gell-Mann-Low function. In the low energy domain that we will be interested in (below the threshold of the charm production) the number of active flavors is $N_f = 3$, and, hence, $b = 9$.

In perturbation theory the effective coupling is calculated order by order. The result is known up to three loops, and for $N_c = N_f = 3$ can be conveniently written as [6]

$$a(\mu^2) = \frac{1}{\ln\frac{\mu^2}{\Lambda^2}} - 0.79\frac{\ln\ln\frac{\mu^2}{\Lambda^2}}{\ln^2\frac{\mu^2}{\Lambda^2}} + \frac{(0.79)^2}{\ln^3\frac{\mu^2}{\Lambda^2}}\left[\left(\ln\ln\frac{\mu^2}{\Lambda^2}\right)^2 - \ln\ln\frac{\mu^2}{\Lambda^2} + 0.415\right] + \mathcal{O}\left(\frac{1}{\ln^4\frac{\mu^2}{\Lambda^2}}\right), \tag{3}$$

where we define

$$a(\mu^2) \equiv \frac{b}{4}\frac{\alpha_s(\mu^2)}{\pi}\,, \tag{4}$$

and Λ is the scale parameter of QCD introduced in a standard way [6]. In the third and higher loops the law of the running of α_s becomes scheme-dependent. The third term in Eq. (3) refers to the so-called modified minimal subtraction ($\overline{\text{MS}}$) scheme. More exactly, since Eq. (3) describes running with three flavors, the parameter Λ is actually $\Lambda^{(3)}_{\overline{\text{MS}}}$. Below we will deal exclusively with $\Lambda^{(3)}_{\overline{\text{MS}}}$; therefore, not to make the notation too clumsy, we will suppress the sub(super)scripts.

Let us ignore for a while the issue of the effective coupling constant, with the intention of returning to it later. The only lesson one should remember at this stage is that the effects caused by perturbative gluon exchanges are logarithmic.

Equations (1) or (2) imply that the effective color interaction becomes stronger as the separation between the probe color charges increases. Being remarkable by itself, this phenomenon carries the seeds of another, even more remarkable property of QCD. When the distance becomes larger than some

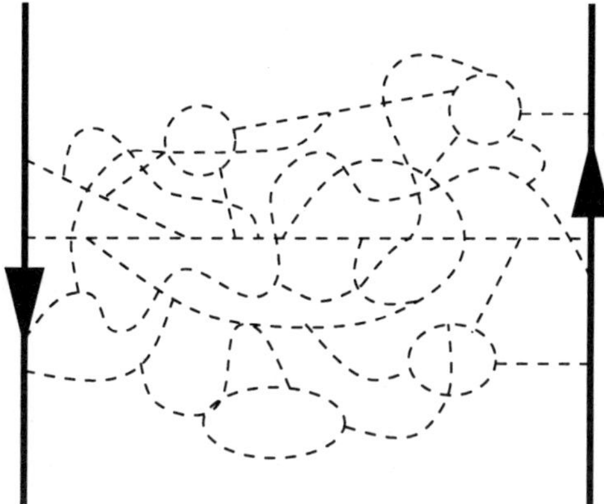

Fig. 3. When the separation between the probe color charges exceeds the critical value the nonlinearity becomes so strong that it makes no sense to speak about individual gluons. The color fields between the probe charges form a flux tube with transverse dimensions $\sim \Lambda^{-1}$.

number times Λ^{-1}, and exceeds a critical one, the branchings of gluons become so intensive (Fig. 3) that it makes no sense to speak about individual gluons.

Rather, we should phrase our consideration in terms of the chromoelectric and chromomagnetic fields. In "normal" theories, like QED, the field induced by the probe charges separated by a large distance, is dispersed all over space. In QCD it is conjectured that a specific organization of the QCD vacuum makes such a dispersed configuration energetically inexpedient [7]. Rather, the chromoelectric field between the probe charges squeezes itself into a sausage-like configuration. The situation is reminiscent of the Meissner effect in superconductivity. As is well-known, the superconducting media do not tolerate the magnetic field. If one imposes, as an external boundary condition, a certain flux of the magnetic field through such a medium, the magnetic field will be squeezed into a thin tube carrying all the magnetic flux; the superconducting phase is destroyed inside the tube, and there is no magnetic field outside.

Superconductivity is caused by the condensation of the Cooper pairs — pairs of electric charges. A phenomenologically acceptable picture of the infrared dynamics in QCD requires chromoelectric flux tubes, not magnetic ones. If something like condensation of "chromomagnetic monopoles" had taken place in the QCD vacuum then this would naturally explain nonproliferation of the chromoelectric field from the external probe color charges in the entire space — a dual Meissner effect would force the field to form flux tubes in the QCD vacuum, in this way ensuring color confinement [7].

Nobody has ever succeeded in proving that the dual Meissner effect does indeed take place in QCD. There are reasons to believe, however, that it does. First, some evidence came from the numerical study in the lattice QCD (see e.g. Ref. [8]). Quite recently, a breakthrough [9] was achieved in the understanding of the infrared dynamics in a theory which might be considered as a relative of QCD, the so-called $N = 2$ supersymmetric (SUSY) Yang-Mills theory, in which it was analytically shown that the monopoles do condense in the strong coupling regime, thus providing a basis for the dual Meissner mechanism.

Even if, as a result of a breakthrough sometime in the future, it becomes clear how a similar mechanism might develop in QCD, we have a long way to go from a qualitative picture of the phenomenon to a quantitative approach allowing one to exactly *calculate* all the variety of the hadronic properties from the first principles. It may be possible that such string-based exact approach will never emerge. While the issue of the exact solution is still under investigation, let us see what analytic QCD can offer as an approximate solution.

The basic idea lying behind the SVZ method is quite transparent. The quarks comprising the low-lying hadronic states, e.g. classical mesons or baryons, are not that far from each other, on average. The distance between them is of order Λ^{-1}. Under the circumstances, the string-like chromoelectric flux tubes, connecting well-separated probe color charges, hardly have a chance to be developed. Moreover, the valence quark pair injected in the vacuum, in a sense, perturbs it only slightly. Then we do not need the full machinery of the QCD strings, whatever it might mean, to approximately describe the properties of the low-lying states. Their basic parameters depend on how the valence quarks of which they are built, interact with typical *vacuum* field fluctuations.

It is established that the QCD vacuum is characterized by various condensates [1]. Half-dozen of them are known: the gluon condensate $G^2_{\mu\nu}$, the quark condensate $\bar{q}q$, the mixed condensate $\bar{q}\sigma Gq$, and so on. The task is to determine the regularities and parameters of the classical mesons and baryons from a few simple condensates.

Of course, without invoking the entire infinite set of condensates one can hope to capture only gross features of the vacuum medium, at best. Correspondingly, any calculation of the hadronic parameters of this type is admittedly approximate. Since hadronic physics is deprived of small parameters, in the vast majority of problems high accuracy is by far not the most desired requirement to calculation, however. Does it make any difference if, say, the proton magnetic moment is predicted theoretically to two or to three digits?

I do not think so. Rather, it is the high reliability of predictions in a *wide range of problems where the answer is not known a priori* and the theoretical control over qualitative aspects that are the primary goals of the theory of hadrons. Below I will try to demonstrate that the SVZ method is both a reliable and controllable approach which has enjoyed enormous success in dozens and dozens of instances. There are a few cases where it fails [3], but we do understand why. And the very fact of the failure of the standard strategy teaches us a lot. This happens for specific, "nonclassical" hadrons, with a very strong coupling to the vacuum fluctuations, which are very different from say the ρ meson or nucleon. So, we can reverse the argument and convert the failure into success by saying that the method predicts that *not all hadrons are alike.*

2.2. *Getting Started/Playing With Toy Models*

The basic microscopic degrees of freedom of quantum chromodynamics are the quarks and gluons. Their interaction is described by the Lagrangian

$$\mathcal{L} = -\frac{1}{4} G^a_{\mu\nu} G^a_{\mu\nu} + \sum_f \bar{q} i \not{D} q \tag{5}$$

where f is the flavor index, and the color index of the quark fields q is suppressed. For simplicity we will assume that the quark mass terms vanish. In other words, we will work in the chiral limit. In this limit the pion is massless, $m_\pi^2 = 0$. This is known to be quite a good approximation to the real world.

Neither quarks nor gluons are asymptotic states. Experimentally observed are hadrons — color-singlet bound states. In order to study the properties of the classical hadrons it is convenient to start from the empty space — the vacuum — inject a quark–antiquark pair there, and then follow the evolution of the valence quarks injected in the vacuum medium. The injection is achieved by external currents. The most popular are the vector and axial currents. Their popularity is due to the fact that they actually exist in nature: virtual photons and W bosons couple to the vector and axial quark currents. Therefore, they are experimentally accessible in the e^+e^- annihilation into hadrons or hadronic τ decays.

Thus, the objects we will work with are the correlation functions of the quark currents. More concretely, let us consider the vector current with the isotopic spin $I = 1$,

$$J_\mu = \frac{\bar{u}\gamma_\mu u - \bar{d}\gamma_\mu d}{\sqrt{2}}, \qquad \text{or} \quad J_\mu = \bar{u}\gamma_\mu d\,. \tag{6}$$

The first current shows up in the e^+e^- annihilation while the second is relevant to the τ decays. The two-point function $\Pi_{\mu\nu}$ is defined as

$$\Pi_{\mu\nu} = i \int e^{iqx} d^4x \langle 0|T\{J_\mu(x) J_\nu^\dagger(0)\}|0\rangle \tag{7}$$

where q is the total momentum of the quark–antiquark pair injected in the vacuum. Due to the current conservation $\Pi_{\mu\nu}$ is transversal and, hence,

$$\Pi_{\mu\nu} = (q_\mu q_\nu - q^2 g_{\mu\nu})\Pi(q^2)\,. \tag{8}$$

For historical reasons $\Pi(q^2)$ is often called the *polarization operator*; we will use this nomenclature in what follows. It is easy to count that the function $\Pi(q^2)$ is dimensionless. The imaginary part of $\Pi(q^2)$ at positive values of q^2 (i.e. above the physical threshold of the hadron production) is called *the spectral density*,

$$\rho(s) = \frac{12\pi}{N_c} \mathrm{Im}\,\Pi(s)\,, \qquad s \equiv q^2\,. \tag{9}$$

It coincides up to normalization with the cross-section of e^+e^- annihilation into hadrons (measured in the units $\sigma(e^+e^- \to \mu^+\mu^-)$) or the τ decay distribution function. The numerical factor in the definition of $\rho(s)$ in Eq. (9) is introduced for convenience, as will become apparent shortly. In the real world $N_c = 3$ but we will keep this factor explicit for a while, to keep track of the N_c dependence.

The spectral density carries full information about the spectrum and the widths of hadrons with given quantum numbers. Every QCD practitioner dreams of the exact calculation of the spectral density. Later on we will see how the SVZ sum rules constrain the parameters of the lowest-lying states, the ρ meson in the case at hand. But first, prior to submerging into the technical aspects of the SVZ approach, let us make an educated guess of how the spectral density might look, in gross features, to get an idea of what is expected for $\Pi(q^2)$.

To begin with, consider the limit $N_c \to \infty$. As is well-known [10], in this limit all hadrons are infinitely narrow, since all decay widths are suppressed by powers of $1/N_c$. Correspondingly, $\rho(s)$ is a sum of delta functions.

Moreover, there are good reasons to believe that these delta functions must be approximately equidistant, at least asymptotically, for highly excited states. A string-like picture of color confinement naturally leads to (approximately) linear Regge trajectories, and, hence, $m_n^2 \propto n$ at large n where m_n is the mass of the nth ρ-meson excitation. In the world of purely linear Regge trajectories there are an infinite number of daughter trajectories associated with each Regge trajectory. The daughter trajectories are parallel to the parent trajectory and are shifted by integers. Thus, in the old Veneziano model (a review is given e.g. in Ref. [11]) $m_n^2 = m_\rho^2 + n/\alpha'$ (α' is the slope of the Regge trajectory). In some of the later versions, which are more "QCD-friendly", $m_n^2 = m_\rho^2 + 2n/\alpha'$, see e.g. Ref. [12]. Since for now the focus is on a qualitative picture, the distinction between these two scenarios is not essential for our illustrative purposes. For definiteness, let us accept that the distance between two consecutive excitations contributing to $\rho(s)$ is $2/\alpha' \approx 2\ \mathrm{GeV}^2$.

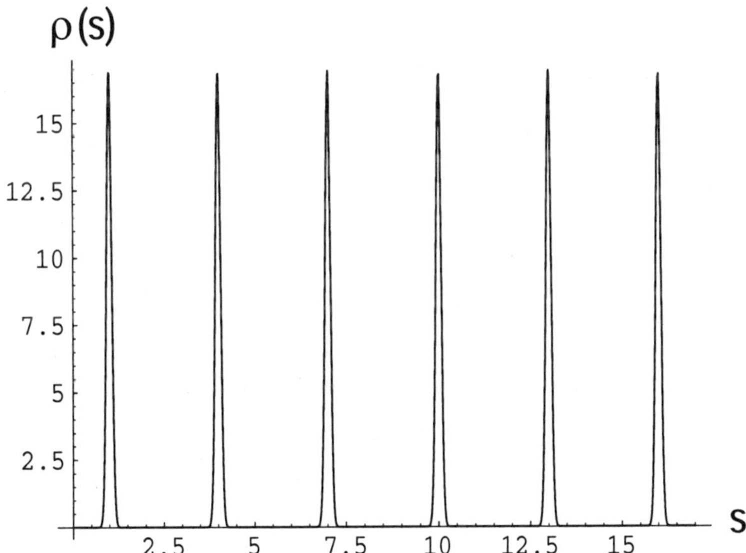

Fig. 4. The spectral density in the toy model. For clarity I gave a tiny width to the δ functions in Eq. (10).

Assembling all these elements together one obtains a sketch of the spectral density (Fig. 4),

$$\rho(s) = 3 \sum_{n=0}^{\infty} \delta(s - 1 - 3n)\,. \tag{10}$$

Here m_ρ^2 is set equal to unity; all dimensional quantities are measured in these units, for instance, $2/\alpha' \approx 3$. The couplings of all mesons to the current J_μ are chosen to be equal. This choice is represented by the overall numerical factor 3 in the sum (10). The explanation will become clear momentarily.

Now, given the imaginary part (10), it is not difficult to reconstruct the polarization operator Π itself,

$$\Pi(Q^2) = -\frac{N_c}{12\pi^2}\psi(z) + \text{Const.}\,, \tag{11}$$

$$Q^2 \equiv -q^2\,, \qquad z = \frac{Q^2+1}{3}\,,$$

where ψ is the logarithmic derivative of the Γ function,

$$\psi(z) = -C + \sum_{k=0}^{\infty}\left[\frac{1}{k+1} - \frac{1}{z+k}\right].$$

The constant on the right-hand side of Eq. (11) is irrelevant, since it can be always eliminated by an appropriate subtraction. It will be omitted hereafter.

Although our desired target is the spectral density on the physical cut, i.e. at positive values of q^2, all theoretical calculations in quantum chromodynamics are carried out off the physical cut, say, at negative values of q^2 (positive Q^2). The reason is obvious: the QCD Lagrangian (5) is formulated in terms of quarks and gluons, not hadrons. In the absence of the final solution of QCD we can deal only with the quark and gluon fields. By working in the Euclidean domain, off the physical cuts, we can calculate in terms of quarks and gluons. This is a common feature of the SVZ sum rules and lattice strategies.

At positive values of Q^2 an asymptotic representation exists for the ψ function,

$$\psi(z) = \ln z - \frac{1}{2z} - \sum_{n=1}^{\infty}\frac{B_{2n}}{2n}z^{-2n}\,, \tag{12}$$

where B_{2n} stand for the Bernoulli numbers,

$$B_{2n} = (-1)^{n-1}\frac{2(2n)!}{(2\pi)^{2n}}\zeta(2n)\,; \tag{13}$$

here ζ is the Riemann function. (In some textbooks $(-1)^{n+1}B_{2n}$ is called the nth Bernoulli number and is denoted by B_n.)

Equations (11) and (12) at large positive Q^2 imply that in the leading approximation

$$\Pi(Q^2) \to -\frac{N_c}{12\pi^2}\ln Q^2\,. \tag{14}$$

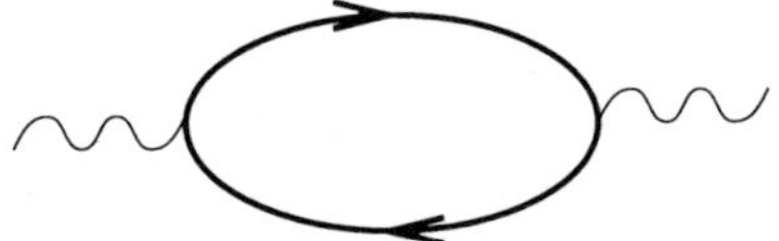

Fig. 5. The two-point function of the vector currents in the free-quark approximation. The external current injecting the quark–antiquark pair in the vacuum (and then annihilating it) is denoted by wavy lines.

This logarithmic formula for the polarization operator exactly matches what we expect from perturbation theory. Indeed, at large Q^2, in the deep Euclidean domain, the points of injection and annihilation of the quark pair in the vacuum are separated by a small space-time interval. The quarks have no time to interact with the vacuum medium. They propagate as free objects (Fig. 5). For free quarks, obviously,

$$\Pi_{\mu\nu}(q) = i \int e^{iqx} d^4x \mathrm{Tr}\left\{\gamma_\mu S_0(x,0)\gamma_\nu S_0(0,x)\right\}, \tag{15}$$

where $S_0(x,y)$ is the free-quark Green function describing propagation from the point y to the point x,

$$S_0(x,y) = \frac{1}{2\pi^2}\frac{\not{\Delta}}{(\Delta^2)^2}, \qquad \Delta \equiv x - y\,. \tag{16}$$

It is trivial to obtain

$$\mathrm{Tr}\left\{\gamma_\mu S_0(x,0)\gamma_\nu S_0(0,x)\right\} = -\frac{N_c}{\pi^4}\frac{2x_\mu x_\nu - x^2 g_{\mu\nu}}{x^8}\,. \tag{17}$$

Note that the expression on the right-hand side is automatically transversal. Substituting Eq. (17) in Eq. (15) and doing the Fourier transformation, we arrive at Eq. (14).

Thus, the leading asymptotic behavior of the polarization operator stemming from the free-quark graph in Fig. 5 is purely logarithmic. An infinite comb of infinitely narrow resonances, taken with one and the same residue, see Eq. (10), yields the same logarithm in the deep Euclidean domain. This explains why all resonance coupling constants are set equal in Eq. (10). I hasten to add that the spectral density in the real world is much more contrived. The toy spectral density (10) is supposed to be a caricature, only roughly reminding the actual one. Nevertheless, the exercise with the toy spectral density is instructive in several respects.

It is clear that the leading logarithmic term (14) carries absolutely no information on the mass of the lowest-lying state or on the spacing between the resonances: it has no built-in dimensional parameter. This information is encoded in the power corrections.

Examining the $1/Q^2$ expansion in Eq. (11),[2] we conclude that the high-order terms have factorially divergent coefficients. The factorial divergence at large n is due to the factorial growth of the Bernoulli numbers B_{2n} in Eq. (12). In QCD the expansion in powers of $1/Q^2$ is nothing but the condensate expansion. The example considered teaches us that the power expansion of $\Pi(Q^2)$ has zero radius of convergence. By itself, this is no disaster, one can work with asymptotic series as long as the expansion parameter is small. In our case the expansion parameter is $\tau \equiv 1/Q^2$, and we are interested in the expansion in the vicinity of $Q^2 \sim 1$. (Remember that we put $m_\rho^2 = 1$. If Q^2 is substantially larger than 1, we get very little information on the ρ meson from $\Pi(Q^2)$.)

In the ideal world we would calculate $\Pi(Q^2)$, and, hence, the spectral density proportional to $\operatorname{Im}\Pi(s)$, exactly. In reality we must settle for $\Pi(Q^2)$, calculated in the Euclidean domain (positive Q^2), in the form of a *truncated* series. The dispersion relation provides a bridge between the Euclidean calculation and the spectral density,

$$\Pi(Q^2) = \frac{1}{\pi}\int ds \frac{\operatorname{Im}\Pi(s)}{s+Q^2} + \text{Const.} \tag{18}$$

This explains why the method is referred to as the sum rules.

The first five terms in the power expansion of the polarization operator (11) are

$$\Pi(Q^2) = -\frac{N_c}{12\pi^2}\left[\ln Q^2 - \frac{\tau}{2} + \frac{\tau^2}{4} + \frac{\tau^3}{3} - \frac{13}{40}\tau^4 - \tau^5 + \dots\right], \qquad \tau \equiv 1/Q^2 . \tag{19}$$

From the structure of the series it is seen that at $\tau \sim 1$ the polarization operator $\Pi(Q^2)$ is determined by its expansion with very poor accuracy, up to a factor of 2, at best. This accuracy is obviously insufficient to allow one to estimate, say, m_ρ^2 with a reasonable precision.

One can drastically improve the accuracy by considering the Borel-transform of $\Pi(Q^2)$ [1]. The Borel transformation can be defined in various ways.

---

[2] In the case at hand we deal with pure powers of $1/Q^2$. Our toy model is too rude to reproduce the logarithmic corrections and logarithmic anomalous dimensions of the condensates typical of QCD.

For the functions obeying dispersion relations, the most convenient definition is through the following limiting procedure:

$$\hat{\mathcal{B}} = \lim \frac{1}{(n-1)!}(Q^2)^n \left(-\frac{d}{dQ^2}\right)^n , \qquad Q^2 \to \infty , \quad n \to \infty ,$$
$$\frac{Q^2}{n} \equiv M^2 \text{ fixed} . \tag{20}$$

M^2 is called the *Borel parameter.* It is not difficult to show that applying $\hat{\mathcal{B}}$ to $(1/Q^2)^n$ we get

$$\hat{\mathcal{B}} \left(\frac{1}{Q^2}\right)^n = \frac{1}{(n-1)!} \left(\frac{1}{M^2}\right)^n , \tag{21}$$

which entails, in turn

$$\hat{\mathcal{B}} \left(\frac{1}{s+Q^2}\right) = \frac{1}{M^2} e^{-s/M^2} . \tag{22}$$

Another trivial but useful relation we will need is

$$\hat{\mathcal{B}}(\ln Q^2) = -1 . \tag{23}$$

The Borel-transformed dispersion relation (18) takes the form

$$\tilde{\Pi}(M^2) \equiv \{\hat{\mathcal{B}}\Pi(Q^2)\} = \frac{1}{\pi M^2} \int ds \, \mathrm{Im}\, \Pi(s) e^{-s/M^2} . \tag{24}$$

The advantages gained in dealing with the Borel-transformed sum rules are obvious. First, we improve, factorially, the convergence of the power series. Thus, in the toy model we are playing with — an infinite comb of infinitely narrow peaks presented in Eq. (10) — the $1/Q^2$ expansion for $\Pi(Q^2)$ is asymptotic, while the $1/M^2$ expansion of $\tilde{\Pi}(M^2)$ is *convergent*! Indeed,

$$\tilde{\Pi}(M^2) = \frac{N_c}{4\pi^2} \frac{1}{M^2} \frac{e^{-1/M^2}}{1 - e^{-3/M^2}} , \tag{25}$$

and the radius of convergence of the $1/M^2$ expansion is

$$|M_*^2| = \frac{3}{2\pi} . \tag{26}$$

The factor 2π in the denominator ensures that we can go down to a remarkably low value of M^2; the point $M^2 \sim 1$ is well inside the convergence radius. Let us examine the first five terms of the truncated series,

$$\tilde{\Pi}(M^2) = \frac{N_c}{12\pi^2}\left[1 + \frac{\tau}{2} - \frac{\tau^2}{4} - \frac{\tau^3}{6} + \frac{13\tau^4}{240} + \frac{\tau^5}{24} + \ldots\right], \qquad \tau \equiv 1/M^2 . \tag{27}$$

It is seen that in the vicinity of $\tau \sim 1$ our "theoretical prediction" is expected to carry an error of order $\sim 2\%$. Such accuracy is already good enough to get the mass and residue of the lowest-lying state with a reasonable "resolution".

On the phenomenological side, by proceeding to the Borel-transformed dispersion relations we automatically kill possible subtraction constants. What is even more important, the exponential weight function in Eq. (24) makes the integral over the imaginary part well-convergent. Thanks to the improved convergence the relative role of the ρ-meson contribution is strongly enhanced, as we will prove shortly by quantitative estimates.

To see how it all works let us take a closer look at $\tilde{\Pi}(M^2)$; to avoid cumbersome numerical factors we will change the overall normalization and will deal with

$$I(M^2) = \frac{12\pi^2}{N_c}\tilde{\Pi}(M^2) = \frac{1}{M^2}\int ds e^{-s/M^2}\rho(s) . \tag{28}$$

In the toy model under consideration everything is known explicitly: the exact expression for $I(M^2)$, its power expansion, and the ρ-meson contribution to $I(M^2)$. The corresponding plots are displayed in Fig. 6. The exact curve has a typical shape: at small M^2 it is exponentially suppressed, approaching zero as $\exp(-m_\rho^2/M^2)$; at large M^2 it begins to flatten and slowly approaches its asymptotic value, which is equal to unity thanks to a smart choice of the normalization factor in Eq. (28). This unity is in one-to-one correspondence with the free quark result (14). It is convenient to normalize the sum rules to the free-quark result at asymptotically large M^2, and we are always going to do that. The exact curve has a steep (left) shoulder, and a shallow (right) one, with a maximum at M^2 slightly above m_ρ^2. Remember this pattern since we will encounter a very similar picture more than once in real QCD.

The ρ-meson mass and residue follow immediately from the consideration of $I(M^2)$ in the small M^2 domain (the left shoulder). Indeed, in this domain all higher states are exponentially suppressed (for instance, the contribution of the first excitation relative to that of ρ is $\sim \exp(-3/M^2)$. Therefore, $I(M^2)$ is fully saturated by ρ in the limit $M^2 \to 0$. Fitting the left shoulder by $C_1 \exp(-C_2/M^2)$ we would find $C_{1,2}$, the ρ-meson residue and mass, respectively.

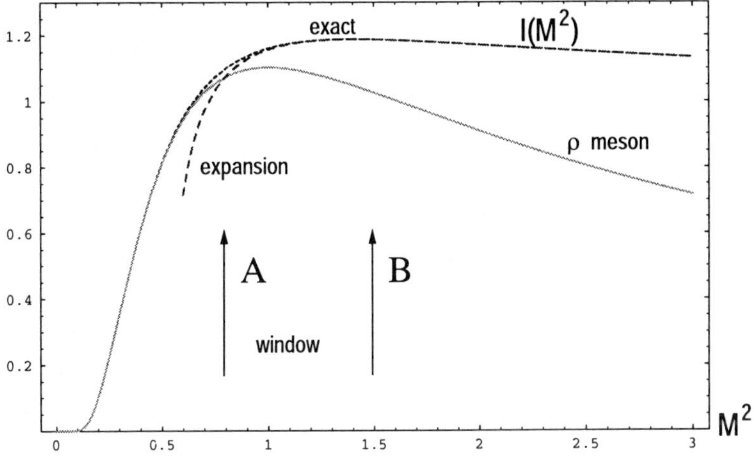

Fig. 6. The sum rule for the ρ-meson in the toy model.

Alas... in the real world $I(M^2)$ is not known at small M^2. All that is known is the large-M^2 asymptotics, in the form of expansion in the coupling constant $\alpha_s(M^2)$, and a (truncated) condensate expansion. Therefore, let us pretend that in our toy model we have the same: a truncated power series (27). The corresponding curve is shown in Fig. 6 (where I also have actually included the $O(\tau^6)$ term). The expansion is convergent only to the right from $M^2 \approx 0.5$. If we want the truncated series, with a few first terms included, to accurately represent the theoretical curve, we have to stop at $M^2 \approx 0.7$ (arrow A in Fig. 6). This is the left boundary of what the sum-rule practitioners usually call *the window* or *working window.*

The right boundary of the window (arrow B) is provided by the requirement that the first and higher excitations show up at the level not exceeding, say, 20 to 30%. In this way we ensure sufficient sensitivity to the ρ-meson parameters.

The larger the M^2, the weaker the ρ-meson dominance will be. The requirement of the ρ-meson dominance forces us to move to smaller values of M^2, while keeping control over the power expansion suggests that we move to larger values of M^2. Thus, these two requirements are contradictory. *A priori,* in any given problem, it is not evident that the AB window exists at all. If it does, we can obviously use this fact to approximately calculate the mass and the residue of the lowest-lying state. The reasons why the window exists in the problems of the classical mesons and baryons will be discussed later.

To get an idea of the accuracy one can expect from the sum rule calculations of the ρ-meson parameters we must investigate how sensitive our results are to various assumptions regarding the spectral density. The spectral density presented in Fig. 4 is not fully realistic, for many reasons. If minor details affect the results too strongly, the SVZ method would be useless. The right edge of the window (arrow B) is chosen in such a way as to minimize the impact of the (unknown) details of the spectral density above the ρ-meson peak. Yet, it is important to obtain a quantitative measure of the corresponding uncertainty.

The most obvious unrealistic feature of the spectral density (10) is the infinitely narrow width of all resonances from the comb. It is true that in the multicolor QCD the width-to-mass ratio of all hadrons built from quarks is proportional to $1/N_c$ [10] and, thus, vanishes at $N_c \to \infty$. If N_c is fixed, however, and we are interested in the widths of the excited states, there is another large parameter, the excitation number, which compensates for the effect of large N_c. Moving to the right along the s-axis in Fig. 4, very soon we find ourselves in the situation with the overlapping resonances — the δ functions in the comb representing the spectral density are smeared and the spectral density becomes perfectly smooth. Let us discuss the consequences of such *dynamical smearing*.

First of all, we must establish the dependence of the width Γ_n on the excitation number n. If one believes that a string-like picture of color confinement develops in QCD, an asymptotic estimate of Γ_n at large n becomes a simple exercise. Unlike the ρ meson and other low-lying classical states, for highly excited states the string-like picture of the chromoelectric tubes is expected to be fully relevant.

When a highly excited meson state is created by a local source, it can be considered, quasiclassically, as a pair of (almost free) ultrarelativistic quarks; each of them with energy $m_n/2$. These quarks are created at the origin, and then fly back-to-back, creating behind them a flux tube of the chromoelectric field. The length of the tube $L \sim m_n/\Lambda^2$ where Λ^2 represents the string tension. The decay probability is determined, to order $1/N_c$, by the probability of producing an extra quark–antiquark pair. Since the pair creation can happen anywhere inside the flux tube, it is natural to expect that

$$\Gamma_n \sim \frac{1}{N_c} L\Lambda^2 = \frac{B}{N_c} m_n \tag{29}$$

where B is a dimensionless coefficient of order one. I used here the fact that within the quasiclassical picture the meson mass $m_n \sim L\Lambda^2$. Thus, the width of the nth excited state is proportional to its mass which, in turn, is proportional to $\sqrt{n}$ for the linear Regge trajectories.[3]

This simple estimate was obtained in Ref. [13] long ago. Since the argument bears a very general nature, the square root dependence of Γ_n should take place in all models with linear confinement. And it does, indeed! Recently, the square root formula for Γ_n was obtained numerically [14] in the 't Hooft model [15].

If Eq. (29) does indeed take place, it is not difficult to find the impact of the resonance widths on the spectral density. The infinitely narrow pole is substituted with

$$\frac{1}{Q^2+m_n^2} \to \frac{1}{Q^2+m_n^2-im_n\Gamma_n} \to \frac{1}{Q^2+m_n^2-im_n^2B/N_c}$$
$$\to \frac{1}{Q^2+m_n^2-\gamma Q^2\ln Q^2} \to \frac{1}{(Q^2)^{1-\gamma}+m_n^2}, \tag{30}$$

where

$$\gamma = \frac{B}{\pi N_c}, \tag{31}$$

and I used the fact that $1/N_c$ is a small parameter. This explains the last transition in Eq. (30). For the same reason, in the third transition I replaced m_n^2/N_c with $N_c^{-1}\pi^{-1}Q^2\ln Q^2$. Near the pole (i.e. at $Q^2=-m_n^2$) both expressions give the same in the leading $1/N_c$ approximation, and are equally legitimate.[4]

As a result, all poles, which lay previously at positive real q^2, are now shifted onto unphysical sheets, away from the physical cut (Fig. 7). The δ functions in the spectral density are replaced by peaks with finite widths. Correspondingly, the toy model number two, which replaces Eq. (11) is

$$\Pi(Q^2) = -\frac{N_c}{12\pi^2}\frac{1}{1-\gamma}\psi(z) + \text{Const.}, \tag{32}$$

where now z is given by the following formula

$$z = \frac{(Q^2)^{1-\gamma}+1}{3}. \tag{33}$$

[3] Let us note in passing that the $1/N_c^2$ corrections due to the creation of two quark pairs are of order L^2/N_c^2 within this picture. Since $L \sim m_n \sim \sqrt{n}$, the expansion parameter is $\sqrt{n}/N_c$.

[4] The logarithm introduces not only the imaginary part, i.e. the width, but a shift in the real part, equivalent to a shift in the resonance mass. Both effects are of order $1/N_c$.

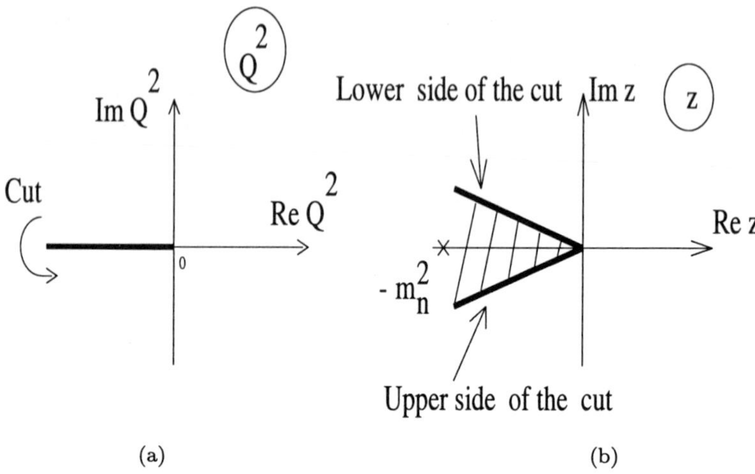

(a) (b)

Fig. 7. Analytical structure of the polarization operator. (a) The polarization operator must be analytic everywhere in the complex Q^2 plane, except for the cut running on the negative real semiaxis of Q^2 (positive real semiaxis of q^2). The imaginary part of the polarization operator must be positive at the upper side of the q^2 cut; (b) The mapping of the Q^2 plane onto the z plane, Eq. (33). The physical sheet on the Q^2 plane corresponds to the z plane with the shaded sector removed. The boundaries of the sector correspond to the lower and upper sides of the q^2 cut. The angle of the removed sector is proportional to γ.

The additional factor $(1-\gamma)^{-1}$ in the overall normalization is chosen to reproduce the free quark result (14) at large Euclidean Q^2. Note that the polarization operator defined by Eqs. (32) and (33) has the correct analytical structure: it is non-singular in the whole complex Q^2 plane, apart from the cut at real positive q^2. The singularities associated with the poles of the ψ function lie on the unphysical sheet (Fig. 7).

The spectral density (i.e. the imaginary part of Π at $q^2 = s + i\varepsilon$) takes the form of the sum of (modified) Breit-Wigner peaks

$$\rho(s) = \frac{3}{\pi(1-\gamma)} \sum_{k=0}^{\infty} \frac{s^{1-\gamma}\sin\pi\gamma}{[s^{1-\gamma}\cos\pi\gamma - (1+3k)]^2 + [s^{1-\gamma}\sin\pi\gamma]^2} \,. \tag{34}$$

It is depicted in Fig. 8 for $B = 0.6$ and $N_c = 3$. Although the choice of the parameter B is rather arbitrary, it is not unreasonable; it corresponds to $\Gamma/m \approx 0.2$. Experimentally the ρ meson has a close width-to-mass ratio.

We can see on the picture a pronounced ρ-meson peak, accompanied by a deep dip, and then an almost smooth curve which approaches the asymptotic value (unity) in an oscillating mode. The smearing of the δ functions into a

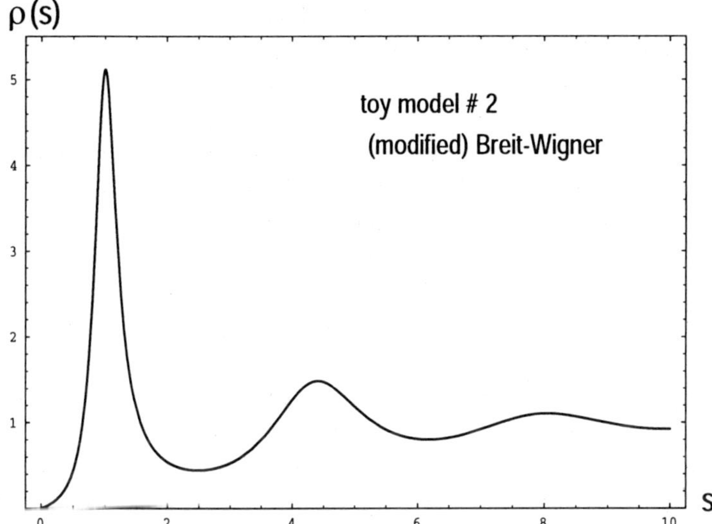

Fig. 8. The spectral density corresponding to a sum of (modified) Break-Wigner peaks with one and the same width-to-mass ratio 0.2. All residues are set equal, and $\gamma \approx 0.06369$, see Eq. (34). The s-axis is in the units m_ρ^2.

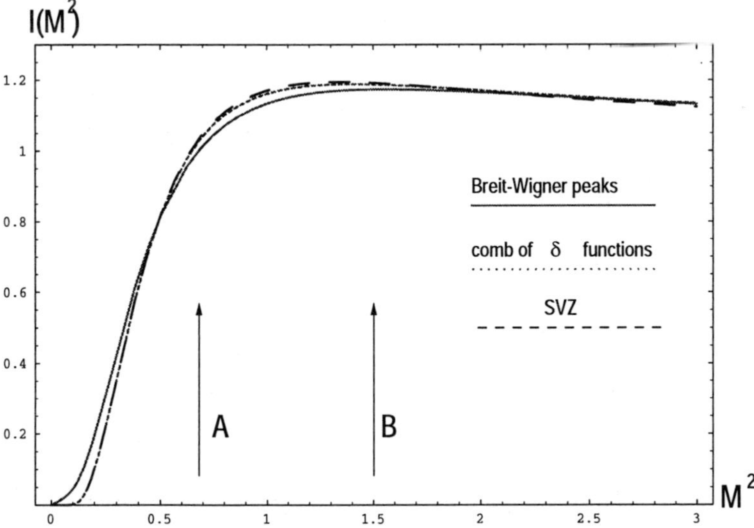

Fig. 9. $I(M^2)$ for three models of the spectral density.

smooth curve occurs because the width of the excited states is proportional to m_n while the distance between the two neighboring states $\Delta m_n \sim 1/m_n$. It is assumed, of course, that N_c is fixed. The larger the value of N_c, the further we must go, to higher excitation numbers, for the resonances to overlap. Figure 8 shows that at $N_c = 3$ a smooth curve starts right after the ground state. The first excitation already belongs to the smooth curve. Remember this pattern of the spectral density — a conspicuous first peak accompanied by a relatively smooth (and oscillating) curve approaching its constant asymptotic value. The gross features of this picture follow from very general theoretical arguments [16]. We will see shortly that experimental data, being different in fine details, exhibit the very same behavior.

The spectral densities shown in Figs. 4 and 8 at first sight do not produce an impression of close relatives. Let us integrate them over, with the exponential weight function, and compare $I(M^2)$ obtained in this way. The comparison is shown in Fig. 9. The dotted curve corresponds to the comb of δ functions. It was already presented in Fig. 6. The solid curve corresponds to Eq. (34) and Fig. 8. Finally, the dashed curve is obtained with an extremely crude model of the spectral density depicted in Fig. 10,

$$\rho(s) = 3\delta(s-1) + \theta(s-s_0)\,, \qquad s_0 = 2.7\,. \tag{35}$$

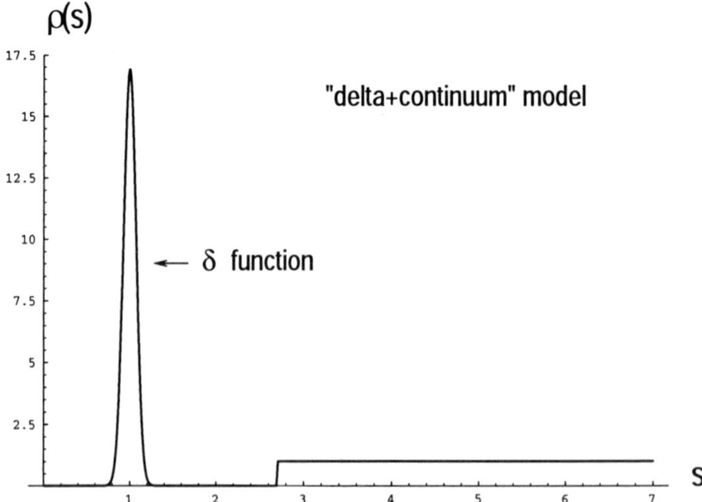

Fig. 10. The original SVZ model of the spectral density.

It presents an infinitely narrow ρ, plus a gap, plus all higher excitations fused into "continuum" which starts abruptly at $s = s_0$ and coincides (at $s > s_0$) with the asymptotic free-quark expression for the spectral density. This is the original SVZ model [1]. It, obviously, captures only gross features, and misses all details: the fact that the gap at $s > m_\rho^2$ is not quite empty, the onset of continuum is not so abrupt, and the limit $\rho(s) \to 1$ at $s \to \infty$ is achieved through oscillations. Nevertheless, it is seen that $I(M^2)$ in all three models are very close to each other. The Breit-Wigner model of Eq. (34) gives a slightly less steep fall to the left of the window, at $M^2 \to 0$. This is quite understandable, since in this model there is a nonvanishing spectral density at $s < m_\rho^2$, due to the ρ-meson width.

The comparison of three curves in Fig. 9 teaches us a lesson: by examining $I(M^2)$, calculated approximately inside the window, one cannot expect to predict, with any reasonable accuracy, such fine structure of the spectral density as the width of the ρ meson, and the shape of the curve in the continuum (i.e. higher excitations). As Fig. 9 shows, it is hardly possible to distinguish between the three models under discussion, which, in a sense, represent extreme situations. The best one can hope for is estimating m_ρ^2, its residue, and, to a lesser extent, an effective value of s where the dip following the ρ-meson peak ends (i.e. s_0).

In a broader context, this message is instructive for the lattice practitioners too. Doing calculations in the Euclidean domain — a common feature of the sum rules and lattice technology — derives virtually no information about the spectral density beyond the ground state in the given channel. Indeed, the original SVZ model and the comb of the δ functions, representing drastically different $\rho(s)$ point-by-point, lead to $I(M^2)$ coinciding with each other to a high accuracy. To distinguish between these two scenarios one needs exponential precision going far beyond the level so far achieved in the lattice calculations.

Now we are ready to explore the sensitivity of $I(M^2)$ to the ρ-meson parameters. To this end we will take the SVZ model in the form

$$\rho(s) = C_1 \delta(s - C_2) + \theta(s - s_0) \tag{36}$$

and let the parameters float by, say, 10% around their "reference" values, $C_1 = 3, C_2 = 1$ and $s_0 = 2.7$. We then compare $I(M^2)$ obtained in this way with $I(M^2)$ emerging in our (most realistic) Breit-Wigner toy model, see Eq. (34).

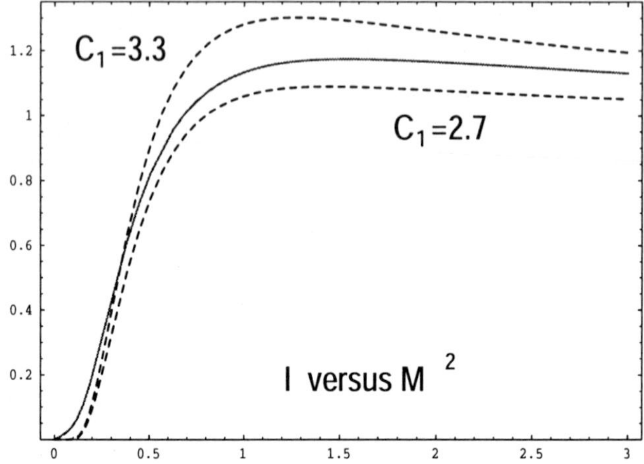

Fig. 11. Changing the ρ-meson residue. The solid curve corresponds to a "reference" spectral density presented in Eq. (34). The dashed curves correspond to Eq. (36) with $C_1 = 3.3$ and 2.7, respectively. The reference value is $C_1 = 3$.

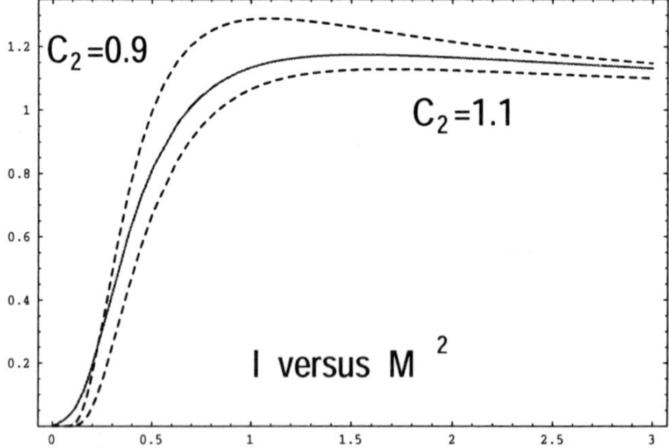

Fig. 12. Changing the ρ-meson mass. The dashed curves correspond to the 10% variation of m_ρ^2, around its reference value ($m_\rho^2 = 1$).

Figure 11 illustrates the sensitivity of the sum rule to the ρ-meson residue (coupling constant). The upper and lower dashed curves correspond to $C_1 = 3.3$ and 2.7, respectively. Figure 12 illustrates the sensitivity to the ρ-meson mass. The upper and lower dashed curves correspond to $m_\rho^2 = 0.9$ and 1.1

of the experimental value, respectively. It is quite realistic to expect to get g_ρ^2 and M_ρ^2 with the accuracy $\sim 10\%$ by inspecting $I(M^2)$ in the window and fitting the theoretical prediction by the model (36). The estimates of s_0 are less accurate. One should not be surprised, of course, since the sum rules were designed to be most sensitive to the ground-state parameters and are relatively insensitive to the details of the spectral density in the continuum. The exponential weight in the definition of $I(M^2)$ takes care of this feature. Deviations of s_0 from its reference value at the level of 10% lead to rather insignificant changes in $I(M^2)$. Deviations of s_0 at the level of 20% are quite noticeable.

I hasten to add, however, that the 10% accuracy is not a typical outcome of the sum rule analysis. The ρ-meson channel is most favorable from the point of view of applications of the SVZ sum rules. In a sense, this is a dream case: the role of continuum with respect to ρ is as tempered as it can possibly be, and higher (unknown) condensates in the truncated condensate expansion show up at remarkably low values of M^2 so that the working window is comfortably wide. In many other channels, mesonic and baryonic, we have to deal with a narrower window. The general rule is: the narrower the window, the worse the accuracy. In some channels, as we will see later, the window shrinks to zero. Then the SVZ method fails. The reasons why it is successful in some cases and fails in others, when properly understood, give us a unique hint as to the structure of the QCD vacuum (see Sec. 8).

3. Vacuum Condensates

It is time to abandon our baby version of QCD and proceed to the real thing. What can be said about $I(M^2)$ now?

Needless to say that no analytic calculation of $I(M^2)$ at $M^2 \lesssim \Lambda^2$, that would be based entirely on the first principles, exists. QCD at large distances has not yet been solved analytically. Under the circumstances it seems reasonable to start at short distances and advance to larger distances "step by step", staying on the solid ground — working with the microscopic variables, quarks and gluons, where it is fully legitimate. At short distances the quark–gluon interactions are adequately described by perturbation theory. Certainly, as we have just learned by inspecting the toy models, the (truncated) *perturbative* series is not going to yield us any estimates relevant to the ρ-meson parameters. To get these parameters, we should add, at least, some information regarding the large distance dynamics.

In the ideal world this information would be obtained from the theory *per se.* In the real world we may try to parametrize the effects caused by the vacuum fields. If the quark–antiquark pair injected in the vacuum by the current J_μ does not propagate too far, its impact on the vacuum fields is, hopefully, not drastic. This means that the polarization operator can be well approximated by the interaction of the valence quarks with a few vacuum condensates. For the validity of this assumption it is necessary that the characteristic frequencies ω of the valence quarks inside the ρ meson be larger than the characteristic scale parameter of the vacuum medium μ. Certainly, we can count only on an interplay of numbers: since the only dimensional parameter of QCD is Λ, all quantities of dimension of mass are of order Λ. It may well happen, however, that, say $\omega \sim 3\Lambda$ while $\mu \sim \Lambda/2$. In a sense, the success of the SVZ sum rules confirms this assumption *a posteriori.*

On general grounds it is expected that all gauge invariant Lorentz singlet local operators built of the quark and/or gluon fields develop nonvanishing vacuum expectation values (VEV's). It is convenient to order all operators (and their VEV's) according to their normal dimensions. The higher the dimension, the higher the power of $1/M$ of the corresponding coefficient. This fact allows one to control the condensate expansion inside the working window.

The lowest dimension (zero) belongs to the unit operator. The unit operator is trivial. When one calculates the perturbative contribution to $I(M^2)$, one actually calculates the coefficient of the unit operator.

The operator of the next lowest dimension (3) is the quark density operator,

$$\mathcal{O}_q = \bar{q}q\,. \tag{37}$$

No other gauge and Lorentz invariant operators of dimension 3 exist.

There is one operator of dimension 4, the gluon operator

$$\mathcal{O}_G = \frac{\alpha_s}{\pi} G^a_{\mu\nu} G^a_{\mu\nu}\,. \tag{38}$$

The factor α_s/π appears in the definition naturally. Since both operators, (37) and (38), play a very special role we will interrupt our excursion before passing to higher dimensions, to make several important remarks.

In the chiral limit (i.e. when the quark mass term in the Lagrangian is put to zero) $\mathcal{O}_q$ is the order parameter — its VEV signals the spontaneous breaking of the axial $SU(N_f)$ symmetry and the occurrence of the corresponding massless pions. In perturbation theory the chiral symmetry remains unbroken, of course, $\langle \mathcal{O}_q \rangle$ vanishes identically. It is tempting to say then that $\langle \mathcal{O}_q \rangle \neq 0$ measures deviations from the perturbation theory, and the same refers to all other condensates.

In the early days of nonperturbative QCD such an understanding was widely spread. I hasten to say that this is a wrong understanding. It is impossible to define the condensates as "truly nonperturbative" residues obtained after subtracting a "perturbative part" from them [17]. A heated debate took place in the literature in the eighties regarding the possibility of defining the condensates, in a rigorous way, by subtracting a "perturbative part". Glimpses of the debate can be seen in Ref. [18] where it was clearly emphasized, for the first time, that the procedure of isolating "purely nonperturbative" quantities does not (and must not) work.[5] I will outline the proper procedure below using the gluon condensate as the most instructive example.

Unlike $\mathcal{O}_q$, the gluon operator $\mathcal{O}_G$ is not an order parameter. There is no known symmetry whose spontaneous breaking would generate a vacuum expectation value $\langle \mathcal{O}_G \rangle$. Correspondingly, $\langle \mathcal{O}_G \rangle$ is generated in perturbation theory. Physically, $\langle \mathcal{O}_G \rangle$ measures the vacuum energy density ε_{vac}. To see that this is indeed the case let us use the fact that the trace of the energy-momentum tensor

$$\theta^\mu_\mu = -\frac{b}{8}\frac{\alpha_s}{\pi} G^a_{\mu\nu} G^a_{\mu\nu} \, . \tag{39}$$

This expression is nothing but the scale anomaly of QCD [20]. Strictly speaking, the coefficient on the right-hand side contains higher orders in α_s; they are not essential for our purposes, however, and will be ignored. The vacuum expectation value of the left-hand side is obviously $4\varepsilon_{\text{vac}}$. Hence

$$\varepsilon_{\text{vac}} = -\frac{b}{32} \langle \mathcal{O}_G \rangle \, . \tag{40}$$

Everybody knows that the vacuum energy density badly diverges in field theory, as the fourth power of the cut-off.[6] If so, how can one make sense of the gluon condensate?

---

[5] Surprisingly, the debate was revived recently, over a decade after the issue had been seemingly settled, in connection with the heavy quark theory. This theory, as it exists today, is also based on the operator product expansion and is a close relative [19] of the SVZ sum rules, see Chapter I. One can ask there the very same questions that are usually raised in connection with the condensate expansion in the SVZ method. Later on, in Sec. 10.2, we will return to the issue and discuss some elements of the heavy quark theory.

[6] Let me parenthetically note that in supersymmetric glyodynamics the vacuum energy density vanishes, and so does the gluon condensate.

The vacuum fields fluctuate; these fluctuations contribute to the vacuum energy density. High frequency modes of the fluctuating fields belong to the weak coupling regime; their contribution in the correlation functions is described by perturbation theory, giving rise to the standard perturbative expansions. What we are interested in are the low frequency modes, soft vacuum fields that are responsible for the peculiar properties of the vacuum medium. By saturating the condensate by the *soft modes only* we get a consistent definition of the condensates that automatically solves the problem of divergencies.

Specifically, we must introduce a somewhat artificial boundary, μ: all fluctuations with frequencies higher than μ are supposed to be hard, those with frequencies lower than μ soft. The parameter μ is usually referred to as the *normalization point.* Only soft modes are to be retained in the condensates. Thus, the separation principle is "soft versus hard" rather than "perturbative versus nonperturbative". This is the basic principle of Wilson's operator product expansion, which, in turn, is the foundation of the SVZ sum rules. Being defined in this way the condensates are explicitly μ dependent. All physical quantities are certainly μ independent; the normalization point dependence of the condensates is compensated by that of the coefficient functions.

Generalities of Wilson's approach and its particular implementation in QCD will be further discussed in Secs. 5 and 6. Here we complete our consideration of the gluon condensate.

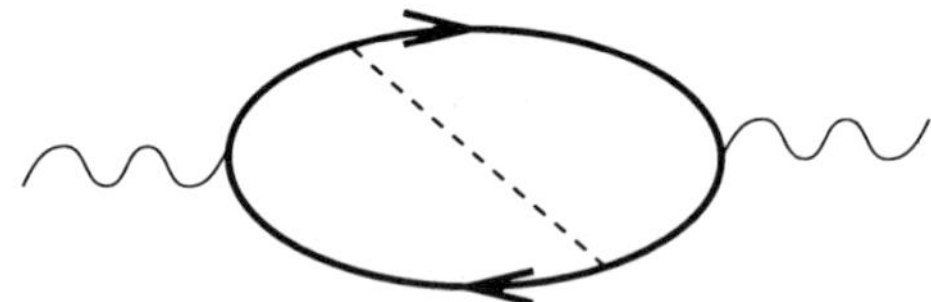

Fig. 13. The one-gluon correction to the D function.

It is instructive to trace how the gluon condensate emerges from a slightly different perspective. A natural starting point is the diagram in Fig. 13 presenting the one-gluon correction to $\Pi(Q^2)$ (two other similar graphs with the one-gluon exchange are not shown). For convenience we will deal with the so-called D function,

$$D(Q^2) = -(4\pi^2)Q^2(d\Pi/dQ^2)\,, \tag{41}$$

rather than with the polarization operator Π itself. In terms of $D(Q^2)$ the bare quark loop of Fig. 5 becomes just unity.

If the gluon momentum is denoted by k the one-gluon term in $D(Q^2)$ takes the form

$$\Delta D(Q^2) = \alpha_s \int k^2 dk^2 \frac{1}{k^2} F(k^2, Q^2) \tag{42}$$

where it is implied that the angular integration over the orientations of the vector k with respect to q is already carried out. The factor $1/k^2$ comes from the gluon propagator, while $F(k^2, Q^2)$ represents the rest of the graph. The function $F(k^2, Q^2)$ was calculated in [21] (it can also be extracted from App. B of Ref. [22]),

$$F(k^2, Q^2) = \frac{8}{3\pi}\frac{1}{Q^2}\left\{\left(\frac{7}{4} - \ln\tau\right)\tau + (1+\tau)\left[L_2(-\tau) + \ln\tau\ln(1+\tau)\right]\right\}, \tag{43}$$

where

$$\tau = \frac{k^2}{Q^2},$$

and

$$L_2(x) = -\int_0^x dy y^{-1}\ln(1-y)$$

is the dilogarithm function. The analytic continuation of the function (43) to $\tau > 1$ can be more conveniently rewritten as

$$\begin{aligned} F(k^2, Q^2) = \frac{8}{3\pi}\frac{1}{Q^2}\Bigg\{ & 1 + \ln\tau + \left(\frac{3}{4} + \frac{1}{2}\ln\tau\right)\frac{1}{\tau} \\ & + (1+\tau)\left[L_2(-1/\tau) - \ln\tau\ln(1+\tau^{-1})\right]\Bigg\}. \end{aligned} \tag{44}$$

Substituting Eqs. (43) and (44) in Eq. (42) and doing the integral over k^2 in the most straightforward manner one gets $\Delta D = \alpha_s/\pi$, the standard well-known one-gluon correction to the D function.

The plot of the function $F(k^2, Q^2)$ is shown in Fig. 14. Although it has a clear-cut peak at $\tau \sim 0.5$, it presents a distribution spanning the entire range from $k^2 = 0$ to $k^2 = \infty$. The gauge coupling constant α_s in Eq. (42) is literally a constant provided one limits oneself to the graph in Fig. 13. In higher orders, however, α_s starts running, and it is evident that $\alpha_s \to \alpha_s(k^2)$. Then $\alpha_s(k^2)$ should be placed inside the integral in Eq. (42) rather than outside. By doing so we effectively resum a subset of the multigluon graphs.

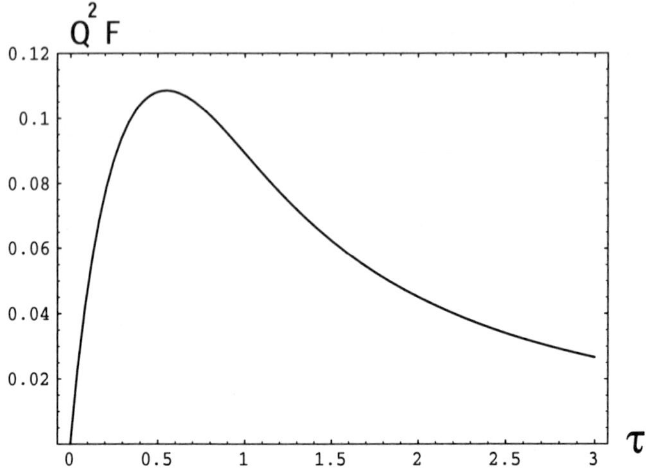

Fig. 14. The distribution function $Q^2F(k^2, Q^2)$ defined in Eq. (42) versus τ.

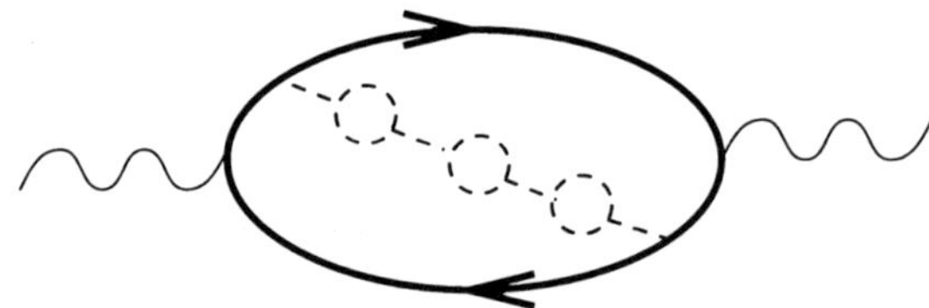

Fig. 15. One of the graphs from the bubble chain set.

Certainly, putting

$$\alpha_s(k^2) = \alpha_s(Q^2)\left[1 + \frac{b\alpha_s(Q^2)}{4\pi}\ln\frac{k^2}{Q^2}\right]^{-1} \tag{45}$$

inside the integral in Eq. (42) we do not account for *all* higher order corrections. Only those are included that are responsible for the running of α_s in the leading logarithmic approximation. Usually they say that the corresponding multigluon graphs are of the bubble chain type (Fig. 15). Strictly speaking, in the covariant gauges the gauge coupling renormalization (45) in QCD (unlike QED) does not reduce to the bubble insertions in the gluon propagator depicted in Fig. 15. Various graphs of a more complicated structure are involved too; they combine together to produce a gauge-invariant expression (45). The bubble chain saturation of Eq. (45) takes place in the physical gauges, e.g. in the Coulomb gauge. In the covariant gauges one can artificially reduce the problem to the bubble chains, without explicit identification of the full set of graphs, by exploiting the so-called "large negative N_f trick". We will not dwell on this technical issue, since it is irrelevant for our purposes. The interested reader is referred to Refs. [21, 22].

Even though the bubble chain is a very specific subset of graphs, and the majority of the multigluon graphs are left aside in this procedure, it still makes sense to substitute the running $\alpha_s(k^2)$ inside the integral in Eq. (42), since in this way we take into account an essential part of the underlying dynamics: the fact that the quark–gluon interaction becomes weak if the gluon is a far off shell, and, on the contrary, becomes stronger as the gluon virtuality decreases. If we follow this strategy, then

$$\Delta D(Q^2) = \int_0^\infty \alpha_s(k^2)dk^2 F(k^2,Q^2)$$

$$= \alpha_s(Q^2)\int_0^\infty dk^2 \left[1+\frac{b\alpha_s(Q^2)}{4\pi}\ln\frac{k^2}{Q^2}\right]^{-1} F(k^2,Q^2)\,. \tag{46}$$

We immediately run into a problem here. The running α_s has the Landau pole at small k^2, the square bracket in the integrand explodes right inside the integration interval. The explosion occurs at

$$k^2 = Q^2 \exp\left[-\frac{4\pi}{b\alpha_s(Q^2)}\right] = \Lambda^2\,. \tag{47}$$

Although the function $F(k^2,Q^2)$ is suppressed at small k^2, i.e. $\tau \to 0$, (see Fig. 14), it by no means vanishes at $\tau = \Lambda^2/Q^2$, rendering the integration impossible. In order to do the k^2 integral literally, as it is given in Eq. (46), we must say how the singularity at $k^2 = \Lambda^2$ is to be treated. Some people say: "let us take the principal value", others bypass the singularity by shifting the integration contour in the complex k^2 plane, still others do something else. All these prescriptions are arbitrary and physically meaningless for obvious reasons. Indeed, let us introduce the normalization point μ =several units $\times\,\Lambda$ and split the integration over k^2 in two parts: from zero to μ^2 in the first integral and from μ^2 to ∞ in the second. Then at $k^2 > \mu^2$, in the weak coupling domain, we do trust Eq. (46) while below μ^2 *we do not.* At best, below μ^2 one can use Eq. (46) for the purpose of orientation. At small k^2

$$F(k^2,Q^2) \to \frac{2}{\pi}\frac{k^2}{Q^4}, \tag{48}$$

implying that

$$\Delta_{\{k^2<\mu^2\}}D(Q^2) = \frac{2\alpha_s(Q^2)}{\pi}\frac{1}{Q^4}\int_0^{\mu^2} k^2dk^2\left[1+\frac{b\alpha_s(Q^2)}{4\pi}\ln\frac{k^2}{Q^2}\right]^{-1}$$
$$\sim \frac{8}{b}\frac{1}{Q^4}\int_0^{\mu^2} k^2dk^2\frac{\Lambda^2}{k^2-\Lambda^2} \sim \frac{8\pi}{b}\frac{\Lambda^4}{Q^4}. \tag{49}$$

This is a very crude estimate; I substituted the square bracket by the corresponding pole, and then evaluated the integral by equating it to its imaginary part, i.e. substituting $(k^2-\Lambda^2)^{-1} \to \pi\delta(k^2-\Lambda^2)$.

From this exercise we learn the following lesson. The domain $k^2 < \mu^2$ gives a contribution to $D(Q^2)$ which scales as $1/Q^4$. The coefficient in front of $1/Q^4$ cannot be reliably calculated and must be parametrized by the gluon condensate — that is the best we can do. It is not accidental that the k^2 expansion of $F(k^2,Q^2)$ starts from k^2, while the term of the *zero*th order is absent. If it were not the case, the condensate series would have to start from $\mathcal{O}(Q^{-2})$. This is impossible, however, since no gauge invariant local operator of dimension two exists in QCD. The absence of such operators and the absence of the terms k^0 in the distribution function $F(k^2,Q^2)$ are in one-to-one correspondence.

At first sight, one could try to circumvent the problem of the Landau pole inside the integration domain by expanding the square bracket in Eq. (46) in the series in $\alpha_s(Q^2)$. This is one of the "remedies" sometimes cited in this context. I put the word remedy in quotation marks since in reality it does not work. True, in every given order of the α_s expansion the integral (46) becomes well-defined. It is easy to see, however, that the expansion is going to be factorially divergent in high orders. Indeed,

$$\int_0^{\mu^2} k^2dk^2\left(\ln\frac{Q^2}{k^2}\right)^n \sim 2^{-n}n!, \qquad n \gg \ln\frac{Q^2}{\mu^2}. \tag{50}$$

It is seen that perturbation theory *per se* carries the seeds of the gluon condensate. The perturbative series cannot be consistently defined unless the gluon condensate is introduced. Once we cut off the $k^2 < \mu^2$ tail of the integral from the perturbative expression obtained from Eq. (46), the factorial divergence of the α_s series at high n disappears, the series becomes well-defined, and so is the gluon condensate.

The factorial divergence of the perturbative series of the type that we have just discussed is called the *infrared renormalon* [23, 24] for reasons which need not concern us here. It is associated with the bubble chain graphs of Fig. 15 and is inevitable if one forces the perturbative integrals to run all the way down to $k^2 = 0$. By bounding the integration interval in the perturbative part from below by μ^2 we get rid of the infrared renormalon altogether. The domain below μ^2 is not lost; it is fully represented by the gluon condensates. Higher condensates appear as higher order terms in the k^2 expansion of $F(k^2, Q^2)$. The intimate relationship between the infrared renormalons and condensates was first revealed by Mueller [25].

I hasten to warn that one should not literally equate the infrared renormalons to the condensate expansion. Even if one sticks to a certain definition for summation of the bubble chains, that eliminates ambiguities, say the principal value prescription in the integral (46), one typically gets a numerical value of the Q^{-4} term which is grossly off compared to the gluon condensate, let alone ambiguities associated with the various possible choices of the regularization prescription. For instance, in the case at hand,

$$D(Q^2) = 1 + \frac{2\pi^2}{3Q^4}\langle \mathcal{O}_G \rangle + \dots , \tag{51}$$

and the $\mathcal{O}(1/Q^4)$ term is 0.08 GeV^4/Q^4 under the standard choice of the gluon condensate (see below). Equation (49) yields, on the other hand, only 0.004 GeV^4/Q^4 at $\Lambda = 0.2$ GeV (0.01 GeV^4/Q^4 at $\Lambda = 0.25$ GeV). Within the SVZ method, based on Wilson's expansion, the role of the infrared renormalons is purely illustrative. They show, in a very straightforward manner, that without the condensates the perturbative series cannot be defined and, on the contrary, by introducing the condensates one eliminates ambiguities associated with the high order tails of the perturbative series.[7]

[7]The infrared renormalon is not the only source of the factorial divergence of the α_s series in QCD. The α_s expansion of Eq. (46) contains also the so-called ultraviolet renormalon — factorial divergence of high-order α_s terms coming from the large k^2 domain. This factorial divergence is readily summable, however, as is perfectly clear from the unexpanded expression (46). Indeed, the integrand has no singularities at large k^2 and is well convergent. Below, the ultraviolet renormalons will never be mentioned again. Those readers who want to familiarize themselves with this subject are referred to an excellent review [26]. A factorial divergence of a different nature showing up in the coefficient functions is briefly discussed in Sec. 5.

On the other hand, in some (perturbatively infrared stable) processes that have *no* operator product expansion, most notably in the jet physics, the infrared renormalons, in spite of all their obvious limitations, become a poor man's substitute for the condensate expansion. Although, unlike OPE, the renormalon analysis does not provide us with the proper numerical coefficients, we may infer from the renormalons what powers of $1/E$ (or $1/Q$) appear in the nonperturbative parts of such quantities as, say, thrust. Whether the nonperturbative corrections are $\mathcal{O}(1/E)$ or $\mathcal{O}(1/E^2)$ is clearly a question of paramount importance for phenomenology. Using the infrared renormalons for counting the powers of $1/E$ in the processes without OPE is a totally fresh idea which can be considered as a distant spin off of the SVZ method, see Ref. [27] for a review. Previously the nonperturbative corrections were just ignored since nobody knew how to approach the issue scientifically.

Now, when the meaning of the condensates appearing in the Wilson expansion is hopefully clear, we can proceed to a brief discussion of their numerical values. In principle, two alternative approaches are conceivable: (i) calculating the condensates from first principles (e.g. on the lattices); (ii) extracting the condensates from the sum rules themselves. Attempts of the lattice calculation of the condensates were reported in the literature (for a review see Ref. [28]). The main difficulty, which, to a large extent, devaluates this work is the fact that the strategy of "isolating a nonperturbative residue from the perturbative background" was adopted. As we already know, this strategy is doomed to failure. Instead, one should have consistently implemented Wilson's procedure. This has never been done, however. Some initial ideas as to how Wilson's procedure can be implemented on the lattices are presented in Ref. [29].

Therefore at present, as twenty years ago, one has to exploit the sum rule themselves in order to determine the condensates. One picks up certain channels where sufficient experimental information on the corresponding spectral densities is available. These channels are sacrificed, i.e. the analysis goes in the direction opposite to conventional: the values of the condensates are extracted from the Euclidean correlation functions obtained through dispersion integrals. The first determination of the gluon condensate was carried out exactly in this way, from the sum rules in the J/ψ channel [30, 1]. The corresponding value will be referred to as standard,

$$\langle \mathcal{O}_G \rangle = 0.012\ \text{GeV}^4\,.$$

Since then this estimate was subject to multiple tests, sometimes with conflicting results. A brief discussion can be found in the beginning of Sec. 2 in Ref. [2]. The overall conclusion is that certain deviations from the standard value (say, at the level of $\sim 30\%$) are not ruled out.[8] One of the reasons limiting the accuracy is the fact that the gluon condensate was determined within a simplified (the so-called practical) version of the operator product expansion, which is strictly speaking somewhat ambiguous. We will dwell on the issue in Sec. 6. At the present level of understanding it is highly desirable to repeat the analysis within the framework of consistent Wilson's procedure. This has not been done so far. To take into account a certain degree of numerical uncertainty it is reasonable to accept that

$$\langle \mathcal{O}_G \rangle = 0.012 x_G \ \mathrm{GeV}^4 \,, \tag{52}$$

where a dimensionless numerical factor x_G is allowed to float in the vicinity of unity. This parametrization will be used below in the ρ-meson sum rule.

Let us pass now to the quark condensate $\langle \mathcal{O}_q \rangle$. As was mentioned, this condensate is the order parameter for the spontaneous breaking of the chiral symmetry. Hence, its value can be independently determined from the corresponding phenomenology. In particular, the celebrated Gell-Mann-Oakes-Renner formula relates the product of the light quark masses and $\langle \mathcal{O}_q \rangle$ to observable quantities,

$$(m_u + m_d)\langle \bar{u}u + \bar{d}d \rangle = -M_\pi^2 f_\pi^2 \,, \tag{53}$$

where f_π is the pion constant ($f_\pi \approx 133$ MeV) and M_π is the pion mass. The original estimate of $\langle \mathcal{O}_q \rangle$ used in Ref. [1] was obtained by substituting the crude estimates of the light quark masses that existed at that time [31, 32],

$$m_u + m_d \approx 11 \ \mathrm{MeV} \,.$$

In this way one arrives at

$$\langle \mathcal{O}_q \rangle = -(250 \ \mathrm{MeV})^3 \,. \tag{54}$$

Since then considerable work has been carried out to narrow down the theoretical uncertainty. First of all, chiral corrections to the Gell-Mann-Oakes-Renner

[8] I leave aside extremist, and to my mind unfounded statements in the literature, that the standard value underestimates the gluon condensate by a factor of 2 to 5.

formula were derived and analyzed. Moreover, much effort has been invested in perfecting our knowledge of the quark masses. The progress is summarized in Ref. [33], where an extensive list of references can be found. No dramatic changes occurred. The value of $m_u + m_d$ extracted from phenomenology of the chiral symmetry breaking went up by about 30%, with an error at the level of 15%. Thus, one is tempted to say that the actual value of the quark condensate is close to its "standard" value quoted in Eq. (54); perhaps, somewhat suppressed.[9]

Events took a dramatic turn in 1996, after results of the lattice calculations of the light quark masses (with two dynamical quarks) became known. The lattice result for $m_u + m_d$ lies significantly lower (by a factor of two to three, see e.g. Ref. [34]) than any of the reasonable analytic estimates! Were these results confirmed, this would mean that the quark condensate is significantly larger than the number given in Eq. (54).

I would caution against hasty conclusions at this point. The discrepancy between the analytic and lattice estimates may well be an artifact of the lattice procedure. Putting dynamical almost massless quarks on the lattice is a notoriously difficult task. On the other hand, the discrepancy, if persists, may prove to be a signature of a very interesting physical phenomenon — strong dependence of physical quantities on the number of light dynamical quarks. The surprisingly low lattice result for $m_u + m_d$ was obtained under the assumption that the number of light quarks is two, while in fact it is three. We will return to the issue of possible strong dependence on N_f in Sec. 12.

Under the circumstances it seems reasonable, analogously to the gluon condensate, to parametrize the quark condensate as

$$\langle \mathcal{O}_q \rangle = -(250\ \text{MeV})^3 x_q\,, \tag{55}$$

and let x_q vary around unity.

The full catalog of all relevant operators up to dimension six, was worked out in Ref. [1]. It is not as large as one might think at first sight. There is only one operator of dimension 5, the mixed quark–gluon operator

[9]The quark mass is definition dependent. Usually the $\overline{\text{MS}}$ scheme and a "reference" normalization point $\mu = 1$ GeV are implied. Not to confuse the reader we will follow this convention, in spite of shortcomings of the $\overline{\text{MS}}$ scheme. One cannot avoid explicitly introducing the normalization point μ in the case of the quark condensate: $\mathcal{O}_q$ has a nonvanishing anomalous (logarithmic) dimension. The $\bar{q}q$ vertex dressed by gluons is logarithmically suppressed. Including the gluons with off-shellness from M_0 down to μ suppresses the vertex by a factor $\propto [\alpha_s(\mu)/\alpha_s(M_0)]^{4/b}$. This formula explicitly demonstrates why μ cannot be put to zero in the case at hand: the logarithm explodes.

$$\mathcal{O}_{qG} = ig\bar{q}\sigma_{\mu\nu}G_{\mu\nu}q \tag{56}$$

where $G_{\mu\nu} \equiv G^a_{\mu\nu}T^a$ (T^a are the color generators). This operator plays no role in the ρ-meson sum rules, to be considered below: because of its "wrong" chirality it can enter only being multiplied by the quark mass m_q. It is very important, however, in a wide range of problems involving baryons [35], and mesons built from one light and one heavy quarks [36].

At the level of dimension 6, there is one operator built from three gluon field strength tensors (symbolically GGG) and several four-quark operators of the type

$$\mathcal{O}_{4q} = (\bar{q}_1\Gamma_1 q_2)(\bar{q}_3\Gamma_2 q_4) \tag{57}$$

where $\Gamma_{1,2}$ denote certain combinations of the Lorentz and color matrices and q_i ($i = 1, \ldots, 4$) are the light quark fields of different flavors, u, d and s. The overall flavor must be conserved, of course, but q_1 need not coincide with q_2 and so on.

The three-gluon operator is expected to have a significant impact in heavy quarkonium [37]; it does not appear in the ρ-meson sum rules, as was first shown in Ref. [38], by virtue of a very elegant theorem.

As for the four-quark operators, their vacuum matrix elements are not known independently. It became standard to evaluate them in the factorization approximation, which was first applied in this context in Ref. [1]. Note that using the Fierz identities for the color and Lorentz matrices one can always arrange the operators $\mathcal{O}_{4q}$ to have a natural color flow. By natural I mean that $\Gamma_{1,2}$ do not contain color matrices, and the color indices of quarks are contracted inside each bracket in Eq. (57). The color of q_1 flows to q_2 and that of q_3 flows to q_4. If the number of colors N_c is large, then the expectation value of $\mathcal{O}_{4q}$ would be totally saturated by the vacuum intermediate state,

$$\langle \mathcal{O}_{4q} \rangle = \langle \bar{q}_1\Gamma_1 q_2\rangle\langle \bar{q}_3\Gamma_2 q_4\rangle \,. \tag{58}$$

Corrections to this equation are formally of order of $1/N_c^2$. It is clear that in the factorization approximation, after the natural color flow is achieved by the Fierz rearrangement, the only surviving structure is that with $\Gamma_{1,2} = 1$.

In the real world $N_c = 3$, and we certainly do expect deviations from factorization at a certain level. As far as we can tell today, factorization works surprisingly well at least for those four-quark operators that appear in the ρ-meson sum rule. All attempts to detect deviations, both on the lattices [39] and in the sum rules themselves [40], gave results consistent with zero, within errors that typically lie in the 10% ballpark. This is quite nontrivial, considering the fact that quite a few examples are known where large numerical coefficients neutralize formal $1/N_c$ suppression factors. Whatever the reasons might be, we will accept Eq. (58) in what follows.

4. ρ Meson in QCD

The experience accumulated in Sec. 2 will now be applied for explorations in actual QCD. As we learned from the toy models, the polarization operator $\Pi(Q^2)$ (or its Borel transform $\tilde{\Pi}(M^2)$) in the deep Euclidean domain has an expansion in $\ln Q^2$ (perturbation theory) plus power terms $(1/Q^2)^k$ or $(1/M^2)^k$ (the condensates). If in the toy models the power terms (truncated in certain order) can be made as large or as small as we want, in QCD they are determined dynamically; through interactions of the valence quark pair with the soft vacuum fields. This interaction gives rise to the condensate expansion. The first power term is due to the gluon condensate, while the next one is associated with the four-quark condensate. In principle, higher condensates were analyzed in the literature too, but we will not discuss them in this lecture.

The quark–antiquark pair $q\bar{q}$ with the total energy $\sqrt{s}$ is injected in the QCD vacuum either by the virtual photon (e^+e^- annihilation) or W boson (τ decays). The $q\bar{q}$ pair, being injected, starts evolving according to the dynamical laws of QCD. At first the quarks do not feel the impact of the vacuum "medium". As separation between them grows, the effects of the medium become more and more important, so that eventually it prevents quarks from appearing in the detectors. The injected quarks get dressed and materialize themselves in the form of hadrons. In the sum rule approach we control only the beginning of this process.

In the sum rule framework, the condensates become important in the vicinity of the left edge of the window. To the left of the arrow A they explode, and theoretical control is lost. On the other hand, the value of $I(M^2)$ near the right edge of the window is determined mostly by ordinary perturbation theory. All perturbation theory sits in the unit operator. Technically it is convenient to formulate the corresponding calculation in terms of the perturbative corrections to the spectral density. We take the graph in Fig. 5, and attach to it various gluon and quark loops (e.g. Figs. 13 and 15). In this way we get the perturbative part of $\rho(s)$; at present in the vector–isovector channel it is known up to third order in $\alpha_s(s)$ [41]. In the $\overline{\text{MS}}$ scheme the result takes the form

$$\rho(s) = 1 + k_1 a + k_2 a^2 + k_3 a^3 \tag{59}$$

where $a(s)$ is given in Eq. (3). In the low-energy domain the number of active flavors is $N_f = 3$. For $N_f = N_c = 3$ the parameter $b = 9$, and the values of k_i are

$$k_1 = \frac{4}{9}, \qquad k_2 = 0.729 k_1, \qquad k_3 = -2.03 k_1. \tag{60}$$

Equation (59) can be immediately translated into the prediction for the perturbative part of $I(M^2)$. The perturbative expansions for $I(M^2)$ and $\rho(s)$ do not coincide. To get the former, one must integrate the latter with the exponential weight function. The transition is readily carried out with the aid of the formula [42]

$$\int \frac{ds}{M^2} \frac{\exp(-s/M^2)}{\left(\ln \frac{s}{\Lambda^2}\right)^\nu} = \frac{1}{\left(\ln \frac{M^2}{\Lambda^2 e^\gamma}\right)^\nu} \left[1 + \frac{\nu(\nu+1)}{2} \frac{\pi^2/6}{\left(\ln \frac{M^2}{\Lambda^2 e^\gamma}\right)^2} + \mathcal{O}\left(\frac{1}{\left(\ln \frac{M^2}{\Lambda^2 e^\gamma}\right)^4} \right) \right], \tag{61}$$

where γ denotes the Euler constant, $\gamma = 0.577$. By expanding this expression near integer values of ν one also gets necessary expressions for the integrals containing logarithm of logarithm (ln ln).

In this way we arrive at [42]

$$I_{\rm pc}(M^2) = 1 + k_1 a\left(\frac{M^2}{e^\gamma}\right) + k_2 \left[a\left(\frac{M^2}{e^\gamma}\right)\right]^2 + \left(k_3 + \frac{\pi^2}{6} k_1\right) \left[a\left(\frac{M^2}{e^\gamma}\right)\right]^3. \tag{62}$$

Similar expressions were obtained in Ref. [43], where $a(M^2/e^\gamma)$ was re-expressed in terms of $a(M^2)$. It is more convenient to work directly with the expansion (62).

The quark mass corrections in $\rho(s)$ and $I(M^2)$ are also known, they are proportional to $m_{u,d}^2/M^2$ and are negligibly small due to the smallness of the current quark masses. They will be discarded since the corresponding uncertainty is invisible in the background of other uncertainties.

Now we return to the condensates. The modern perturbative calculations of the coefficients C_n are based on the background field technique. This is an important aspect of the SVZ approach as it exists today. The method is quite elegant, to say nothing of being much more economic than the original "brute force" calculations in Ref. [1]. We shall not discuss the technique *per se*, however. Conceptually everything is clear here. If you need to do a calculation, you have to just go and learn the corresponding technique using, for instance, the review paper [44].

The coefficients C_G and C_{4q}, to the leading order, are known from the ancient times [1],

$$I_{\rm npc}(M^2) = 1 + \frac{\pi^2}{3M^4}\langle \mathcal{O}_G\rangle - \frac{8\pi^3}{M^6}\langle \alpha_s(\bar{u}\gamma_\alpha\gamma_5 T^a u - \bar{d}\gamma_\alpha\gamma_5 T^a d)^2\rangle$$
$$- \frac{16\pi^3}{9M^6}\left\langle \alpha_s\left(\bar{u}\gamma_\alpha T^a u + \bar{d}\gamma_\alpha T^a d\right)\sum_q (\bar{q}\gamma_\alpha T^a q)\right\rangle . \tag{63}$$

The four-quark structures appearing in the angle brackets, as well as the factor α_s in front of them, are normalized at Q. The anomalous (logarithmic) dimension of this particular combination nearly vanishes [1]; we shall ignore this in evolving the four-quark operators in Eq. (63) down to μ. Applying factorization, as was explained in Sec. 3, we get for the four-quark term in $I_{\rm npc}(M^2)$

$$C_{4q}\langle\mathcal{O}_{4q}\rangle = -\frac{448\pi^3}{81M^6}\alpha_s(\mu)\langle\bar{q}q(\mu)\rangle^2 \to -\frac{1}{M^6}0.03x_{4q}\ \mathrm{GeV}^6\,, \tag{64}$$

where, as previously, we have introduced a factor x_{4q} to allow for the possible theoretical uncertainties. Under the standard choice of parameters [1] this factor is equal to unity. If we are on the right track it is reasonable to expect that x_{4q} is close to unity.

A huge amount of work was carried out in calculating two-loop corrections in the coefficients of various condensates. The results can be inferred, say, from Ref. [45] or from the reviews [46]. Unfortunately, the existing uncertainty in the condensates themselves precludes us from taking advantage of the precision achieved in perturbative corrections. Therefore, we will limit ourselves to the leading order results for C_G and C_{4q}.

Now, the stage is set and we can finally present the sum rule for $I(M^2)$,

$$I(M^2) = 1 + \frac{4}{9}a\left(\frac{M^2}{e^\gamma}\right)\left\{1 + 0.729\left[a\left(\frac{M^2}{e^\gamma}\right)\right] - 0.386\left[a\left(\frac{M^2}{e^\gamma}\right)\right]^2\right\}$$
$$+ 0.1x_G\left(\frac{0.6}{M^2}\right)^2 - 0.14x_{4q}\left(\frac{0.6}{M^2}\right)^3 , \tag{65}$$

where M in the second line of Eq. (65) is measured in GeV. The term M^{-4} is due to the gluon condensate, while the term M^{-6} is due to the quark

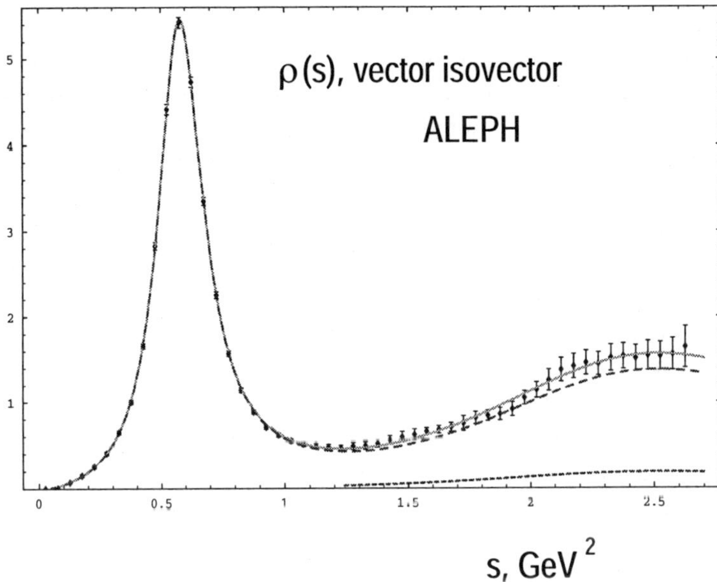

Fig. 16. The spectral density in the vector–isovector channel measured in τ decays. The data points belong to ALEPH. The solid curve is a best fit. The dashed curve is the best fit minus $KK\pi$ (the latter contribution is shown by the dotted curve) which, I think, should not have been included in the vector spectral density, see text.

condensate. With the "standard" numerical values of the gluon and quark condensates in the the nonperturbative part $x_G = x_{4q} = 1$; the dimensionless constants x_G and x_{4q} allow the gluon and four-quark condensates to "breathe" to a certain degree. For instance, $x_G > 1$ would imply that the actual value of the gluon condensate is larger than the "standard" value, and so on.

We are ready to examine the situation in the ρ-meson channel. Figure 16 shows experimental data on the spectral density in the vector–isovector channel, measured in the τ decays. The data points are from Ref. [47]. Note a remarkably close resemblance with our toy model number two. If we take the beginning in Fig. 8 and expand it to make the energy scales coincide[10] the curves will essentially repeat each other. There exist a few extra points in the τ decays, spanning the interval of s from 2.7 to 3 GeV^2, and some data above 3 GeV^2 from the e^+e^- annihilation, but the corresponding error bars are so large that plotting these points would just obscure the picture.

[10]Warning: the energy scale (the horizontal axis) in Fig. 8 presents s in the units of m_ρ^2 while that in Fig. 13 is in GeV^2.

There is a subtle point which I need to discuss here. Directly measurable in the hadronic τ decays is the sum of the vector and axial spectral densities. To obtain the spectral densities separately one has to sort out all decays by assigning specific quantum numbers to each given final hadronic state. In the majority of cases such an assignment is unambiguous. For instance, two pions (whose contribution is the largest) can be produced only by the vector current. Some processes, however, can occur in both channels, for instance, the $KK\pi$ production. Using certain theoretical arguments it was decided in Ref. [47] that around 3/4 of all $KK\pi$ yield should be ascribed to the vector channel. Other theoretical arguments [42], which seem to be much more convincing to me, say that virtually all $KK\pi$ production should take place in the axial channel. Therefore, in dealing with the data, I will subtract the $KK\pi$ yield from the ALEPH data points. I hasten to add that the subtracted quantity is small (see Fig. 16), and this subtraction is not very essential for a general picture I draw here by broad touches.

The solid curve in Fig. 16 is a best fit by a smooth curve representing the sum of two Breit-Wigner peaks (the first one is a modified Breit-Wigner taking into account threshold effects important for the ρ meson). The dashed curve is what I believe the actual spectral density is. The data points above 1 GeV^2 carry a noticeable uncertainty. Since we will be using the spectral density only in the integrals, where many data points are summed over, these individual errors can be neglected, since they will be statistically insignificant in the integrals. Systematic uncertainty might be important, but since nobody knows how to estimate it, I will ignore it for the time being. For our limited purposes we can consider the dashed curve in Fig. 16 as an exact experimental result for the spectral density.

How would the experimental spectral density look above 2.7 GeV^2? This question is irrelevant for calculating the ρ-meson parameters. It is very instructive, however, to have a broader perspective. I will show you what I call a "theoretical experimental" spectral density, and then will make a short digression explaining where the answer comes from.

My expectations are summarized in Fig. 17. The curve to the left of the arrow is just the fit to the experimental data, as explained above (it identically coincides with the dashed curve in Fig. 16). To the right of the arrow I have attached a "tail" which approaches a smooth asymptotic prediction for $\rho(s)$ (the dashed line) given in Eq. (59), in an oscillating manner. The tail and the experimental data are smoothly matched at 2.65 GeV^2.

Why do I think that the actual spectral density, when and if it is measured, will approach the smooth asymptotic curve with oscillations rather than monotonously? The reason is of a very general nature [16]. In a nut shell, there are exponential terms that are not seen in the truncated OPE series for $\Pi(Q^2)$ in the Euclidean domain. When these terms are analytically continued to the Minkowski domain, where $\mathrm{Im}\,\Pi(s)$ belong, they show up as oscillating terms. The smooth asymptotic prediction is obtained from the truncated series and, thus, carries no traces of the exponential/oscillating terms. We will return to the issue later on (see Sec. 5).

Although the very existence of oscillations in the spectral densities may be considered firmly established, the question of how rapidly they die off is highly nontrivial and has no unambiguous answer in today's theory. In resonance-inspired models the damping factor is $\sim \exp(-\text{const.}\, s)$ (see Sec. 2.2), in the instanton-based models [48] it is $\sim \exp(-\text{const.}\, \sqrt{s})$. The tail added in Fig. 17 is taken to be proportional to $\exp(-\pi\sqrt{s})\cos(\pi s + \text{const.})$. This rather eclectic formula is chosen for sophisticated reasons which need not concern us here. The task which we address is illustrative, anyway. I certainly cannot guarantee that

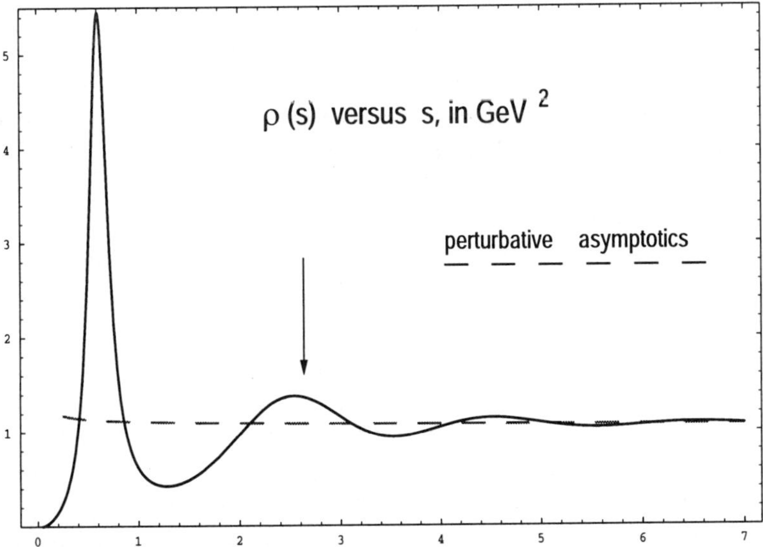

Fig. 17. "Theoretical experimental" spectral density in the ρ-meson channel. The solid curve presents experiment (see text), while the dashed curve is the perturbative result for $\rho(s)$, including terms up to $O(\alpha_s^3)$, with $\Lambda = 0.2$ GeV.

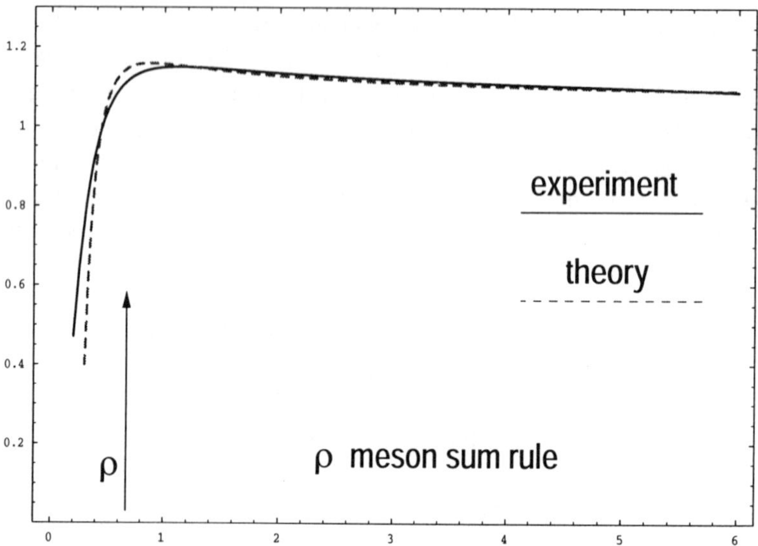

Fig. 18. I versus M^2 (in GeV^2) in the ρ-meson channel: confronting experiment and theory. The theoretical curve corresponds to $x_G = 0.8$, $x_{4q} = 1.3$ and $\Lambda = 0.2$ GeV.

around 3.5 GeV^2 the value of $\rho(s)$ is 0.97, as it is shown in Fig. 17, but I could bet that a shortage of the spectral density in this domain will be observed in precision measurements, so that $\rho(s)$ at 3.5 GeV^2 will lie somewhere around unity. I emphasize again that the precise form of the tail does not affect the sum rule calculation of the ρ-meson parameters, and is discussed only for completeness of the picture.

With the experimental spectral density in hands, we can calculate the "experimental" value of the integral $I(M^2)$. The corresponding result is presented in Fig. 18 by the solid curve.

The agreement is excellent; it is even better than one would expect *a priori*. We see that the conspicuous ρ peak and all further twiddles characteristic to $\rho(s)$ are washed out. If we descend from larger to smaller values of M, the behavior of $I(M)$ is flat down to $M \sim 0.8$ GeV, i.e. down to the ρ-meson mass. At $M \sim 0.8$ GeV the regime smoothly changes, the curve dives down, and at still lower values of M, approaches the exponential asymptotics (this domain is not shown in the figure). The boundary value of M where the regime changes is correlated with the mass of the lowest state, the ρ meson in the case at hand. This is practically the only characteristic dimensional parameter in $I(M)$.

Let us pretend now that we do not know the spectral density and want to use the sum rules to determine the ρ-meson mass. It is quite clear that the truncated condensate series does not allow one to go to the limit $M \to 0$ where this determination would be exact: the series explodes. We must stay inside the window where the expansion is still under control. Correspondingly, our determination of M_ρ is going to be approximate. The strategy can be formulated as follows. Let us assume that somebody shows us a sketch of the spectral density presented in Fig. 10 with the numbers along the horizontal and vertical axes erased. This sketch is known to correctly reproduce the basic features of the actual spectral density but leaves us ignorant as to where the ρ-meson peak lies and what is its height. We insert this sketch in our sum rule, do the integral

$$\frac{1}{\pi M^2} \int ds \rho(s) e^{-s/M^2} ,$$

fit the numbers in such a way as to be as close as possible to the theoretical prediction (65) inside the window — click, click — the scales are restored, and there come out the ρ mass and the coupling constant. Since inside the window the ρ meson saturates the integral at the level $\sim 90\%$, the fact that our continuum model is a caricature (minor details are lost) is unimportant. Even if we are off by a factor of two in this model this will affect our estimates referring to the ρ meson at the level of $\sim 10\%$.

This example is quite typical. A similar situation takes place for all classical low-lying hadrons: those built from the light quarks, heavy, and light and heavy. I do not define here precisely what the "classical meson" means but will return to this point later, after considering some nonclassical channels. I must admit that the ρ meson is an example where the SVZ method demonstrates its best facets. The window is sufficiently broad, everything is clean. As we proceed to higher spins, the mesons become larger in size. A snapshot of a high-spin meson would show a string-like picture, with a longitudinal size of the "sausage" much larger than its transverse size. Under the circumstances one could hardly expect that the SVZ method would work. And, indeed, it ceases to be informative for spins higher than two [49].

5. Basic Theoretical Instrument —Wilson's OPE

The theoretical basis of any calculation within the SVZ method is the operator product expansion (OPE). It allows one to consistently define and build the (truncated) condensate series for any amplitude of interest in the Euclidean domain. The physical picture lying behind OPE was described above: consistent separation of short and large distance contributions. The former are then represented by the vacuum condensates while the latter are accounted for in the coefficient functions. Thus, the operator product expansion in the form engineered by Wilson [50] is nothing but a book-keeping procedure. Wilson's idea was adapted to the QCD environment in Ref. [18].

Although OPE is used in an innumerable amount of works since the mid-1970's, there are no good textbooks or reviews devoted to this issue. Moreover, one typically encounters a lot of confusion in the literature. Below I will try to explain the Wilsonean procedure using, as an example, the T product of the two vector currents defined in Eq. (7).

Technically it is more convenient to deal with the D function (41) rather than $\Pi_{\mu\nu}$,

$$D(q^2) = \sum_n C_n(q;\mu)\langle \mathcal{O}_n(\mu)\rangle \tag{66}$$

where the normalization point μ is indicated explicitly. The sum in Eq. (66) runs over all possible Lorentz and gauge invariant local operators built from the gluon and quark fields. The operator of the lowest (zero) dimension is the unit operator $\mathbf{I}$, followed by the gluon condensate $G^2_{\mu\nu}$, of dimension 4. The four-quark condensate gives an example of dimension-6 operators.

At short distances QCD is well approximated by perturbation theory. The coefficient functions C_n absorb the short-distance contributions. Therefore, as a first approximation, it is reasonable to calculate them perturbatively, in the form of expansion in $\alpha_s(Q) \sim (\ln Q)^{-1}$. This certainly does not mean that the coefficients C_n are free from nonperturbative, nonlogarithmic terms of the type $\sim Q^{-\gamma}$ where γ is a positive number.

To substantiate the point let us consider the coefficient of the unit operator. The Feynman graphs for $D(q^2)$ of the lowest order are depicted in Figs. 5 and 13. Assume that the momentum flowing through the graph is Euclidean and large, $Q \to \infty$. If all subgraphs in these and similar graphs are renormalized at Q, the final result, being expressed in terms of the running coupling constant $\alpha_s(Q)$, is finite. The virtual momenta saturating the loop integrals scale with Q as the first power of Q. At any finite order the perturbative series is well-defined.

At the same time, if the number of loops n becomes very large, in a random graph (Fig. 19) each loop carries momentum of the same order of magnitude, and the total momentum Q is shared between many lines, so that the

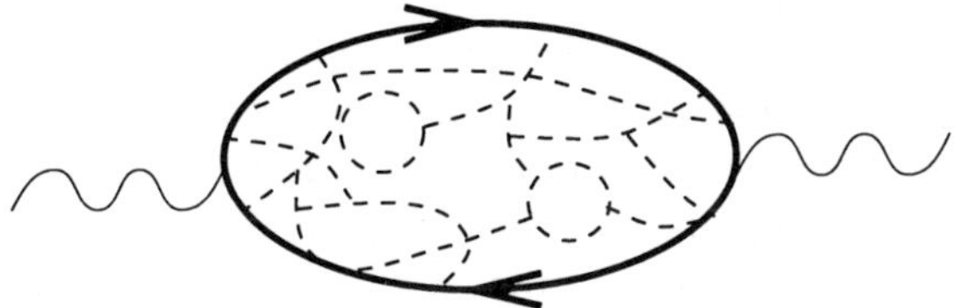

Fig. 19. A typical multiloop graph contributing to the D function. The number of such graphs grows factorially with the number of loops.

characteristic virtual momentum is proportional to Q/n. The contribution of each graph to C_I is of order $[\alpha_s(Q)]^n$, but the number of distinct diagrams ν grows factorially, $\nu \sim n!$, so that the expansion in $\alpha_s(Q)$ is asymptotic (for an exhaustive discussion of this phenomenon see Ref. [51]).

Note that this factorial divergence has nothing to do with the renormalon divergence considered in Sec. 3. There, we had one specific graph of the nth order whose contribution was proportional to $n!$. This factorial was a spurious artifact of an improper calculation where the domain $k \sim \Lambda$ was included. Introducing a lower cut-off at μ = several units times Λ and discarding the contribution coming from $k^2 < \mu^2$, as it should be done in the consistent calculation of $C_I(\mu)$, we would eliminate the renormalon divergence. At the same time, the factor $n!$ counting the number of graphs of the nth order certainly cannot be eliminated in this way.

If n is not too large, so that $Q/n > \mu$, the graphs of Fig. 19 present a legitimate contribution to $C_I(\mu)$. It is quite clear that the tail of the α_s series with $n \gtrsim 1/\alpha_s(Q) \sim \ln Q$ generates a nonperturbative contribution of the type

$$\Delta C_I \sim \exp\left(-2\pi\gamma/\alpha_s(Q)\right) \sim \left(\frac{\kappa\Lambda}{Q}\right)^{b\gamma} . \tag{67}$$

Here γ and κ are numerical constants. An explicit example of such terms is provided by the so-called direct instantons [3]. Integration over the sizes ρ of the direct instantons is saturated at $\rho \sim Q^{-1}$. The constant $\gamma = 1$ for one instanton, 2 for two and so on.

The correspondence between the Feynman graphs with $1/\alpha_s(Q)$ loops and the small-size instantons is not straightforward. I will not go further in this issue, referring the interested reader to Ref. [51]. At the qualitative level the inevitability of occurrence of the hard nonperturbative terms (67) is evident. At the moment the only semiquantitative framework that we have at our disposal for their evaluation is the instanton mechanism.

Equation (67) gives a noncondensate power term. Generically we will call such terms *hard nonperturbative.* They may or may not be important numerically depending on the particular value of Q under consideration. The value of $b\gamma$ need not be integer, generally speaking; numerically it is very large. This means that once the boundary is crossed where the argument in the brackets in Eq. (67) becomes less than one, $Q > \kappa\Lambda$, these terms immediately become totally unimportant. On the other hand, below this boundary they are so large that no expansion is possible.

It is assumed that inside the window the terms (67) are unimportant; they are completely disregarded in the *practical version* of OPE (Sec. 6) which, thus, applies to the values of Q above the critical point. Note that the large value of $b\gamma$ results in a large degree of suppression of such terms after borelization.

Although the assumption above works very well in many "classical" channels, it seems to fail in exceptional cases of "nonclassical" mesons. This issue will be addressed in more detail in Sec. 8.

I pause here to make a remark concerning nondynamical power terms in $C_I(\mu)$ whose origin is related to the introduction of the normalization point μ. Let us return to the one-loop graph of Fig. 13 and the corresponding expression (42). One should not forget that in doing the loop integrations in $C_I(\mu)$ we *must* discard the domain of virtual momenta below μ, by definition. Subtracting this domain from the perturbative loop integrals we introduce power corrections of the type $(\mu^2/Q^2)^n$ in $C_I(\mu)$, by hand. For instance, in the first order in α_s

$$D(Q^2) = 1 + \left[\frac{\alpha_s(Q)}{\pi} - \frac{\alpha_s}{\pi}\frac{\mu^4}{Q^4}\right] + \frac{2\pi^2}{3Q^4}\langle \mathcal{O}_G(\mu)\rangle + \dots , \tag{68}$$

where I combined Eqs. (48) and (51). The expression in the square brackets is the α_s part of the coefficient $C_I(\mu)$. The explicit μ dependence of this expression conspires with an implicit μ dependence residing in $\langle \mathcal{O}_G(\mu)\rangle$ to ensure the μ independence of the physical quantity (i.e. the D function). If higher-order terms in the k^2 expansion of the F function (43) were taken into account, we would get higher powers of μ/Q in the square brackets, which then should have been combined with higher condensates, e.g. GD^2G and so on.

If μ can be chosen sufficiently low, the explicit and implicit μ dependent terms in Eq. (68) may be numerically insignificant and can be ignored.

The coefficient functions that are in front of other operators generally speaking have the same structure as C_I. The general situation is quite similar.

For higher dimensional operators it may so happen (and, in fact, happens) that some of the loop integrations are not saturated at Q or μ but are, rather, logarithmic. Logarithms $[\ln(Q/\mu)]^\gamma$ occurring in this way are associated with the anomalous dimensions of the operators at hand.

Let us take a closer look at the graph depicted in Fig. 13. As we already know, some of the lines can be in a special regime — the corresponding virtual momenta are soft and their off-shellness does not scale with Q. If the gluon line is soft, this gives rise to the gluon condensate, if some of the quark lines are soft we deal with the four-quark condensate. Graphically one distinguishes the soft lines by cutting them; then the remainder of the graph shrinks to a point while those lines that are cut form a local operator. The question is: what happens if we pass to the multiloop graph in Fig. 19 and start cutting more and more lines?

Naturally, in the power series we proceed to the operators of higher and higher dimensions. When the number of the soft lines becomes very large (of order of Q/Λ) it is conceivable that there are no hard lines at all: the external momentum Q is transferred from the initial to the final vertex through a very large number of quanta (growing like a power of Q), and none of the quanta carries momentum scaling with Q. Of course, in this situation one cannot speak of individual quanta, one should rather use the language of typical field fluctuations transmitting the momentum Q without having any Fourier components with frequencies of order Q. Contributions of this type are not seen in OPE truncated at any finite order. They correspond to the high-order tail of the condensate series which, analogously to the α_s series, is factorially divergent. In this way we obtain the so-called *exponential terms* [16].

The fact that the condensate series is factorially divergent in high orders is rather obvious from the analytic structure of the polarization operator $\Pi(Q^2)$. In a nut shell, since the cut in $\Pi(Q^2)$ runs all the way to infinity along the positive real semiaxis of q^2, the $1/Q^2$ expansion cannot be convergent. The actual argument is more subtle than that, but I would not like to go into details here, referring the reader to Ref. [16] where a careful consideration of the issue is carried out. The final conclusion is perfectly transparent. It is intuitively clear that the high-order tail of the (divergent) power series gives rise to exponentially small corrections $\sim \exp(-Q^\sigma)$ where σ is some critical index.

The numerical value of σ is correlated with the rate of divergence of high orders in the power series. At the moment very little is known about this rate from first principles, if at all. The best we can do is to rely on toy models.

The simplest example is again provided by instantons. This time one has to fix the size of the instanton ρ by hand, $\rho = \rho_0$. Then the fixed-size instanton contribution is $\mathcal{O}(\exp(-Q\rho_0))$. The exponential factor is the price we pay for transmitting the large momentum Q through a soft field configuration whose characteristic frequencies are of order ρ_0^{-1}.

One can look at this example from a broader perspective. Consider the polarization operator Π in the coordinate rather than the momentum space. If the regular terms in OPE are in one-to-one correspondence with singularities of $\Pi(x)$ at $x = 0$, the exponential terms, invisible in the truncated OPE, are related to the singularities of $\Pi(x)$ located at a finite distance from the origin [52]; the distance from the origin plays the same role as the instanton radius ρ_0. For our illustrative purposes it is sufficient to consider the singularities of the simplest possible structure, namely,

$$\frac{1}{x^2+\rho_0^2}\,, \qquad \ln\left(x^2+\rho_0^2\right), \qquad \left(x^2+\rho_0^2\right)\ln\left(x^2+\rho_0^2\right), \qquad \text{and so on}\,. \tag{69}$$

The Fourier transforms of these expressions have the generic form

$$(Q\rho_0)^{-n}K_n(Q\rho_0)\,, \qquad n = 1, 2, \ldots \tag{70}$$

where K_n is the McDonald function. At large Euclidean Q^2 the corresponding contribution dies off exponentially,

$$\Delta\Pi = (Q\rho_0)^{-n}K_n(Q\rho_0) \propto (Q\rho_0)^{-n-1/2}e^{-Q\rho_0}\,, \tag{71}$$

in full accord with the intuition regarding transmitting a large momentum through a soft field fluctuation.

The terms that are exponentially small for Euclidean values of Q^2 become oscillating upon the analytic continuation to the Minkowski domain, $Q \to iQ$. Analytically continuing $\Delta\Pi$ given in Eq. (71) we arrive at

$$\begin{aligned} \operatorname{Im}\Delta\Pi &= (-1)^{n+1}\frac{\pi}{2}(\sqrt{s}\rho_0)^{-n}J_n(\sqrt{s}\rho_0) \\ &\propto (-1)^{n+1}\frac{\pi}{2}(\sqrt{s}\rho_0)^{-n-1/2}\cos(\sqrt{s}\rho_0-\delta_n)\,, \end{aligned} \tag{72}$$

where J_n is the Bessel function. The oscillating character at $q^2 > 0$ is the most important signature of the exponential terms representing the high-order tail of the power series.

Having fixed ρ_0 is unrealistic. We would do much better than that by allowing ρ_0 to vary around a typical value of order of Λ^{-1}. In other words, any reasonable model will introduce a distribution over ρ_0, with a weight function $w(\rho_0)$ peaked at Λ^{-1},

$$\langle \mathrm{Im}\,\Delta\Pi\rangle \propto \int (\sqrt{s}\rho_0)^{-n} J_n(\sqrt{s}\rho_0) w(\rho_0) \frac{d\rho_0}{\rho_0} \tag{73}$$

where the angle brackets denote ρ_0 smearing. The weight function $w(\rho_0)$ is model-dependent. Particular details of the result will certainly depend on $w(\rho_0)$, but the general pattern — the exponential fall-off of $\mathrm{Im}\,\Delta\Pi$ modulated by oscillations — remains the same under any reasonable choice of $w(\rho_0)$. This is the reason why the experimental spectral density in Sec. 4 is extrapolated beyond the range of the direct measurements in an oscillating mode (Fig. 17). The exponential suppression factor at large E takes the form

$$\exp(-kE^{\sigma}), \tag{74}$$

where the critical index σ in various models lies in the interval $0 < \sigma < 2$.

It remains to be added that the exponential terms are usually ignored in the sum rule analyses one encounters with in practical applications.

6. Practical Version of OPE

In practical calculations one can hardly go beyond several lowest-order terms in the α_s expansion of the coefficient functions and in the condensate expansion. For instance, in the ρ-meson channel, which is the most advanced, four terms of the perturbative series are known in the unit operator.

Thus, the approximation we make at the very first stage of building OPE is truncating both the perturbative and condensate series. By doing so we expect that those few terms that are kept, represent $D(Q^2)$ or $I(M^2)$ inside the window with sufficient accuracy.

Besides truncating the series, one usually ignores the hard nonperturbative terms in the coefficient functions. The reason why these terms are neglected is quite obvious: no framework allowing one to reliably calculate them has been worked out so far. From the instanton studies we know [3] that such terms must have an extremely steep M dependence, so that above a critical value of M their impact is expected to be totally negligible. It is assumed that the critical value of M, where the hard nonperturbative terms die off, lies to the left of the window.

This is not the end of the story, however. The third key element inherent to the practical OPE is a simplified treatment of the μ dependence in the coefficient functions and the condensates. The perturbative parts of the coefficient functions are often borrowed from earlier calculations of relevant Feynman graphs which, sometimes, date back to QED. These calculations make no distinction between small and large virtual momenta. Moreover, in analytic QCD it is technically not always easy to explicitly carry out the Wilsonean separation of virtual momenta (higher than μ and lower than μ). At the very least, such a separation requires dedicated analyses.

This is an obvious task for the future. A simplified strategy of the present times is as follows. Assume that we do not plan to go beyond the one-loop correction in Fig. 13 in the perturbative part of C_I and beyond the leading-order result for C_G (and higher condensates). The corresponding proper expression for the D function is

$$D = 1 + \alpha_s \int_{\mu^2}^{\infty} dk^2 F(k^2, Q^2) + \frac{\pi^3 F'}{3} \langle O_G(\mu) \rangle \tag{75}$$

(cf. Eq. (68)). Here F' is dF/dk^2 at $k^2 = 0$. The dimension-6 condensate GD^2G enters with the coefficient proportional to F'' at $k^2 = 0$, and so on.

Now, let us add and subtract to the right-hand side the integral

$$\alpha_s \int_0^{\mu^2} dk^2 F(k^2, Q^2)\,. \tag{76}$$

This integral looks exactly as the one-gluon contribution naively continued below μ^2. Of course, below μ^2 the gluon propagator has nothing to do with $1/k^2$, but since the transformation is identical, we are doing nothing wrong. Equation (75) then takes the form

$$D = 1 + \alpha_s \int_0^{\infty} dk^2 F(k^2, Q^2) + \frac{\pi^3 F'}{3} \left[\langle O_G(\mu) \rangle - \frac{3\alpha_s}{\pi^3} \int_0^{\mu^2} k^2 dk^2 \right] \tag{77}$$

where in the square brackets, I expanded the integrand in k^2 and kept only the leading term. The next term proportional to k^4 would be important at the level of dimension-6 condensate, etc.

The first integral on the right-hand side is the full perturbative one gluon correction to C_I, with no Wilsonean separation (it is equal to α_s/π). The term in the square brackets can be called *an effective one-loop* gluon condensate. It is obtained by subtracting from the genuine condensate its perturbative one-loop expression. By construction, the μ dependence of the effective one-loop gluon condensate cancels, provided we do not go beyond the accuracy specified above.

Adding the integral (76) to the right-hand side we achieve the desired goal. Now the coefficient C_I, to order $\mathcal{O}(\alpha_s)$, is given by full perturbative graphs in Fig. 13. To avoid double counting we subtract the same contribution from the condensates. More exactly, we subtract *almost* the same contribution. Since the sum over condensates is truncated at some finite order n_0, the best we can do is to subtract from the condensate part the integral

$$\alpha_s \int_0^{\mu^2} dk^2 \sum_{n=1}^{n_0} \frac{1}{n!} F^{(n)}(k^2 = 0, Q^2)(k^2)^n \,. \tag{78}$$

In this way we subtract from each condensate its "one-loop perturbative value".

It is quite clear that this strategy of converting the genuine Wilson OPE into the practical version is admittedly approximate and works only if we limit ourselves to a given number of loops and a given number of terms in the condensate expansion. The effective condensates obtained in this way should be supplemented by a superscript indicating the number of loops in the subtracted part. Their numerical value is μ independent but does depend on the number of loops.

This approximate procedure (i) leads to a loss of factorization of short and large distance contributions inherent to Wilson's OPE; (ii) cannot be systematically generalized to arbitrary number of loops since there is no way to unambiguously define the integrand to all orders in the small k^2 domain;[11] (iii) only approximately avoids double counting due to the necessity of truncating the condensate series at a finite order (the lower the order we truncate, the larger the error). Moreover, the effective condensates are not universal, strictly speaking; they become process-dependent. If this nonuniversality were numerically significant, the SVZ method, as it exists now, would be undermined.

Therefore, the practical version is useful in applications only provided μ^2 can be made small enough to ensure that the "one-loop perturbative" contributions to the condensates are much smaller than their genuine values and, at the same time, $\alpha_s(\mu^2)/\pi$ is small enough for the expansion to make sense. The existence of such "μ^2 window" is not granted *a priori* and is a very fortunate feature of QCD. We do observe this feature empirically although the roots of the phenomenon remain unclear.

[11]Within practical OPE we automatically forbid questions to ourselves concerning all aspects of the high-order behavior.

Although the origin of the μ^2 window, ensuring the applicability of practical OPE[12] is not understood, one can find similar situations in simplified settings. The best example of which I am aware is the two-dimensional $O(N)$ sigma model in the limit of large N. In this limit, the model is exactly solvable; therefore any question of interest can be exhaustively answered. The model bears remarkable parallels to QCD — it is asymptotically free, and the mass scale is generated dynamically. A surprising result was established some time ago [53]. In the large N limit all vacuum condensates in this model are μ independent, and the practical version of OPE becomes exact; μ dependence in the condensates and in the coefficient functions comes only at the level of $1/N$. If N is large these terms are parametrically small. (It would be great to find a hidden parameter which would explain the suppression of the μ^2/Q^2 terms in QCD in the same way $1/N$ explains it in the sigma model.)

It is often asked whether it is legitimate to retain logarithmic and power terms simultaneously in the truncated expansions.[13] Indeed, formally any $1/(\ln Q)^n$ term is parametrically larger that, say, $1/Q^4$ (moreover, any power term is parametrically larger than the exponential ones). If so, the theoretical uncertainty due to truncation of the logarithmic series is formally larger than even the lowest-order condensate correction.

The logarithmic and power terms in practical OPE have distinct physical nature. In some instances power terms describe effects that do not show up in perturbation theory. This is valid in all cases where the chiral quark condensate is involved. In other cases, when the power terms are not different in their structure from those occurring in perturbation theory, there is a strong numeric enhancement of the power terms (see Sec. 3). This mysterious fact — the numerical enhancement of the vacuum condensates — explains why Wilson's OPE can be substituted in QCD by the practical version, at least in the classical channels.

In the early days of QCD, in 1970's, it was common to assume that the coefficient functions in the operator product expansion are saturated by perturbation theory, with no separation of virtual momenta. Although the difference between Wilson's OPE and the practical version was realized long ago [18], very little has been done to investigate this difference and perfect the procedure.

---

[12] A natural abbreviation for practical OPE would be POPE, but I will not risk to put this abbreviation into circulation.

[13] In discussing the issue of duality violations one deals with the logarithmic, power, and exponential terms simultaneously, see Ref. [48].

Only now are we witnessing attempts in this direction, mostly in connection with the heavy quark theory (e.g. Refs. [19, 48, 54]).

7. Low-Energy Theorems

Although the low-energy theorems are independent of the sum rules, at least some of them were derived in connection with the development of the SVZ method, and were later combined with the sum rules giving rise to an intriguing observation of the hadronic nonuniversality. This observation is the topic of the next section. Here we will dwell on the low-energy theorems related to the trace anomaly of QCD.

In the limit of massless quarks (and we will never leave the limit $m_q = 0$) the classical Lagrangian of QCD is free from dimensional parameters. The scale parameter appears at the quantum level,

$$\Lambda = M_0 \exp\left(-\frac{8\pi^2}{bg_0^2}\right) \tag{79}$$

where M_0 is the ultraviolet cut off, g_0 is the bare coupling constant, and I ignored the second- and higher-order terms in the β function. (The corresponding modifications are quite trivial and are suggested to the reader as an exercise.)

The occurrence of the scale parameter (79) is in one-to-one correspondence with the trace anomaly (39). Let us denote the trace operator

$$\theta_\mu^\mu \equiv \sigma\,. \tag{80}$$

Then, it is not difficult to show [3] that for arbitrary local operator $\mathcal{O}$

$$\lim_{q\to 0}\left\{i\int d^4x e^{iqx}\langle T\{\mathcal{O}(0),\sigma(x)\rangle_c\right\} = -d\langle\mathcal{O}\rangle\,, \tag{81}$$

where d is the (normal) dimension of the operator $\mathcal{O}$, and the subscript c denotes the connected part (it will be omitted hereafter). If

$$\mathcal{O} = \sigma \propto G^2\,,$$

then the dimension $d = 4$, so that

$$i\int e^{iqx}d^4x\left\langle T\left\{\frac{-b\alpha_s}{8\pi}\,G^2(x),\,\frac{-b\alpha_s}{8\pi}G^2(0)\right\}\right\rangle = (-4)\left\langle\frac{-b\alpha_s}{8\pi}G^2\right\rangle \text{ at } q^2 = 0\,. \tag{82}$$

For the quark operator $\bar{q}q$ the dimension $d = 3$, and so on.

The derivation is straightforward. First, we rescale the gluon field, $\mathcal{G}^a_{\mu\nu} = g_0 G^a_{\mu\nu}$, so as to factor out the g_0 dependence in the Lagrangian. Instead of Eq. (5) we now have

$$\mathcal{L} = -\frac{1}{4g_0^2}\mathcal{G}^a_{\mu\nu}\mathcal{G}^a_{\mu\nu} + \text{quark term}\,.$$

Then we observe that

$$i\int d^4x \langle T\{\mathcal{O}(0), \mathcal{G}^2(x)\}\rangle = -\frac{\partial}{\partial(1/4g_0^2)}\langle \mathcal{O}\rangle\,.$$

Since $\langle \mathcal{O}\rangle = \Lambda^d$, we use Eq. (79) to perform differentiation on the right-hand side. In this way we arrive at the formula (81).

In fact, the simple derivation outlined above should be supplemented by some regularization since the matrix elements $\langle \mathcal{O}\rangle$, as well as the two-point functions (81), are divergent unless the cut-off at μ is introduced. In Ref. [3] (see Appendix B) it is shown that the regularization procedure does not affect Eq. (81) provided the cut-off is introduced in a concerted way.

Denote the two-point function on the left-hand side of Eq. (82) as $\Pi_s(Q^2)$; the subscript s marks the scalar glueball channel. If $\Pi_s(0)$ is known, this knowledge can be immediately traded for a power term in the Borel-transformed sum rule. Indeed, in this case, instead of Borelizing the dispersion relation for $\Pi_s(Q^2)$, one can deal with the dispersion relation for

$$\frac{\Pi_s(Q^2) - \Pi_s(0)}{Q^2}\,.$$

After Borelization one gets a prediction for the integral

$$\frac{1}{\pi M^2}\int \frac{ds}{s}\text{Im}\,\Pi_s(s)e^{-s/M^2}\,.$$

Note the occurrence of an extra factor s^{-1} in the weight function in the integrand (cf. Eq. (24)). At the very end

$$\Pi_s(0) = \frac{b}{2}\langle \mathcal{O}_G\rangle$$

is transferred to the right-hand side of the sum rule where it acts as a coefficient in the corresponding power correction.

In this way one arrives at

$$\frac{32\pi^3}{b^2M^4}\int\frac{ds}{s}\mathrm{Im}\,\Pi_s(s)e^{-s/M^2}=\alpha_s^2(M)\left[1+\frac{16\pi^4}{b\alpha_s^2(M)}\frac{1}{M^4}\langle\mathcal{O}_G\rangle+\ldots\right], \quad (83)$$

where the first term in the square brackets corresponds to the free gluon loop, the second term is due to $\Pi_s(0)$, while the ellipses stand for all other corrections, perturbative and nonperturbative. Comparing the gluon condensate correction to that in the ρ-meson sum rule, Eq. (63), we observe, with astonishment, a very strong numerical enhancement of the relative coefficient, by a factor

$$\frac{48}{b}\left(\frac{\alpha_s}{\pi}\right)^{-2}\sim 10^2\,.$$

The standard condensate expansion, with a few first terms kept, does not match this huge low-energy constant. Attempts to smoothly match this gigantic power term with other nonperturbative corrections and to saturate the sum rule in the scalar glueball channel lead us [3] to the conclusion of extremely violent distortions of the spectral density up to the scale $s_0\sim 25$ GeV^2, to be compared with $s_0\sim 1.5$ GeV^2 in the ρ-meson sum rule, see Eq. (36) and Fig. 10. The scale up to which strong violations of asymptotic freedom extend in the scalar glueball channel is unconventionally large.

Thus, a new (numerically) large scale was discovered, that does not show up in the classical channels. It was predicted to have a major impact on the properties of the "nonclassical" mesons. The physical picture lying behind this phenomenon is discussed in Sec. 8. The quantum numbers of the mesons affected by the "superstrong" interaction with the vacuum medium were found to be in one-to-one correspondence with the presence of the direct instantons. This observation served as an initial impetus for the development of the instanton liquid model of the QCD vacuum [55] (for a review see Ref. [56]).

The topic of the QCD low-energy theorems is very vast; it reaches out to such profound issues as the subtleties of the θ dependence, the η' problem, effective dilaton Lagrangians, the 't Hooft matching condition, etc. It has never been reviewed in full, although it certainly deserves attention. This endeavor goes far beyond the scope of the present lecture, and I will limit myself to fragmentary remarks on the literature.

The low-energy limit of the n-point functions generated by $G\tilde{G}$ was studied in Ref. [57]. A celebrated mass formula for η' was obtained in Refs. [58, 59]. An elegant low-energy theorem for the coupling of the Goldstone mesons to the gluonic sources was derived in Ref. [60]. The couplings of photons to the gluonic sources were elaborated in Refs. [3, 61]. Low-energy theorems instrumental in the sum-rule analyses, additional to those considered in Ref. [3], were found in Ref. [62]. Selected aspects of the subject are reviewed in Ref. [63].

8. Are All Hadrons Alike?

The first, superficial, impression of the hadronic family may leave one quite bored. There are no obvious small or large parameters, all quantities are of order one in the appropriate units, and nothing special attracts attention. In particular, one might think that the characteristic hadronic scale is given in all cases by the slope of the Regge trajectory, i.e. universal $\alpha' \sim 1$ GeV^{-2}. The masses of the "most" typical hadrons, nucleons and ρ mesons, are of this order of magnitude. Roughly at the same momentum transfers, scaling sets in deep inelastic scattering heralding the onset of asymptotic freedom — the observer sees (almost) free quarks with modest corrections due to the gluon exchanges.

A careful reflection shows, however, that the hadronic family is far from being that monotonous. Messages of its nonuniversality have been coming from various sides, although it was not so easy to decipher them.

Many years ago Witten posed a question [58]:

"There is an obvious troublesome question. If $m^2_{\eta'} \sim 1/N_c$ and the $1/N_c$ expansion is usually a good approximation, why is $m_{\eta'}$ not much smaller than its actual value of almost one GeV?

This is not merely a problem of $1/N_c$ expansion. It is a more general phenomenological problem. The $\eta' - \pi$ mass splitting violates Zweig's rule, since it involves $q\bar{q}$ annihilation. Since Zweig's rule is usually rather good, why is the $\eta' - \pi$ splitting so large?"

A number of the glueball candidates have been experimentally observed in the last decade. Still, none was unambiguously proved to be a glueball. Let us ask ourselves:

Why, unlike the quark mesons, the masses of the glueballs are unusually heavy or they have an unusually large mixing with the quark states which makes their widths large enough to escape clear experimental identification, in defiance with the $1/N_c$ counting?

The answers to all these questions naturally come with the observation [3] of large scales in the nonclassical channels. Yes, $m^2_{\eta'}$ and the Zweig rule violating parameters are suppressed by $1/N_c$ and are small, but they are small compared to the intrinsic scale $s_0 \sim 25$ GeV^2, rather than 1 GeV^2.

The existence of the diverse scales in the hadronic family is due to the fact that various currents injecting the valence quarks/gluons in the vacuum medium interact with this medium nonuniversally. If in the ρ meson (and similar classical mesons) the impact of the vacuum fields on the valence quark–antiquark pair is modest and is reasonably well described by the average densities of the vacuum fields, in the glueballs the coupling to the vacuum fields is much stronger, so that they actually feel fine details (a grain structure) of the vacuum medium. The strongest coupling to the vacuum fields takes place in the channels with the total spin zero, both quark and gluon (scalar and pseudoscalar quarkonia and glueballs). Phenomenologically deviations from the Zweig rule are most drastic here.

The occurrence of the "superstrong" coupling manifests itself in an abnormally large critical scale which may or may not materialize as the mass of the lowest-lying state with the given quantum numbers. By critical scale I mean the energy at which the asymptotic perturbative regime sets in the spectral density.

A combined analysis of the sum rules and the low-energy theorems has led us [3] to a number of qualitative (and, in many instances, semiquantitative) predictions. First of all, the hierarchy of glueball masses was established. The scalar glueball was predicted to be the lightest, with mass $M(0^+) \sim$ 1.5 GeV (see Ref. [64]). Its coupling to the two-pion state was shown to be very large, with no trace of the Zweig suppression. The pseudoscalar and tensor 2^+ glueballs were predicted to lie in the vicinity of 2 GeV.

In both scalar and pseudoscalar channels the valence gluons injected in the vacuum are affected by the vacuum medium in the most drastic way, and the critical scale is the largest. In the tensor glue channel the critical scale was expected to be smaller than in the scalar/pseudoscalar channels, although larger than for the classical mesons.

Why are the $0^\pm$ channels so special, with the strongest reaction of the medium? It was natural to assume that the phenomenon had to do with the structure of the dominant vacuum fluctuations. It was noted [3] that instantons would do the job, since the direct instantons show up exactly in the $0^\pm$ channels. This was sufficient for a qualitative classification. Quantitative instanton-based models of the QCD vacuum appeared later [55, 56]; they automatically incorporate all relevant features of this peculiar picture. Hunting for instantons in the vacuum is the *vogue* of the day in the lattice community. There are indications that they are abundant and almost saturate the string tension.

It remains to be added that, many years after these predictions, we are witnessing how other approaches started recovering the very same pattern of nonuniversality. Recent calculations in the instanton liquid model produce — not surprisingly — a very close hierarchy of masses and mixing parameters [65]. What is interesting is that this model explains where the large nonperturbative scale in the scalar glueball channel is hidden. The violent vacuum fluctuations squeeze this state to an unusually small radius because of a very strong attraction. The radius of the 0^+ glueball in this model is only 0.2 fm, to be compared with 0.6 fm in the ρ-meson case [65]. The ratio of the radii squared is $\sim 1/10$. It is clear that no constituent model compatible with quantum mechanics can accommodate a state light and narrow simultaneously (see the end of this section).

This part of the hadronic family is being intensively explored on the lattices too. Of course, the light quarks are usually nondynamical on today's lattices, and a typical lattice site of 0.2 fm is comparable with the radius of the 0^+ glueball. With all these reservations in mind it is still instructive to compare what the lattice community learned about the nonclassical hadrons. To this end I borrowed relevant numbers from the talks [66] and original publications to be cited below:

(i) The lightest glueball is scalar, and its mass is close to 1.6 GeV;
(ii) The tensor glueball is significantly heavier; its mass is slightly above 2 GeV; the pseudoscalar glueball is found at approximately the same place, with larger errors;
(iii) The sizes of scalar and tensor glueballs were found to be drastically different. This can be inferred from the different magnitudes of the finite size effects (see e.g. Ref. [67]), or seen directly, in the glueball Bethe-Salpeter amplitudes (or "wave functions") [68, 69]. The radius of the scalar glueball was found to be 0.2 fm, while that of the tensor one is much larger, approximately 0.8 fm. A similar lattice measurement of the ρ-meson size yields approximately 0.5 fm.

Needless to say that the lattice calculations provide us only with final numbers, with no insight as to the mechanisms of the nonuniversality. To the best of my knowledge, no lattice publication on this topic mentions the previous sum rule predictions. True, the lattice results have a potential to be perfected in the future, while the accuracy of the sum rule method is admittedly limited.

To emphasize the highly nontrivial nature of our findings, let us conclude this section by comparing them to the expectations of conventional models, say, the bag model. The quark and (electric) gluon modes in the spherical cavity have energies $2.04/R$ and $2.7/R$, respectively, where R is the cavity radius. Thus, the glueball states are expected to be only marginally heavier than the quark states. In particular, if the spin-dependent forces are neglected the model predicts that $M(2^{++}) \approx M(0^{++}) \approx 1$ GeV and $M(0^{-+}) \approx 1.3$ GeV. The inclusion of the spin-dependent forces changes these numbers rather insignificantly [70]. The complex picture of the vacuum medium, which most strongly affects the scalar and pseudoscalar mesons, especially glueballs, is completely missed.

9. Ecological Niche

In the 1980's the original strategy of the SVZ method was tested, with remarkable success, in analyzing practically every static property of all established low-lying hadronic states. The method was developed in various directions — three-point functions, inclusion of external electromagnetic and other auxiliary fields, light-cone modifications and so on — which allowed one to expand the range of applicability to such advanced problems as magnetic moments [71], form factors [72] at intermediate momentum transfers, weak decays [73] and structure functions [74] of deep inelastic scattering at intermediate x. At the initial stage the method was essentially unchallenged since the lattice calculations were lagging far behind, burdened by the multiple internal problems of lattice QCD.

Now the situation has changed. The machinery of lattice QCD has been improved, and it has become a powerful tool in many problems where qualitative insight is unimportant, and the prime emphasis is on numbers. Let us remember that the SVZ method is admittedly approximate: typically, theoretical accuracy is at the level of 20%. In some cases it does not work at all.

In order to survive, the approach based on OPE and the sum rules has to find a range of applications in the hadronic physics, where it can successfully compete. I must say that it copes with the challenge. Its strongest advantage is the analytic character of calculations which can be continued to the Minkowski space, term by term. Thus, the most promising directions of growth are those where the Minkowski kinematics is entangled in this or that way and/or the processes under consideration develop in more complicated media, rather than in the vacuum medium. (Such problems are very difficult for lattice QCD which operates with vacuum correlation functions numerically evaluated in the Euclidean space.) Let me give three examples.

(1) *Particle decays with a large spatial momentum carried away by produced hadrons, for instance, $B \to \rho \ell \nu$.* If the invariant mass of the lepton pair is small, $q^2 \sim 0$, the spatial momentum of the ρ meson is of order $M_B/2 \gg \Lambda$. Treating the problem in the Euclidean domain we will have to face the necessity of an extremely remote analytic continuation. At the same time the light-cone sum rule, to be discussed in Sec. 10, are perfectly fit to deal with this problem.
(2) *Heavy flavor sum rules.* They allow one to determine basic parameters of the heavy quark theory — the masses of the heavy quarks, form factors at zero recoil, and so on — in terms of directly measurable quantities. This is a close relative of the SVZ sum rules. The main distinction is that instead of the vacuum condensates one deals with the expectation values of various local operators over the heavy hadron state, for instance the B meson. This approach will be also considered in brief in Sec. 10.
(3) *Pre-asymptotic effects in inclusive processes at high energies.* A typical problem from this class is the determination of the lifetime differences in the b quark family. Although the task is somewhat different from those usually treated in the SVZ sum rules, the method for its solution is exactly the same. The key technical element is the operator product expansion. In the given context it is imperative to address, additionally, an issue going beyond OPE (more exactly, its practical version) — duality violations. The topic of the deviations from duality is of paramount practical importance; it came under renewed scrutiny recently [16].

In all these and other instances the OPE-based methods remain to be the only theoretical tool available on the market today. We see that the approach is sufficiently rich and flexible to remain viable in the future. I am sure it will continue to play a key role in solving many applied problems in the hadronic physics in the next decade. The potential for further growth exists, the active stage is not yet over.

10. New Developments

To give you a flavor of what is going on in this field I will pick up two or three subjects which come to my mind first. My selection by no means presents the full picture.

The heavy quark mass $1/m_Q$ expansions that flourished in the 1990's revived interest in the SVZ sum rules for hadrons containing one heavy quark[14] (in practice, b quark). It was realized — and this is a new element — that (i) the hybrid logarithms occur in the theoretical part of the sum rules, and (ii) they may result in a strong enhancement of the gluon radiative corrections. Hybrid logarithms depend on the ratio m_Q/μ, they were unknown before 1988. A typical example where summation of the hybrid logarithms significantly shifts the answer is f_B. I single out this problem because this parameter is fundamentally important in the heavy quark physics and because of an instructive story attached to it. Early sum rule calculations of f_B, which had missed previously undiscovered hybrid logarithms, yielded a number close to 130 MeV [76]. The main impact of the hybrid logarithms is a change of the argument of the running coupling constant in the sum rule: instead of $\alpha_s(m_b)$ the radiative corrections, as they turn out [77], are governed by α_s (1 GeV). As a result, the sum rule prediction jumped up to $f_B = 180$ MeV, with the theoretical uncertainty 20 to 30 MeV [77].

Meanwhile, f_B was the object of the multiple lattice studies. The first lattice measurements produced an unbelievably huge value, somewhere around 250 MeV. With the elimination of various sources of the systematic uncertainties the lattice number has been systematically decreasing. The modern number, quoted at recent conferences, is close to 180 MeV, with the error in the same ballpark as the sum rule uncertainty. Thus, the initial contradiction between the two methods faded away giving place to perfect agreement.

Along with the technical developments of the type that I have just mentioned the method experienced ideological developments too. Below we will discuss some of them.

10.1. *Light-Cone Sum Rules*

This formalism was designed to overcome difficulties in the traditional sum rules for three-point functions. Assume we want to calculate a cubic coupling constant corresponding to the amplitude $A \to B + C$, where A, B and C

[14]The $1/m_Q$ expansion of the SVZ sum rules *per se* dates back to the work of Shuryak [75].

stand for either hadrons or external currents, say, electromagnetic. Typical examples are $D^* \to D\pi$ or $\gamma^* + \gamma^* \to \pi$ (here γ^* denotes a virtual photon). In many instances in order to get the desired constant it is necessary to apply Borelization with respect to two momenta, p_A and p_B, independently. Then the components of the third momentum, p_C, cannot be small compared to $p^2_{A,B}$. Indeed, $p_C(p_A + p_B) = p_A^2 - p_B^2 \sim p^2_{A,B}$. The standard condensate expansion contains a series of operators with derivatives which, eventually, give rise to the expansion parameter of the type $p_C(p_A + p_B)/p^2_{A,B} \sim 1$. In other words, all terms in this subseries must be summed over.

This (partial) summation is carried out automatically if, instead of the three-point function $\langle j_A, j_B, j_C \rangle$ one considers the correlation function of the currents j_A and j_B sandwiched between the vacuum and the state $|C\rangle$. The vacuum expectation values of local operators (condensates) in the SVZ sum rules are substituted by the *light-cone wave functions* (*distribution amplitudes* in the American literature) describing the momentum fraction distribution of the components in the Fock wave function of the hadron C. If the SVZ approach is based on the short-distance expansion of the current T products in terms of local operators, and the expansion runs in *dimensions* of the local operators, the light-cone sum rules exploit the expansion in *nonlocal* "string" operators on the light-cone; the expansion runs in *twists*, not dimensions. From the viewpoint of the standard SVZ sum rules, in the light-cone approach one performs a partial summation of an infinite chain of operators of arbitrary dimension, but given twist. In effect, the light-cone sum rules combine the SVZ approach with the technique used in the description of hard exclusive processes (see Ref. [78]). The price we have to pay for the partial summation is rather high: the large distance dynamics is parametrized not by numbers, as in the SVZ condensates, but by functions — the leading twist, next-to-leading twist, and so on distribution functions.

A Russian proverb says that it is better to see once than to hear hundred times. Let us take a closer look at a specific example. Below I will outline the calculation of the $D^*D\pi$ constant. This example, as well as the explanations below, are borrowed from Ref. [79].

The $D^*D\pi$ coupling constant g is defined as follows:

$$\langle D^*(p)\pi(q)|D(p+q)\rangle = -g q_\mu \varepsilon^\mu ,$$

where the hadrons' momenta are indicated in the parentheses, and ε^μ is the D^* polarization vector. To determine this constant from the light-cone sum rules one considers the correlation function

$$F_\mu(p,q) = i\int d^4x e^{ipx}\langle\pi(q)|T\{\bar{d}(x)\gamma_\mu c(x), \bar{c}(0)i\gamma_5 u(0)\}|0\rangle\,. \tag{84}$$

The first operator in the T product is the interpolating current for D^* and the second is the interpolating current for D.

The correlation function (84) depends on two variables, p^2 and $(p+q)^2$ (note that $q^2 = 0$ since the pion is on the mass shell). The contribution proportional to the $D^*D\pi$ amplitude has two poles, in p^2 and $(p+q)^2$, corresponding to the ground state mesons both in the vector and pseudoscalar channels. Namely,

$$F_\mu(p,q) = \frac{M_D^2 M_{D^*} f_D f_{D^*}}{m_c(p^2 - M_{D^*}^2)((p+q)^2 - M_D^2)} g q_\mu + \ldots\,. \tag{85}$$

The ellipses on the right-hand side denote a term proportional to p_μ irrelevant for our analysis, as well as terms with single pole and nonpole continuum contributions, that will be exponentially suppressed under double Borelization (with respect to p^2 and $(p+q)^2$). All other notations are self-explanatory.

On the theoretical side we assume that $p^2 - M_{D^*}^2$ and $(p+q)^2 - M_D^2$ are Euclidean and large enough to allow for the operator product expansion of the correlation function (84). As usual, the corresponding Borel parameters will lie in the window. The simplest graph pertinent to OPE that has to be built is depicted in Fig. 20. The large distance part of the process is denoted by the shaded area. The light quark lines form the pion, rather than the quark condensate, that was discussed in Sec. 4. More exactly, we deal here with the leading twist pion wave function whose definition will be given shortly.

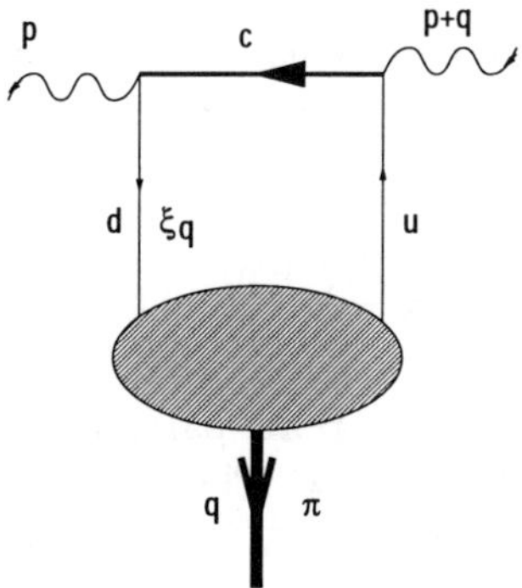

Fig. 20. Light-cone sum rule for $D^*D\pi$ in the leading order. The current labeled by $p+q$ interpolates D while that labeled by p interpolates D^*.

If ξ is the fraction of the pion momentum carried by the d quark, the graph of Fig. 20 gives rise to the following expression:

$$F_\mu \equiv q_\mu F(p^2,(p+q)^2) = m_c f_\pi q_\mu \int_0^1 d\xi \frac{\phi_\pi(\xi)}{m_c^2 - (p+\xi q)^2} + \dots , \tag{86}$$

where $\phi_\pi(\xi)$ is the distribution function, and the ellipses denote corrections, both perturbative (the α_s expansion) and nonperturbative (due to higher twists). The c quark Green's function is substituted, in the leading approximation, by that of a free quark. Emission of gluons from this line would result either in α_s corrections or in the distribution functions of nonleading twists.

For very small values of q one could have substituted the denominator in Eq. (86) by $m_c^2 - p^2$. Then the integral $\int d\xi\phi_\pi(\xi) \to 1$, and F_μ would reduce to the matrix element of the operator $\bar{d}\gamma_\mu\gamma_5 u$ sandwiched between the pion state and vacuum.

Expanding the denominator in Eq. (86) in q we readily identify higher dimension operators that are summed over: the term linear in q originates from $\bar{d}\overleftarrow{D}_\alpha\gamma_\mu\gamma_5 u$, quadratic in q from $\bar{d}\overleftarrow{D}_\alpha\overleftarrow{D}_\beta\gamma_\mu\gamma_5 u$, etc. The actual expansion parameter is

$$\frac{pq}{p^2 - m_c^2} \sim 1 .$$

The (leading twist) light-cone wave function $\phi_\pi(\xi)$ is formally defined as

$$\langle\pi(q)|\bar{d}(x)\gamma_\mu\gamma_5 \exp\left\{ig\int_0^x A_\alpha(y)dy_\alpha\right\}u(0)|0\rangle = -iq_\mu f_\pi \int_0^1 d\xi e^{i\xi qx}\phi_\pi(\xi) . \tag{87}$$

It parametrizes the large-distance dynamics of the $\bar{d}u$ pair.

If we do the double Borel transformation, with respect to p^2 and $(p+q)^2$ the denominator in Eq. (86) becomes

$$\frac{M_1^2 M_2^2}{M_1^2+M_2^2}\exp\left(-m_c^2\frac{M_1^2+M_2^2}{M_1^2M_2^2}\right)\delta\left(\xi - \frac{M_1^2}{M_1^2+M_2^2}\right) , \tag{88}$$

where $M_{1,2}$ are the corresponding Borel parameters, that are expected to lie close to each other inside the window. The sum rule is obtained by matching the double Borel transform of Eq. (85) with the double Borel transform of Eq. (86). Thus, we see that the theoretical part of the sum rule is determined by $\phi_\pi(0.5)$. Of course, there are corrections of higher order in α_s and from higher twist distribution functions, but as usual inside the window they should be kept at a modest level. In addition we must check the stability of the result inside the window. Provided all this is properly done, one obtains [79] g in terms of $\phi_\pi(0.5)$. The latter quantity must be extracted from independent sources.

(The situation we face here is not quite generic. *A priori* one could expect the coupling g to be expressible in terms of $\phi_\pi(\xi)$, rather than one constant, $\phi_\pi(0.5)$.)

Information on $\phi_\pi(\xi)$ is quite abundant; there is some limited information on the higher-twist wave functions too. The issue of $\phi_\pi(\xi)$ is almost as old as the SVZ sum rules themselves. The key element of the corresponding theory, which was studied in great detail in connection with hard exclusive processes, is the (approximate) conformal symmetry of QCD. Other elements are provided by the sum rule and lattice calculations and models. The issue is not completely settled, although it seems that a heated debate of the 1980's gradually wanes. Out of two competing scenarios — a double hump distribution function by Chernyak and Zhitnitsky [78] and a narrow distribution by Radyushkin and collaborators [80] that is close to the asymptotic form of $\phi_\pi(\xi)$ — the second is gaining more recognition now. Experts lean towards a narrow almost asymptotic distribution. As usual, the best strategy is sacrificing one of the sum rules in order to extract $\phi_\pi(\xi)$ from data. A recent attempt based on the $\gamma^* + \gamma \to \pi^0$ data has been undertaken in Ref. [81]. I will not go into further details here, referring the reader to a very rich original literature devoted to this topic.

It remains to be added that (i) the term "light-cone sum rules" first appears in Ref. [82]; (ii) the idea of the approach dates back to Ref. [83]; and (iii) a review of this topic, with a representative list of references was published recently [84].

If the partial summation of the higher-dimension operators turns out to be so successful in the light-cone sum rules, it is natural to ask [85] why it is not applied in the vacuum correlation functions appearing in the SVZ sum rules. In this case, in order to partially sum, say, the quark operators we must substitute the quark condensate $\langle \bar{q} q \rangle$ by a "string expectation value" in the vacuum,

$$\left\langle \bar{q}(x) \exp\left\{ ig \int_0^x A_\alpha(y) dy_\alpha \right\} q(0) \right\rangle , \tag{89}$$

which is a function of x.

In the light-cone sum rules one can ascribe certain twist to each given distribution function. Each distribution function sums up a tower of the local operators, with all dimensions but given twist. Therefore, the distribution functions can be systematically ordered in twist. Twist is obviously irrelevant in the vacuum correlation functions. Therefore, the partial summation implemented by the "string expectation values" (89) is hard to justify theoretically. A principle that would allow us to order distinct "string expectation values" has to be elaborated.

Another major difference with the light-cone sum rules is that the pion distribution function is known at least at a certain level, while next to nothing is known about the vacuum function (89). Until recently even the large x asymptotic behavior of this function was erroneously assumed to be Gaussian. In fact, at large $|x|$ this behavior is of the type $\exp(-\mu|x|)$, see Ref. 16. We are at the initial stage, when first observations are being made regarding properties of the function (89) following from the general structure of QCD [16]. I do not rule out that, as our understanding progresses, the corresponding modification of the sum rule approach will be elaborated.

10.2. *Heavy Flavor Sum Rules*

Although the character of problems in which this formalism is instrumental is somewhat different from the classical applications of the SVZ sum rules, the basic techniques — OPE, matching of the theoretical expansions with the phenomenological expressions represented by dispersion integrals over the observable spectral densities, etc. — are the same. This topic was discussed in some detail in Chapter I. The vacuum condensates are replaced by the expectation values of local operators over the heavy flavor states, say, B mesons. All questions that can be asked in connection with the vacuum condensates exist here too; the answers are more transparent, however. We will consider below a sample application. Our main goal is discussing subtle nuances associated with the condensate expansion (Wilsonean OPE versus practical version, see Sec. 6).

The heavy flavor sum rules were engineered [19] to treat the inclusive heavy flavor decays, for instance the semileptonic decays $B \to X_c \ell \nu$ where X_c is the arbitrary hadronic state containing c quark. The lowest-lying c-containing state is the D meson, then comes D^* and then higher excitations/continuum. By varying the momentum of the lepton pair one can measure various spectra in the transition $b \to c$ induced by the weak current. The object of the theoretical

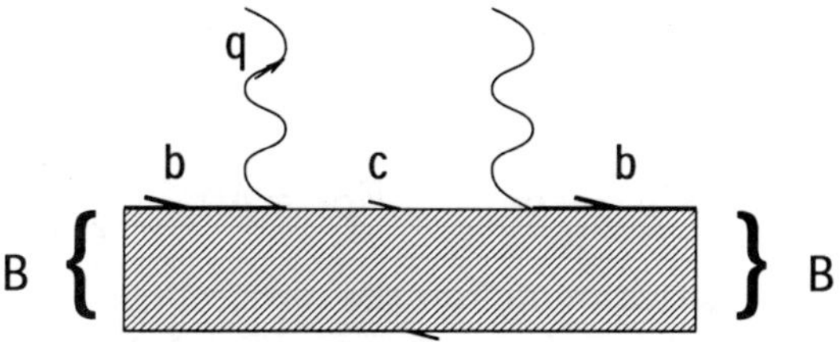

Fig. 21. The operator product expansion for h_{ab}. For large negative ε the c quark is far off shell, and the points of the current emission and absorption are close. The B meson in the initial and final states is assumed to be at rest.

study is the *transition operator*

$$\hat{T}_{ab}(q) = i \int d^4x \, e^{iqx} \, T\{j_a^\dagger(x) j_b(0)\} \, , \tag{90}$$

that will be eventually sandwiched between the states $|B\rangle$ and $\langle B|$ (see Fig. 21). Here j_a denotes a current of the type $\bar{c}\Gamma_a b$ with an arbitrary Dirac matrix Γ_a; q is the momentum carried away by the lepton pair.

The average of $\hat{T}_{ab}$ over the heavy hadron state B with the momentum p_B represents a forward scattering amplitude (the so-called hadronic tensor),

$$h_{ab}(p_B, q) = \frac{1}{2M_B} \langle B|\hat{T}_{ab}|B\rangle \, . \tag{91}$$

The structure functions w_{ab} are obtained from the hadronic tensor by taking its imaginary part,

$$w_{ab} = (1/i) \, \text{disc} \, h_{ab} \, .$$

All observable spectra are represented as certain integrals over w_{ab}. The hadronic tensor h_{ab} and the structure functions w_{ab} can be expanded in terms of various kinematic structures. For each set of the currents generically we may have up to five distinct kinematic structures; these details need not bother us here. For illustrational purposes it is sufficient to limit ourselves to the axial currents, so that $\Gamma_{a,b} \to \gamma_\mu \gamma_5$, and to the first function,

$$h_{\mu\nu} = -h_1 g_{\mu\nu} + \dots \, , \qquad w_1 = (1/i) \, \text{disc} \, h_1 \, .$$

The lowest-lying state produced by the axial current from B is D^*. To end up with the kinematical aspects we must specify the lepton pair momentum q. It will be assumed that the spatial momentum $\mathbf{q}$ is small but nonvanishing, $\Lambda \ll |\mathbf{q}| \ll M_D$. The terms $\mathcal{O}(\mathbf{q}^2)$ will be kept while those of higher order in $|\mathbf{q}|$ will be neglected. Since $\mathbf{q}$ is fixed, h_1^{AA} becomes a function of one (complex) variable, q_0. Instead of q_0 it is more convenient to work with ε defined as

$$\varepsilon = q_{0\,\max} - q_0 \tag{92}$$

where

$$q_{0\,\max} = M_B - E_{D^*}\,, \qquad E_{D^*} = M_{D^*} + \frac{\mathbf{q}^2}{2M_{D^*}}\,. \tag{93}$$

When ε is real and positive we are on the physical cut where the imaginary part of h_1^{AA} is measurable. For negative ε we are below the cut, in the Euclidean domain, where the T product (90) can be computed as an expansion in $1/m_{c,b}$. The leading operator in this expansion is $\bar{b}b$. It has dimension 3. No operators of dimension 4 exist. At the next level, of dimension 5, there are two operators, $\bar{b}(i\mathbf{D})^2 b$ and $(ig/2)\bar{b}\gamma^\mu\gamma^\nu G_{\mu\nu}b$. The first operator (sometimes called the *kinetic energy operator*) is not a Lorentz scalar. In the problem considered the operators to be retained in the expansion need not necessarily be Lorentz scalars since, unlike the vacuum case, the very presence of the B meson in the initial/final state singles out a reference frame.

The whole procedure is perfectly analogous to the $1/Q^2$ expansion of the polarization operator discussed in Sec. 4. The operator $\bar{b}b$ occupies the same position in the hierarchy as the unit operator in the expansion of the vacuum correlators. The operators of dimension 5 and higher generate $1/m_{c,b}$ corrections analogous to the condensate corrections. In the problem at hand all local operators appearing in the expansion of the T product (90) are averaged over the B-meson state. In this way we get a theoretical expression for h_1^{AA} off the cut.

A bridge between h_1^{AA} at negative ε and w_1^{AA} at positive ε is provided by the dispersion relation. Expanding the denominator in the dispersion relation in $1/\varepsilon$ we obtain an infinite set of the sum rules. The third such rule is as follows [86]:

$$\frac{1}{2\pi}\int d\varepsilon\varepsilon^2 w_1^{AA}(\varepsilon) = \frac{1}{3}\mu_\pi^2\mathbf{v}^2 + \dots\,. \tag{94}$$

Here $\mathbf{v}$ is the velocity of the produced heavy hadron (in the B rest frame), and μ_π^2 is defined as the expectation value

$$\mu_\pi^2 = \frac{1}{2M_B}\langle B|\bar{b}(i\mathbf{D})^2 b|B\rangle\,. \tag{95}$$

This parameter, μ_π^2, has the meaning of the average spatial momentum squared of the heavy quark inside B meson, and is one of the fundamental parameters of the heavy quark theory. One encounters μ_π^2 in a large number of applied problems. It is fair to say that in the heavy quark physics μ_π^2 plays the same role as the gluon condensate in classical applications of the SVZ method.

The ellipses in Eq. (94) denote corrections suppressed by powers of $1/m_{c,b}$. Since the weight function in Eq. (94) is proportional to ε^2, the "elastic" $B \to D^*$ transition drops out, and the integral on the left-hand side is saturated by "inelastic" transitions, i.e. decays of the B meson in the excited X_c states. For such transitions $w_1^{AA}(\varepsilon)$ is proportional to $\mathbf{v}^2$ at $|\mathbf{v}| \ll 1$. It is convenient then to introduce a spectral function $\sigma(\varepsilon)$ which does not vanish in the limit $|\mathbf{v}| \to 0$,

$$\sigma(\varepsilon) = |\mathbf{v}|^{-2} w_1^{AA}(\varepsilon)\,, \qquad \varepsilon > 0\,. \tag{96}$$

Our brief excursion in the heavy flavor sum rules is almost over. In terms of the new spectral density the sum rule we are interested in takes the form

$$\frac{1}{2\pi}\int d\varepsilon\varepsilon^2\sigma(\varepsilon) = \frac{1}{3}\mu_\pi^2\,. \tag{97}$$

As with any sum rule it can be read in both directions: if μ_π^2 is known this is a prediction for the spectral density. On the other hand, if the spectral density is measured we can take the integral and get μ_π^2. Below we will pursue the latter strategy assuming that $\sigma(\varepsilon)$ is measured. In actuality this is not the case, but we have a pretty good idea of how $\sigma(\varepsilon)$ will look like when it is measured. First of all, it is positive-definite. At small values of ε, of order of a few units times Λ, there is one, perhaps, two resonances; then at higher values of ε the spectral density must approach its perturbative asymptotics which is known too: $\sigma(\varepsilon) \to C\alpha_s/\varepsilon$, where C is a constant which was calculated in Ref. [19]. The numerical value of C is unimportant for our purposes. The asymptotic regime is achieved through oscillations — the reasons are the same as in Secs. 2.2 and 4. A schematic plot of $\varepsilon^2\sigma(\varepsilon)$ is depicted in Fig. 22.

If you think about Eq. (97) you will, probably, find it remarkable: it gives a condensate-like nonperturbative parameter in terms of an integral over the observable spectral density. Further examination raises a question. Indeed, if the integral extends to infinity, it badly diverges.

One need not be concerned, however. This infinity reflects the fact that the proper definition of μ_π^2 requires a normalization point, just in the same way as the for the gluon condensate. In the case at hand this normalization point has a very transparent physical meaning — μ is the maximal excitation energy of the hadronic state belonging to the class of "soft excitations" (Fig. 22). Higher excitations are considered to be hard, and are well approximated by

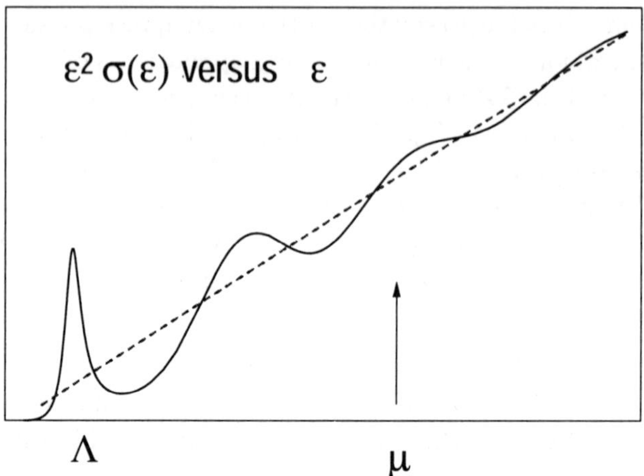

Fig. 22. A sketch of $\varepsilon^2\sigma(\varepsilon)$ in the $b \to c$ transition at small $|\mathbf{v}|$ (solid curve). The spectral density $\sigma(\varepsilon)$ is defined in Eq. (96). The dashed curve represents $\varepsilon^2\sigma(\varepsilon)$ in perturbation theory at one loop.

perturbation theory. I remind that the hard contributions do not belong to μ_π^2; rather, they determine the coefficient functions in OPE. In the perturbative calculation of the coefficient functions one must remove the part with $\varepsilon < \mu$. In fact, the cut off in the excitation energy as a normalization point was suggested in the given context long ago in Ref. [87]. Thus, in the Wilsonean approach $\mu_\pi^2(\mu)$ is determined as

$$\mu_\pi^2(\mu) = \frac{3}{2\pi}\int_0^\mu d\varepsilon\varepsilon^2\sigma(\varepsilon)\,. \tag{98}$$

In practical OPE we deal, instead, with subtracted "surrogates". Say, if one limits oneself to one-loop accuracy,

$$\text{“}\mu_\pi^2\text{”} = \frac{3}{2\pi}\int_0^\mu d\varepsilon\varepsilon^2\left[\sigma(\varepsilon) - \sigma_{1-\text{loop}}(\varepsilon)\right]\,.$$

The right-hand side in this expression is μ independent at one loop. Its μ dependence shows up only at two loops. If we agree to discard all effects $\mathcal{O}(\alpha_s^2)$ and higher, *just totally ignore them*, then the integration in "μ_π^2" can be extended to infinity. Correspondingly, the coefficient functions must be calculated at one loop, following the standard Feynman rules, with no removal of the $\varepsilon < \mu$ part.

11. Sum Rules and Lattices

In this lecture it has been already noted, more than once, that certain aspects of the condensate approach, as well as a rich experience in the analysis of various spectral densities, might be useful for lattice practitioners. Remarks to this effect are scattered here and there. Of particular importance are qualitative observations, such as an enhanced sensitivity to the vacuum fermion loops in certain channels (this effect is neglected in the quenched approximation), the existence of new, abnormally large scales in the $0^{\pm}$ glueball channels, and so on. Surprisingly though, all these hints, which might have produced a strong impact on the lattice theory, are so far almost totally ignored.[15] The reaction of rejection of any ideas coming from outside the lattice theory *per se* plagues the lattice community. This happens even in those cases where the overlap is quite obvious.

Let me give an almost anecdotal example. As was mentioned in Sec. 7, in 1981 a wide class of "QCD scale anomaly" low-energy theorems was obtained [3] in connection with the SVZ sum rules. These theorems relate to each other n-point functions with arbitrary number of insertions of the operator $\sigma(x) = \theta^\mu_\mu(x) \propto G^2$ (see Eq. (39)) at vanishing momentum. One of these theorems can be cast in the form of a prediction for the expectation value of $G^2_{\mu\nu}$ in the presence of the Wilson loop operator W,

$$\langle W\rangle^{-1}\left\langle\int d^3x\sigma(x)W\right\rangle_{\rm E,c} = V(R) + R\frac{\partial V(R)}{\partial R} \tag{99}$$

where $V(R)$ is the static potential between the heavy quarks separated by distance R,

$$W = P\oint_C \mathrm{Tr}\,\exp(igA_\mu(x)dx_\mu)\,,$$

the contour C in the definition of the Wilson loop is depicted in Fig. 23, and the subscript E reminds us that all definitions and calculations refer to the Euclidean space. Moreover, the integration in Eq. (99) is performed in three directions perpendicular to the Euclidean time; the subscript c refers to the connected part, i.e. the disconnected part of the correlation function (99) must be discarded. One can interpret the operator $\sigma(x)$ in Eq. (99) as a probe of the energy density of the vacuum medium at the point x in the presence of a

[15] I am aware of two dedicated works were the sum rule results were analyzed in parallel with the lattice calculations [88]. This seems to be a rare exception.

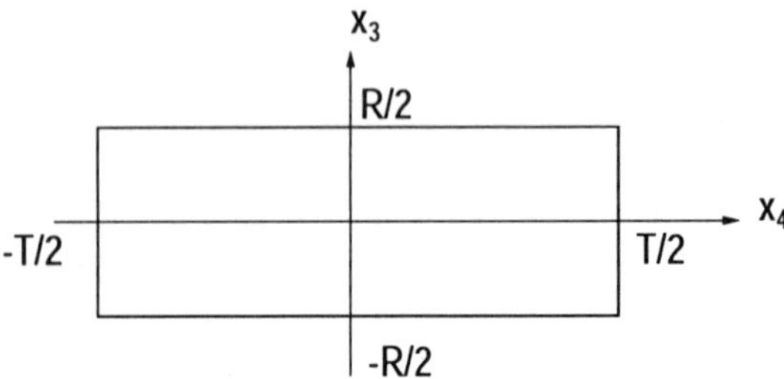

Fig. 23. Rectangular contour in the Wilson loop operator. One can consider it as a trajectory of infinitely heavy quarks (in the imaginary time). R is the distance between the quark and antiquark, $T \to \infty$.

pair of heavy static quark sources separated by distance R [3]. In the case of purely linear potential the right-hand side reduces to $2V(R)$. In this form the theorem (99) was derived in Ref. [3] (see Appendix B). The generalization to the case of arbitrary potentials is due to Dosch *et al.* [89].

Derivation of the theorem (99) is very similar to other theorems discussed in Sec. 7. The functional-integral representation for $\langle W \rangle$ is differentiated with respect to $1/g_0^2$. We then use the fact that

$$\frac{\partial V(R)}{\partial(1/4g_0^2)} = -\frac{32\pi^2}{b}\left(V(R) + R\frac{\partial V(R)}{\partial R}\right). \tag{100}$$

The latter equality, in turn, follows from consideration similar to that presented in Sec. 7. Thus, the low-energy theorem (99) is a reflection of the scale anomaly of QCD, and as such, represents a fundamental aspect of the theory.

Now I come to the culmination of the story. In 1987 the theorem (99) was rediscovered in the lattice community [90], where it goes under the name of the "action Michael sum rule". The original derivation in Ref. [90] was erroneous, the second term on the right-hand side of Eq. (99) was omitted. The error was corrected in [89] which was specifically devoted to the issue. Nevertheless, the theorem continues to circulate in the lattice community as the "Michael sum rule", see e.g. Ref. [91].

This example of a "lattice xenophobia" is by no means unique. Needless to say, I consider the situation as absolutely unhealthy. The problems we face in the hadronic world are too complicated to afford neglecting the work done in related areas. The analytic and numerical approaches should complement each other. I sincerely hope that the attitude of the lattice community will start changing after recent breakthrough discoveries in supersymmetric gauge theories. Supersymmetry is a powerful tool which allows us to reveal features of strong coupling gauge dynamics that we never suspected of. One of the lessons which seems undeniable is the proof that massless (or light) quarks play a more significant role than is universally believed among the lattice practitioners. In the quenched approximation, the most wide spread today, the dynamical quarks are neglected altogether. Even the most advanced lattice investigations, which do not rely on the quenched approximation, routinely use extrapolations in the number of the massless quarks. Supersymmetric gauge theories teach us that increasing the number of massless quarks from two to three or changing the color representation of the fermion fields from fundamental to adjoint is not necessarily a smooth process. Dynamics may change drastically!

Penetration of ideas and insights obtained analytically, in the practice of lattice calculations, where they can be used, at the very least, to test reliability of various approximations, is unavoidable. The first steps in this direction have been already reported [92].

12. Vacuum Fluctuations Are Subtle Creatures

The rich SVZ phenomenology confirms the wide-spread belief that the structure of the hadronic family, in all its gross features and intricacies, derives from a complicated organization of the vacuum medium. The vacuum fluctuations are violent, but fine-tuned in a very specific way. Within the method, as it exists today, we do not calculate the vacuum condensates. This does not mean, of course, that the origin of the condensates, and their relative values, are of no interest. I merely wanted to say that the problem of the vacuum structure in analytic QCD has not been solved in full. There are indirect methods, however, which show that this structure is very subtle — much more subtle than one might think of *a priori* — and depends on seemingly insignificant details of the theory. For instance, changing the number of colors from three to two dramatically changes the pattern of the spontaneous chiral symmetry breaking resulting in a drastic rearrangement in the Goldstone sector of the hadron family. Instead of $N_f^2 - 1$ QCD "pions" (N_f stands for the number of the light quark flavors) we get $2N_f^2 - N_f - 1$ massless Goldstone hadrons.[16] This conclusion can be achieved [93] from the consideration of the 't Hooft

[16]It is curious to note that the existence of the extra massless states was discovered in the lattice SU(2) theory, which has been used as a theoretical laboratory for decades, only recently. The result is known in analytic QCD since the mid-eighties. This is another manifestation of the "lattice xenophobia", see Sec. 11.

matching conditions [94] (for earlier analyses based on less rigorous arguments see [95]).

Let us dwell on this issue. Assume that we have N_f massless (light) quarks but the color gauge group is SU(2). The quarks belong to the fundamental representation of SU(2). Unlike QCD, in which the color triplets and antitriplets are independent representations, the SU(2) doublets are essentially the same as antidoublets. In other words, all representations of SU(2) are (quasi)real.

This simple observation has far reaching consequences. Indeed, the flavor symmetry of the Lagrangian is $\mathrm{SU}(2N_f)$ now, rather than $\mathrm{SU}(N_f) \times \mathrm{SU}(N_f)$ we deal with in QCD. Out of $4N_f^2 - 1$ currents $2N_f^2 - N_f - 1$ are axial; all correspond to the spontaneously broken symmetries since there is no way one can achieve the matching of the triangular AVV anomalies otherwise [93]. This means that the pattern of the chiral symmetry breaking is $\mathrm{SU}(2N_f) \to \mathrm{Sp}(2N_f)$ rather than the $\mathrm{SU}(N_f) \times \mathrm{SU}(N_f) \to \mathrm{SU}(N_f)$ we got used to in QCD. The very gross features of the vacuum structure do depend on the number of colors!

Not only do they depend on the number of colors, they are also sensitive to the fermion content of the theory.

Indeed, let us substitute the standard quarks of QCD with one Majorana fermion λ in the adjoint representation of the color group. The theory thus obtained is nothing but supersymmetric gluodynamics. Superficially it looks very similar to conventional QCD. At first sight, one can hardly expect any conspicuous deviations from the conventional pattern of behavior.

It has been known for many years that the supersymmetric gauge theories posses some miraculous properties. In particular, in supersymmetric gluodynamics the Gell-Mann-Low function is known *exactly* [96]

$$\beta(\alpha_s) = -\frac{\alpha_s^2}{2\pi}\frac{3N}{1-(N\alpha_s/2\pi)}, \tag{101}$$

where the SU(N) gauge group is assumed.

As QCD, supersymmetric gluodynamics is believed to be confining in the infrared domain. Unlike QCD, however, the condensate $\langle\lambda^2\rangle$ may or may not develop. It was argued recently [97] that two distinct phases coexist in this theory — the conventional chirally asymmetric phase with $\langle\lambda^2\rangle \neq 0$, that reminds QCD, and a very unusual chirally symmetric phase where $\langle\lambda^2\rangle = 0$. The arguments leading to this conclusion are too technical to be reproduced here. They would lead us far astray. The interested reader is referred to Ref. [97].

Instead, let us dwell on another example which is pretty close to actual QCD. In fact, we will consider just conventional QCD with more than three massless flavors.

The Gell-Mann-Low function in QCD has the form [6]

$$\beta(\alpha_s) = -b\frac{\alpha_s^2}{2\pi} - b_1\frac{\alpha_s^3}{4\pi^2} - \ldots\,,$$

$$b = 11 - \frac{2}{3}N_f\,, \qquad b_1 = 51 - \frac{19}{3}N_f\,. \tag{102}$$

At small α_s it is negative (asymptotic freedom!) since the first term always dominates. When the scale μ decreases the running gauge coupling constant grows, and the second term becomes important. Generically the second term takes over the first one at $\alpha_s/\pi \sim 1$, when all terms in the α_s expansion are equally important, i.e. in the strong coupling regime. We do not know what exactly happens at $\alpha_s/\pi \sim 1$. Assume, however, that for some reasons the first coefficient b is abnormally small, and this smallness does not propagate to higher orders. Then the second term catches up with the first one when $\alpha_s/\pi \ll 1$, we are in the weak coupling regime, and higher-order terms are inessential. Inspection of Eq. (102) shows that this happens when N_f is close to 33/2, say 16 or 15 (N_f has to be less than 33/2 to ensure asymptotic freedom). For these values of N_f the second coefficient b_1 turns out to be negative. This means that the β function develops a zero in the weak coupling regime, at

$$\frac{\alpha_s^*}{2\pi} = \frac{b}{-b_1} \ll 1\,. \tag{103}$$

(Say, if $N_f = 15$ the critical value is at 1/44 and is indeed small.) This zero is nothing but the infrared fixed point of the theory. At large distances $\alpha_s \to \alpha_s^*$, and $\beta(\alpha_s^*) = 0$, implying that the trace of the energy-momentum vanishes. Then the theory is in the conformal regime. There are no localized particle-like states in the spectrum of this theory; rather we deal with massless unconfined interacting quarks and gluons; all correlation functions at large distances exhibit a power-like behavior. In particular, the potential between two heavy static quarks at large distances R will behave as $\sim \alpha_s^*/R$, i.e. we have a pure Coulomb behavior. The situation is not drastically different from conventional QED. As long as α_s^* is small, the interaction of the massless quarks and gluons in the theory is weak at all distances, short and large, and is amenable to the standard perturbative treatment (renormalization group, etc.). The chiral symmetry is not broken spontaneously, the quark condensates do not develop. QCD becomes a fully calculable theory.

The fact that at N_f close to 16, QCD becomes conformal and weakly coupled in the infrared limit is known for about 20 years [98, 99]. If the conformal regime takes place at $N_f = 16$ and 15, let us ask ourselves how far can we descend in N_f without ruining the conformal nature of the theory in the infrared? Certainly, when N_f is not close to 16 the gauge coupling α_s^* is not small. The theory is *strongly coupled*, as conventional QCD, and, simultaneously, conformal in the infrared. The chiral symmetry presumably is unbroken, $\langle\bar{q}q\rangle = 0$. The vacuum structure is totally different, and so are all properties of the theory.

The range of N_f where this phenomenon occurs is called the *conformal window*. The right edge of the conformal window is at $N_f = 16$. There are indications that the left edge of the conformal window N_f^* lies not too far from the actual QCD (i.e. not too far from $N_f = 3$).

First, in the instanton liquid model it was found [100] that at $N_f = 5$ the chiral condensate disappears, $\langle\bar{q}q\rangle = 0$. Although nothing is said about the onset of the conformal regime in this work, the vanishing of $\langle\bar{q}q\rangle$ may be a signal. More definite is the conclusion of the lattice investigation [101]. It is claimed that at $N_f = 7$ not only the quark condensate disappears, $\langle\bar{q}q\rangle = 0$, but the decay law of the correlation functions at large distances changes from the exponential to power-like, i.e. the theory switches from the confining regime to the conformal one. One could speculate that at $N_f = 5$ and 6 we are in the intermediate phase, when confinement is still operative at large distances but the chiral symmetry is unbroken. It was believed previously that confinement of color in QCD implies the chiral symmetry breaking [102]. Today's wisdom tells us that this need not be necessarily the case. At least in some supersymmetric gauge theories we have both: confinement and unbroken chiral symmetry (see e.g. Ref. [97]). In any case, we see that ascending from $N_f = 3$ to $N_f = 5$ or 7 — quite a modest variation — we drastically change the vacuum structure of the theory, with the corresponding abrupt change of the picture of the hadron world.

An interesting question is what happens slightly below the left edge of the conformal window N_f^*. If N_f were a continuous parameter, and the phase transition in N_f were of the second order, slightly below N_f^* the string tension σ would be parametrically small in its natural scale given by Λ^2. In actuality N_f changes discretely. Still it may well happen that at $N_f = N_f^* - 1$ the ratio

σ/Λ^2 is numerically small. This would mean that the string and its excitations, whose scale is set by σ, are abnormally light. Under the circumstances one might hope to build effective low-energy approaches analogous to the chiral Lagrangians of actual QCD. In the limit $\sigma/\Lambda^2 \to 0$ the string becomes, in a sense, classic; the quantum corrections are unimportant. I do not rule out that the string representation of QCD — the holy grail of two generations of theorists — is easier to construct in this limit. One could even dream of an expansion in $N_f^* - 3$. At this moment, this is a pure speculation, however, and I have to wind up.

13. Instead of Conclusions

As time passes, the hope that the full analytic solution of QCD will be found, fades away. After all, the problem is with us for over a quarter of the century, and none of the numerous theoretical attacks have reached the goal. Many theorists whose philosophy is "all or nothing" abandoned the field.[17] Does it mean that the theory of hadrons is an aging science at the verge of submerging into a permanently dormant state?

I do not think so. Those theorists who stayed in the field can claim many partial successes. The field is messy and down-to-earth but — what can we do? — this is the only world God gave to us. The theory of hadrons is alive. Our understanding of the vacuum structure of quantum chromodynamics and the various dynamical regimes that can be supported continues to grow. Especially fruitful were the last years, with many breakthrough discoveries in supersymmetric gauge theories. These discoveries are to be used as hints and insights in actual QCD.

OPE-based methods — and the SVZ sum rules is one of them — continue to grow too. They are admittedly approximate, but their analytic nature and a graphic physical interpretation make them indispensable in several important practical problems. In some issues they share their role with other methods, which were developed later, for instance, lattice QCD. Since the theory of hadrons is so difficult it is in our best interests to combine all sources of information. We are in no position to neglect one of them in favor of the other. Many ideas and theoretical devices surfaced in connection with the

[17]The "all or nothing" philosophy is wide-spread and, unfortunately, not only in theoretical physics. This is a favorite child of the so-called revolutionaries in all times and in all countries. The misfortunes it brought to our world are innumerable. Needless to say, I personally think this is a rotten philosophy.

SVZ method (e.g. the low-energy theorems, Sec. 7); the method also produced a number of variations and spin-offs. This is a healthy process which will hopefully continue.

Acknowledgments

I am grateful to Prof. A. Zichichi for his kind invitation to lecture at XXXIII International School of Subnuclear Physics *Vacuum and Vacua: The Physics of Nothing*, in July 1995, Erice, Italy, and for the kind hospitality extended to me at the Ettore Majorana Center for Scientific Culture.

It is my pleasure to thank Prof. T. Suzuki and Prof. T. Kunihiro who were my hosts during my 1997 visit to Japan, for their kind hospitality. Prof. T. Suzuki kindly arranged this lecture in the framework of the International Workshop *NonPerturbative QCD — Structure of the QCD Vacuum*, Yukawa Institute for Theoretical Physics, Kyoto, December 2–12, 1997.

Useful communications with V. Braun, H. G. Dosch, B. L. Ioffe, A. Kaidalov, A. Khodjamirian, A. Radyushkin and E. Shuryak are acknowledged. I would like to thank A. Vainshtein for numerous illuminating discussions.

This work was supported in part by DOE under the grant number DE-FG02-94ER40823.

14. Appendix. Some Useful Definitions and Formulae

The Gell-Mann-Low function is defined as

$$\frac{\partial \alpha_s}{\partial \ln \mu} \equiv \beta(\alpha_s) = -b\frac{\alpha_s^2}{2\pi} - b_1\frac{\alpha_s^3}{4\pi^2} + \dots\,. \tag{A.1}$$

This definition is standard except that in some sources the first coefficient is called β_0, the second β_1, and so on (cf. Ref. [6]). At two loops the β function is scheme independent.

The coefficients are

$$b = 11 - \frac{2}{3}N_f\,, \qquad b_1 = 51 - \frac{19}{3}N_f\,. \tag{A.2}$$

The running coupling $\alpha_s(\mu)$ is parametrized as follows

$$\alpha_s(\mu) = \frac{4\pi}{b\ln\frac{\mu^2}{\Lambda^2}}\left(1 - \frac{2b_1}{b^2}\frac{\ln\ln\frac{\mu^2}{\Lambda^2}}{\ln\frac{\mu^2}{\Lambda^2}} + \dots\right)\,. \tag{A.3}$$

A more convenient form of the very same expression is

$$\frac{2\pi}{\alpha_s(\mu)} = b \ln \frac{\mu}{\Lambda} + \frac{b_1}{b} \ln \ln \frac{\mu^2}{\Lambda^2} + \ldots . \tag{A.4}$$

When expanded, at two loops, the latter expression is equivalent to the former; Eq. (A.4) effectively sums up some higher-order terms.

The scale parameter Λ following from Eq. (A.3) is

$$\Lambda^b = \left(\frac{2}{b}\right)^{b_1/b} \mu^b \exp\left\{-\left[\frac{2\pi}{\alpha_s(\mu)} - \frac{b_1}{b} \ln \frac{2\pi}{\alpha_s(\mu)}\right]\right\} . \tag{A.5}$$

The first factor on the right-hand side is an (inconvenient) artifact of the definition. It could have been easily avoided. Since the definition is standard, we will keep it not to cause confusion.

Perturbative calculations in QCD, as a rule, are carried out in the so-called modified minimal subtraction ($\overline{\mathrm{MS}}$) scheme [103], using dimensional regularization. For massless quarks and gluons it works nicely; the treatment of the heavy quark mass thresholds in this approach is ugly. The most common is the step-function approximation [104]: immediately above the threshold the quark is declared massless, while below the threshold it is treated as infinitely heavy, so that it is frozen out and does not participate in loops. This is equivalent to treating N_f as a function of μ constructed from several step functions, $N_f = 3$ below m_c, then it jumps to $N_f = 4$ between m_c and m_b, etc. Matching conditions at thresholds require equivalence of one effective theory with N_f massless quarks to another effective theory with $N_f - 1$ massless quarks.

If the step-function approximation is applied, at one loop the coupling constant is obviously continuous, while the first μ derivative experiences a jump. At two and three loops the coupling itself becomes discontinuous if the matching is done at the quark masses [105]. Several suggestions as to how one can smooth out the running coupling constant using various "physical" definitions were presented in the literature [106]. I propose a somewhat different approach inspired by supersymmetry. In supersymmetric theories the quark mass thresholds can be accounted for *exactly*, to all orders [107]. For instance, consider supersymmetric QCD with one flavor, with the mass term m. Assume that we start our evolution at a high normalization point M_0 (the corresponding coupling constant is α_{s0}), pass the threshold and descend down to $\mu \ll m$. A continuously running α_s can be obtained from the formula

$$\frac{2\pi}{\alpha_s(\mu)} = \frac{2\pi}{\alpha_{s0}} - 3N \ln \frac{M_0}{\mu(\alpha_{s0}/\alpha_s(\mu))^{1/3}} + \ln \frac{M_0}{\mu Z(\mu)} , \tag{A.6}$$

for SU(N) gauge group. Here $Z(\mu)$ is the Z factor of the matter fields and μ is arbitrary: larger or smaller than m. At $\mu = m$ and below, the second logarithm freezes at $\ln M_0/m_0$ where $m_0 = mZ$ is the mass parameter normalized at M_0. The one loop Z factor implies $\alpha_s(\mu)$ at two loops. Needless to say that $Z(\mu)$ varies continuously at one loop. At the two-loop level the high- and low-energy scale parameters are related as follows:

$$\Lambda_{\rm LOW}^{3N} = \Lambda_{\rm HIGH}^{3N-1}\left(\frac{2}{3N}\right)^{N}\left(\frac{3N-1}{2}\right)^{N-\frac{2C_2}{3N-1}}\left[m_0\left(\frac{2\pi}{\alpha_{s0}}\right)^{2C_2/(3N-1)}\right], \qquad (A.7)$$

where

$$C_2 = \frac{N^2-1}{2N}.$$

At this level of accuracy m_0 and α_{s0} in the square brackets can be treated in the leading logarithmic approximation. In this approximation the expression in the square brackets is renormalization-group invariant. In principle, the procedure can be extended to all loops, but we will not pursue this goal here.

In non-supersymmetric QCD the exact treatment of the mass thresholds seems impossible. However, numerically the situation is quite close to what we have in supersymmetric QCD. One can work out compact formulae applicable at two loops. Say, for the charm threshold

$$\frac{2\pi}{\alpha_s(\mu)} = \frac{2\pi}{\alpha_{s0}} - 9\ln\frac{M_0}{\mu(\alpha_{s0}/\alpha_s(\mu))^{32/81}} + \frac{2}{3}\ln\frac{M_0}{\mu Y(\mu)}, \qquad (A.8)$$

where

$$Y = \left[\frac{\alpha_{s0}}{\alpha_s(\mu)}\right]^{3y/25}$$

and

$$y = \frac{107}{18}.$$

Correspondingly,

$$\Lambda_{\rm LOW}^{9} = \Lambda_{\rm HIGH}^{25/3}\left(\frac{2}{9}\right)^{32/9}\left(\frac{25}{6}\right)^{77/25}\left[m_0\left(\frac{2\pi}{\alpha_{s0}}\right)^{12/25}\right]^{2/3}\left[\frac{2\pi}{\alpha_s(m)}\right]^{7/45}. \qquad (A.9)$$

If it were not for the last factor (which is quite close to unity) the relation between the low- and high-energy scales would be similar to that in supersymmetric QCD.

Matching at two thresholds, charm and beauty, yields

$$\Lambda^9_{\text{LOW}} = \Lambda^{23/3}_{\text{HIGH}} \left(\frac{2}{9}\right)^{32/9} \left(\frac{23}{6}\right)^{58/23} [m_1^* m_2^*]^{2/3}$$
$$\times \left[\frac{2\pi}{\alpha_s(m_2)}\right]^{21/115} \left[\frac{2\pi}{\alpha_s(m_1)}\right]^{7/45} , \tag{A.10}$$

where

$$m_i^* = m_{i0} \left[\frac{2\pi}{\alpha_{s0}}\right]^{12/23} .$$

References

[1] M. Shifman, A. Vainshtein and V. Zakharov, *Nucl. Phys.* **B147** (1979) 385, 448.

[2] *Vacuum Structure and QCD Sum Rules*, Ed. M. Shifman (North-Holland, Amsterdam, 1992).

[3] V. Novikov, M. Shifman, A. Vainshtein and V. Zakharov, *Nucl. Phys.* **B191** (1981) 301.

[4] D. Gross and F. Wilczek, *Phys. Rev. Lett.* **30** (1973) 1343; H. D. Politzer, *Phys. Rev. Lett.* **30** (1973) 1346.

[5] V. Novikov *et al.*, *Phys. Rep.* **41** (1978) 1.

[6] See e.g. the review paper I. Hinchliffe, *Phys. Rev.* **D54** (1996) 77 and references therein.

[7] G. 't Hooft, *The Confinement Phenomenon in Quantum Field Theory*, 1981 Cargese Summer School Lecture Notes on Fundamental Interactions, Eds. M. Levy and J.-L. Basdevant, NATO Adv. Study Inst. Series B: Phys. vol. 85, p. 639 [reprinted in G. 't Hooft, *Under the Spell of the Gauge Principle* (World Scientific, Singapore, 1994) p. 514]; S. Mandelstam, *Phys. Rep.* **23** (1976) 245.

[8] For a review see e.g. A. Di Giacomo, *Nucl. Phys. Proc. Suppl.* **47** (1996) 136 [hep-lat/9509036]; *Monopole Condensation in Gauge Theory Vacuum*, in *Confinement 95*, Proc. Int. Workshop on Color Confinement and Hadrons, Japan, March 1995, Eds. H. Toki, Y. Mizuno, H. Suganuma, T. Suzuki, O. Miyamura (World Scientific, Singapore, 1995) [hep-lat/9505006]; *Mechanisms for Color Confinement*, in *Selected Topics in Nonperturbative QCD*, Proc. Int. School of Physics "Enrico Fermi", Eds. A. Di Giacomo and D. Diakonov (IOS Press, 1996) p. 147.

[9] N. Seiberg and E. Witten, *Nucl. Phys.* **B426** (1994) 19; (E) **B430** (1994) 485; *Nucl. Phys.* **B431** (1994) 484.

[10] G. 't Hooft, *Nucl. Phys.* **B72** (1974) 461; E. Witten, *Nucl. Phys.* **B160** (1979) 57; for a brief review see Chap. 8 in [2].

[11] A. B. Kaidalov, *Usp. Fiz. Nauk* **105** (1971) 97 [*Sov. Phys. Uspekhi* **14** (1972) 600]; P. D. B. Collins, *An Introduction to Regge Theory and High Energy Physics* (Cambridge Univ. Press, 1977).

[12] A. Yu. Dubin, A. B. Kaidalov and Yu. A. Simonov, *Phys. Lett.* **B323** (1994) 41.

[13] A. Casher, N. Neurberger and S. Nussinov, *Phys. Rev.* **D20** (1979) 179.

[14] B. Blok, M. Shifman and D.-X. Zhang, *Phys. Rev.* **D57** (1998) 2691; Erratum, *Phys. Rev.* **D59** (1999) 019901.

[15] G. 't Hooft, *Nucl. Phys.* **B75** (1974) 461 [Reprinted in G. 't Hooft, *Under the Spell of the Gauge Principle* (World Scientific, Singapore, 1994) p. 443]; see also F. Lenz, M. Thies, S. Levit and K. Yazaki, *Ann. Phys.* (N.Y.) **208** (1991) 1; C. Callan, N. Coote and D. Gross, *Phys. Rev.* **D13** (1976) 1649; M. Einhorn, *Phys. Rev.* **D14** (1976) 3451; M. Einhorn, S. Nussinov and E. Rabinovici, *Phys. Rev.* **D15** (1977) 2282; I. Bars and M. Green, *Phys. Rev.* **D17** (1978) 537.

[16] M. Shifman, *Theory of Preasymptotic Effects in Weak Inclusive Decays*, in Proc. Workshop on *Continuous Advances in QCD*, Ed. A. Smilga (World Scientific, Singapore, 1994) p. 249 [hep-ph/9405246]; *Recent Progress in the Heavy Quark Theory*, in *Particles, Strings and Cosmology*, Proc. XIX Johns Hopkins Workshop on Current Problems in Particle Theory and V PASCOS Interdisiplinary Symposium, Baltimore, March 1995, Ed. J. Bagger (World Scientific, Singapore, 1996), p. 69 [hep-ph/9505289].

[17] F. David, *Nucl. Phys.* **B209** (1982) 433; **B234** (1984) 237.

[18] V. Novikov, M. Shifman, A. Vainshtein and V. Zakharov, *Nucl. Phys.* **B249** (1985) 445.

[19] I. Bigi, M. Shifman, N. Uraltsev and A. Vainshtein, *Phys. Rev.* **D52** (1995) 196.

[20] R. Crewther, *Phys. Rev. Lett.* **28** (1972) 1421; M. Chanowitz and J. Ellis, *Phys. Lett.* **B40** (1972) 397; *Phys. Rev.* **D7** (1973) 2490; J. Collins, L. Duncan and S. Joglekar, *Phys. Rev.* **D16** (1977) 438.

[21] M. Neubert, *Phys. Rev.* **D51** (1995) 5924.

[22] P. Ball, M. Beneke and V. Braun, *Nucl. Phys.* **B452** (1995) 563.

[23] G. 't Hooft, *Can We Make Sense out of "Quantum Chromodynamics"?* in *The Whys Of Subnuclear Physics*, Erice 1977, Ed. A. Zichichi (Plenum, New York, 1977) p. 943 [reprinted in G. 't Hooft, *Under the Spell of the Gauge Principle* (World Scientific, Singapore, 1994) p. 547].

[24] B. Lautrup, *Phys. Lett.* **69B** (1977) 109; G. Parisi, *Phys. Lett.* **76B** (1978) 65; *Nucl. Phys.* **B150** (1979) 163; A. Mueller, *Nucl. Phys.* **B250** (1985) 327.

[25] A. H. Mueller, in Proc. Int. Conf. *QCD — 20 Years Later*, Aachen 1992, Eds. P. Zerwas and H. Kastrup (World Scientific, Singapore, 1993), vol. 1, p. 162.

[26] R. Akhoury and V. Zakharov, *The Physics of the Ultraviolet Renormalon*, *Nucl. Phys. Proc. Suppl.* **64** (1998) 350 [hep-ph/9710257].

[27] R. Akhoury and V. Zakharov, *Nucl. Phys.* Proc. Suppl. **54A** (1997) 217; M. Beneke, hep-ph/9706457; V. Braun, *Proc. $\bar{V}$ Int. Conf. on Physics Beyond*

the Standard Model, eds. G. Eigen, P. Osland, and B. Stugu (Woodbury, AIP, 1997), p. 341 [hep-ph/9708386]; B. R. Webber [hep-ph/9712236].

[28] A. Di Giacomo, *Lattice Gauge Theories and SVZ Expansion*, in *Nonperturbative Methods*, Proc. 1985 Montpellier Int. Workshop, Ed. S. Narison (World Scientific, Singapore, 1986), p. 135.

[29] X. Ji, hep-ph/9506216 (unpublished).

[30] M. Shifman, A. Vainshtein, M. Voloshin and V. Zakharov, *Phys. Lett.* **B77** (1978) 80.

[31] J. Gasser and H. Leutwyler, *Nucl. Phys.* **B94** (1975) 269.

[32] S. Weinberg, in *A Festschrift for I. I. Rabi*, Ed. L. Motz, *Trans. New York Acad. Sci.* Ser. II **38** (1977) 185.

[33] H. Leutwyler, in *Proc. Cargése Summer School on Masses of Fundamental Particles*, eds. M. Levy *et al.* (Plenum, New York, 1997), p. 149 [hep-ph/9609467] and a forthcoming book.

[34] T. Bhattacharya and R. Gupta, *Nucl. Phys. Proc. Suppl.* **63** (1998) 95 [hep-lat/9710095].

[35] For a review see B. L. Ioffe, *Acta Phys. Polon.* **B16** (1985) 543.

[36] V. Novikov *et al.*, in *Neutrinos 78*, Proc. VIII Int. Conference on Physics and Neutrino Astrophysics, West Lafayette, April 1978, Ed. E. C. Fowler (Purdue University, 1978) p. C278.

[37] M. Voloshin, *Nucl. Phys.* **B154** (1979) 365; L. J. Reinders, H. Rubinstein and S. Yazaki, *Phys. Rep.* **127** (1985) 1; S. N. Nikolaev and A. V. Radyushkin, *Nucl. Phys.* **B213** (1983) 285; *Phys. Lett.* **B124** (1983) 243.

[38] M. Dubovikov and A. Smilga, *Nucl. Phys.* **B185** (1981) 109.

[39] M. Crisafulli *et al.*, *Phys. Lett.* **B369** (1996) 325.

[40] K. G. Chetyrkin *et al.*, *Phys. Lett.* **B174** (1986) 104; L. Reinders and S. Yazaki, *Nucl. Phys.* **B288** (1987) 789; K. Chetyrkin and A. Pivovarov, *Nuov. Cim.* **A100** (1988) 899. Literally speaking, these authors test factorization for the four-quark operators averaged over the Goldstone mesons, π, K, etc. rather than over the vacuum state. Within the soft pion technique the former matrix elements are reducible to the latter.

[41] The two-loop correction was found by K. Chetyrkin, A. Kataev and F. Tkachev, *Phys. Lett.* **B85** (1979) 277; M. Dine and J. Sapirstein, *Phys. Rev. Lett.* **43** (1979) 668; W. Celmaster and R. Gonsalves, *Phys. Rev. Lett.* **44** (1980) 560. Three-loop calculations are due to S. Gorishny, A. Kataev and S. Larin, *Phys. Lett.* **B259** (1991) 144; L. Surguladze and M. Samuel, *Phys. Rev. Lett.* **66** (1991) 560; (E) **66** (1991) 2416; K. Chetyrkin, *Phys. Lett.* **B391** (1997).

[42] S. Eidelman, L. Kurdadze, M. Shifman and A. Vainshtein, unpublished.

[43] S. Gorishny, A. Kataev and S. Larin, *Pisma ZhETF* **53** (1991) 121 [*JETP Lett.* **53** (1991) 127].

[44] V. Novikov, M. Shifman, A. Vainshtein and V. Zakharov, *Fortsch. Phys.* **32** (1984) 585.

[45] L. Surguladze and F. Tkachev, *Nucl. Phys.* **B 331** (1990) 35.

[46] D. Broadhurst, A. G. Grozin, in Proc. Int. Workshop on Software Engineering and Artificial Intelligence and Expert Systems for High Energy and Nuclear Physics *New Computing Techniques in Physics Research IV*, Pisa, Italy, April 1995, Eds. B. Denby and D. Perret-Gallix (World Scientific, Singapore, 1995), p. 217 [hep-ph/9504400]; L. Surguladze and M. Samuel, *Rev. Mod. Phys.* **68** (1996) 259.

[47] R. Barate *et al.* (ALEPH Collab.), *Z. Phys.* **C76** (1997) 15.

[48] B. Chibisov, R. D. Dikeman, M. Shifman and N. Uraltsev, *Int. J. Mod. Phys.* **A12** (1997) 2075.

[49] M. Shifman, *Yad. Fiz.* **36** (1982) 1290 [*Sov. J. Nucl. Phys.* **36** (1982) 749].

[50] K. Wilson, *Phys. Rev.* **179** (1969) 1499; K. Wilson and J. Kogut, *Phys. Rep.* **12** (1974) 75.

[51] *Large-Order Behaviour of Perturbation Theory*, Eds. J. C. Le Guillou and J. Zinn-Justin, (North-Holland, Amsterdam, 1990).

[52] M. S. Dubovikov and A. V. Smilga, *Yad. Fiz.* **37** (1983) 984 [*Sov. J. Nucl. Phys.* **37** (1983) 585].

[53] V. Novikov, M. Shifman, A. Vainshtein and V. Zakharov, *Phys. Rep.* **116** (1984) 103.

[54] X. Ji, hep-ph/9506216, hep-ph/9507322 (unpublished).

[55] E. Shuryak, *Nucl. Phys.* **B198** (1982) 83; D. Diakonov and V. Petrov, *Nucl. Phys.* **B245** (1984) 259.

[56] E. Shuryak, *The QCD Vacuum, Hadrons and the Superdense Matter* (World Scientific, Singapore, 1988) [World Scientific Lecture Notes in Physics, v. 8]; T. Schäfer and E. Shuryak, *Rev. Mod. Phys.* **70** (1998) 323 and references therein.

[57] R. Crewther, *Phys. Lett.* **B70** (1977) 349; for a review see G. A. Christos, *Phys. Rep.* **116** (1984) 251.

[58] E. Witten, *Nucl. Phys.* **B156** (1979) 269.

[59] G. Veneziano, *Nucl. Phys.* **B159** (1979) 213.

[60] M. Voloshin, *Yad. Fiz.* **44** (1986) 738 [*Sov. J. Nucl. Phys.* **44** (1986) 478]; *Yad. Fiz.* **45** (1987) 190 [*Sov. J. Nucl. Phys.* **45** (1987) 122].

[61] H. Leutwyler and M. Shifman, *Phys. Lett.* **B221** (1989) 384.

[62] I. Balitsky, D. Diakonov and A. Yung, *Phys. Lett.* **B112** (1982) 71; *Z. Phys.* **C33** (1986) 265; for a review see D. Diakonov, in Proc. XXI Winter School of Physics (LINP, Leningrad, 1986), p. 3.

[63] M. Shifman, *Phys. Rep.* **209** (1991) 341.

[64] M. Shifman, *Z. Phys.* **C9** (1981) 347.

[65] T. Schäfer and E. Shuryak, *Phys. Rev. Lett.* **75** (1995) 1707.

[66] C. Michael, in Proc. X Les Rencontres de Physique de la Vallee d'Aoste: Results and Perspectives in Particle Physics, La Thuile, Italy, March 1996; *Results and Perspectives in Particle Physics*, Ed. M. Greco (Istituto Naz. Fis. Nucl., 1996) [hep-ph/9605243]; T. DeGrand, hep-th/9610132.

[67] P. van Baal and A. Kronfeld, *Nucl. Phys.* **B** (Proc. Suppl.) **9** (1989) 227.

[68] K. Ishikawa, G. Schierholz, H. Schneider and M. Teper, *Nucl. Phys.* **B227** (1983) 221; T. DeGrand, *Phys. Rev.* **D36** (1987) 176; R. Gupta *et al.*, *Phys. Rev.* **D43** (1991) 2301.
[69] P. de Forcrand and K.-F. Liu, *Phys. Rev. Lett.* **69** (1992) 245.
[70] C. E. Carlson, T. H. Hansson and C. Peterson, *Phys. Rev.* **D27** (1983) 1556; (E) **D28** (1983) 2895; M. Chanowitz and S. Sharpe, *Nucl. Phys.* **B222** (1983) 211.
[71] B. L. Ioffe and A. V. Smilga, *JETP Lett.* **37** (1983) 298; *Nucl. Phys.* **B232** (1984) 109; I. Balitsky and A. Yung, *Phys. Lett.* **B129** (1983) 328; see also Chap. 7 in [2].
[72] V. Nesterenko and A. Radyushkin, *Phys. Lett.* **B115** (1982) 410; B. L. Ioffe and A. V. Smilga, *Nucl. Phys.* **B216** (1983) 373; see also Chap. 6 in [2].
[73] B. Blok and M. Shifman, *Sov. J. Nucl. Phys.* **45** (1987) 135; 301; 522; *Sov. J. Nucl. Phys.* **46** (1987) 1310; for a review and a representative list of references see A. Khodjamirian and R. Rückl, hep-ph/9801443.
[74] V. Belyaev and B. L. Ioffe, *Nucl. Phys.* **B130** (1998) 548; *Int. J. Mod. Phys.* **A6** (1991) 1533; B. L. Ioffe and A. Khodjamirian, *Phys. Rev.* **D51** (1995) 3373.
[75] E. Shuryak, *Nucl. Phys.* **B198** (1982) 83.
[76] T. Aliev and V. Eletsky, *Sov. J. Nucl. Phys.* **38** (1983) 936.
[77] E. Bagan, P. Ball, V. Braun and H. G. Dosch, *Phys. Lett.* **B278** (1992) 457.
[78] V. L. Chernyak and A. R. Zhitnitsky *JETP Lett.* **25** (1977) 510; A. V. Efremov and A. V. Radyushkin, *Phys. Lett.* **B94** (1980) 245; G. P. Lepage and S. J. Brodsky, *Phys. Lett.* **B87** (1979) 359; *Phys. Rev.* **D22** (1980) 2157; for a review see V. L. Chernyak and A. R. Zhitnitsky, *Phys. Rep.* **112** (1984) 173.
[79] V. M. Belyaev, V. M. Braun, A. Khodjamirian and R. Rückl *Phys. Rev.* **D51** (1995) 6177.
[80] A. V. Radyushkin, *Nucl. Phys.* **A532** (1991) 141 in *Perspectives in Hadronic Physics*, eds. S. Boffi *et al.* (World Scientific, Singapore, 1997) [hep-ph/9707335].
[81] A. Khodjamirian *Eur. Phys. J.* **C6** (1999) 477.
[82] P. Ball, V. M. Braun and H. G. Dosch, *Phys. Rev.* **D44** (1991) 3567.
[83] N. S. Craigie and J. Stern, *Nucl. Phys.* **B216** (1983) 209; I. Balitsky, V. Braun and A. Kolesnichenko, *Nucl. Phys.* **B312** (1989) 509 and earlier works of the same authors cited therein.
[84] V. Braun, *Proc. Int. Workshop Progress in Heavy Quark Physics*, eds. M. Beyer, T. Mannel and H. Schröder (Univ. of Rostock Press, 1998) p. 105 [hep-ph/9801222].
[85] S. V. Mikhailov and A. V. Radyushkin, *Phys. Rev.* **D45** (1992) 1754; A. V. Radyushkin, *Phys. Lett.* **B271** (1991) 218; A. P. Bakulev and A. V. Radyushkin, *Phys. Lett.* **B271** (1991) 223.
[86] I. Bigi *et al.*, *Phys. Lett.* **B339** (1994) 160.
[87] N. Isgur and M. Wise, *Phys. Rev.* **D43** (1991) 819.
[88] M. C. Chu, *et al.*, *Nucl. Phys.* Proc. Suppl. **30** (1993) 495; J. W. Negele, M. Burkardt and J. M. Grandy, *Ann. Phys.* (N.Y.) **238** (1995) 441.

[89] H. G. Dosch, O. Nachtmann and M. Reuter, hep-ph/9503386 (unpublished).
[90] C. Michael, *Nucl. Phys.* **B280** (1987) 13.
[91] H. Rothe, *Phys. Lett.* **B364** (1995) 227.
[92] N. Evans, S. Hsu and M. Schwetz, hep-th/9707260.
[93] M. Vysotsky, I. Kogan and M. Shifman, *Yad. Fiz.* **42** (1985) 504 [*Sov. J. Nucl. Phys.* **42** (1985) 318].
[94] G. 't Hooft, in *Recent Developments in Gauge Theories*, Eds. G. 't Hooft *et al.* (Plenum Press, New York, 1980).
[95] S. Dimopoulos, *Nucl. Phys.* **B168** (1980) 69; M. Peskin, *Nucl. Phys.* **B175** (1980) 197.
[96] V. Novikov, M. Shifman, A. Vainshtein and V. Zakharov, *Nucl. Phys.* **B229** (1983) 381; *Phys. Lett.* **B166** (1986) 329; M. Shifman and A. Vainshtein, *Nucl. Phys.* **B296** (1988) 445.
[97] A. Kovner and M. Shifman, *Phys. Rev.* **D56** (1997) 2396.
[98] A. Belavin and A. Migdal, *Pis'ma ZhETF* **19** (1974) 317 [*JETP Lett.* **19** (1974) 181]; *Scale Invariance and Bootstrap in the Non-Abelian Gauge Theories*, Landau Institute Preprint-74-0894, 1974 (unpublished).
[99] T. Banks and A. Zaks, *Nucl. Phys.* **B196** (1982) 189.
[100] T. Schäfer and E. Shuryak, *Phys. Rev.* **D53** (1996) 6522.
[101] Y. Iwasaki *et al.*, *Z. Phys.* **C71** (1996) 343.
[102] A. Casher, *Phys. Lett.* **B83** (1979) 395.
[103] W. A. Bardeen *et al.*, *Phys. Rev.* **D18** (1978) 3998.
[104] W. Marciano, *Phys. Rev.* **D29** (1984) 580; R. M. Barnett, H. E. Haber and D. E. Soper, *Nucl. Phys.* **B306** (1988) 697; G. Rodrigo and A. Santamaria, *Phys. Lett.* **B313** (1993) 441.
[105] W. Wetzel, *Nucl. Phys.* **B196** (1982) 259; W. Bernreuther and W. Wetzel, *Nucl. Phys.* **B197** (1982) 228; W. Bernreuther, *Ann. Phys.* (N.Y.) **151** (1983) 127; *Z. Phys.* **C20** (1983) 331; S. Larin, T. van Ritbergen and J. A. M. Vermaseren, *Nucl. Phys.* **B438** (1995) 278; K. Chetyrkin, B. A. Kniehl and M. Steinhauser, *Phys. Rev. Lett.* **79** (1997) 2184.
[106] S. Brodsky, M. Gill and J. Rathman, *Phys. Rev.* **D58** (1998) 116006, and references therein.
[107] M. Shifman, *Int. J. Mod. Phys.* **A11** (1996) 5761.

Recommended Literature

Two review papers B. L. Ioffe, *Acta Phys. Polon.* **B16** (1985) 543; L. J. Reinders, H. Rubinstein and S. Yazaki, *Phys. Rep.* **127** (1985) 1 were published in the mid-1980's. They may serve for an initial exposure but they do not provide an overview of modern developments. Neither do they reflect modern understanding of conceptual issues. A dedicated comprehensive review on this subject is long overdue.

A brief survey of the light-cone sum rules is given in V. Braun, Proc. Int. Workshop *Progress in Heavy Quark Physics*, Eds. M. Beyer, T. Mannel, and H. Schröder (Univ. of Rostock Press, 1998), p. 105 [hep-ph/9801222].

SVZ sum rules in the heavy quark theory are discussed in M. Neubert, *Phys. Rep.* **245** (1994) 259.

For a review of the sum rule applications in weak decays see A. Khodjamirian and R. Rückl, in *Heavy Flavors*, second edition, eds. A. Buras and M. Linder (Would Scientific, Singapore, 1998) p. 345 [hep-ph/9801443].

Some topics related to the SVZ sum rules are covered in M. Shifman, *Phys. Rep.* **209** (1991) 341; T. Schäfer and E. Shuryak, *Rev. Mod. Phys.*, **70** (1998) 323; *Annu. Rev. Nucl. Part. Sci.* **47** (1997) 359.

A collection of the original papers with an extended commentary reflecting the state of the art in the late 1980's can be found in *Vacuum Structure and QCD Sum Rules*, Ed. M. Shifman (North-Holland, Amsterdam, 1992).

Some useful formulae are compiled in S. Narison, *QCD Spectral Sum Rules*, (World Scientific, Singapore, 1989). I do not agree with the treatment of many conceptual issues and technical details in this book.

Chapter III

ABC of Instantons

V. A. Novikov,[a] M. A. Shifman,[b] A. I. Vainshtein[b] and V. I. Zakharov[c]

[a] *Institute of Theoretical and Experimental Physics, Moscow 117259, Russia*
[b] *Theoretical Physics Institute, University of Minnesota, Minneapolis, MN 55455, USA*
[c] *Max-Planck Institute für Physik, 80805 München, Germany*

This lecture was first delivered at XVI Winter School of Physics of Leningrad Institute for Nuclear Physics in spring 1981. Revised April 1994. First published in Russian in *Usp. Phys. Nauk* **136** (1982) 553.

Abstract

These lectures present an extended introduction to instantons in gauge theories. The lectures consist of several distinct parts. To reveal the physical meaning of instantons we consider in detail the simplest quantum-mechanical problem where they appear: tunneling in the double-well potential. This pedagogical example was suggested by Polyakov. Then we proceed to quantum chromodynamics (QCD). The discovery of instantons was instrumental in the understanding of the vacuum structure of QCD. The θ vacuum is described from the quasiclassical perspective. The second part is devoted to the instanton formalism. We discuss various aspects of the instanton calculations: the solution *per se* in different gauges, the instanton measure in QCD and in the Higgs phase, the impact of external background fields. A related topic we dwell on is the sphaleron and its interpretation. Finally, the last part deals with the massless fermions in the instanton transitions. Their impact is drastic both at the conceptual and technical levels. We explain how the tunneling interpretation changes in the presence of the massless fermions. If the fermions are chiral rather than Dirac, under certain conditions the theory becomes ill-defined (Witten's global anomaly). Although these lectures are self-contained, they are best read in conjunction with Coleman's lecture *The Uses of Instantons* [S. Coleman, *Aspects of Symmetry* (Cambridge University Press, London, 1985), p. 265].

Contents

Introduction

It appears that all fundamental interactions in nature are of the gauge type. The modern theory of hadrons — quantum chromodynamics (QCD) — is no exception. It is based on local gauge invariance with respect to the color group SU(3), which is realized by an octet of massless gluons. The idea of gauge invariance, however, is much older and derives from quantum electrodynamics, which was historically the first field-theoretical model in which successful predictions were obtained. By the end of the forties, theoreticians had already learned how to calculate all observable quantitites in electrodynamics in the form of series in $\alpha = 1/137$. The first steps in QCD in the mid-1970's were also made in the framework of perturbation theory. However, it gradually became clear that, in contrast to electrodynamics, quark–gluon physics is not exhausted by perturbation theory. The most interesting phenomena — the confinement of colored objects and the formation of the hadron spectrum — are associated with nonperturbative (i.e. not describable in the framework of perturbation theory) effects. The latter, in turn, are due to complicated structure of the QCD vacuum, which is filled with fluctuations of the gluon field.

It is now clear that the construction of the complete analytical "wave function" of the vacuum is a very difficult problem. Despite numerous attacks by theoreticians it still remains unsolved. Nevertheless, quite a lot is already known. The study of "old," traditional hadrons gives information about the fundamental properties of the vacuum. In turn, having obtained this information, we can make a number of nontrivial predictions about gluonium and other poorly investigated aspects of hadron phenomenology.

The corresponding approach has been developed by the authors over a number of years, but it will not be discussed here. We note only that the main element is the introduction of several vacuum expectation values. For example, the intensity of gluon fields in vacuum is obviously measured by the quantity [1]

$$\langle 0|\, G^a_{\mu\nu} G^a_{\mu\nu}\, |0\rangle\,,$$

where $G^a_{\mu\nu}$ is the gluon field strength tensor ($a = 1, \ldots, 8$ is the color index). Similarly, the quark condensate expectation value $\langle 0|\bar{q}q|0\rangle$ serves as a measure of the quark fields.

In the "final theory," if such is constructed, it will be possible to calculate all phenomenological matrix elements on the basis of the Lagrangian of QCD.

It can already be said that this will require knowledge of nonperturbative fluctuations in the physical vacuum. Here, phenomenology makes contact with the purely theoretical development, which as yet has not had great applications, though it has made possible the reexamination of a number of problems.

In 1975 one of the most beautiful phenomena in quantum chromodynamics was discovered, instantons, classical solutions of the field equations with nontrivial topology [2]. The beauty of the theoretical constructions has attracted the interest of many physicists and mathematicians, and it is difficult to overestimate the popularity of instantons. The importance of instantons as the first example of fluctuations of the gluon field not encompassed by perturbation theory is undoubted. Therefore, it appears appropriate to explain the physical essence of the phenomenon and derive the basic formulas to enable the reader to find his (or her) way about the literature.

The original Belavin-Polyakov-Schwarz-Tyupkin solution [3] (BPST instanton) may or may not be the fluctuation which is dominant in the vacuum wave function. Although there are some numerical evidence in favor of the instanton dominance [4] the arguments are far from being conclusive. The instanton-based models of the QCD vacuum do exist, but the last word in this line of research is yet to be said. Therefore, we will not dwell on this issue. Instead, we will focus on those aspects of the instanton calculus which are completely settled and will stay with us forever.

We begin with a simple quantum-mechanical problem that illustrates the role of nonperturbative fluctuations. This example was analyzed in detail by Polyakov [5], who made a major contribution to the development of the entire subject.

A double-well potential will be considered, and the famous problem of the level splittings will be solved by exploiting an instanton approach, which is rather awkward in this particular problem, but has an important advantage over the standard WKB method: it can be directly extended to field theory, while the standard method cannot. All technical elements of the instanton calculus (the Euclidean time, classical solutions, zero modes and determinants) which we will encounter later in QCD are introduced in this setting. Having dealt with the toy model we proceed to QCD. General arguments are presented revealing a nontrivial topology in the space of the gauge fields. The existence of distinct classical minima of the "potential" is demonstrated. Classical trajectories interpolating between these distinct minima ("pre-vacua") are BPST instantons. We discuss the explicit form of the instanton, and calculate, in a pedagogical manner, the instanton density. The notion of the vacuum angle θ is introduced.

We then briefly consider an applied aspect of the instanton calculus. Instantons submerged in background fields, produced by other fluctuations, deform. As a result of this deformation the instanton density changes. The change in the density caused by the gluon condensate is considered in some detail.

The second part of the lecture is devoted to the role of fermions. Massless fermions have a drastic impact both on interpretation of the instanton as a tunneling trajectory, and on all technical aspects of the instanton calculus. We first consider the Dirac fermions and explain how the instanton calculations must be modified. Then a more subtle problem of chiral fermions is addressed. Here we have to reanalyze anew the very foundations of the procedure, such as the Euclidean continuation. The chiral fermions are an indispensible element of supersymmetric guage theories. A brief excursion in the topic of supersymmetric instantons concludes the lecture.

1. Quantum Mechanics, Imaginary Time, Path Integrals

In this section, we consider the problem of the one-dimensional motion of a spinless particle in a potential $V(x)$. This problem is usually treated in all textbooks on quantum mechanics, but we shall use a somewhat unusual method to solve it. The reader may find it inconvenient, just as sum rules [1] are "inconvenient" for finding the eigenvalues of the Schrödinger equation. But — and this is the most important property — the method can be directly generalized to field theory.

If we take the mass of the particle equal to unity, $m = 1$, then the Lagrangian of the system has the simple form

$$\mathcal{L} = \frac{1}{2}\left(\frac{\mathrm{d}x}{\mathrm{d}t}\right)^2 - V(x)\,. \tag{1}$$

Suppose that the particle at the initial time $(-t_0/2)$ is at the point x_i and at the final time $(+t_0/2)$ at the point x_f. An elegant method of expressing the amplitude of such a process was invented by Feynman [6]. The prescription is that the amplitude is equal to the sum over *all* paths joining the world points $(-t_0/2, x_i)$ and $(t_0/2, x_f)$ taken with weight

$$e^{iS}\,.$$

The action, which we shall denote by the letter S in what follows, is related to the Lagrangian by

$$S = \int_{-t_0/2}^{t_0/2} \mathrm{d}t\, \mathcal{L}(x, \dot{x})\,. \tag{2}$$

Thus, the transition amplitude is

$$\langle x_f | e^{-iHt_0} | x_i \rangle = N \int [Dx] e^{iS[x(t)]}\,, \tag{3}$$

where H is the Hamiltonian and $\exp(-iHt_0)$ is the ordinary evolution operator of the system. The factor N on the right-hand side is a normalization factor, to the discussion of which we shall return below. $[Dx]$ denotes integration over all functions $x(t)$ with boundary conditions $x(-t_0/2) = x_i$ and $x(t_0/2) = x_f$.

Before we consider dynamical questions, we examine the left-hand side. If we pass from states with a definite coordinate to states with a definite energy,

$$H|n\rangle = E_n |n\rangle\,,$$

then, obviously,

$$\langle x_f | e^{-iHt_0} | x_i \rangle = \sum_n e^{-iE_n t_0} \langle x_f | n \rangle \langle n | x_i \rangle\,, \tag{4}$$

and we obtain a sum of oscillating exponentials. If we are interested in the ground state (and in field theory we are always interested in the lowest state — the vacuum), it is much more convenient to transform the oscillating exponentials into decreasing exponentials. To this end, we make the substitution $t \to -i\tau$. Then in the limit $\tau_0 \to \infty$ only a single term survives in the sum (Eq. (4)), and this directly tells us what are the energy E_0 and the wave function $\psi_0(x)$ of the lowest level, $e^{-E_0\tau_0}\psi_0(x_f)\psi_0^*(x_i)$.

In the literature, the transition to the imaginary time is frequently called the Wick rotation, and the corresponding version of the theory is referred to as the Euclidean version. Below, we shall see that the substitution $t \to -i\tau$ is in a certain sense not only a matter of convenience, since it gives a new language for describing a very important aspect of the theory.

We now turn to the right-hand side of Eq. (3). In the Euclidean formulation, the action takes the form

$$iS[x(t)] \to \int_{-\tau_0/2}^{\tau_0/2} \left[-\frac{1}{2}\left(\frac{\mathrm{d}x}{\mathrm{d}\tau}\right)^2 - V(x) \right] \mathrm{d}\tau\,, \tag{5}$$

where we assume the boundary condition $x(-\tau_0/2) = x_i$, $x(\tau_0/2) = x_f$, and the origin of the energy is chosen such that $\min V(x) = 0$.

We call

$$S_E = \int_{-\tau_0/2}^{\tau_0/2} \left[\frac{1}{2}\left(\frac{\mathrm{d}x}{\mathrm{d}\tau}\right)^2 + V(x)\right] \mathrm{d}\tau \tag{6}$$

the Euclidean action. Since $S_E \geq 0$, we have acquired an exponentially decreasing weight on the right-hand side of Eq. (3). In the present lecture, we shall remain in the Euclidean space and shall not return to the Minkowski space (i.e. to real time) until Sec. 13; therefore, in what follows we shall omit the subscript E.

The Euclidean variant of Eq. (3) is

$$\langle x_f | e^{-H\tau_0} | x_i \rangle = N \int [Dx] e^{-S} \,. \tag{7}$$

It is now time to make the next important step and explain what integration over all paths actually means. Let $X(\tau)$ be some function satisfying the boundary conditions. Then an *arbitrary* function with the same boundary conditions can be represented in the form

$$x(\tau) = X(\tau) + \sum_n c_n x_n(\tau) \,, \tag{8}$$

where $x_n(\tau)$ is a complete set of orthonormal functions that vanish at the boundary:

$$\int_{-\tau_0/2}^{\tau_0/2} \mathrm{d}\tau x_n(\tau) x_m(\tau) = \delta_{nm} \,, \qquad x_n\left(\pm \frac{\tau_0}{2}\right) = 0 \,.$$

The measure $[Dx]$ can be chosen in the form

$$[Dx] = \prod_n \frac{dc_n}{\sqrt{2\pi}} \,. \tag{9}$$

The coefficient of proportionality in this relation does not in general have in itself a particular meaning until the normalization factor N has been fixed.

Now suppose that in the problem under consideration the characteristic value of the action is large for certain reasons. Well-known is the situation when the quasiclassical approximation, or in other words, the method of steepest descent (the latter, "mathematical" term may be more readily understood by some readers), "works." In other words, the entire integral in (7) is accumulated from regions near the extremum (minimum) of S. The path corresponding to the least action, which we denote by $X(\tau)$, is known in the literature as an extremal path, an extremal, or a stationary point. If there is one extremal and $S[X(\tau)] = S_0$, then

$$N \int [Dx] e^{-S} \sim e^{-S_0} . \tag{10}$$

Thus, to find the principal, *exponential factor* in the result, it is sufficient to put in information about a single, extremal path. (If there are several stationary points, we have in general the sum of the contributions of all the stationary points.)

There exists a standard procedure which enables us to take the next step and fix the pre-exponential factor. This operation is already somewhat more laborious. Suppose for simplicity that there is a single stationary point, $X(\tau)$. The following formula expresses, in mathematical language, the fact that $X(\tau)$ realizes a minimum of the action:

$$\delta S = S[X(\tau) + \delta x(\tau)] - S[X(\tau)] = \int_{-\tau_0/2}^{\tau_0/2} d\tau \delta x(\tau) \left[-\frac{\mathrm{d}^2 X}{\mathrm{d}\tau^2} + V'(X) \right] = 0 ,$$

where $V' = \mathrm{d}V/\mathrm{d}x$. The equation

$$\frac{\mathrm{d}^2 X}{\mathrm{d}\tau^2} = V'(X) , \tag{11}$$

is of course well-known to the reader from school days (we recall that "the mass multiplied by the acceleration is equal to the force"). It is the *classical* equation of motion of a particle in the potential *minus* $V(x)$.[1]

We shall shortly return to this circumstance, but first recall how the pre-exponential factor in (10) is calculated. It is determined by an entire "beam" of paths near the extremal path, i.e. by the paths with action that differs little from S_0. In other words, we take into account only the quadratic deviation:

$$S[X(\tau) + \delta x(\tau)] = S_0 + \int_{-\tau_0/2}^{\tau_0/2} \mathrm{d}\tau \delta x \left[-\frac{1}{2} \frac{\mathrm{d}^2}{\mathrm{d}\tau^2} \delta x + \frac{1}{2} V''(X) \delta x \right] \tag{12}$$

(as the reader will recall, there is no term linear in the deviation).

[1]The minus sign is due to the fact that the Euclidean formulation is considered [see Ref. [6]].

Suppose we know a complete set of eigenfunctions and eigenvalues of the equation

$$-\frac{\mathrm{d}^2}{\mathrm{d}\tau^2}x_n(\tau) + V''(X)x_n(\tau) = \varepsilon_n x_n(\tau)\,. \tag{13}$$

Then we can choose these functions as the orthonormalized system which occurs in (8), and the action (12) is transformed to the simple *diagonal* form

$$S = S_0 + \frac{1}{2}\sum_n \varepsilon_n c_n^2\,.$$

Recalling the definition (9) and the rule of Gaussian integration

$$\int_{-\infty}^{+\infty} \mathrm{d}c \exp\left(-\frac{1}{2}\varepsilon c^2\right) = \frac{\sqrt{2\pi}}{\sqrt{\varepsilon}}$$

(it is important that after the diagonalization each such integration can be performed independently of the others), we obtain

$$\langle x_f|e^{-H\tau_0}|x_i\rangle = e^{-S_0}N\prod_n \varepsilon_n^{-1/2}\,. \tag{14}$$

Sometimes, instead of the product of eigenvalues one uses the notation

$$\prod_n \varepsilon_n^{-1/2} = \left[\det\left(-\frac{\mathrm{d}^2}{\mathrm{d}\tau^2} + V''(X(\tau))\right)\right]^{-1/2}, \tag{15}$$

which, of course, derives from the theory of ordinary finite-dimensional matrices. In fact, the relation (15) can be regarded as the definition of the determinant of a differential operator. It is here appropriate to make three comments. First, the result (14) does not depend on the explicit form of the eigenfunctions but only on the eigenvalues. Second, we have assumed that all the ε_n are positive. In most cases, this is so, but in the instanton example several eigenvalues vanish. The resulting infinity has a simple physical meaning. The problem of how it should be handled is the subject of the next section. The third and final comment is the following. The normalization factor N has yet to be fixed. We shall not attempt to give a general prescription but consider a simple example, which will serve us in the future too. Suppose the original particle with mass $m = 1$ is placed in the potential $V(x)$ shown in Fig. 1. We do not need the actual form of this potential, but to achieve "normalization" to the harmonic oscillator (in which the potential is usually taken to be $m\omega^2x^2/2$),

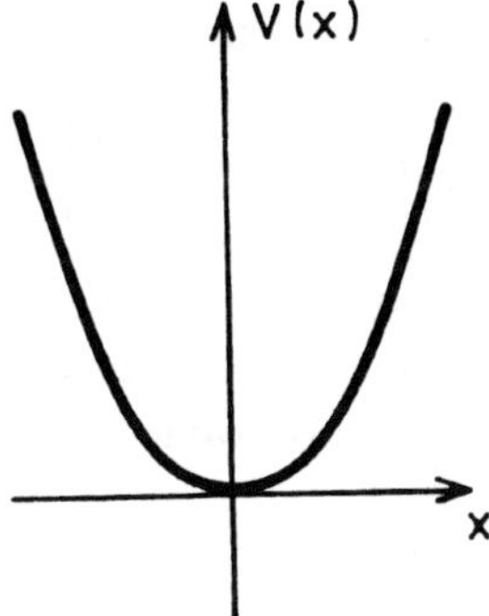

Fig. 1. The quantum-mechanical problem with the potential of the oscillator type.

we set $V''(x=0)=\omega^2$. As the initial and final points of the motion we choose $x_i = x_f = 0$.

The rich physical intuition that we each have for potential mechanical motion enables us to find the extremal from Eq. (11) without knowing the explicit form of $V(x)$. Indeed, this equation describes the motion of a ball on the profile shown in Fig. 2. At the time $-\tau_0/2$, the ball is displaced from the upper point, to which it returns at the time $+\tau_0/2$. It is entirely clear that there exists only one path with such properties: $X(\tau) \equiv 0$. Any other path corresponds to an infinite motion with the ball going away to plus or minus ∞. It is also clear that the action on the path $X(\tau) = 0$ vanishes.

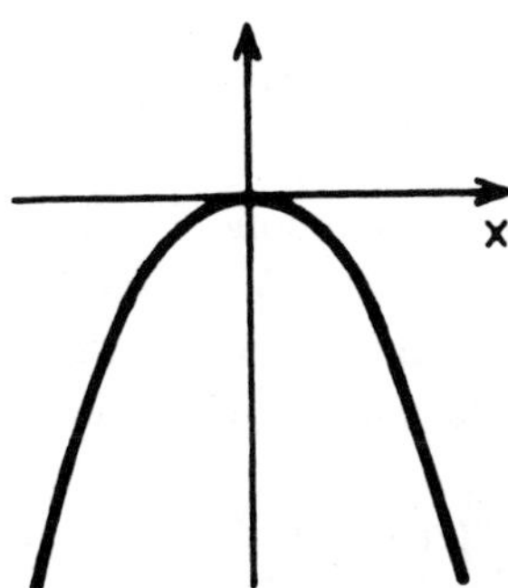

Fig. 2. The potential appearing in the same problem in the Euclidean time.

Thus, in the given particular problem the general formula (14) becomes

$$\langle x_f = 0|e^{-H\tau_0}|x_i = 0\rangle = N\left[\det\left(-\frac{\mathrm{d}^2}{\mathrm{d}\tau^2} + \omega^2\right)\right]^{-1/2} (1 + \text{subleading terms}),$$

The eigenfunctions of the operator in the square brackets are the sine and cosine functions. For instance, the lowest eigenfunction is $\cos(\pi\tau/\tau_0)$, the next is $\sin(2\pi\tau/\tau_0)$, and so on. All the eigenvalues ε_n are immediately fixed by the boundary conditions $x_n(\pm\tau_0/2) = 0$:

$$\varepsilon_n = \frac{\pi^2 n^2}{\tau_0^2} + \omega^2, \qquad n = 1, 2, \ldots .$$

We have now arrived at the point at which it is impossible to advance further without saying what is the value of N. To avoid the necessity of explicit determination of N we split the determinant into two factors:

$$N\left[\det\left(-\frac{\mathrm{d}^2}{\mathrm{d}\tau^2} + \omega^2\right)\right]^{-1/2} = \left[N\left(\prod_{n=1}^{\infty}\frac{\pi^2 n^2}{\tau_0^2}\right)^{-1/2}\right] \times\left[\prod_{n=1}^{\infty}\left(1 + \frac{\omega^2\tau_0^2}{\pi^2 n^2}\right)\right]^{-1/2} . \tag{16}$$

Obviously, the first factor corresponds to *free* motion of the particle, and therefore, it must, of course, reproduce the free result:

$$N\left(\prod_{n=1}^{\infty}\frac{\pi^2 n^2}{\tau_0^2}\right)^{-1/2} = \langle x_f = 0|e^{-\hat{p}^2\tau_0/2}|x_i = 0\rangle = \sum_n |\langle p_n|x = 0\rangle|^2 e^{-p_n^2\tau_0/2}$$
$$= \int_{-\infty}^{+\infty}\frac{\mathrm{d}p}{2\pi}e^{-p^2\tau_0/2} = \frac{1}{\sqrt{2\pi\tau_0}} . \tag{17}$$

Of course, Eq. (17) is somewhat symbolic, but it can be regarded as the definition of the normalization factor N. We now consider the second, less trivial factor in Eq. (16). For the infinite product which occurs in it we have the well-known formula [see e.g. formula (1.431.2) in Ref. [7]

$$\pi y\prod_{n=1}^{\infty}\left(1 + \frac{y^2}{n^2}\right) = \sinh \pi y ,$$

where in our case $y = \omega\tau_0/\pi$.

We now collect all the factors together, take into account (16) and (17), and write down the final result:

$$\langle x_f = 0|e^{-H\tau_0}|x_i = 0\rangle = N\left[\det\left(-\frac{\mathrm{d}^2}{\mathrm{d}\tau^2} + \omega^2\right)\right]^{-1/2}$$
$$= \frac{1}{\sqrt{2\pi\tau_0}}\left(\frac{\sinh\omega\tau_0}{\omega\tau_0}\right)^{-1/2}$$
$$= \left(\frac{\omega}{\pi}\right)^{1/2}(2\sinh\omega\tau_0)^{-1/2}\,. \tag{18}$$

Going to the limit $\tau_0 \to \infty$, we find

$$\langle x_f = 0|e^{-H\tau_0}|x_i = 0\rangle \xrightarrow[\tau_0\to\infty]{} \left(\frac{\omega}{\pi}\right)^{1/2} e^{-\omega\tau_0/2}\left(1 + \frac{1}{2}e^{-2\omega\tau_0} + \cdots\right),$$

from which it follows that for the lowest state $E_0 = \omega/2$ and $[\psi_0(0)]^2 = (\omega/\pi)^{1/2}$. The next term in the expansion corresponds to the level of the harmonic oscillator with $n = 2$ [the odd n do not contribute, since for them $\psi_n(0) = 0$]. The results are exact for the harmonic oscillator and serve as a zeroth approximation for a potential with small anharmonicity, say $(\omega^2/2)x^2 + \lambda x^4$.

2. Double-Well (Two-Humped) Potential. Tunneling

In the previous section, we reformulated in the language of Euclidean space and path integrals one of the most fundamental problems — an oscillator system near the equilibrium position. This problem provides the basis of all field theory. In fact, we have taken into account small oscillations — small deviations from the equilibrium position — and have made the first step to ordinary perturbation theory. For more than 20 years, right up to the middle of the seventies, all field-theoretical models (apart from the small exception of exactly solvable two-dimensional models) were developed in this, and only this, direction. The field variables were regarded as a system of an infinitely large number of oscillators coupled to each other and each possessing zero-point oscillations; one then considered small deviations, with respect to which perturbation theory was constructed successively. In this sense, the "infant" period of quantum chromodynamics, when quark–gluon perturbation theory was created, did not introduce anything fundamentally new. It was only the discovery of instantons which showed that QCD contains effects which cannot be described if one does not go beyond the framework of small deviations from the equilibrium position. It is in principle impossible to describe these effects by expansions in the coupling constant. Here, we again turn to a simple quantum-mechanical analogy, in which, however, all the main features are already present.

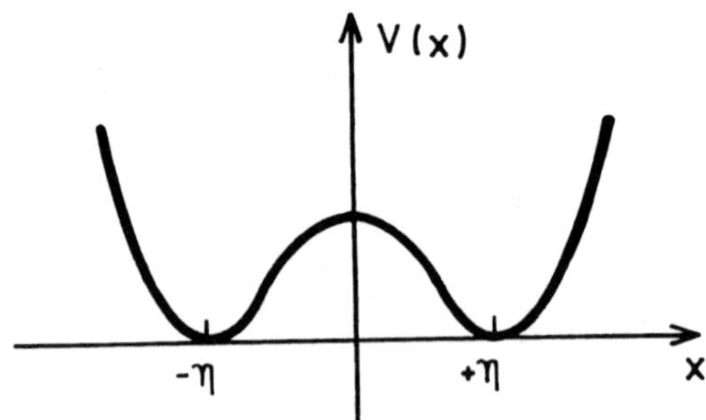

Fig. 3. The double-well potential.

Thus, we again consider the one-dimensional potential motion of a spinless particle with unit mass. The potential

$$V(x) = \lambda(x^2 - \eta^2)^2 \tag{19}$$

is shown in Fig. 3. We fix the parameters λ and η in such a way that

$$8\lambda\eta^2 = \omega^2 \,,$$

where ω is the frequency introduced in the previous section. Then near each minima which are indicated by the symbols $\pm\eta$, the curve is identical to the potential of the previous section. If $\lambda \ll \omega^3$, then the wall separating the two minima is high. Its height is $\omega^4/64\lambda$. Suppose for a moment that it is actually equal to infinity. Then the lowest state of the system has a twofold degeneracy — the particle may be in the right-hand well or in the identical left-hand well, i.e. it executes small oscillations near the point $+\eta$ or $-\eta$. At first glance, the solution to our problem should be constructed in exactly the same way. The expectation value of the coordinate in the ground state should be

$$\langle x\rangle_0 = +\eta\ (1 + \text{corrections}) \quad \text{or} \quad \langle x\rangle_0 = -\eta\ (1 + \text{corrections})\,,$$

the original symmetry of the system with respect to the substitution $x \to -x$ is broken, $E_0 = (\omega/2)(1 + \text{corrections})$ in both cases, and at small λ the corrections are small. In fact, it is known from courses of quantum mechanics that this picture is *qualitatively* incorrect. The symmetry is *not* broken, the expectation value of x for the ground level is *exactly* zero, and there is *no* degeneracy:

$$E_0 = \frac{\omega}{2} - \sqrt{\frac{2\omega^3}{\pi\lambda}} e^{-\omega^3/12\lambda} \frac{\omega}{2} ,$$
$$E_1 = \frac{\omega}{2} + \sqrt{\frac{2\omega^3}{\pi\lambda}} e^{-\omega^3/12\lambda} \frac{\omega}{2} . \tag{20}$$

We note the fact that $E_1 - E_0 \sim \exp(-\omega^3/12\lambda)$ and this quantity cannot be expanded in a series in λ. [It is assumed that $\omega^3/\lambda \gg 1$. In reality, Eqs. (20) begin to "work" when $\omega^3/12\lambda \gtrsim 6$.]

Thus, we have gone wrong and failed to take into account an important element that leads to qualitative changes. What is this element? Everyone knows the standard answer given in courses of quantum mechanics. If at the initial time the particle is concentrated in, say, the left-hand minimum, it nevertheless feels the existence of the right-hand well despite the fact that the latter is inaccessible according to the classical equations of motion. Quantum-mechanical tunneling transfers the wave function from one well to the other and, in Polyakov's terminology, "smears" the ground states. The correct wave function of the ground state is an even superposition of the wave functions in each well.

We now consider how this phenomenon appears in the imaginary time and how the technique presented in the previous section is changed. It turns out — and this is a great good fortune — that all fundamental technical elements remain unchanged. It is only necessary to take into account the fact that the classical equations of motion in the imaginary time have not only the trivial solutions $X(\tau) = \text{const}$ considered earlier but also additional topologically nontrivial solutions which extend far from both the minima. These solutions connect the points $\pm\eta$, and they are entirely responsible for the phenomenon under discussion. We emphasize that in real time there are no additional classical solutions, since the transition from one minimum to the other occurs below the barrier and is classically forbidden.

The solutions arise only after the Euclidean rotation. The double-well potential becomes a two-humped potential of Fig. 4.

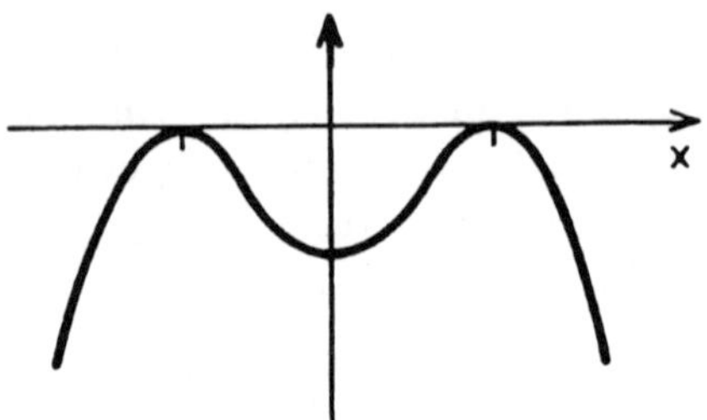

Fig. 4. The two-hump potential relevant to the motion in the Euclidean time.

We consider the calculation of the amplitudes

$$\langle\eta|e^{-H\tau_0}|-\eta\rangle \qquad \text{and} \qquad \langle-\eta|e^{-H\tau_0}|\eta\rangle\,.$$

The first step consists of solving Eq. (11). The "mechanical profile" for this equation is shown in Fig. 4. We are interested in solutions of Eq. (11) that have finite action in the limit $\tau_0 \to \infty$, since such solutions are important in the quasiclassical approximation that is under discussion. Most paths correspond to trajectories on which $x \to \infty$ as $\tau \to \infty$, and they have infinite action.

A finite action in the limit $\tau_0 \to \infty$ is obviously obtained when the particle stays at the top of a hump, i.e. $X(\tau) = \eta$ and $X(\tau) = -\eta$. The contribution of these trajectories was considered above. Another interesting motion leading to a finite action as $\tau_0 \to \infty$ corresponds to the particle sliding from one hump and stopping on the other. Thus, we are interested in a path which begins at $-\tau_0/2$ at the point $-\eta$ and ends at the point η at the time $\tau_0/2$.[3] Physical intuition suggests that such trajectories exist, though their explicit form for finite τ_0 is complicated. We are always interested in only the lowest state, and therefore we can directly assume that $\tau_0 \to \infty$. In this limit, the solution is very simple:

$$X(\tau) = \eta \tanh \frac{\omega(\tau - \tau_c)}{2} \tag{21}$$

(it corresponds to mechanical motion with zero energy, $E = (1/2)\dot{x}^2 - V(x) = 0$, so that the equation can be rewritten as a first-order equation

$$\dot{X} = -\sqrt{2\lambda}(X^2 - \eta^2)$$

and readily integrated).

---

[3]Here we have allowed a slight inaccuracy. If τ_0 is large but not infinite, the path begins just to the right of $-\eta$ and ends just to the left of $+\eta$. It is only in the limit $\tau_0 \to \infty$ that the end points coincide with $\pm\eta$.

Such a solution is called *instanton* (Polyakov proposed the name "pseudoparticle," which can also be found in the literature); the arbitrary parameter τ_c indicates its center. Of course, there also exist antiinstantons, which begin at $+\eta$ and end at $-\eta$. They are obtained from (21) by the substitution $\tau \to -\tau$.

Since all the integrals can be calculated, it is easy to obtain a closed expression for the action of the instanton (we recall that for the instanton $\frac{1}{2}\dot{x}^2 = V(x)$):

$$S_0 = S[X(\tau)]_{\text{inst}} = \int_{-\infty}^{+\infty} d\tau \dot{X}^2 = \int_{-\eta}^{\eta} (-\sqrt{2\lambda})(X^2 - \eta^2) dx = \frac{\omega^3}{12\lambda}. \qquad (22)$$

We recall that the principal exponential factor in the amplitude is $e^{-\text{action}}$ (see Eq. (10)). The exponential which occurs in (20) has emerged. Of course, we still have a long way to go before we can reproduce the complete answer.

We draw attention to an additional property of the instanton, which has far reaching consequences. The center of the solution may be at any point, and the action of the instanton does not depend on the position of the center. This circumstance obviously reflects the symmetry of the original problem. Namely, the Lagrangian of the system is invariant with respect to shifts in time, and the time origin can be chosen arbitrarily. Each concrete solution (21) has a definite position with respect to the origin, and thus there exists an infinite family of solutions distributed arbitrarily with respect to the origin. Intuitively, it is clear that the instanton must occur in any physical quantity in the form of an integral over the position of its center. How does this integral arise formally and what weight is then obtained? Answers to these questions are given in the following section.

3. Determinant and Zero Modes

In this section, we find the one-instanton contribution to $\langle -\eta | e^{-H\tau_0} | \eta \rangle$. We shall not, of course, be concerned with the exponential factor, which has actually been found already, but rather the pre-exponential factor, whose calculation presents a more laborious problem. It is true that in the case under consideration one can employ various devices that significantly simplify the problem and are sometimes discussed in the literature [8]. However, we shall proceed in a "brute force" manner, which is the closest approximation to the method used by 't Hooft [9] to calculate the instanton determinant in QCD. We hope that this will subsequently enable the reader to reproduce for himself (herself) all details of 't Hooft's work, which is central for the entire instanton problem.

The original formula (14) is conveniently rewritten as

$$\langle -\eta | e^{-H\tau_0} | \eta \rangle = N \left[\det \left(-\frac{d^2}{d\tau^2} + \omega^2 \right) \right]^{-1/2} \times \left\{ \frac{\det[-(d^2/d\tau^2) + V''(X)]}{\det[-(d^2/d\tau^2) + \omega^2]} \right\}^{-1/2} e^{-S_0}(1 + \text{corrections}) .$$

We have multiplied and divided by a known number — the determinant for the harmonic oscillator see Eq. (18). The harmonic oscillator will serve as a "point of reference" for manipulations with the more complicated determinant in the numerator. Substituting the explicit expression $X(\tau) = \eta \tanh(\omega\tau/2)$ in $V''(X)$, we arrive at the eigenvalue equation

$$-\frac{d^2}{d\tau^2} x_n(\tau) + \left(\omega^2 - \frac{3}{2}\omega^2 \frac{1}{\cosh^2(\omega\tau/2)} \right) x_n(\tau) = \varepsilon_n x_n(\tau) . \qquad (23)$$

It can be regarded as a certain Schrödinger equation, which, fortunately, is very well studied. Indeed, Eq. (23) is described in detail in, for example, the textbook of Landau and Lifshitz ([10], pp. 73 and 80), and we shall use this source. We recall that the boundary conditions are $x_n(\pm\tau_0/2) = 0$ and $\tau_0 \to \infty$. These conditions are automatically satisfied with exponential accuracy for bound levels, i.e. for the truly discrete spectrum.[4]

There are two such levels in Eq. (23). One of them corresponds to the eigenvalue $\varepsilon_1 = (3/4)\omega^2$, and the other to

$$\varepsilon_0 = 0 .$$

The wave function of the latter, normalized to unity, is

$$x_0(\tau) = \sqrt{\frac{3\omega}{8}} \frac{1}{\cosh^2(\omega\tau/2)} . \qquad (24)$$

The vanishing of the eigenvalue may discourage the reader, since the answer contains $\varepsilon_n^{-1/2}$! However, this result, $\varepsilon_0 = 0$, cannot be regarded as a surprise.

[4] With our boundary conditions, the complete spectrum is, in fact, discrete. The genuine discrete levels can however be readily distinguished from the quasidiscrete levels formed from the continuum after the system has been enclosed in the "box" $x(\pm\tau_0/2) = 0$. The former are separated by intervals of order ω^2, while the latter are at a distance of order $1/\tau_0^2$ from their neighbors.

Indeed, Eq. (23) actually describes the response of the dynamical system under consideration to small perturbations imposed on $X(\tau)$. Since $X(\tau)$ is a solution which realizes a "local" minimum of the action, a generic perturbation of $X(\tau)$ increases the action. Accordingly, ε_n is positive. However, we already know that there is one direction in the functional space along which the solution can be perturbed without changing the action. We have in mind a shift of the center. By virtue of the translational invariance,

$$S[X(\tau,\tau_c)] - S[X(\tau,\tau_c+\delta\tau_c)] = 0\,.$$

The so-called zero mode (i.e. the mode with $\varepsilon = 0$) is obviously proportional to $X(\tau,\tau_c) - X(\tau,\tau_c+\delta\tau_c)$. The correctly normalized zero mode has the form

$$x_0(\tau) = S_0^{-1/2}\left(-\frac{\mathrm{d}}{\mathrm{d}\tau_c}\right)X(\tau,\tau_c)\,,$$

or, which is the same as

$$x_0(\tau) = S_0^{-1/2}\frac{\mathrm{d}}{\mathrm{d}\tau}X(\tau)\,. \tag{25}$$

The fact that the normalization factor reduces to $S_0^{-1/2}$ follows from the expression (22). It is readily seen that (25) is identical to (24), and we now see that this agreement is not fortuitous but a consequence of the translational invariance.

Thus, integration with respect to the coefficient c_0 corresponding to the zero mode (see Eqs. (8) and (9)) is non-Gaussian, and the integral between infinite limits does not exist at all. The way out of the dilemma is simple. We shall not calculate this integral explicitly. It is clear that the integration over $\mathrm{d}c_0$ is the same as the integration over $\mathrm{d}\tau_c$, apart from a coefficient of proportionality. We have here the same integral over the position of the center of the instanton whose appearance our intuition required. In the literature, this trick is sometimes called the introduction of a collective coordinate.

We determine the coefficient of proportionality. If c_0 changes by Δc_0, then $x(\tau)$ changes by

$$\Delta x(\tau) = x_0(\tau)\Delta c_0$$

(see Eq. (8)). On the other hand, the change $\Delta x(\tau)$ under a shift $\Delta\tau_c$ of the center is

$$\Delta x(\tau) = \Delta X(\tau) = \frac{\mathrm{d}x}{\mathrm{d}\tau_c}\Delta\tau_c = -\sqrt{S_0}\,x_0(\tau)\Delta\tau_c\,.$$

Equating the two increments, we obtain

$$\mathrm{d}c_0 = \sqrt{S_0}\,\mathrm{d}\tau_c\,. \tag{26}$$

(In Eq. (26), we have not inserted the minus sign to ensure that as c_0 varies from $-\infty$ to $+\infty$ the parameter τ_c changes in the same interval.) This is not yet the end of the story, since we agreed to normalize the result to the ordinary oscillator (we recall that we are interested in the ratio of determinants). In the oscillator problem, the minimal eigenvalue is $\omega^2 + \pi^2/\tau_0^2 \to \omega^2$ in the limit $\tau_0 \to \infty$. Finally,

$$\left\{\frac{\det[-(\mathrm{d}^2/\mathrm{d}\tau^2) + V''(X)]}{\det[-(\mathrm{d}^2/\mathrm{d}\tau^2) + \omega^2]}\right\}^{-1/2}$$

$$= \sqrt{\frac{S_0}{2\pi}}\omega\mathrm{d}\tau_c\left\{\frac{\det'[-(\mathrm{d}^2/\mathrm{d}\tau^2) + V''(X)]}{\omega^{-2}\det[-(\mathrm{d}^2/\mathrm{d}\tau^2) + \omega^2]}\right\}^{-1/2}, \tag{27}$$

where det′ denotes the reduced determinant with the zero mode removed.

We emphasize that although we analyzed only a single specific example with the simplest instanton $\eta\tanh(\omega\tau/2)$, the method of dealing with the zero modes is in fact general. Thus, in the BPST instanton any invariance will generate a zero mode, and the integration with respect to the corresponding coefficient must be replaced by integration with respect to some collective coordinate. We have already learned how to find the Jacobian of the transformation.

We now consider nonzero modes. It is easiest to deal with the second discrete level, whose eigenvalue is $(3/4)\omega^2$. If we denote by Φ the ratio

$$\Phi = \frac{\det'[-(\mathrm{d}^2/\mathrm{d}\tau^2) + V''(X)]}{\omega^{-2}\det[-(\mathrm{d}^2/\mathrm{d}\tau^2) + \omega^2]}, \tag{28}$$

then the contribution of this level to Φ as $\tau_0 \to \infty$ is obviously

$$\frac{3}{4}\,. \tag{29}$$

We now turn to other modes, with $\varepsilon > \omega^2$. If we did not have the boundary condition $x(\pm\tau_0/2) = 0$, Eq. (23) in this region would have a continuous spectrum. Let us forget the boundary conditions for a moment. The general solution of (23) is given in the book of Landau and Lifshitz; however, we do not need its explicit form. It is sufficient to know the following. First, the solutions with $\varepsilon > \omega^2$ are labeled by a continuous index p. This index is related to the eigenvalue ε by $p = \sqrt{\varepsilon_p - \omega^2}$ and ranges over the entire interval $(0, \infty)$. Second, for the values of the parameters that occur in (23) there is no reflection. In other words, choosing one of the linearly independent solutions in such a way that

$$x_p(\tau) = e^{ip\tau} \qquad \text{as} \qquad \tau \to +\infty\,,$$

we have in the other asymptotic region the same exponential:

$$x_p(\tau) = e^{ip\tau + i\delta_p} \qquad \text{as} \qquad \tau \to -\infty\,.$$

The second exponential, $e^{-ip\tau}$, which should in principle arise, is absent, and the entire dynamical effect is reduced to the phase

$$e^{i\delta_p} = \frac{1 + (ip/\omega)}{1 - (ip/\omega)} \frac{1 + (2ip/\omega)}{1 - (2ip/\omega)} \tag{30}$$

(we have used here the formula from [10] on p. 81). The second linearly independent solution can be chosen in the form $x_p(-\tau)$. The general solution is $Ax_p(\tau) + Bx_p(-\tau)$, where A and B are arbitrary constants.

This information is already sufficient to find the spectrum if we recall the boundary condition $x(\pm\tau_0/2) = 0$. The equations for A and B,

$$Ax_p\left(\frac{\tau_0}{2}\right) + Bx_p\left(-\frac{\tau_0}{2}\right) = 0, \qquad Ax_p\left(-\frac{\tau_0}{2}\right) + Bx_p\left(-\frac{\tau_0}{2}\right) = 0$$

have nontrivial solutions if and only if

$$\frac{x_p(\tau_0/2)}{x_p(-\tau_0/2)} = \pm 1\,.$$

This gives an equation for p:

$$e^{ip\tau_0 - i\delta_p} = \pm 1\,,$$

or, which is the same as,

$$p\tau_0 - \delta_p = \pi n, \qquad n = 0, 1, \ldots\,. \tag{31}$$

We denote the nth solution by $\tilde{p}_n$. In the case of $\det[-(\mathrm{d}^2/\mathrm{d}\tau^2) + \omega^2]$, by which we normalize, the equation is $p\tau_0 = \pi n$ and the nth solution $p_n = \pi n/\tau_0$. We need to calculate the product[5]

$$\prod_{n=1}^{\infty} \frac{\omega^2 + \tilde{p}_n^2}{\omega^2 + p_n^2}.$$

For any preassigned n, the ratio $(\omega^2+\tilde{p}_n^2)/(\omega^2+p_n^2)$ is arbitrarily close to unity as $\tau_0 \to \infty$. Only the multiplication of a very large number of factors with $n \sim \omega\tau_0$, each of them differing from 1 by an amount of order $1/\omega\tau_0$, gives an effect. (For $n \gg \omega\tau_0$, the difference between $\omega^2+\tilde{p}_n^2$ and $\omega^2+p_n^2$ again becomes unimportant, in complete agreement with our physical intuition.) Under these conditions, we can write

$$\prod \frac{\omega^2 + \tilde{p}_n^2}{\omega^2 + p_n^2} = \exp\left(\sum_n \ln \frac{\omega^2 + \tilde{p}_n^2}{\omega^2 + p_n^2}\right) \approx \exp\left[\sum_n \frac{2p_n(\tilde{p}_n - p_n)}{\omega^2 + p_n^2}\right],$$

where we have made an expansion with respect to the small difference $\tilde{p}_n - p_n$. Going over from summation over n to integration over p_n and using (31) for $\tilde{p}_n - p_n$, we obtain on the right-hand side

$$\exp\left[+\frac{1}{\pi}\int_0^{\infty} \frac{\delta_p \cdot 2p\mathrm{d}p}{p^2 + \omega^2}\right] = \exp\left[-\frac{1}{\pi}\int_0^{\infty} \frac{\mathrm{d}\delta_p}{\mathrm{d}p} \ln\left(1 + \frac{p^2}{\omega^2}\right)\mathrm{d}p\right].$$

Differentiating the phase by means of (30) and introducing the dimensionless variable $y = p/\omega$, we transform this expression identically to

$$\exp\left[-\frac{2}{\pi}\int_0^{\infty} \mathrm{d}y \left(\frac{1}{1+y^2} + \frac{2}{1+4y^2}\right) \ln(1+y^2)\right] = \frac{1}{9}. \tag{32}$$

Finally, combining (32) and (29), we find that

$$\Phi = \frac{1}{12}. \tag{33}$$

We have now made all the necessary preparations, namely, we have derived formulas (33), (28), (27), (22), and (18), and we write down the result for the

[5]The reader may recall that we have already "taken up" in the denominator two eigenvalues, $\omega^2+\pi^2/\tau_0^2$ and $\omega^2+4\pi^2/\tau_0^2$, in calculating the contribution of the discrete modes with $\varepsilon = 0$ and $\varepsilon = 3\omega^2/4$. Therefore, it would be more correct in the denominator to write $\omega^2 + p_{n+2}^2$. However, as we shall see very shortly, it is the region of very large n of order $\omega\tau_0$ that is important, so that the difference between p_{n+2} and p_n is immaterial.

one-instanton contribution:

$$\langle -\eta | e^{-H\tau_0} | \eta \rangle_{\text{one-inst}} = \left(\sqrt{\frac{\omega}{\pi}} e^{-\omega\tau_0/2} \right) \left(\sqrt{\frac{6}{\pi}} \sqrt{S_0} e^{-S_0} \right) \omega d\tau_c \,. \tag{34}$$

This result can be trusted as long as

$$\sqrt{S_0} e^{-S_0} \omega \tau_0 \ll 1 \,.$$

At large τ_0, when this condition is violated, it is necessary to take into account paths constructed from many instantons and antiinstantons, and this will be done in the following section.

It is appropriate here to make some comments. The factor in the first brackets corresponds to a simple harmonic oscillator. By separating it, we have been able to normalize, or regularize, the instanton calculations. A similar device for regularization is used in quantum chromodynamics. The factor in the second brackets can naturally be called the instanton density. Besides the exponential factor e^{-S_0}, the density contains the pre-exponential $\sqrt{S_0}$, which is associated with the existence of the zero mode. This circumstance is also of a general nature. In quantum chromodynamics too, each zero mode is associated with $\sqrt{S_0}$. Finally, the existence of the zero mode leads to the appearance of a regularization frequency and of integration over the collective coordinate $\omega d\tau_c$.

We wish to emphasize that is is worth remembering the lessons we have learned, since they can be directly transferred to the BPST instanton. The only thing specific in the present case is the number, $\sqrt{6/\pi}$. If this number is not particularly important (and in QCD, as we shall see below, this is indeed the case), all the remaining results can be reconstructed almost at once, without calculations. We have given so much attention to the relatively simple determinant for a pedagogical reason — to avoid greater boredom in the case of the BPST instanton.

4. Instanton Gas

It remains for us to make the final, small step to reproduce formula (20). The energy of the lowest state is determined by the transition to the limit $\tau_0 \to \infty$. We cannot go to this limit directly in Eq. (34). At very large τ_0, paths constructed of many instantons and antiinstantons are important. If the distance between their centers is large, such a path is also a classical solution.

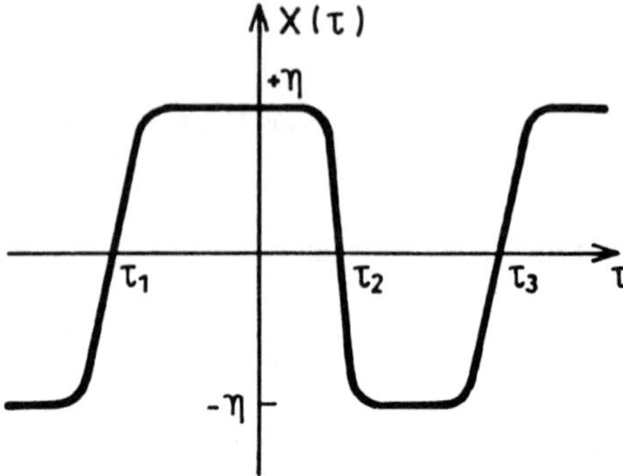

Fig. 5. The chain of n well-separated instantons (antiinstantons).

Suppose we have n instantons or antiinstantons with centers $\tau_1, \tau_2, \ldots, \tau_n$ (Fig. 5). The points τ_i satisfy the condition

$$-\frac{\tau_0}{2} < \tau_1 < \tau_2 < \cdots < \tau_n < \frac{\tau_0}{2},$$

and otherwise can be distributed arbitrarily. If the characteristic intervals satisfy $|\tau_i - \tau_j| \gg \omega^{-1}$ (we shall verify the condition *a posteriori*), then the action corresponding to such a configuration is nS_0, where S_0 is the action of one instanton. With regard to the determinant, it is obvious that if we did not have the n narrow transition regions (near $\tau_1, \tau_2, \ldots, \tau_n$) we should obtain the same result as in the case of the harmonic oscillator, $\sqrt{\omega/\pi}e^{-\omega\tau_0/2}$. The transition regions lead to a correction, and we now know in what way:

$$\sqrt{\frac{\omega}{\pi}}e^{-\omega\tau_0/2} \to \sqrt{\frac{\omega}{\pi}}e^{-\omega\tau_0/2}\left(\sqrt{\frac{6}{\pi}}\sqrt{S_0}e^{-S_0}\right)^n \prod_1^n (\omega\, \mathrm{d}\tau_i)\,.$$

Finally, the contribution of the n-instanton configuration can be written in the form

$$\sqrt{\frac{\omega}{\pi}}e^{-\omega\tau_0/2}d^n \int_{-\tau_0/2}^{\tau_0/2} \omega \mathrm{d}\tau_n \int_{-\tau_0/2}^{\tau_n} \omega \mathrm{d}\tau_{n-1} \cdots \int_{-\tau_0/2}^{\tau_2} \omega \mathrm{d}\tau_1$$

$$= \sqrt{\frac{\omega}{\pi}}e^{-\omega\tau_0/2}d^n \frac{(\omega\tau_0)^n}{n!},$$

where d denotes the instanton density,

$$d = \sqrt{\frac{6}{\pi}}\sqrt{S_0}e^{-S_0}\,. \tag{35}$$

The amplitudes $\langle -n|e^{-H\tau_0}|\eta\rangle$ and $\langle\eta|e^{-H\tau_0}|\eta\rangle$ are obtained by summation over n. In the first case, we start from $-\eta$ and arrive at $+\eta$ and therefore the number of pseudoparticles is odd. In the second case, conversely, only an even number of pseudoparticles works:

$$\begin{aligned}\langle -\eta|e^{-H\tau_0}|\eta\rangle &= \sum_{n=1,3,\ldots}\sqrt{\frac{\omega}{\pi}}e^{-\omega\tau_0/2}\frac{(\omega\tau_0 d)^n}{n!}\\ &= \sqrt{\frac{\omega}{\pi}}e^{-\omega\tau_0/2}\sinh(\omega\tau_0 d)\,,\end{aligned}\tag{36}$$

$$\langle\eta|e^{-H\tau_0}|\eta\rangle = \sum_{n=0,2,\ldots}\sqrt{\frac{\omega}{\pi}}e^{-\omega\tau_0/2}\frac{(\omega\tau_0 d)^n}{n!} = \sqrt{\frac{\omega}{\pi}}e^{-\omega\tau_0/2}\cosh(\omega\tau_0 d)\,.$$

Going to the limit $\tau_0\to\infty$, we immediately reproduce formula (20) for the energy of the lowest state. Denoting the ground state of the system by $|0\rangle$, we see that $\langle\eta|0\rangle = \langle -\eta|0\rangle = (\omega/4\pi)^{1/4}$, i.e. the symmetry between the right- and left-hand wells is, indeed, not broken.

We now return to the assumption that the characteristic distances between the centers of the instantons are large,

$$|\tau_i - \tau_j| \gg \omega^{-1}\,,$$

and consider how well it works. It is clear that the sums in (36) converge well, and all terms with number $n \gg d\omega\tau_0$ are unimportant. Thus, $n_{\rm char}\sim d\omega\tau_0$ and $|\tau_i-\tau_j|_{\rm char}\sim d^{-1}\omega^{-1}$. Having at our disposal the free parameter λ, we can achieve an arbitrary smallness of d, since $d\to 0$ as $e^{-\omega^3/12\lambda}$ in the limit $\lambda\to 0$.

Thus, for $\lambda \ll 1$ we are fully justified in "stringing" instantons and antiinstantons on one another, forming thereby a chain of noninteracting pseudoparticles. Noninteracting in the sense that they are all far from one another, know nothing about the remaining partners, and the total weight function is obtained by multiplying the individual weight functions [d^n in formulas (36)].

Such an approximation is called a dilute instanton gas. This has been exploited particularly by Callan, Dashen and Gross [11] in quantum chromodynamics. Unfortunately, in QCD we do not have free parameters like λ that can be kept small. Therefore, a dilute instanton gas is not suitable from the quantitative point of view in QCD, and the most we can extract from it are heuristic indications. Further details regarding attempts to improve the instanton gas approximation (the so-called, instanton liquid) can be found in Ref. [4].

To conclude the section, we note that a somewhat more extensive exposition of the instanton approach to the double-well, potential problem is contained in Coleman's lecture [8]. The reader interested in special questions, for example, situations not covered by the gas approximation in quantum mechanics, must consult Ref. [12].

5. Tunneling in Quantum Chromodynamics

5.1. *Nontrivial Topology in the Space of Fields in the Yang-Mills Theories*

Now, when we are done with the toy model, we can pass to real QCD. The Lagrangian of the theory has the form

$$L = -\frac{1}{4} G^a_{\mu\nu} G^a_{\mu\nu} + \sum \bar{\psi}(iD_\mu \gamma^\mu - M)\psi$$

where the sum runs over all quark flavors, $G_{\mu\nu}$ is the gluon field strength tensor,

$$G^a_{\mu\nu} = \partial_\mu A^a_\nu - \partial_\nu A^a_\mu + g f^{abc} A^b_\mu A^c_\nu \, ,$$

g is the gauge coupling constant, f^{abc} stands for the structure constants of the gauge group. In QCD the gauge group is SU(3); the quarks are described by the Dirac fields ψ^i transforming according to the fundamental (triplet) representation of SU(3). The issue to be discussed in this section is independent of the particular choice of the gauge group and the presence (absence) of the quark fields. To make the picture as transparent as possible we will disregard, for the time being, the quarks, and consider the simplest non-Abelian group, SU(2). Of course, later on we will include quarks and pass from SU(2) to SU(3).

If tunneling is important for understanding the structure of QCD, the first question to be asked is from where to where does the system tunnel.

At first glance it is not obvious at all that the Lagrangian of gluodynamics has a discrete set of degenerate classical minima. In the double-well potential problem this was evident. The main distinction is due to the fact that in the field theory the number of degrees of freedom is infinite, while in the toy example of the previous sections we dealt with only one degree of freedom. The space of fields in QCD is infinitely-dimensional. Most of these field-theoretical degrees of freedom are oscillator-like and, thus, "uninteresting." Our task is to single out such degree(s) of freedom which tunnel and are delocalized in the space of fields, much in the same way as the genuine ground state wave function in the double-well potential is not localized in any of the two minima, but is rather smeared along the x-axis.

Below, we will demonstrate that in QCD there exists one such direction in the infinitely-dimensional space of fields. If we forget for a while about all other degrees of freedom, and focus on this chosen degree of freedom, we will see that the corresponding dynamics, being somewhat more complicated than that of the double-well potential system, is similar in one aspect: it calls for the consideration of tunneling. The closest analogy one can keep in mind in this context is quantum mechanics of a particle living on a vertically oriented circle subjected to a constant gravitational force (Fig. 6). Classically the particle with the lowest possible energy (the ground state of the system) just stays at rest at the bottom of the circle. Quantum-mechanically, the zero-point oscillations come into play. Within the perturbative treatment, we deal exclusively with small oscillations near the equilibrium point at the bottom of the circle. For such small oscillations, the existence of the upper part of the circle plays no role. It could have been eliminated altogether with no impact on the zero-point oscillations.

From the courses of quantum mechanics it is known, however, that the genuine ground-state wave function is different. The particle oscillating near the

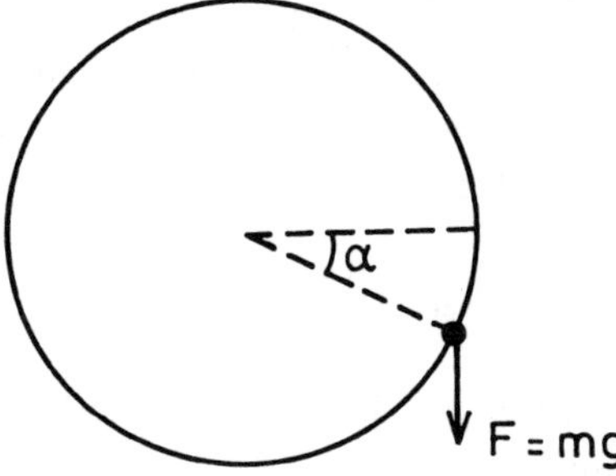

Fig. 6. Quantum mechanics of a particle (•) on a one-dimensional topologically nontrivial manifold, circle.

origin "feels" that it can wind around the circle to which it belongs, tunneling under the potential barrier it experiences at the top of the circle.

To single out the relevant degree of freedom in the infinitely-dimensional space of the gluon fields, it is necessary to proceed to the Hamiltonian formulation of the Yang-Mills theory which implies, of course, that the time component of the four-potential has to be gauged away, $A_0 = 0$. Then,

$$\mathcal{H} = \frac{1}{2} \int d^3x \{ \mathbf{E}^a \mathbf{E}^a + \mathbf{B}^a \mathbf{B}^a \}$$

where $\mathcal{H}$ is the Hamiltonian and $E_i^a = \dot{A}_i^a$ are to be treated as canonical momenta.

Two subtle points are to be mentioned in connection with this Hamiltonian. First, the equation $\operatorname{div} \mathbf{E} = \rho$, inherent to the original Yang-Mills theory, does not stem from this Hamiltonian *per se*. This equation must be imposed as a constraint on the states from the Hilbert space by hand. Second, the gauge freedom is not fully eliminated. Gauge transformations which depend on $\mathbf{x}$ but not t are still allowed. This freedom is reflected in the fact that, instead of two transverse degrees of freedom $\mathbf{A}_\perp$, the Hamiltonian above has three (three components of $\mathbf{A}$). Imposing, say, the Coulomb gauge condition,

$$\partial_i A_i = 0$$

we could get rid of the "superfluous" degree of freedom, a procedure quite standard in perturbation theory (in the Coulomb gauge). Alas! If we want to keep and reveal the topologically nontrivial structure of the space of fields, the Coulomb gauge condition can *not* be imposed. We have to work, with certain care, with the "undergauged" Hamiltonian.

Quasiclassically the state of the system described by this Hamiltonian at any given moment of time is characterized by the field configuration $A_i^a(\mathbf{x})$. Since we are interested in the zero-energy states (classically, this is obviously the minimal energy) A_i^a must be pure gauge,

$$A_i^a(\mathbf{x})|_{\text{vac}} = \frac{i}{g} U(\mathbf{x}) \partial_i U^{-1}(\mathbf{x})$$

where U is a matrix belonging to SU(2) and depending on the *spatial* components of the four-coordinate.

Moreover, we are interested only in those zero-energy states which might be connected with each other by tunneling transitions, i.e. the corresponding classical action must be finite. The latter requirement results in the following boundary condition (see e.g. Ref. [11] for a detailed discussion)

$$U(|\mathbf{x}| \to \infty) = 1\,,$$

or any other constant matrix U_0 independent of the direction in the three-dimensional space along which $\mathbf{x}$ tends to infinity. This boundary condition compactifies our *three-dimensional* space which becomes topologically equivalent to three-dimensional sphere.

On the other hand, the group space of SU(2) is also a three-dimensional sphere. Indeed, any matrix belonging to SU(2) can be parametrized as

$$M = A + i\mathbf{B}\boldsymbol{\sigma}, \qquad M \in \mathrm{SU}(2)\,.$$

Here A and $\mathbf{B}$ are four real parameters; $\boldsymbol{\sigma}$ are the Pauli matrices. Both conditions, $M^+M = 1$ and $\det M = 1$, are met provided

$$A^2 + \mathbf{B}^2 = 1\,.$$

Since $U(\mathbf{x})$ is a matrix from SU(2), and the space of all coordinates $\mathbf{x}$ is topologically equivalent to a sphere (after compactification $U(|\mathbf{x}| \to \infty) = 1$), the function $U(\mathbf{x})$ realizes a mapping of the sphere in the coordinate space onto a sphere in the group space. Intuitively it is quite obvious that all continuous mappings $S_3 \to S_3$ are classified according to the number of coverings. This number will be the number of times we sweep the group sphere S_3 when the coordinate $\mathbf{x}$ sweeps the sphere in the coordinate space once. The number of coverings can be zero (topologically trivial mapping), one, two, and so on. The number of coverings can be negative too, since the mappings $S_3 \to S_3$ are orientable [13]. This fact is especially transparent for the mappings $S_1 \to S_1$, i.e. circle onto circle. If one circle is swept in the clockwise direction and the other one in the anticlockwise, we say that the number of coverings is -1.

In other words, the matrices $U(\mathbf{x})$ can be sorted out in distinct classes labeled by an integer number, $U_n(\mathbf{x})$, $n = 0,\ \pm1,\ \pm2,\ \ldots$, which is referred to as the *winding number*. All matrices belonging to a given class $U_n(\mathbf{x})$ are reducible to each other by a continuous $\mathbf{x}$-dependent gauge transformation. At the same time, no continuous gauge transformation can transform $U_n(\mathbf{x})$ into $U_{n'}(\mathbf{x})$ if $n \neq n'$. The unit matrix represents the class $U_0(\mathbf{x})$. For $n = 1$ one can take, for instance,[6]

[6]Let us note in passing that exactly the same topological classification is the basis of the theory of Skyrmions, see Ref. [14] for a review.

$$U_1(\mathbf{x}) = \exp\left[i\pi\frac{\mathbf{x}\boldsymbol{\sigma}}{(\mathbf{x}^2+\rho^2)^{1/2}}\right],$$

where ρ is an arbitrary parameter. An example of the matrix from U_n is U_1^n.

Any field configuration $A_i^a(\mathbf{x})|_{\rm vac} = (i/g)U_n(\mathbf{x})\partial_i U_n^{-1}(\mathbf{x})$, being pure gauge, corresponds to the lowest possible energy — the zero energy. As a matter of fact, the set of points $\{U_n\}$ in the space of fields, obviously consists simply of the gauge images of one and the same physical point (analogous to the bottom of the circle in Fig. 6). The fact that the matrices U_n from different classes are not continuously transformable to each other indicates the existence of a "hole" in the space of fields, with a noncontractible loop winding around this "hole".

We are finally ready to identify the degree of freedom corresponding to the motion along this circle. Let us consider the vector K_μ,

$$K_\mu = 2\varepsilon_{\mu\nu\alpha\beta}\left(A_\nu^a\partial_\alpha A_\beta^a + \frac{g}{3}f^{abc}A_\nu^a A_\alpha^b A_\beta^c\right).$$

The vector K_μ is called the *Chern-Simons current*; it plays an important role in the instanton calculus. We will encounter it more than once in what follows. Now, define the charge $\mathcal{K}$ corresponding to the Chern-Simons current,

$$\mathcal{K} = \frac{g^2}{32\pi^2}\int K_0(x)d^3x\,.$$

It is not difficult to show that for any pure gauge field $A_i^a(\mathbf{x})$ the Chern-Simons charge $\mathcal{K}$ measures the winding number. For any field $A_i(\mathbf{x}) = (i/g)U_n(\mathbf{x})\partial_i U_n^{-1}(\mathbf{x})$ we have[7]

$$\mathcal{K} = n\,.$$

Summarizing, moving in the "direction of $\mathcal{K}$" in the space of fields we observe that this particular direction has the topology of circle. The points $\mathcal{K}$ and $\mathcal{K}+1$, and $\mathcal{K}-1,\ldots$ are physically one and the same point. The integer values of $\mathcal{K}$ correspond to the bottom of the circle in Fig. 6.

[7] *Exercise:* Prove the assertion formulated in the above paragraph. In case of difficulties consult Coleman's lecture [8].

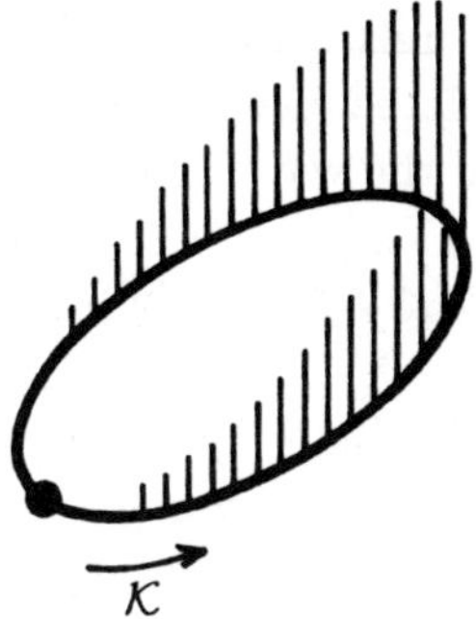

Fig. 7. Nontrivial topology in the space of gauge fields in the $\mathcal{K}$ direction. The length of the circle is 1. The vertical lines indicate the strength of a potential acting on the effective degree of freedom living on the circle.

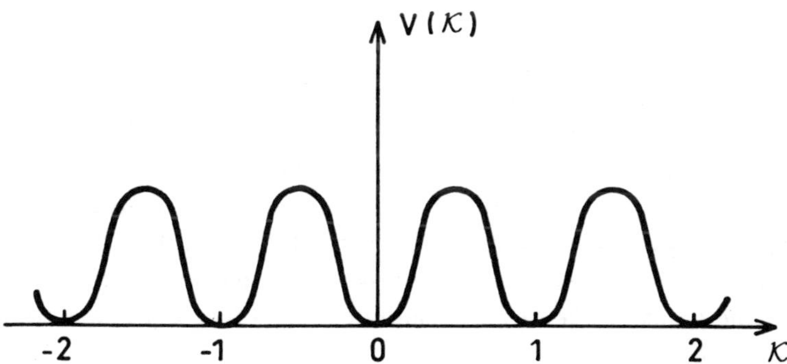

Fig. 8. If we unwind the circle of Fig. 7 onto a line we get a periodic potential.

It is convenient to visualize dynamics of the Yang-Mills system in the "direction of $\mathcal{K}$" as shown in Fig. 7. The vertical lines indicate the potential energy — the higher the line the larger the energy. It is well-known (see e.g. Ref. [15]) that the only consistent way of treating the quantum-mechanical systems living on a circle (angle-type degrees of freedom) is to cut the circle and map it many times onto a straight line. In other words, we pretend that the variable $\mathcal{K}$ lives on the line (Fig. 8). Any integer value of $\mathcal{K}$ in Fig. 8 corresponds to a pure gauge configuration with zero energy. On the other hand, if $\mathcal{K} \neq n$ the field strength tensor is nonvanishing and the energy of the field configuration is positive. Viewed as a function on the line, the potential energy $V(\mathcal{K})$ is, of course, periodic — with the unit period. To take into account the fact that the original problem is formulated on the circle, we impose the (quasi)periodic Bloch boundary condition on the wave function Ψ,

$$\Psi(\mathcal{K}+1) = e^{i\theta}\Psi(\mathcal{K}) .$$

The phase θ appearing in the Bloch quasiperiodic boundary condition is a hidden parameter of QCD, the *vacuum angle.* We will return to the issue of the vacuum angle later on. The classical minima of the potential of Fig. 8 can be called *pre-vacua.* The correct wave-function of the quantum-mechanical vacuum state of the Bloch form is built as a linear combination of these pre-vacua.

We would like to emphasize here a subtle point which in many presentations remains fogged. It might seem that the systems depicted in Figs. 7 and 8 (a particle on a circle and that in the periodic potential) are physically identical. This is not quite the case. In the periodic potentials, say in crystals, one can always introduce impurities that would slightly violate periodicity. For a system on a circle this cannot be done. The correct analog system for gluodynamics, where the gauge invariance is a sacred principle, is that of Fig. 7.

Assume that at $t=-\infty$ and at $t=+\infty$ our system is at the classical minimum (zero-energy state). Assume also that at $t=-\infty$ the winding number $\mathcal{K}=n$ while at $t=+\infty$ $\mathcal{K}=n\pm 1$. In Fig. 7 this means that our system tunnels from the point marked by the closed circle to the very same point under the hump of the potential energy.

Consider now a field configuration $A_\mu(t,\mathbf{x})$ continuously interpolating (with the minimal action) between these two states in the Euclidean time, the least-action tunneling trajectory. This is the BPST instanton.

The fact that the BPST instantons describe tunnelings in the space of fields, which possesses noncontractible paths, was realized by V. Gribov shortly after the discovery of the BPST instantons.[8] Almost simultaneously the tunneling picture was revealed in Refs. [16] and [17]. A very pedagogical and illustrative discussion of the tunneling interpretation given in the *Minkowski* space is presented in Ref. [18] which may be recommended to the reader just beginning to study instanton calculus.

---

[8]See e.g. a remark in Ref. [5] where Polyakov acknowledges Gribov's suggestion of the tunneling interpretation.

The analysis outlined above (the one based on the Hamiltonian formulation) is convenient for establishing the existence of a nontrivial topology, nonequivalent vacuum states and, hence, the existence of nontrivial interpolating field configurations corresponding to tunneling. As we already know, the minimal action is achieved for the solutions of the classical equations of motion (in Euclidean time) with the given boundary conditions. In practice, however, the Hamiltonian gauge $A_0 = 0$ is rarely used in constructing these solutions. This gauge is extremely inconvenient for this purpose.

Below we will describe a standard procedure based on a specific *ansatz* for $A_\mu(x)$ in which all four Lorentz components of A_μ are nonvanishing. This *ansatz* entangles the color and Lorentz indices; the field configurations emerging in this way are, following Polyakov, generically referred to as "hedgehogs".

Since the solutions we are going to deal with are those of the Euclidean equations of motion, the first question to be asked is what needs to be done with QCD in order to pass to the Euclidean time. One can choose two alternative routes. In pure Yang-Mills theory with no fermions, it is advantageous, from the very beginning, to formulate a Euclidean version of the theory, and work only with this version. With the instanton calculations we will be never required to return to the original Minkowski version of the theory. The Euclidean formulation can be also developed in the presence of fermions, provided all fermions in the theory are described by the Dirac fields, i.e. are nonchiral. We will follow this route almost up to the very end of the lecture. The approach does not work, however, for chiral fermions, and in many supersymmetric field theories. For such problems one must choose the second route, which will be discussed in Sec. 13.

5.2. θ *Vacuum*

The existence of a noncontractable loop in the space of fields A_μ leads to drastic consequences for the vacuum structure in non-Abelian gauge theories. Let us take a closer look at the potential of Fig. 8. The argument presented below is formulated in quasiclassical language. One should keep in mind, however, that the general conclusion is valid even though the quasiclassical approximation is inappropriate in quantum chromodynamics where the coupling constant becomes large at large distances.

The lowest-energy state of the system depicted in Fig. 8, classically, is in one of the minima of the potential. Quantum-mechanically the zero point oscillations arise. The wave function[9] corresponding to oscillations near the nth zero-energy state, Ψ_n, is localized near the corresponding minimum. The genuine wave function is delocalized, however,

$$\Psi_\theta = \sum_{n=0,\pm 1,\pm 2,\ldots} e^{in\theta}\Psi_n\,,$$

where θ is a parameter,

$$0 \le \theta \le 2\pi\,,$$

analogous to the quasimomentum in the physics of crystals [15]. The nth term in the sum can be called a *pre-vacuum* while the total sum represents the θ vacuum of QCD. The vacuum angle θ is a global fundamental constant characterizing the boundary condition on the wave function. It does not make sense to say that in one part of the space θ takes some value, while in another part θ takes a different value, or that θ depends on time. The worlds with different values of θ have orthogonal wave functions; for any operator $\mathcal{O}$ from the Hilbert space of the physical states

$$\langle \Psi_\theta | \mathcal{O} | \Psi_{\theta'} \rangle = 0 \qquad \text{if } \theta \neq \theta'\,.$$

This property is referred to as the *superselection rule.* The energy of Ψ_θ can (and does) depend on θ. From the definition of the vacuum angle it is clear that the θ dependence of all physical quantities, including the vacuum energy, must be periodic, with the period 2π.

Since all states Ψ_n are degenerate in energy the question is often raised on why one should form a linear combination corresponding to the θ vacuum. Is it possible to take, say, Ψ_0 as the vacuum wave function?

This question can be answered at different levels. Purely theoretically, if we want to implement the full gauge invariance of the theory, including the invariance under "large" gauge transformations, we must pass from Ψ_n to Ψ_θ. At a more pragmatic level one can say that introduction of Ψ_θ is necessary to maintain the property of the cluster decomposition, which must take place in any sensible field theory. (We are reminded that the cluster decomposition means that the vacuum expectation value of the T product of any two operators, $\mathcal{O}_1(x_1)$ and $\mathcal{O}_2(x_2)$, at large separations $|x_1 - x_2| \to \infty$

[9] In application to QCD we should rather use the term wave functional; nevertheless, we will continue referring to the wave function.

must tend to $\langle \mathcal{O}_1 \rangle \langle \mathcal{O}_2 \rangle$.) If the vacuum wave function is chosen to be Ψ_n, this property would not be valid, as we will see below. Finally, by proceeding to Ψ_θ we ensure that the vacuum state is stable under small perturbations. This is not the case if the vacuum wave function is Ψ_n. A small mass term of the quark fields would then cause a drastic restructuring of the vacuum wave function.

Although the physical meaning of the parameter θ is absolutely transparent within the Hamiltonian formulation, when we speak practically of instantons in field theory we keep in mind the Lagrangian formulation based on the path-integral formalism. In the Lagrangian formalism the vacuum angle is introduced as the θ *term* in the Lagrangian,

$$L = -\frac{1}{4} G^a_{\mu\nu} G^a_{\mu\nu} + L_\theta \,, \qquad L_\theta = \theta \frac{g^2}{32\pi^2} G^a_{\mu\nu} \tilde{G}^a_{\mu\nu} \,,$$

where

$$\tilde{G}^a_{\mu\nu} = \frac{1}{2} \varepsilon_{\mu\nu\alpha\beta} G^a_{\alpha\beta} , \qquad \varepsilon^{0123} = 1 \,.$$

Note that if $\theta \neq 0$ or π, the θ term violates P and T invariance.

Before the discovery of the instantons it was believed that QCD naturally conserves P and CP. Indeed, the only gauge invariant Lorentz scalar operator one could construct from the A_μ fields of dimension 4 violating P and T is $G\tilde{G}$. This operator is a full divergency, $G\tilde{G} = \partial_\mu K_\mu$ where K_μ is the Chern-Simons current. It was believed that full divergencies have no impact on the action.

In the instanton field, however, the integral of $G\tilde{G}$ does not vanish. The reasons for that will be explained below. What is important for us now is the fact that adding the θ term to the QCD Lagrangian we do break P and CP in the strong interactions if $\theta \neq 0$. Since it is known experimentally, that P and CP symmetries are conserved in the strong interactions, to a very high degree of accuracy, this means that in nature the vacuum angle is fine-tuned, and is very close to zero. (As a matter of fact, estimates show that $\theta \leq 10^{-9}$ [19].) Thus, with the advent of instantons the naturalness of QCD is gone. Can this fine-tuning be naturally explained? There exist several suggestions as to how one could solve the problem of P and CP conservation in QCD in a natural way. One of the most popular is the axion conjecture [20]. This topic, however, definitely lies outside the scope of the present lecture. Interested readers are referred to Ref. [21] for a pedagogical review. We will simply assume that $\theta = 0$, although theoretically it could take any value from the interval $[0, 2\pi]$.

6. Euclidean Formulation of QCD

As we said above, we are concerned with the solution of classical equations in *Euclidean* space. Therefore, we first formulate the Euclidean version of QCD. We give the formulas for the transition from Minkowski to Euclidean space. In Minkowski space one distinguishes between the contravariant and covariant vectors, v^μ and v_μ, respectively. ($\mu = 0, 1, 2, 3$.) The spatial vector $\mathbf{v}$ coincides with the components of the contravariant four-vector, $\mathbf{v} = \{v^1, v^2, v^3\}$. In Euclidean space the distinction between the lower and upper vectorial indices is immaterial, we consider just one vector $\hat{v}_\mu (\mu = 1, 2, 3, 4)$. (In this section the caret is used to denote all quantities on Euclidean space.) In transition to Euclidean the spatial coordinates are not changed, $\hat{x}_i = x^i$, $i = 1, 2, 3$. For the time coordinate x_0, we make the substitution

$$x_0 = -ix_4 \,. \tag{37}$$

Clearly, when x_0 is continued to imaginary values the zeroth component of the vector potential A_μ also becomes imaginary.

We define the Euclidean vector potential $\hat{A}_\mu$ as follows:

$$A^m = -\hat{A}_m \; (m = 1, 2, 3), \; A_0 = i\hat{A}_4 \tag{38}$$

With this definition, the quantities $\hat{A}_\mu$ ($\mu = 1, \ldots, 4$) form a Euclidean vector. The difference between formulas (38) and the corresponding relations for the vector x^μ is introduced for convenience in the expression of the following formulas.[10]

Thus, for the operator of covariant differentiation

$$D_\mu = \partial_\mu - igA_\mu^a T^a \,, \tag{39}$$

where T^a are the matrices of the generators in the representation being considered, we obtain

$$\begin{aligned} D^m &= -\hat{D}_m, \qquad D_0 = i\hat{D}_4 \,, \\ \hat{D}_\mu &= \frac{\partial}{\partial \hat{x}_\mu} - ig\hat{A}_\mu^a T^a \,. \end{aligned} \tag{40}$$

For the field strength tensor $G_{\mu\nu}$ we obtain the formulas

$$G_{mn}^a = \hat{G}_{mn}^a \; (m, n = 1, 2, 3), \; G_{0n}^a = i\hat{G}_{4n}^a \,, \tag{41}$$

---

[10] If we use the definition $\hat{A}_m = A^m$ ($m = 1, 2, 3$), then in all the following connection formulas it is necessary to make the substitution $g \to -g$.

where the Euclidean field strength tensor $\hat{G}^a_{\mu\nu}$,

$$\hat{G}^a_{\mu\nu} = \frac{\partial}{\partial \hat{x}_\mu}\hat{A}^a_\nu - \frac{\partial}{\partial \hat{x}_\nu}\hat{A}^a_\mu + g f^{abc}\hat{A}^b_\mu \hat{A}^c_\nu \qquad (\mu, \nu = 1, \ldots, 4) \tag{42}$$

can be expressed in terms of $\hat{A}_\mu$ and $\partial/\partial\hat{x}_\mu$ in the same way as the Minkowskian $G^a_{\mu\nu}$.

To complete the transition to the Euclidean space, what remain are the formulas for the Fermi fields. We begin with the definition of four Hermitian γ matrices $\hat{\gamma}_\mu$:

$$\begin{aligned} &\hat{\gamma}_4 = \gamma_0, \ \hat{\gamma}_m = -i\gamma^m \qquad && (m = 1, 2, 3), \\ &\{\hat{\gamma}_\mu, \hat{\gamma}_\nu\} = 2\delta_{\mu\nu} && (\mu, \nu = 1, \ldots, 4), \end{aligned} \tag{43}$$

where γ_0 and γ^m are the ordinary Dirac matrices.

The fields ψ and $\bar{\psi}$ are regarded as independent anticommuting variables, with respect to which integration is performed in the functional integral. On the transition to the Euclidean space, it is convenient to define the variables $\hat{\psi}$ and $\hat{\bar{\psi}}$ as

$$\psi = \hat{\psi}, \qquad \bar{\psi} = -i\hat{\bar{\psi}}. \tag{44}$$

Note that under rotations of the pseudo-Euclidean space, $\bar{\psi}$ transforms as $\psi^*\gamma_0$. In the Euclidean space, $\hat{\bar{\psi}}$ transforms as $\hat{\psi}^+$. Indeed, under infinitesimal rotations of the *pseudo*-Euclidean space characterized by the parameters $\omega_{\mu\nu}$ $(\mu, \nu = 0, 1, \ldots, 3)$ the spinor ψ varies as follows:

$$\delta\psi = -\frac{1}{4}(\gamma_\mu\gamma_\nu - \gamma_\nu\gamma_\mu)\omega^{\mu\nu}\psi .$$

For the change in $\bar{\psi} = \psi^*\gamma_0$ we deduce from this

$$\delta(\psi^+\gamma_0) = -\frac{1}{4}\psi^+\gamma_0\gamma_0(\gamma^+_\nu\gamma^+_\mu - \gamma^+_\mu\gamma^+_\nu)\gamma_0\omega^{\mu\nu} = \frac{1}{4}(\psi^+\gamma_0)(\gamma_\mu\gamma_\nu - \gamma_\nu\gamma_\mu)\omega^{\mu\nu} ,$$

so that $\psi_1^+\gamma_0\psi_2$ is a scalar and $\psi_1^+\gamma_0\gamma_\mu\psi_2$ a vector.

On the transition to the Euclidean space, the parameters ω_{mn} $(m, n = 1, 2, 3)$ do not change, and $\omega_{0n} = i\omega_{4n}$ (because of the substitution $x_0 = -ix_4$).

For the variations of $\hat{\psi}$ and $\hat{\psi}^+$ under rotations, we obtain

$$\delta\hat{\psi} = \frac{1}{4}(\hat{\gamma}_\mu\hat{\gamma}_\nu - \hat{\gamma}_\nu\hat{\gamma}_\mu)\hat{\omega}_{\mu\nu}\hat{\psi}\,, \qquad \delta\hat{\psi}^+ = -\frac{1}{4}\psi^+(\hat{\gamma}_\mu\hat{\gamma}_\nu - \hat{\gamma}_\nu\hat{\gamma}_\mu)\hat{\omega}_{\mu\nu}\,,$$

so that $\psi_1^+\psi_2$ and $\psi_1^+\gamma_\mu\psi_2$ are a scalar and vector, respectively.

Finally, we can write down an expression for the Euclidean action:

$$iS = -\hat{S}\,,$$

$$S = \int \mathrm{d}^4x\Big[-\frac{1}{4}G^a_{\mu\nu}G^a_{\mu\nu} + \bar{\psi}(i\gamma_\mu D_\mu - M)\psi + \theta\frac{g^2}{32\pi^2}G^a_{\mu\nu}\tilde{G}^a_{\mu\nu}\Big]\,, \qquad (45)$$

$$\hat{S} = \int \mathrm{d}^4\hat{x}\Big[\frac{1}{4}\hat{G}^a_{\mu\nu}\hat{G}^a_{\mu\nu} + \bar{\hat{\psi}}(-i\hat{\gamma}_\mu\hat{D}_\mu - iM)\hat{\psi} + i\theta\frac{g^2}{32\pi^2}\hat{G}^a_{\mu\nu}\tilde{\hat{G}}^a_{\mu\nu}\Big]\,,$$

where it is assumed that $\hat{\psi}$ is a column in the space of flavors (with color index), and M is a matrix in this space. Note that in the Euclidean space the Levi-Civita tensor $\varepsilon_{\mu\nu\alpha\beta}$ is defined in such a way that $\varepsilon_{1234} = 1$.

Below, we shall use the Euclidean space and omit the caret. The formulas given above make it possible to relate the quantities in the pseudo-Euclidean and Euclidean spaces.

To conclude the section, we note that if we are considering quantities such as the vacuum expectation values of the time-ordered products of currents for space-like external momenta, i.e. when the sources do not produce real hadrons from the vacuum, the Euclidean formulation is not only merely possible but in fact is more adequate than the pseudo-Euclidean. The region of timelike momenta, where there are singularities, can be reached by means of analytic continuation. Such an approach is particularly necessary for quantum chromodynamics, for which the fundamental objects of the theory — the quarks and gluons — have meaning only in the Euclidean domain, and the real singularities corresponding to hadrons have to be obtained.

7. BPST Instantons General Properties

7.1. *Finiteness of the Action and the Topological Charge*

We have learned that in quantitative description of tunneling an important part is played by solutions that give a minimum of the Euclidean action in the limit $\tau_0 \to \infty$. In general, the action increases unboundedly in the limit $\tau_0 \to \infty$, and the condition that it be finite imposes strong restrictions on the paths.

Thus, in the quantum-mechanical example we have analyzed in Sec. 2, the finite-action condition means that the function $x(\tau)$ as $\tau \to \pm\infty$ must have the limits $\pm\eta$. In this way there arises naturally a topological classification of functions giving a finite action on the basis of their limiting values. Formally, a topological charge can be introduced as follows:

$$Q = \frac{1}{2\eta} \int_{-\infty}^{+\infty} \mathrm{d}t\, \dot{x}(t) = \frac{x(+\infty) - x(-\infty)}{2\eta}.$$

It is obvious that Q can take on the values $0, +1, -1$. Functions with different Q cannot be deformed into one another by a continuous deformation that leaves the action finite. Therefore, in each of the classes $Q = 0, +1, -1$ there exists a corresponding minimum of the action and corresponding functions that realize it. The instanton and antiinstanton realize minima for $Q = \pm 1$.

We now turn to gluodynamics — the theory of non-Abelian vector fields — and consider first the case of the group SU(2). We pose the same question: what must be the behavior of the vector fields A_μ^a as $x \to \infty$ if the action is to be finite? (We have in mind the Euclidean action $\hat{S}$; see Eq. (45).) It is clear that the field strength tensor $G_{\mu\nu}^a$ must decrease more rapidly than $1/x^2$. But this by no means implies that the fields A_μ^a must decrease faster than $1/x$. Indeed, suppose A_μ^a in the limit $x \to \infty$ has the form

$$A_\mu = g\frac{\tau^a}{2} A_\mu^a \xrightarrow[x \to \infty]{} iS\partial_\mu S^+ \tag{46}$$

where we have introduced matrix notation: S is a unitary unimodular matrix that depends on the angles in the Euclidean space. Although the angular components of A_μ are proportional to $1/x$, it is clear that in the region in which the expression (46) holds the field strength tensor $G_{\mu\nu}^a$ vanishes, since A_μ^a has a purely gauge form.

Thus, the behavior of A_μ^a at large x is determined by the matrix S, which depends on the angles. Under a gauge transformation of A_μ defined[11] by the matrix $U(x)$:

$$A_\mu \to U^+ A_\mu U + iU^+ \partial_\mu U\,,$$

the matrix S is replaced by $U^+(x \to \infty)S$. It would appear that one can always choose $U(x)$ such that $U(x \to \infty) = S$ and thus remove the terms $1/x$

[11]Warning: Although one and the same letter U is used here and in Sec. 5, these are different gauge matrices which depend on different variables: on the spatial coordinate $\mathbf{x}$ in Sec. 5, and on the Euclidean four-coordinate x_μ here.

from A_μ. However, this argument is correct only if the matrix $U(x)$ does not have singularities at any value of x. Otherwise, the problem of the behavior of $A_\mu(x)$ is transferred from the point at infinity to the position of the singularity of $U(x)$.

As a result, the problem of classifying the fields A_μ^a which give finite action reduces to the topological classification of the matrices S. We shall not present this classification, which was obtained in the pioneering paper of Ref. [3], but rather give examples of nontrivial (not reducible to the unit matrix) matrices S. For example, we have the matrix

$$S_1 = \frac{x_4 + i\mathbf{x}\boldsymbol{\tau}}{\sqrt{x^2}}\,. \tag{47}$$

It corresponds to unit topological charge (there is a one-to-one correspondence between the space of unitary unimodular matrices and the points of the hypersphere in Euclidean space). To topological charge n there corresponds a matrix of the form

$$S_n = (S_1)^n, \qquad n = 0, \pm 1, \pm 2, \ldots\,. \tag{48}$$

Of course, one could choose a different form of the matrix S corresponding to the charge n, but the difference between it and S_n reduces to a topologically trivial gauge transformation.

For the careful reader it should be clear already that there exist two related, but not identical topological arguments. The first argument, discussed in detail in Sec. 5, reveals the existence of the distinct topologically inequivalent zero-energy states. Outlined here is a four-dimensional topological aspect; it refers to the topology of the trajectories connecting (in the Euclidean space-time) the distinct zero-energy states discussed in Sec. 5. The field configurations $A_\mu(x_4, \mathbf{x})$ satisfying Eq. (46) with $S = S_1$ interpolate between the state with the winding number $\mathcal{K}$ and that with the winding number $\mathcal{K}+1$. (To see that this is indeed the case we must, of course, transform the instanton into the $A_0 = 0$ gauge,see Sec. 8.4.) For $S = S_2$ we deal with the trajectory $A_\mu(x_4, \mathbf{x})$ connecting $\mathcal{K}$ and $\mathcal{K}+2$, etc. The topological charge Q of any given field configuration $A_\mu(x_4, \mathbf{x})$ satisfying Eq. (46) is actually nothing but

$$Q = \mathcal{K}' - \mathcal{K}\,,$$

where the prime marks the distant (Euclidean) future while the unprimed quantity refers to the distant (Euclidean) past. A gauge-invariant integral representation exist for the topological charge Q (making unnecessary the transition to the $A_0 = 0$ gauge):

$$Q = \frac{g^2}{32\pi^2} \int d^4x G^a_{\mu\nu} \tilde{G}^a_{\mu\nu} , \tag{49}$$

where

$$\tilde{G}^a_{\mu\nu} = \frac{1}{2} \varepsilon_{\mu\nu\gamma\delta} G^a_{\gamma\delta}, \qquad \varepsilon_{1234} = 1 . \tag{50}$$

The validity of Eq. (49) can be verified by using the fact that $G^a_\mu \tilde{G}^a_{\mu\nu}$ can be represented in the form of a total derivative,

$$G_{\mu\nu}\tilde{G}_{\mu\nu} = \partial_\mu K_\mu ,$$

$$K_\mu = 2\varepsilon_{\mu\nu\gamma\delta} \left(A^a_\nu \partial_\gamma A^a_\delta + \frac{1}{3} g \varepsilon^{abc} A^a_\nu A^b_\gamma A^c_\delta \right) ,$$

so that the volume integral (49) can be transformed into an integral over the surface of a large sphere embedded in four-dimensional space, where A^a_μ has the form (46).

7.2. *The Distinguished Role of the Group* SU(2)

Hitherto, we have discussed the group SU(2). For groups different from SU(2), the construction of instanton solutions with $n = 1$ reduces to the case of SU(2) by means of separation of SU(2) subgroups. Why is the group SU(2) distinguished? We shall attempt to explain this without using topological terminology.

The possibility of deformation of the matrices S is determined by the gauge invariance discussed above. To fix the gauge we represent an arbitrary field A_μ in the form

$$A_\mu(x) = S(x)\tilde{A}_\mu(x)S^+(x) + iS(x)\partial_\mu S^+(x) , \tag{51}$$

where the field $\tilde{A}_\mu$ satisfies definite gauge conditions (for example, $\tilde{A}_0 = 0$ or $\partial_m \tilde{A}_m = 0$ $(m = 1, 2, 3)$). This fixing does not completely determine the transition to the new fields $\tilde{A}_\mu(x)$ and $S(x)$, since A_μ is invariant under global transformations of the form

$$S(x) \to S(x)U_2^+, \qquad \tilde{A}_\mu(x) \to U_2 \tilde{A}_\mu U_2^+ \tag{52}$$

with matrix U_2 that does not depend on x.

In addition, even after fixing the gauge, the theory is still invariant with respect to global color rotations for A_μ, which in terms of the new fields $\tilde{A}_\mu(x)$ and $S(x)$ is equivalent to the transformations

$$S(x) \to U_1 S, \qquad \tilde{A}_\mu(x) \to \tilde{A}_\mu(x)\,. \tag{53}$$

Thus, the color SU(2) invariance of the theory together with the gauge invariance reduce to the set of *global* transformations (52) and (53), which obviously form the group SU(2) $\times$ SU(2). The field $S(x)$ transforms in accordance with the representation $(1/2, 1/2)$, and $\tilde{A}_\mu(x)$ in accordance with the representation $(1,0)$.

On the other hand, the group of rotations of four-dimensional Euclidean space is again, as is well-known, SU(2) $\times$ SU(2), and the generators of the SU(2) subgroups have the form

$$\begin{aligned} I_1^a &= \frac{1}{4}\,\eta_{a\mu\nu} M_{\mu\nu}\,, \\ I_2^a &= \frac{1}{4}\,\bar{\eta}_{a\mu\nu} M_{\mu\nu} \end{aligned} \qquad \begin{pmatrix} a = 1,2,3 \\ \mu,\nu = 1,\ldots,4 \end{pmatrix}, \tag{54}$$

where $M_{\mu\nu} = -ix_\mu \partial/\partial x_\nu + ix_\nu \partial/\partial x_\mu$ + spin part, are the operators of infinitesimal rotations in the (μ,ν) plane, and $\eta_{a\mu\nu}$ are the numerical symbols

$$\eta_{a\mu\nu} = \begin{cases} \varepsilon_{a\mu\nu}, & \mu,\nu = 1,2,3, \\ -\delta_{a\nu}, & \mu = 4, \\ \delta_{a\mu}, & \nu = 4, \\ 0 & \mu = \nu = 4\,. \end{cases} \tag{55}$$

(The symbols $\bar{\eta}_{a\mu\nu}$ differ from η by a change in the sign of δ.) η and $\bar{\eta}$ are called the 't Hooft symbols. The coordinate vector x_μ transforms in accordance with the representation $(1/2, 1/2)$. This is conveniently seen by considering transformations of the matrix

$$x_4 + i\mathbf{x}\boldsymbol{\tau} = i\tau_\mu^+ x_\mu\,, \tag{56}$$

where we have introduced the notation

$$\tau_\mu^\pm = (\boldsymbol{\tau}, \mp i)\,. \tag{57}$$

For $\tau_\mu^\pm$, we have

$$\tau_\mu^+\tau_\nu^- = \delta_{\mu\nu} + i\eta_{a\mu\nu}\tau^a, \qquad \tau_\mu^-\tau_\nu^+ = \delta_{\mu\nu} + i\bar{\eta}_{a\mu\nu}\tau^a. \tag{57'}$$

It is not difficult to find the law of transformation of the matrix (56),

$$e^{i\varphi_1^a I_1^a + i\varphi_2^a I_2^a} i\tau_\mu^+ x_\mu = e^{-i\varphi_1^a(\tau^a/2)}(i\tau_\mu^+ x_\mu)e^{i\varphi_2^a(\tau^a/2)},$$

where φ_1^a and φ_2^a are the parameters of rotations. In other words, a rotation of x_μ is equivalent to the multiplication of unitary unimodular matrices both from the left and the right.

The choice of S in the form $S_1 = ix_\mu\tau_\mu^+/\sqrt{x^2}$ distinguishes certain directions in the isotopic and coordinate spaces. However, under rotation by the same angles in the spatial SU(2)×SU(2) group and in the SU(2)×SU(2) group given by the transformations (52) and (53), the matrix S_1 obviously does not change. In other words, if instead of I_1^a and I_2^a we call $I_1^a + T_1^a$ and $I_1^a + T_2^a$, the angular momentum operators, the introduced object has spin zero (here $T_{1,2}^a$ are the operators of the infinitesimal transformations (52) and (53)).

Thus, we see that the group SU(2) is distinguished on account of the dimension of the coordinate space. Further clarifying remarks as to why the group SU(2) is singled out are presented in Sec. 8.5.

7.3. *Value of the Action for Instanton Solutions*

Although we do not yet have the explicit form of the instanton solution, we can nevertheless calculate the value of the action for it. Indeed, for positive values of the topological charge Q, the Euclidean action can be rewritten in the form

$$\begin{aligned} S &= \int d^4x \frac{1}{4} G_{\mu\nu}^a G_{\mu\nu}^a = \int d^4x \left[\frac{1}{4} G_{\mu\nu}^a \tilde{G}_{\mu\nu}^a + \frac{1}{8}(G_{\mu\nu}^a - \tilde{G}_{\mu\nu}^a)^2\right] \\ &= Q\frac{8\pi^2}{g^2} + \frac{1}{8}\int d^4x (G_{\mu\nu}^a - \tilde{G}_{\mu\nu}^a)^2. \end{aligned} \tag{58}$$

It is clear from this formula that in the class of functions with given positive Q the minimum of S is attained for $G_{\mu\nu}^a = \tilde{G}_{\mu\nu}^a$ and is equal to $(8\pi^2 Q/g^2)$. We recall the functions with different Q cannot be related by a continuous deformation if the action is to remain finite. Therefore, minimization of the action can be carried out separately in each class of functions with the given Q. The BPST instanton has $Q = 1$, and the action $S = 8\pi^2/g^2$.

The case of negative Q is obtained from (58) by the reflection $x_{1,2,3} \to -x_{1,2,3}$, under which $G_{\mu\nu}\tilde{G}_{\mu\nu} \to -G_{\mu\nu}\tilde{G}_{\mu\nu}$ and accordingly $Q \to -Q$. Thus, the minimum of the action for negative Q is $(8\pi^2/g^2)|Q|$, and it is attained when $G^a_{\mu\nu} = -\tilde{G}^a_{\mu\nu}$.

As can be seen from this discussion, the fulfillment of the self-duality and antiself-duality conditions $G^a_{\mu\nu} = \pm\tilde{G}^a_{\mu\nu}$ automatically leads to satisfaction of the equations of motion $D_\mu G_{\mu\nu} = 0$. This can also be seen directly; indeed, for a self-dual field, say, we have

$$\begin{aligned} D_\mu G^a_{\mu\nu} &= D_\mu \tilde{G}^a_{\mu\nu} = \frac{1}{2}\varepsilon_{\mu\nu\gamma\delta} D_\mu G^a_{\gamma\delta} \\ &= \frac{1}{6}\varepsilon_{\mu\nu\gamma\delta}(D_\mu G^a_{\gamma\delta} + D_\gamma G^a_{\delta\mu} + D_\delta G^a_{\mu\gamma}) = 0\,, \end{aligned}$$

where we have used the Bianchi identity:

$$D_\mu G_{\gamma\delta} + D_\delta G_{\mu\gamma} + D_\gamma G_{\delta\mu} = 0\,.$$

8. Explicit Form of the BPST Instanton

8.1. *Solution with* $Q = 1$

As discussed in the previous section, the asymptotic behavior of A^a_μ for this solution is

$$\begin{aligned} g\frac{\tau^a}{2}A^a_\mu &\xrightarrow[x\to\infty]{} iS_1\partial_\mu S_1^+\,, \\ S_1 &= \frac{i\tau^+_\mu x_\mu}{\sqrt{x^2}}\,, \end{aligned} \tag{59}$$

where the matrices $\tau^\pm_\mu$ are defined in (57). We shall also use the symbols $\eta_{a\mu\nu}$ and $\bar{\eta}_{a\mu\nu}$ defined by Eqs. (55). These numerical coefficients are frequently called the 't Hooft symbols, and some useful relations for $\eta_{a\mu\nu}$ are given in Sec. 8.3.

The expression for the asymptotic behavior of A^a_μ can be rewritten in terms of the 't Hooft symbols as follows:

$$A^a_\mu \xrightarrow[x\to\infty]{} \frac{2}{g}\eta_{a\mu\nu}\frac{x_\nu}{x^2}\,.$$

For an instanton with its center at the point $x = 0$, it is natural to assume the same angular dependence of the field for all x, i.e. to seek the solution in the form

$$A^a_\mu(x) = \frac{2}{g}\eta_{a\mu\nu}x_\nu \frac{f(x^2)}{x^2}\,, \tag{60}$$

where

$$f(x^2) \xrightarrow[x^2\to\infty]{} 1\,, \qquad f(x^2) \xrightarrow[x^2\to 0]{} \text{const}\cdot x^2\,.$$

The last condition corresponds to the absence of a singularity at the origin. A justification for the assumption (60) will be the construction of a self-dual expression for $G^a_{\mu\nu}$. From (60), we obtain for $G^a_{\mu\nu}$

$$G^a_{\mu\nu} = -\frac{4}{g}\left\{\eta_{a\mu\nu}\frac{f(1-f)}{x^2} + \frac{x_\mu\eta_{a\nu\gamma}x_\gamma - x_\nu\eta_{a\mu\gamma}x_\gamma}{x^4}[f(1-f) - x^2 f']\right\}. \tag{61}$$

Here the prime denotes differentiation with respect to x^2. In deriving (61), we have used the relation for $\varepsilon^{abc}\times\eta_{b\mu\gamma}\eta_{c\nu\delta}$ from the list of formulas at the end in Sec. 8.3. Using the formula for $\varepsilon_{\mu\nu\gamma\delta}\eta_{a\delta\rho}$ from the same list, we obtain for $\tilde{G}^a_{\mu\nu}$ the expression

$$\tilde{G}^a_{\mu\nu} = -\frac{4}{g}\left\{\eta_{a\mu\nu}f' - \frac{1}{x^4}(x_\mu\eta_{a\nu\gamma}x_\gamma - x_\nu\eta_{a\mu\gamma}x_\gamma)[f(1-f) - x^2 f']\right\}.$$

The condition of self-duality, $G^a_{\mu\nu} = \tilde{G}^a_{\mu\nu}$, requires the fulfillment of the equation $f(1-f) - x^2 f' = 0$, which determines the function f:

$$f(x^2) = \frac{x^2}{x^2+\rho^2}\,, \tag{62}$$

where ρ^2 is a constant of integration; ρ is called the instanton size or, the instanton radius. The translational invariance guarantees obtaining the solution with the center at an arbitrary point x_0, for which it is necessary to replace x by $x - x_0$. We will discuss ρ, x_0 and other collective coordinates in more detail later. Note that if $f - \frac{1}{2}$ is denoted as X and $x^2 = e^\tau$ the equation for f becomes identical to the first-order differential equation one obtains in the double-well potential problem, $\dot{X} = \frac{1}{4} - X^2$.

Summarizing, the final expression for the instanton, with its center at the point x_0 and scale ρ, has the form

$$\begin{aligned} A^a_\mu &= \frac{2}{g}\eta_{a\mu\nu}\frac{(x-x_0)_\nu}{(x-x_0)^2+\rho^2}\,, \\ G^a_{\mu\nu} &= -\frac{4}{g}\eta_{a\mu\nu}\frac{\rho^2}{[(x-x_0)^2+\rho^2]^2}\,. \end{aligned} \tag{63}$$

It can now be verified that the action for the instanton is $8\pi^2/g^2$, as was shown in the general form. The antiinstanton is obtained by the substitution $\eta_{a\mu\nu} \to \bar{\eta}_{a\mu\nu}$.

8.2. *Singular Gauge. The 't Hooft Ansatz*

It is frequently convenient to use the expression for A^a_μ in the so-called singular gauge, when the "bad" behavior of A^a_μ is transferred from the point at infinity to the center of the instanton. As was discussed in the previous section, such a transfer can be realized by a gauge transformation with a matrix $U(x)$ which becomes identical with $S(x)$ as $x \to \infty$.[12] We write down the formulas of the gauge transformation,

$$\begin{aligned} g\frac{\tau^a}{2}\bar{A}^a_\mu &= U^+ g\frac{\tau^a}{2}A^a_\mu U + iU^+\partial_\mu U\,, \\ g\frac{\tau^a}{2}\bar{G}^a_{\mu\nu} &= U^+ g\frac{\tau^a}{2}G^a_{\mu\nu}U\,, \end{aligned} \tag{64}$$

and for an instanton with its center at x_0 take a matrix of the form

$$U = \frac{i\tau^+_\mu (x-x_0)_\mu}{\sqrt{(x-x_0)^2}}\,. \tag{64'}$$

Then for the potential $\bar{A}^a_\mu$ and the field strength tensor $\bar{G}^a_{\mu\nu}$ in the singular gauge we obtain

$$\begin{aligned} \bar{A}^a_\mu &= \frac{2}{g}\bar{\eta}_{a\mu\nu}(x-x_0)_\nu \frac{\rho^2}{(x-x_0)^2[(x-x_0)^2+\rho^2]}\,, \\ \bar{G}^a_{\mu\nu} &= -\frac{8}{g}\left[\frac{(x-x_0)_\mu (x-x_0)_\rho}{(x-x_0)^2} - \frac{1}{4}\delta_{\mu\rho}\right]\bar{\eta}_{a\nu\rho}\frac{\rho^2}{[(x-x_0)^2+\rho^2]^2} - (\mu \leftrightarrow \nu)\,. \end{aligned} \tag{65}$$

It is obvious that the quantities $G^a_{\mu\nu}G^a_{\gamma\delta}$ are invariants of the gauge transformation (see, however, the last footnote). Note also the fact that (65) contains the symbols $\bar{\eta}_{a\mu\nu}$ but not $\eta_{a\mu\nu}$. This difference is due to the fact that in the

[12]More precisely, this transformation should be called a quasigauge transformation, since at the point where $U(x)$ has a singularity (and there must be such a singularity) this transformation changes the gauge-invariant quantities, for example, $G^a_{\mu\nu}G^a_{\mu\nu}$. To use such transformations, it is necessary to consider a space with punctured singular points. This we shall do, remembering that the physical quantitites are nonsingular at the singular points.

singular gauge the topological charge (49) is saturated in the neighborhood of $x = x_0$ and not at infinity.

The expression (65) for $\bar{A}^a_\mu$ can be rewritten in the form

$$\bar{A}^a_\mu = -\frac{1}{g}\bar{\eta}_{a\mu\nu}\partial_\nu \ln\left[1 + \frac{\rho^2}{(x-x_0)^2}\right]. \tag{66}$$

As was noted by 't Hooft, this expression can be generalized to a topological charge Q greater than unity. Indeed, if

$$A^a_\mu = -\frac{1}{g}\bar{\eta}_{a\mu\nu}\partial_\nu \ln W(x), \tag{67}$$

then for $G^a_{\mu\nu} - \tilde{G}^a_{\mu\nu}$ we obtain [see the properties of the η symbols in Sec. 8.3]

$$G^a_{\mu\nu} - \tilde{G}^a_{\mu\nu} = \frac{1}{g}\bar{\eta}_{a\mu\nu}\frac{\partial_\gamma\partial_\gamma W}{W}.$$

The self-duality of $G^a_{\mu\nu}$ requires fulfillment of the equation $\partial_\gamma\partial_\gamma W/W = 0$. The solution with topological charge Q has the form

$$W = 1 + \sum_{i=1}^{n}\frac{\rho_i^2}{(x-x_i)^2}, \tag{68}$$

i.e. it describes instantons with centers at the points x_i. The effective scale of an instanton with center at the point x_i is obviously

$$\rho_i^{\text{eff}} = \rho_i\left[1 + \sum_{k\neq i}\frac{\rho_k^2}{(x_k-x_i)^2}\right]^{-1/2}.$$

It should be noted that the choice of A^a_μ in the form (67) did not give the most general solution with charge Q, since all Q instantons have the same orientation in the color space (for the construction of the general solution, see Ref. [22]).

8.3. *Relations for the η Symbols*

We give a list of relations for the symbols $\eta_{a\mu\nu}$ and $\bar{\eta}_{a\mu\nu}$ defined by Eqs. (55):

$$\eta_{a\mu\nu} = \frac{1}{2}\varepsilon_{\mu\nu\alpha\beta}\eta_{a\alpha\beta},$$

$$\eta_{a\mu\nu} = -\eta_{a\nu\mu}, \qquad \eta_{a\mu\nu}\eta_{b\mu\nu} = 4\delta_{ab},$$

$$\eta_{a\mu\nu}\eta_{a\mu\lambda} = 3\delta_{\nu\lambda}\,, \qquad \eta_{a\mu\nu}\eta_{a\mu\nu} = 12\,,$$

$$\eta_{a\mu\nu}\eta_{a\gamma\lambda} = \delta_{\mu\gamma}\delta_{\nu\lambda} - \delta_{\mu\lambda}\delta_{\nu\gamma} + \varepsilon_{\mu\nu\gamma\lambda}\,,$$

$$\varepsilon_{\mu\nu\lambda\sigma}\eta_{a\gamma\sigma} = \delta_{\gamma\mu}\eta_{a\nu\lambda} - \delta_{\gamma\nu}\eta_{a\mu\lambda} + \delta_{\gamma\lambda}\eta_{a\mu\nu}\,,$$

$$\eta_{a\mu\nu}\eta_{b\mu\lambda} = \delta_{ab}\delta_{\nu\lambda} + \varepsilon_{abc}\eta_{c\nu\lambda}\,,$$

$$\varepsilon_{abc}\eta_{b\mu\nu}\eta_{c\gamma\lambda} = \delta_{\mu\gamma}\eta_{a\nu\lambda} - \delta_{\mu\lambda}\eta_{a\nu\gamma} - \delta_{\nu\gamma}\eta_{a\mu\lambda} + \delta_{\nu\lambda}\eta_{a\mu\gamma}\,,$$

$$\eta_{a\mu\nu}\bar{\eta}_{b\mu\nu} = 0\,, \qquad \eta_{a\gamma\mu}\bar{\eta}_{b\gamma\lambda} = \eta_{a\gamma\lambda}\bar{\eta}_{b\gamma\mu}\,.$$

To pass from the relations for $\eta_{a\mu\nu}$ to those for $\bar{\eta}_{a\mu\nu}$ it is necessary to make the substitution

$$\eta_{a\mu\nu} \to \bar{\eta}_{a\mu\nu}\,, \qquad \varepsilon_{\mu\nu\gamma\delta} \to -\varepsilon_{\mu\nu\gamma\delta}\,.$$

8.4. *Instanton in the $A_0 = 0$ Gauge*

In Sec. 7.1 it was mentioned that the relation between the instanton topological charge Q and the winding numbers of the zero-energy states in the distant past and distant future, between which it interpolates,

$$Q = \mathcal{K}' - \mathcal{K}\,,$$

is most transparently seen in the $A_0 = 0$ gauge. Now we can explicitly demonstrate that this relation does indeed take place.

Equations (59) and (62) imply that the instanton field

$$A_\mu = \frac{x^2}{x^2+\rho^2}\, iS_1\partial_\mu S_1^+$$

where $A_\mu = (g/2)\tau^a A_\mu^a$ and the matrix S_1 is defined in Eq. (59). Let us presume that the time component of the gauge-transformed field vanishes identically,

$$U^+ A_\mu U + iU^+\partial_\mu U = 0 \text{ at } \mu = 0\,.$$

Substituting the expression for the instanton field we get the following equation for the gauge matrix U transforming the BPST instanton to the $A_0 = 0$ gauge,

$$\dot{U} + \frac{x^2}{x^2+\rho^2}(S_1\dot{S}_1^+)U = 0\,,$$

where

$$S_1 \dot{S}_1^+ = i\,\frac{\mathbf{x}\boldsymbol{\sigma}}{x^2}\,.$$

The solution of this equation is obvious,

$$U(\tau, \mathbf{x}) = \left[\exp \int_{-\infty}^{\tau} \left(\frac{i\mathbf{x}\boldsymbol{\sigma}}{x^2+\rho^2}\right) d\tau\right] U(\tau = -\infty, \mathbf{x})\,.$$

The instanton field in the $A_0 = 0$ gauge takes the form

$$A_i(\tau, \mathbf{x}) = i\left\{\frac{x^2}{x^2+\rho^2} U^+(\tau, \mathbf{x})(S_1 \partial_i S_1^+) U(\tau, \mathbf{x}) + U^+(\tau, \mathbf{x})\partial_i U(\tau, \mathbf{x})\right\}.$$

In the distant past and distant future

$$A_i \to i(U^+ S_1)\partial_i (U^+ S_1)^{-1}\,.$$

Moreover, $S_1 \to 1$ at $\tau \to= \pm\infty$. As for $U(\tau = +\infty)$, we have

$$\begin{aligned} U(\tau = +\infty, \mathbf{x}) &= \left[\exp \int_{-\infty}^{+\infty} \left(-\frac{i\mathbf{x}\boldsymbol{\sigma}}{x^2+\rho^2}\right) d\tau\right] U(\tau = -\infty, \mathbf{x}) \\ &= \left[\exp\left(-\frac{i\pi\mathbf{x}\boldsymbol{\sigma}}{\sqrt{\mathbf{x}^2+\rho^2}}\right)\right] U(\tau = -\infty, \mathbf{x})\,. \end{aligned}$$

The hedgehog matrix appears on the right-hand side. This obviously concludes the proof that the winding numbers of the filed configurations between which the instanton interpolates differ by unity.

8.5. *Instanton Collective Coordinates*

The instanton solution presented in Eq. (63) has the following collective coordinates: the instanton size ρ (associated with dilatations); and four parameters represented by the instanton center x_0 (associated with translations). The issue of the collective coordinates is important, since each of them gives rise to a zero mode. As we already know (Sec. 3), the latter play a special role in calculating the instanton determinant, and, eventually, the instanton measure. Thus, it is imperative to establish a complete set of the collective coordinates. In this section we will analyze the set of the collective coordinates for the SU(2) instanton.

The action of the pure Yang-Mills theory, Eq. (58), has no dimensional parameters, and is conformally invariant at the classical level. Since the instanton is the solution of the classical equations of motion (which are naturally conformally invariant too), the set of the collective coordinates appearing in the generic instanton solution is determined by the conformal group. Each given instanton solution breaks (spontaneously) some of the invariances. The conformal symmetry is restored only upon consideration of the family of the solutions, as a whole. Those symmetry transformations that act on the instanton solution nontrivially, generate another solution belonging to the same family, with "shifted" values of the collective coordinates. Thus, each symmetry transformation from the conformal group, which does not leave the instanton solution intact, requires a separate collective coordinate.

The conformal group in four dimensions includes 15 transformations (e.g. Ref. [23]): four translations, six Lorentz rotations (in the Euclidean space it is more appropriate to speak of six $O(4)$ rotations), four proper conformal transformations and one dilatation. Moreover, the Yang-Mills action is gauge invariant. We do not need to consider (small) gauge transformations of the instanton, since they produce just the same solution in a different gauge. Global rotations in the color space have to be considered, however. In the SU(2) theory there are three global rotations. Thus, *a priori* one could expect the generic instanton solution to depend on 18 collective coordinates. So far, we have only seen five. Where are the remaining collective coordinates?

The proper conformal transformations can be represented as a combination of translations and inversion. Under inversion

$$x_\mu \to x'_\mu = \frac{x_\mu}{x^2}\,, \qquad A_\mu(x) \to x'^2 A_\mu(x')\,.$$

Translations are already represented by the corresponding collective coordinate, x_0. Now, if we start from the original BPST instanton with the unit radius and make the inversion, we will obviously get an antiinstanton in the singular gauge,

$$\frac{2}{g}\eta_{a\mu\nu}\frac{x_\nu}{x^2+1} \xrightarrow{\text{inv.}} \frac{2}{g}\eta_{a\mu\nu}\frac{x_\nu}{x^2(x^2+1)}$$

(cf. Eqs. (63) and (65)). Thus, no new collective coordinates are associated with the proper conformal transformations.

What remains to be discussed? We must consider six rotations of the Euclidean space and three global color rotations. We will show that only three linear combinations of these nine generators act on the instanton solution nontrivially, resulting in three extra collective coordinates. As a matter of fact, the argument was already outlined in Sec. 7.2 where we explained why the gauge group SU(2) is singled out. Here we rephrase the argument [24] in a more explicit form by passing to a convenient spinorial formalism which we will need later on anyway. This formalism becomes practically indispensable in dealing with the chiral fermions.

For arbitrary vector V_μ from the Euclidean space-time introduce $V_{\alpha\dot\alpha}$,

$$V_{\alpha\dot\alpha} = (\sigma_\mu)_{\alpha\dot\alpha} V_\mu\,, \ (\mu = 1, \ldots, 4, \alpha, \dot\alpha = 1, 2)\,,$$

where

$$\sigma_\mu \equiv \tau_\mu^- = \{\boldsymbol{\sigma}, i\}\,.$$

(The matrices $\tau_\mu^\pm$ were introduced in Eq. (57). Here we find it converient to pass to a more concise notation.) The undotted and dotted indices from the beginning of the Greek alphabet will denote the spinorial indices of the $SU(2)_L$ and $SU(2)_R$ subgroups of the Lorentz group in the Euclidean space ($O(4) = SU(2)_L \times SU(2)_R$). A similar definition refers to the coordinate x_μ itself, $x_{\alpha\dot\alpha} = -(\sigma_\mu)_{\alpha\dot\alpha} x_\mu$. The lowering and raising of the spinorial indices are done by multiplying by antisymmetric tensor (the Levi-Civita tensor) from the left,

$$\chi^\alpha = \varepsilon^{\alpha\beta}\chi_\beta\,, \qquad \chi_\alpha = \varepsilon_{\alpha\beta}\chi^\beta\,,$$

and the same for the dotted indices, where

$$\varepsilon^{\alpha\beta} = -\varepsilon^{\beta\alpha}, \qquad \varepsilon^{12} = -\varepsilon_{12} = 1\,.$$

Then,

$$x_{\dot 1}^1 = -x_1 - ix_2,\ x_{\dot 2}^1 = -ix_4 + x_3,\ x_{\dot 1}^2 = ix_4 + x_3,\ x_{\dot 2}^2 = x_1 - ix_2\,.$$

In this notation the instanton field takes the form

$$A_{\alpha\dot\beta}^{\eta\xi} = -\frac{i}{g}\frac{1}{x^2+\rho^2}(x_{\dot\beta}^\eta \delta_\alpha^\xi + x_{\dot\beta}^\xi \delta_\alpha^\eta)$$

where the color index of A_μ^a is also converted into two spinorial indices according to the formula

$$A^{\eta\xi} = \frac{1}{2}(\tau^a)_\delta^\eta \varepsilon^{\xi\delta}\,,$$

and $\varepsilon^{\xi\delta}$ is the Levi-Civita tensor.

For completeness we give here the expression for the gluon field strength tensor,

$$G^{\gamma\delta}_{\alpha\beta} = \frac{4i}{g}\frac{\rho^2}{(x^2+\rho^2)^2}(\delta^\gamma_\alpha\delta^\delta_\beta + \delta^\gamma_\beta\delta^\delta_\alpha), \quad G^{\gamma\delta}_{\dot\alpha\dot\beta} = 0\,,$$

where the transition from the standard vectorial to spinorial notation is done in the following way

$$(\sigma_\mu)_{\alpha\dot\alpha}(\sigma_\nu)_{\beta\dot\beta}G_{\mu\nu} = \varepsilon_{\dot\alpha\dot\beta}G_{\alpha\beta} + \varepsilon_{\alpha\beta}G_{\dot\alpha\dot\beta}\,.$$

$G_{\alpha\beta}$ and $G_{\dot\alpha\dot\beta}$ are self-dual and anti-self-dual parts of the gluon field strength tensor,[13] so that the equation $G^{\gamma\delta}_{\dot\alpha\dot\beta} = 0$ is nothing but the condition of self-duality of the instanton solution.

Now we are ready to discuss what happens with the instanton when one does the Lorentz and/or color rotations. Note that if in the standard notation the global *color* rotation acts on the four-potential A as $A \to M^+AM$, then within our new convention

$$A^{\alpha\beta} \to (M^+)^\alpha_\gamma(M^+)^\beta_\delta A^{\gamma\delta}\,.$$

Here $M = \exp(i\omega^a\tau^a/2)$, and ω^a are three parameters corresponding to a global rotation in the SU(2) color group.

Let the left-handed rotation act on the vector with the upper undotted index as

$$x^\eta_{\dot\beta} \to L^\eta_\alpha x^\alpha_{\dot\beta}\,,$$

where L is a matrix from $\mathrm{SU}(2)_L$. Then for the vectors with the lower undotted index

$$x_{\alpha\dot\beta} \to (L^+)^\gamma_\alpha x_{\gamma\dot\beta}\,.$$

Transformations from $\mathrm{SU}(2)_R$ (they act on the dotted indices) rotate x and A in the same way. In other words, the form of the instanton solution does not change at all. No collective coordinates corresponding to the $\mathrm{SU}(2)_R$ rotations emerge.

[13] From this expression for $G^{\gamma\delta}_{\alpha\beta}$ it is clear that the field configuration under discussion is a hedgehog in the $\mathrm{SU}(2)_L$ Lorentz subgroup. According to the nomenclature suggested by 't Hooft, it should be actually called *antiinstanton*. Indeed, passing to the standard vectorial notation we get $\bar\eta_{a\mu\nu}$ rather than $\eta_{a\mu\nu}$.

Transformations from $\mathrm{SU}(2)_L$ do change the form of the instanton solution, and so do the global color rotations. Under the combined action of these two transformations the instanton solution becomes

$$A^{\eta\xi}_{\alpha\dot\beta} = \frac{i}{x^2+\rho^2}(M^+)^{\eta}_{\eta'}(M^+)^{\xi}_{\xi'}(L^{\eta'}_{\gamma}L^{\xi'}_{\alpha}x^{\gamma}_{\dot\beta} + L^{\xi'}_{\gamma}L^{\eta'}_{\alpha}x^{\gamma}_{\dot\beta}) \, .$$

We see that if $L = M$ the solution remains intact. This means that out of six transformations (three global color rotations and three $\mathrm{SU}(2)_L$ rotations) only three are independent, giving rise to three collective coordinates. We can choose them to be associated either with the global color rotations (as is usually assumed) or with the rotations from $\mathrm{SU}(2)_L$. In the conventional formalism these three collective coordinates are introduced as

$$\eta_{a\mu\nu} \to O_{ab}\eta_{b\mu\nu} \, , \qquad \bar\eta_{a\mu\nu} \to O_{ab}\bar\eta_{b\mu\nu}$$

where the three-by-three matrix O_{ab} is defined as

$$O_{ab} = \frac{1}{2}\,\mathrm{Tr}(M\tau^a M^+\tau^b) \, .$$

The advantage of the spinorial formalism is obvious — there is no need to introduce the 't Hooft symbols, and the hedgehog nature of instanton is most transparent.

Summarizing, there are eight collective coordinates characterizing the SU(2) instanton. Correspondingly, we will observe eight zero modes. For higher gauge groups the number of the collective coordinates corresponding to global color rotations increases. Altogether, in the $SU(N)$ group the BPST instanton has $4N$ collective coordinates. The counting was first carried out in Ref. [25]. We will return to the discussion of the $SU(N)$ instanton in Sec. 9.5.

8.6. *Instantons in the Higgs Regime*

Quantum chromodynamics is not the only gauge theory of practical importance. The Standard model of the electroweak interactions is a gauge theory too. A drastic distinction in their dynamical behavior is due to the fact that the non-Abelian gauge group is spontaneously broken in the standard model due to the Higgs mechanism, the coupling constant is frozen at the values of momenta of order of the W boson mass. It never becomes strong. Correspondingly, the color confinement and other peculiar phenomena of QCD do not take place. Since the nontrivial topology in the space of the gauge fields is not affected by the Higgs phenomenon, instantons (as the tunneling trajectories) exist in the standard model too, leading to certain nonperturbative effects. The one which was under the most intense scrutiny is the baryon number violations at high energies. We will not dwell on this, applied, aspect of the instanton calculus, referring the reader to Ref. [26] for a detailed review. Instead, we will focus on theoretical issues. As a matter of fact, as we will see below, consideration of instantons in the Higgs regime, even has certain advantages over the QCD instantons. Since the coupling constant never becomes large the quasiclassical approximation in the description of the tunneling phenomena, based on instantons, is always justified, in clear distinction with QCD, where the instanton contribution is dominated by large-size instantons, which are obviously outside the scope of the applicability of quasiclassical methods.

Since we are not concerned with applications, we will limit ourselves to a truncated standard model: SU(2) gauge group, with the minimal Higgs sector — one complex Higgs doublet χ^i, $i = 1, 2$. The $U(1)$ subgroup, as well as fermions, present in the standard model, are discarded.

If the Higgs field is in the fundamental representation of the color group, there is no clear-cut distinction between the confinement phase and the Higgs phase. As the vacuum expectation value (VEV) of the Higgs field χ continuously changes from large values to smaller ones, we continuously flow from the weak coupling regime to the strong coupling one. The spectrum of all physical states, and all other measurable quantities, change smoothly [27].

One can argue that the case is such in many different ways. Perhaps, the most straightforward line of reasoning is as follows. Using the Higgs field in the fundamental representation one can build gauge invariant interpolating operators for *all* possible physical states. The Källén-Lehmann spectral function corresponding to these operators, which carries complete information on the spectrum, depends smoothly on $\langle \chi^* \chi \rangle$. When the latter parameter is large the Higgs description is more convenient, when it is small it is more convenient to think in terms of the bound states. There is no sharp boundary. We deal with a single Higgs/confining phase [27].

It is convenient for our purposes to write the Lagrangian in the Higgs sector in a slightly nonconventional form.

The model at hand has a *global* SU(2) symmetry, associated with the possibility of rotating the doublet χ^i into the conjugated doublet $\varepsilon^{ij}\chi_i^\dagger$. (This global symmetry is responsible for the fact that all three W bosons are degenerate if the $U(1)$ interaction is switched off in the standard model.) The

SU(2) symmetry of the χ sector becomes explicit if we introduce a matrix field

$$X = \begin{pmatrix} \chi^1 & -\chi^{2\dagger} \\ \chi^2 & \chi^{1\dagger} \end{pmatrix},$$

and identically rewrite the standard Higgs Lagrangian in terms of this matrix,

$$\Delta\mathcal{L}_\chi = \frac{1}{2}\operatorname{Tr} D_\mu X^\dagger D_\mu X - \frac{\lambda}{4}\left(\frac{1}{2}\operatorname{Tr} X^\dagger X - \eta^2\right)^2$$

where $D_\mu X = (\partial_\mu - iA_\mu)X$. The complex doublet field χ^i develops a vacuum expectation value η which can be varied continuously.

All physical states form representations of the global SU(2). Consider, for instance, the SU(2) triplets produced from the vacuum by the operators

$$W_\mu^a = -\frac{i}{2}\operatorname{Tr}(X^\dagger \overleftrightarrow{D}_\mu X\tau^a), \qquad a = 1, 2, 3.$$

The lowest-lying states produced by these operators in the weak coupling regime (i.e. when $\langle\chi^\dagger\chi\rangle \gg \Lambda^2$) coincide with the conventional W bosons of the Higgs picture, up to a normalization constant. The mass of the W bosons is $\sim g\eta$. On the other hand, if $\langle\chi^\dagger\chi\rangle \lesssim \Lambda^2$ it is more appropriate to think of the bound states of the χ "quarks" forming vector mesons, triplet with respect to the global SU(2) ("ρ mesons"). Their mass is $\sim \Lambda$. The continuous evolution of η results in the continuous evolution of the mass of the corresponding states. It is easy to check that the complete set of the gauge invariant operators that one can build in this model spans the whole Hilbert space of the physical states.

By the same token, one can treat, in a gauge-invariant manner, elements of the instanton calculus. We will not pursue this line of reasoning in detail, referring the reader to the original publications (e.g. Ref. [24]). Instead, we will dwell on two problems: calculation of the instanton action in the Higgs regime, and the height of the barrier in Fig. 8.

Instanton action

Strictly speaking, if the scalar field develops a vacuum expectation value, the only exact solution of the classical equation of motion is the zero-size instanton. For each given value of ρ one can make the action of the tunneling trajectory smaller by choosing a smaller value of ρ, so that $8\pi^2/g^2$ is achieved asymptotically (see below). Since the nontrivial topology remains intact (one direction in the space of fields forms a circle), for proper understanding of the tunneling phenomena one cannot disregard the trajectories connecting the zero-energy gauge copies (pre-vacua) in the Euclidean time, even though they are not exact solutions any more. Following 't Hooft [9], we will consider *constrained* instantons — trajectories that minimize the action under the condition that the size ρ is fixed. Our analysis will be somewhat heuristic. More rigorously the construction is described, for example, in Ref. [28].

Technically the procedure can be summarized as follows. We first find the solutions of the classical (Euclidean) equations of motion for the gauge field ignoring the scalar field altogether. The solution is of course the familiar instanton. Then we look for the solution of the equations of motion for the χ field in the given instanton background. This solution minimizes the Higgs part of the action. A nonvanishing scalar field, in turn, induces a source term in the equation for the gauge field, which is neglected. This source term will push the instanton towards smaller sizes, in particular, by cutting off the tails of the A_μ field at large distances (where they should become exponentially small). The distance where this occurs is of order $1/(g\eta)$. If we are interested in distances of order $1/\eta$ — and the instanton contributions are indeed saturated at such distances — then we can neglect this effect and continue to disregard the back reaction of the scalar field in considering instantons whose sizes are fixed by hand.

To keep our analysis as simple as possible we will further assume that the scalar self-coupling $\lambda \to 0$. The only role of the scalar self-interaction then is to provide the boundary condition at large distances,

$$\frac{1}{2}\,\mathrm{Tr}(X^+X) \to \eta^2\,.$$

The equation of motion of the scalar field is completely determined by the kinetic term in the Lagrangian,

$$D^2 X = 0\,.$$

If the instanton field is written as

$$A_\mu = \frac{ix^2}{x^2+\rho^2} S_1 \partial_\mu S_1^+$$

(for the antiinstanton $S_1 \leftrightarrow S_1^+$; the matrix S_1 is defined in Eq. (59)), it is not difficult to check that the solution of the equation $D^2X = 0$ takes the form[14]

$$X = \eta\left(\frac{x^2}{x^2+\rho^2}\right)^{1/2} S_1 .$$

This expression is properly normalized. Asymptotically the modulus of the Higgs field approaches its value in the "empty" vacuum, while the "phase" part of the scalar field has a hedgehog winding. At small x the VEV is suppressed.

Using the fact that the Higgs field satisfies the equation of motion we readily rewrite the contribution of the Higgs kinetic term in the action as

$$\int d^4x \partial_\mu \frac{1}{2} \operatorname{Tr}(X^+ D_\mu X) .$$

Moreover,

$$X^+ D_\mu X = \eta^2 \rho^2 \frac{x_\mu}{(x^2+\rho^2)^2} + \eta^2 \rho^2 \frac{x^2}{(x^2+\rho^2)^2} (S_1^+ \partial_\mu S_1) .$$

The last bracket, being an element of the algebra, is proportional to σ^a and, hence, vanishes upon taking the color trace. Therefore, the trace is determined entirely by the first term. Exploiting now the Gauss theorem and rewriting the volume integral as that over the surface of the large sphere dS_μ we arrive at

$$\int d^4x \partial_\mu \frac{1}{2} \operatorname{Tr}(X^+ D_\mu X) = \int dS_\mu \eta^2 \rho^2 \frac{x_\mu}{(x^2+\rho^2)^2} = 2\pi^2 \eta^2 \rho^2 .$$

Summarizing, the extra term in the action induced by a nonvanishing vacuum the expectation value of the Higgs field has the form

$$\Delta S = 2\pi^2 \eta^2 \rho^2 .$$

This term is called the 't Hooft interaction, since 't Hooft was the first to calculate it [9]. The 't Hooft interaction explicitly exhibits the feature we anticipated earlier — the smaller the instanton size ρ the smaller is the instanton action. It is clear that the instanton contribution to physical quantities is determined by an integral over ρ (in Sec. 9 we will calculate the instanton density

---

[14]Exercise: Verify that the expression presented here is indeed a solution.

which will give us the measure of integration). The exponential suppression of the instanton density at large ρ due to the 't Hooft term, $\exp(-2\pi^2\eta^2\rho^2)$, guarantees that $\rho \sim \eta^{-1}$. This, in turn, justifies the approximations made: the back reaction of the scalar field on the gauge field becomes important at much larger distances, $x \sim (g\eta)^{-1}$.

In the SU(2) theory the 't Hooft interaction depends only on one collective coordinate, ρ. In more complicated examples it may acquire dependences on other collective coordinates. For instance, if we consider an SU(3) model with one Higgs triplet (breaking SU(3) down to SU(2)) then the 't Hooft interaction will depend, roughly speaking, on the orientation of the instanton in the color space relatively to the direction of the Higgs VEV. The 't Hooft term becomes $2\pi^2\eta^2\rho^2\cos^2(\alpha/2)$ where α is a certain angle. If the instanton under consideration resides in the corner of SU(3) corresponding to the unbroken SU(2), then $\alpha = \pi$ and the 't Hooft term vanishes. Further details can be found in Ref. [24].

8.7. *The Height of the Barrier. Sphaleron*

Let us return to the tunneling interpretation of instantons discussed in Sec. 5 and ask the question on what is the height of the barrier in Figs. 7 or 8, under which the tunnelings described by the instanton trajectory take place. This issue is not so simple as it might seem at first sight.

Indeed, the QCD Lagrangian at the classical level contains no dimensional parameters. Since the instantons are solutions of the classical equations of motion (in the Euclidean space), they do not carry dimensional constants other than the instanton size which is a variable parameter. The height of the barrier must have a dimension of mass. Therefore, the smaller the size ρ, the higher the barrier the instanton with the given ρ "sees", so that the classical action stays constant, $8\pi^2/g^2$. This is possible, of course, due to the fact that the space of fields is actually infinitely-dimensional. The one-dimensional plot depicted in Fig. 8 is symbolic. Since the infrared limit of QCD is not tractable quasiclassically it is impossible to determine the lowest possible height of the barrier under which the system tunnels. All we can say is that it is of order $\Lambda_{\rm QCD}$.

The situation drastically changes in the Higgs regime considered in the previous section. The vacuum expectation value of the Higgs field provides masses to all gauge bosons. If the vacuum expectation value is much larger than $\Lambda_{\rm QCD}$, the coupling constant always stays small, and the quasiclassical picture is fully applicable. Under the circumstances the question on what is the minimal height of the barrier in Fig. 8 becomes amenable to quantitative analysis. From this figure it is clear that when the system sits right on top of the barrier, this is a solution of the *static* equations of motion, since the position on top of the barrier is an equilibrium. It is also clear that the equilibrium is unstable since it corresponds to a maximum of energy rather than a minimum.

Thus, we will look for the solution of the static equations of motion of the Yang-Mills Lagrangian in the $A_0 = 0$ gauge. By inspecting the structure of these equations it is easy to guess an *Ansatz* which untangles the color and Lorentz indices,

$$A_i^a = \frac{1}{g}\varepsilon_{iak}\frac{x^k}{r}f(r)\,, \qquad X = \frac{\boldsymbol{\sigma}\mathbf{x}}{r}h(r)\,,$$

where $r = \sqrt{\mathbf{x}^2}$ and f, h are profile functions to be determined from the equations. The boundary conditions are quite obvious: at $r \to 0$ both functions f and h must tend to zero to avoid singularities; at $r \to \infty$ the function h tends to η while $f(r) \to -2/r$. The latter condition is necessary to ensure that $A_i(r)$ becomes pure gauge at infinity. Then the energy density of the gauge field vanishes at large r. Simultaneously, the energy density of the scalar field also vanishes in spite of the winding of the field X. The overall energy of the field configuration under consideration can be expected to be finite if both conditions are met.

Technically, instead of solving the equations of motion it is more convenient to write out the energy functional and minimize it with respect to f and h under the given boundary conditions. Substituting our *Ansatz* in the Lagrangian presented in the previous section we readily obtain

$$\mathcal{H} = 4\pi\int_0^\infty r^2 dr\left\{\frac{1}{g^2}\left[f'^2 + \frac{2}{r^2}f^2 + \frac{2}{r}f^3 + \frac{1}{2}f^4\right] + h'^2 + 2h^2\left(\frac{1}{r} + \frac{f}{2}\right)^2\right\}.$$

The terms in the square brackets are from the gauge part (integration by parts is carried out in one of the terms). The second term represents the Higgs part. Since all terms in $\mathcal{H}$ are positive-definite it is clear that a minimum of this functional exists. It can be found numerically; the corresponding profile functions are depicted in Fig. 9.

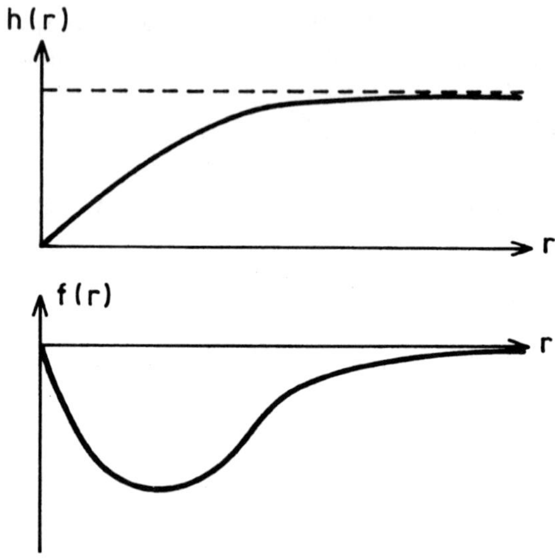

Fig. 9. The solutions for the sphaleron profile functions.

Before minimization it is convenient to rescale the fields and the variable r to make them dimensionless,

$$f = g\eta F, \qquad h = \eta H, \qquad r = R(g\eta)^{-1}\,.$$

In terms of the rescaled fields the energy functional takes the form

$$\mathcal{H} = 4\pi\frac{\eta}{g}\int_0^\infty R^2 dR\Biggl\{\left[F'^2 + \frac{2}{R^2}F^2 + \frac{2}{R}F^3 + \frac{1}{2}F^4\right] + H'^2 + 2H^2\left(\frac{1}{R} + \frac{F}{2}\right)^2\Biggr\},$$

where the prime here denotes differentiation over R. The expression in the braces contains no parameters, neither does the boundary conditions for the dimensionless fields F, H (at $R \to \infty$ the function H approaches unity and the function F tends to $-2/R$). The only parameter of the problem, η/g, is an overall factor. This obviously means that the energy of the solution obtained by minimizing the energy functional $\mathcal{H}$ is

$$E \equiv \mathcal{H}_{\min} = \text{constant of order unity} \times \frac{\eta}{g}\,.$$

The numerical value of the constant is not very important for our illustrative purposes. It can be found in the original papers, see e.g. Ref. [29].

The static solution outlined above, corresponding to the top of the barrier, is called *sphaleron*, from the Greek adjective *sphaleros* meaning unstable, ready to fall. It was found in the SU(2) theory in Ref. [30] and rediscovered later in the context of the standard model by Klinkhamer and Manton ([29]; see also Ref. [31]), who were the first to interpret the sphaleron energy as the height of the barrier separating distinct pre-vacua of the Yang-Mills theory in the Higgs regime. It is instructive to examine the position of the sphaleron on the plot of Fig. 8 directly, by calculating the winding number of the corresponding gauge field. Note that at large distances

$$(A_i)_{\rm sph} \to iU\partial_i U^+\,, \qquad U = \frac{\boldsymbol{\sigma}\mathbf{x}}{r}\,.$$

The matrix U takes different values as we approach infinity from different directions. Thus, the condition of compactification which we impose on the vacuum gauge field, does not hold for the sphaleron. Correspondingly, the winding number $\mathcal{K}[(A_i)_{\rm sph}]$ need not be an integer. A direct calculation (left as an exercise for the reader) readily yields

$$\mathcal{K}[(A_i)_{\rm sph}] = \frac{1}{2}\,,$$

demonstrating that the sphaleron sits right in the middle between two classical minima with $\mathcal{K} = 0$ and $\mathcal{K} = 1$.

To give a well-defined quantitative meaning to the height of the barrier in the absence of the Higgs field we must regularize the Yang-Mills theory in the infrared domain. One of the possible regularizations was suggested in Ref. [32] where the Yang-Mills fields were put on a three-dimensional sphere of a finite radius, instead of the flat space of conventional QCD. The radius of the sphere plays essentially the same role as $(g\eta)^{-1}$ in the Higgs picture. If this radius is small, the quasiclassical consideration becomes closed, and one naturally discovers analogs of the sphaleron solution. The advantage of this regularization over the Higgs field regularization is the existence of analytic expressions. Both, the sphaleron field configuration and its energy can be found analytically [32]. In particular, the sphaleron energy turns out to be $3\pi^2/g^2$ times the inverse radius of the sphere.

9. Calculation of the Pre-Exponential Factor for the BPST Instanton

9.1. *Expansion Near a Saddle Point. Choice of the Gauge and Regularization*

As in the quantum-mechanical example, to calculate the pre-exponential factor in the instanton contribution to the vacuum–vacuum transition, it is necessary to represent the field A^a_μ in the form

$$A^a_\mu = A^{a(\text{ins})}_\mu + a^a_\mu \tag{69}$$

and expand the action $S(A)$ with respect to the deviation a^a_μ from the instanton field $A^{a(\text{ins})}_\mu$:

$$\begin{aligned} S(A) &= S_0 + \frac{1}{2}\int \mathrm{d}^4 x a^a_\mu L^{ab}_{\mu\nu}(A^{\text{ins}}) a^b_\nu \\ &= \frac{8\pi^2}{g^2} - \frac{1}{2}\int \mathrm{d}^4 x a^a_\mu [D^2 a^a_\mu - D_\mu D_\nu a^a_\nu + 2g\varepsilon^{abc} G^b_{\mu\nu} a^c_\nu]\,, \end{aligned} \tag{70}$$

where the instanton field is substituted in D_μ and $G_{\mu\nu}$. As in the one-dimensional case, the integration with respect to the deviations a_μ reduces to the calculation of the determinant of the operator $L^{ab}_{\mu\nu}$. There are however two important differences from the one-dimensional case:

The operator L is degenerate due to the gauge invariance. Indeed, fields a^a_μ of the form $a^a_\mu = (D_\mu \lambda)^a$ with arbitrary function $\lambda^a(x)$ make the quadratic form (70) vanish. In order to have the possibility of working with a degenerate form of this kind, it is necessary to fix the gauge. This can be done conveniently by adding to the action the term

$$\Delta S = \frac{1}{2}\int \mathrm{d}^4 x (D_\mu a^a_\mu)^2 = \frac{1}{2}\int \mathrm{d}^4 x a^a_\mu (\Delta L)^{ab}_{\mu\nu} a^b_\nu\,, \tag{71}$$

which lifts the degeneracy. To avoid changing the content of the theory, we must, as is well-known, simultaneously add the Faddeev-Popov ghosts:

$$\Delta S_{\text{gh}} = -\int \mathrm{d}^4 x \bar{\Phi}^a D^2 \Phi^a = \int \mathrm{d}^4 x \bar{\Phi}^a L^{ab}_{\text{gh}} \Phi^b\,, \tag{72}$$

where Φ^a is a complex anticommuting field. As a result, the instanton contribution can be written in the form

$$\langle 0|0_T\rangle_{\text{ins}} = [\det(L + \Delta L)]^{-1/2} (\det L_{\text{gh}}) e^{-S_0}\,, \tag{73}$$

where $|0_T\rangle$ is the vacuum after time T, $|0_T\rangle = e^{-HT}|0\rangle$, H is the Hamiltonian, $S_0 = 8\pi^2/g^2$, $(L+\Delta L)^{ab}_{\mu\nu}$ is the operator appearing in the quadratic form of the fields a^a_μ, and L_{gh} acts on the ghost fields. The determinant of L_{gh} occurs in a positive power, since $\Phi^a, \bar{\Phi}^a$ are anticommuting fields.

A second difference from the one-dimensional case is the presence in the theory of ultraviolet divergences. By virtue of the renormalizability, all the divergences must be eliminated by a renormalization of the coupling constant, but it is first necessary to regularize the expressions under consideration. The regularization can be done as follows. Instead of the determinant of the operator $L+\Delta L$ we consider the ratio $\det(L+\Delta L)/\det(L+\Delta L+M^2)$, where the introduction of the cut-off parameter M can be interpreted as the addition to the theory of a Pauli-Villars vector field with mass M. The determinant of L_{gh} is regularized similarly. Thus, it is necessary to calculate

$$\langle 0|0_T\rangle^{\text{Reg}}_{\text{ins}} = \left[\frac{\det(L+\Delta L)}{\det(L+\Delta+M^2)}\right]^{-1/2} \frac{\det L_{\text{gh}}}{\det(L_{\text{gh}}+M^2)} e^{-S_0}\,, \tag{74}$$

or, more precisely, the ratio of $\langle 0|0_T\rangle^{\text{Reg}}_{\text{ins}}$ to the corresponding perturbative quantity $\langle 0|0_T\rangle_{\text{p.th}}$, which differs in having $A^a_\mu \equiv 0$ substituted instead of the instanton field. For $A^a_\mu = 0$, it is obvious that $S_0 = 0$, while for the instanton $S_0 = 8\pi^2/g_0^2$, where the subscript in the coupling constant g_0 emphasizes that this is the bare coupling constant normalized at the cut-off parameter M, $g_0 = g(M)$.

We shall not go into a detailed exposition of 't Hooft's calculations for $\langle 0|0_T\rangle/\langle 0|0_T\rangle_{\text{p.th}}$ but obtain the result up to a numerical factor. The study of zero modes plays the main part in obtaining the result.

9.2. *Zero Modes*

As shown in the one-dimensional example, each zero mode leads in $[\det(L+\Delta L)]^{-1/2}$ to a factor proportional to $\sqrt{S_0}$ and an integral with respect to a corresponding collective coordinate. What are the collective coordinates in the case of the BPST instanton in the group SU(2)?

The issue has been discussed in Sec. 8.5. First, there are the four coordinates of the center x_0, then the scale ρ, and, finally, the three Eulerian angles θ, φ, ψ, which specify the orientation of the instanton in the color space. The spatial rotations need not be counted, since they are equivalent to isorotations (see Sec. 8.5).

As a result of the regularization, $[\det(L+\Delta L)]^{-1/2}$ is multiplied by $[\det(L+\Delta L+M^2)]^{1/2}$, i.e. each zero mode gives rise to a factor M. Thus, from all (since we have listed *all* collective coordinates) zero modes, there arises in $\langle 0|0_T\rangle_{\text{ins}}^{\text{Reg}}$ the factor

$$\int \mathrm{d}^4 x_0 d\rho \sin\theta \, \mathrm{d}\theta \, \mathrm{d}\varphi \, \mathrm{d}\psi M^8 (\sqrt{S_0})^8 \rho^3 \,. \tag{75}$$

The factor ρ^3 arises from the Jacobian of the transition to integration over θ, φ, ψ and is recovered on the basis of dimensional considerations.

Using (75) we rewrite $\langle 0|0_T\rangle_{\text{ins}}^{\text{Reg}}/\langle 0|0_T\rangle_{\text{p.th}}$ in the form

$$\frac{\langle 0|0_T\rangle_{\text{ins}}^{\text{Reg}}}{\langle 0|0_T\rangle_{\text{p.th}}} = \text{const} \int \frac{\mathrm{d}^4 x \mathrm{d}\rho}{\rho^5} \left(\frac{8\pi^2}{g_0^2}\right)^4 \exp\left(- \frac{8\pi^2}{g^2} + 8 \ln M\rho + \Phi_1 \right) , \tag{76}$$

where $\exp \Phi_1$ denotes the contribution of the nonzero modes.

9.3. *Nonzero Modes. Effective Coupling Constant*

The quantity Φ_1 depends on the dimensionless parameter $M\rho$, and in the limit $M\rho \gg 1$ can be readily found by means of ordinary perturbation theory. Indeed, calculation of the pre-exponential factor by retaining the terms quadratic in the deviation from the external field corresponds to the calculation of the one-loop corrections in perturbation theory. We are here referring to diagrams of the form

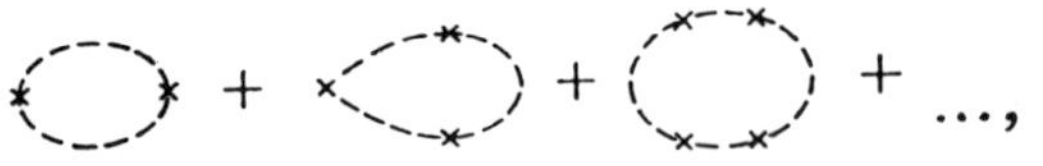
(77)

where the cross denotes vertices of the interaction with the external field, and the broken lines correspond to the propagators of the fields a_μ^a (plus similar loops with the ghosts $\Phi^a, \bar{\Phi}^a$); the external field has the form $A_\mu^{a(\text{ins})}$.

It is clear that complete calculation of the contribution of the zero modes requires summation of a complete chain of diagrams — the zero modes do not appear in any finite order. A manifestation of this nonanalyticity is the presence of the term $\ln(8\pi^2/g_0^2)$ in $\ln\langle 0|0_T\rangle_{\text{ins}}$. It is also clear that there is no nonanalyticity of this kind for the nonzero modes.

In the limit in which we are interested, $M\rho \gg 1$, only the first of the diagrams (77) is important in the calculation, since all the following diagrams are convergent and do not give a dependence on the cut-off parameter M (they change the constant in (76)). Moreover, in the second order in the external field it can be seen that the contribution of the nonzero modes is given by an unsubtracted dispersion relation for the polarization operator $\Pi^{ab}_{\mu\nu}$.

The imaginary part of $\Pi^{ab}_{\mu\nu}$ is obtained by cutting the first diagram (77) and is well defined. In its calculation, it is necessary to take into account only quanta with the spatial transverse polarization states; the unphysical polarizations and ghosts are not necessary. Omitting the details of this simple calculation, we give the result for $\mathrm{Im}\,\Pi_{\mu\nu}$:

$$\mathrm{Im}\,\Pi^{ab}_{\mu\nu} =$$

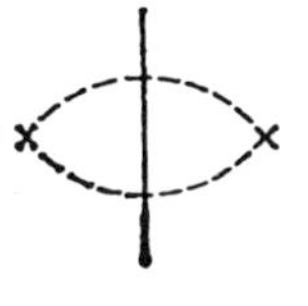

$$= \delta^{ab}(g_{\mu\nu}k^2 - k_\mu k_\nu)\frac{g^2}{16\pi}\cdot\frac{2}{3}\,.$$

Writing down the unsubtracted dispersion representation for $\Pi^{(1)}_{\mu\nu}$ (the part of the polarization operator associated with the nonzero modes), we obtain

$$\begin{aligned}\Pi^{ab(1)}_{\mu\nu} &= \delta^{ab}(g_{\mu\nu}k^2 - k_\mu k_\nu)\frac{1}{\pi}\int\frac{ds}{s-k^2}\cdot\frac{2}{3}\frac{g^2}{16\pi}\\ &= \delta^{ab}(g_{\mu\nu}k^2 - k_\mu k_\nu)\frac{2}{3}\frac{g^2}{16\pi^2}\ln\frac{M^2}{-k^2}\,,\end{aligned} \tag{78}$$

where we cut off the integration over s at M^2, since the regularization involves a subtraction of an analogous contribution with the Pauli-Villars regulators of mass M.

The result (78) for the contribution of the nonzero modes means that the action for the external field acquires from these quantum corrections the effective addition

$$\Delta S^{\mathrm{Mink}} = \frac{2}{3}\frac{g^2}{16\pi^2}\ln M^2\rho^2\int \mathrm{d}^4x\left[-\frac{1}{4}(G^a_{\mu\nu})^2\right], \tag{79}$$

where we use the notation of pseudo-Euclidean space and have replaced $1/(-k^2)$ by the square ρ^2 of the characteristic scale of the field (strictly speaking, we ought to write a differential operator, but for the calculation of the coefficient of $\ln M\rho$ this is not important). Passing to the Euclidean action and substituting the instanton $G^a_{\mu\nu}$, we obtain the result for Φ_1:

$$\Phi_1 = -\frac{2}{3}\ln M\rho\,. \tag{80}$$

Thus, allowance for the zero and nonzero modes has the consequence that $8\pi^2/g_0^2$ in the argument of the exponential (76) is replaced by the effective coupling constant $8\pi^2/g^2(\rho)$:

$$\frac{8\pi^2}{g^2(\rho)} = \frac{8\pi^2}{g_0^2} - 8\ln M\rho + \frac{2}{3}\ln M\rho = \frac{8\pi^2}{g_0^2} - \frac{22}{3}\ln M\rho\,. \tag{81}$$

Of course, this result is a direct consequence of the renormalizability, and we have wasted time on its derivation only to emphasize the very beautiful explanation of the antiscreening of the charge in the non-Abelian theory which arises when the zero modes are considered. In the instanton calculation the antiscreening is entirely due to zero modes.

Indeed, both the sign and the magnitude of the coefficient of the "anti-screening" logarithm in Eq. (76), $8\ \ln M\rho$, are obvious consequences of the consideration above — the coefficient is simply the number of the zero modes.

In the framework of the perturbative calculations, the "antiscreening" result can be most clearly explained in the framework of the ghost-free Coulomb gauge, which was used in calculations by Khriplovich [33] as early as 1969 (see also Ref. [34]). Besides the "dispersion" part, the calculation of which we have discussed above, the polarization operator in this gauge contains a contribution that does not have imaginary part and arises when one of the virtual quanta has the spatial transverse polarization and the second is a Coulomb quantum. The opposite signs of the "nondispersion" and "dispersion" parts of $\Pi_{\mu\nu}$ correspond to the opposite signs of interactions due to the exchange of the Coulomb quantum and the transverse quantum (electric forces repel charges of the same sign, while magnetic forces attract currents of the same type). A more detailed pedagogical discussion of the issue can be found in Ref. [35].

The calculation of the "nondispersion" part in the Coulomb gauge requires care, since it is necessary to use the noncovariant Hamiltonian formalism, and the coefficient of the logarithm is not, of course, known *a priori*. As we have seen, none of these problems arise in the determination of the contribution of the zero modes. With this we conclude our panegyric to the zero modes.

9.4. *Two-Loop Approximation*

The above calculations led to the replacement of the bare coupling constant g_0 in the classical action by the effective constant $g(\rho)$. However, the bare constant still remains in the factor $(8\pi^2/g_0^2)^4$ [see (76)], though it is clear that, because of the renormalizability, it should not occur in the result. The reason for this is that the accuracy obtained by using the one-loop approximation is inadequate to distinguish the factor $(8\pi^2/g_0^2)^4$ from $[8\pi^2/g^2(\rho)]^4$, and we require a two-loop calculation.

From the two-loop calculation we actually only need the expression for the effective coupling constant; such an expression as known from the perturbation theory [36],

$$\frac{8\pi^2}{g^2(\rho)} = \frac{8\pi^2}{g^2(\rho_0)} + N\left[\frac{11}{3}\ln\frac{\rho_0}{\rho} + \frac{17}{11}\ln\left(1 + \frac{11}{3}N\frac{g^2(\rho_0)}{8\pi^2}\ln\frac{\rho_0}{\rho}\right)\right], \tag{82}$$

where we have given the result for the group SU(N) (without the contribution of fermions). The unrenormalized constant is $g_0 = g(\rho_0 = 1/M)$. The instanton contribution to the vacuum–vacuum transition for the group SU(2) has the form

$$\frac{\langle 0|0_T\rangle_{\rm ins}^{\rm Reg}}{\langle 0|0_T\rangle_{\rm p.th}} = \text{const}\cdot\left[\frac{8\pi^2}{g^2(\rho)}\right]^4 e^{-8\pi^2/g^2(\rho)}(1 + O(g^2(\rho)))\,, \tag{83}$$

where $g^2(\rho)$ is given by the expression (82) with $N = 2$. For the factor $[8\pi^2/g^2(\rho)]^4$, we can restrict ourselves to the one-loop expression for $g^2(\rho)$, the difference being of the order of the ignored terms which give relative corrections of order $g^2(\rho)$. Note that the complete two-loop calculation of the instanton contribution would determine these corrections.

The proof of the correctness of (83) is based on the renormalizability of the theory and the method of effective Lagrangians. In the functional integral, we integrate in the spirit of Wilson over fields of small scale (less than ρ_c), i.e. over configurations corresponding to instantons with small $\rho < \rho_c$. As a result, we obtain an effective Lagrangian of the fields with scales greater than ρ_c. In this Lagrangian, the small-scale fluctuations are taken into account in the coefficients of the expansion with respect to the operators.

The calculation of the contribution of the instantons to the vacuum–vacuum transition is equivalent to the determination of their contribution to the coefficient of the unit operator. The calculation of the coefficients of other operators will be considered in Sec. 10. A specific feature of the unit operator is the fact that its matrix elements are independent of the normalization point; i.e. it has the vanishing anomalous dimension. Therefore, the coefficient of it, expressed in terms of $g(\rho)$, cannot contain ρ_c (for operators with a nonvanishing anomalous dimension the factor $[g^2(\rho_c)/g^2(\rho)]^\delta$ arises).

Now, it only remains to express $g^2(\rho)$ in terms of $g^2(\rho_0)$ by means of the renormalization-group equations; the argument above proves that the retention of the two-loop correction in (82) is fully legitimate.

9.5. *Density of Instantons (The Instanton Measure) in the Group SU(N)*

How does the number of zero modes change on the transition to the group SU(N)? We have already said that the instanton field uses only a SU(2) subgroup of the complete group. Suppose this subgroup occupies the top left-hand corner in the $N \times N$ matrix of generators. It is clear that the five zero modes associated with shifts and dilatations remain the same as in the group SU(2), and only the modes associated with group rotations are changed. In SU(2) there were three, and in SU(N) they correspond to three generators in a 2×2 matrix at the top left (Fig. 10). Those of the remaining generators that occur in the $(N-2) \times (N-2)$ matrix at the bottom right obviously do not rotate the instanton field. Thus, to the three SU(2) rotations a further $4(N-2)$ unitary rotations are added. The total number of zero modes is $5 + 3 + 4(N-2) = 4N$. Of course, this number $4N$ exactly corresponds to the coefficient of the "antiscreening" logarithm in the formula for $8\pi^2/g^2(\rho)$. Finally, we write down an expression for the instanton density $d(\rho)$, which is defined as follows:

$$\frac{\langle 0|0_T\rangle_{\rm ins}^{\rm Reg}}{\langle 0|0_T\rangle_{\rm p.th}} = \int \frac{{\rm d}^4x{\rm d}\rho}{\rho^5} d(\rho)\,. \tag{84}$$

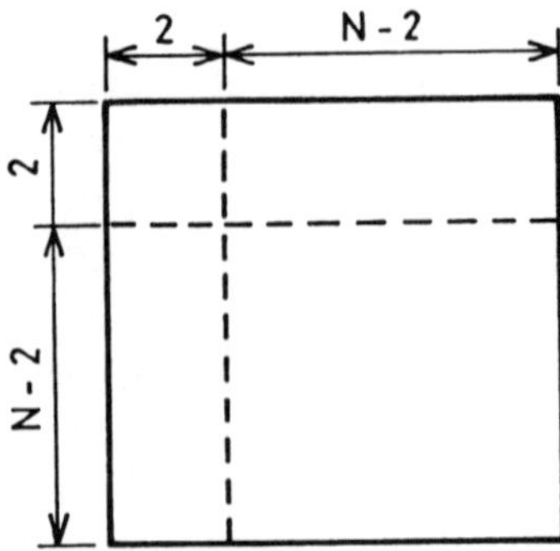

Fig. 10. Counting generators of group rotations in SU(N).

Since the SU(N) group space is finite it is assumed that the integration over the collective coordinates associated with the instanton orientation in the SU(N) group is carried out. The function $d(\rho)$ is equal to

$$d(\rho) = \frac{C_1}{(N-1)!(N-2)!}\left[\frac{8\pi^2}{g^2(\rho)}\right]^{2N} e^{-[8\pi^2/g^2(\rho)]-C_2 N}\,, \tag{85}$$

where $g^2(\rho)$ is expressed in terms of $g_0^2 = g^2(\rho_0 = 1/M)$ by formula (82), and the constants C_1 and C_2 can be found by a certain modification [37] of the 't Hooft's calculations [9]. Concretely, it is necessary to take into account additional $4(N-2)$ vector fields with the above quantum numbers in the contribution of both the zero and nonzero modes. In addition, we need the embedding volume of SU(2) in SU(N); the factor $[(N-1)!(N-2)!]^{-1}$ is associated with it. This part of the modification proved to be the most complicated (see Ref. [37]). The result for C_1 and C_2 has the form

$$C_1 = \frac{2e^{5/6}}{\pi^2} = 0.466\,,$$

$$C_2 = \frac{5}{3}\ln 2 - \frac{17}{36} + \frac{1}{3}(\ln 2\pi + \gamma) + \frac{2}{\pi^2}\sum_{s=1}^{\infty}\frac{\ln s}{s^2} = +1.679\,. \tag{86}$$

Note that the constant C_2 depends on the method of regularization, which actually provides the definition of the bare constant. Equation (86) refers to the Pauli-Villars regularization. Instead of the Pauli-Villars regularization (PV scheme), the so-called dimensional regularization is frequently used. Instead of logarithms of the cut-off parameter, poles with respect to the dimension of space arise in this method, $\ln M \to 1/(4-D)$. Use of the minimal subtraction scheme [38] (MS) for determining the coupling constant leads to an expression of the form (85) with the substitution

$$g(\rho) \to g_{\rm MS}(\rho) \qquad C_2 \to C_{2\rm MS}\,,$$

$$C_{2\rm MS} = C_2 - \frac{5}{36} - \frac{11}{6}(\ln 4\pi - \gamma) = C_2 - 3.721\,. \tag{87}$$

The numerical coefficient in $d(\rho)$ for the MS scheme is $e^{3.72N}$ times greater than in the PV scheme, which for SU(3) gives the factor $\sim 7\cdot 10^4$.

Of course, the relations between the observable amplitudes do not depend on the definition of g^2 — the same conversion constants associated with the change of regularization occur, for example, in the corrections in g^2 to the cross-section of e^+e^- annihilation into hadrons (though there, it is true, the dependence on them is not exponential). We note in this connection that in the perturbation theory the so-called modified minimal subtraction ($\overline{\mathrm{MS}}$) scheme has proved helpful, since in it, too large coefficients of the expansion in g^2 do not arise [39]. The difference between the $\overline{\mathrm{MS}}$ scheme and the MS scheme reduces to

$$\begin{aligned} \frac{8\pi^2}{g^2_{\overline{\mathrm{MS}}}} &= \frac{8\pi^2}{g^2_{MS}} - \frac{11}{6}N(\ln 4\pi - \gamma)\,, \\ C_{2\overline{\mathrm{MS}}} &= C_2 - \frac{5}{36} \approx 1.54\,. \end{aligned} \tag{88}$$

We give finally the explicit form of the dependence on ρ for the function $d(\rho)$ (in the PV scheme)

$$\begin{aligned} d(\rho) = {} & \frac{0.466}{(N-1)!(N-2)!}\left(\frac{\rho}{\rho_0}\right)^{11N/3}\left[1 + \frac{11}{3}N\frac{g^2(\rho_0)}{8\pi^2}\ln\frac{\rho}{\rho_0}\right]^{5N/11} \\ & \times \left[\frac{8\pi^2}{g^2(\rho_0)}\right]^{2N} e^{-[8\pi^2/g^2(\rho_0)]-1.679N}\,. \end{aligned} \tag{89}$$

10. Instantons in the QCD Vacuum

10.1. *Instantons in the Slowly Varying Background Fields*

The quasiclassical methods that have been developed apply to the study of nonperturbative fluctuations of a small scale, among which the instantons are dominant.

In this subsection we take into account the influence on the small-scale instantons of the fields due to the characteristic long-wavelength fluctuations in the QCD vacuum [40].

Since we distinguish fields of two types, namely, the fields of small-scale instantons and the fields of the large-sized vacuum fluctuations, it is convenient to introduce an effective Lagrangian. In it, as usual, the contribution of the rapidly varying fields is included in the coefficients of the various operators that act in the space of the slowly varying fields.

Thus, the effect of a selected instanton with scale ρ and center at x_0 reduces to the following extra term in the effective Lagrangian of the long-wavelength fluctuations:

$$\Delta L(x_0) = \frac{d\rho}{\rho^5} \sum_n C_n(\rho) O_n(x_0) ,$$

where $C_n(\rho)$ are numerical coefficients and $O_n(x_0)$ are local operators constructed from the gluon fields (we consider pure gluodynamics; the changes introduced by fermions are discussed in the following section).

The probability of finding the instanton under consideration in the physical vacuum is given by averaging ΔL over this state. On the other hand, to find the coefficients C_n, it is convenient to consider the matrix elements of ΔL between the perturbation-theory states (with different number of free gluons with momenta $q \ll 1/\rho$). These matrix elements can be calculated by quasiclassical methods.

Concretely, we consider the instanton contribution to the vacuum $\to n$ gluons transition and apply to it the reduction formula

$$\langle n \text{ gluons } |\Delta L|0\rangle = \langle 0| T \prod_{k=1}^{n} \int \mathrm{d}x_k e^{iq_k x_k} \varepsilon_{\mu_k}^{a_k} q_k^2 A_{\mu_k}^{a_k}(x_k)|0\rangle , \tag{90}$$

where q_k and $\varepsilon_{\mu_k}^{a_k}$ are the four-momentum and the polarization of the kth gluon, and $A_\mu^a(k)$ is the operator of the gluon field. For $n = 0$, i.e. for the vacuum–vacuum transition, the right-hand side of (90) was already calculated in Sec. 9 and is equal to $\mathrm{d}\rho\rho^{-5}d(\rho)$; the left-hand side is obviously equal to the coefficient of the unit operator: $C_I \mathrm{d}\rho/\rho^5$.

For $n \neq 0$, the prescription of the quasiclassical calculation of the expression (90) reduces to

(a) the transition to the Euclidean space (see the equations in Sec. 6);

(b) replacement of the Euclidean $\hat{A}_\mu^a(x)$ by the instanton field $\bar{A}_\mu^a(x - x_0)$ given by the formula (65). The singular gauge is used because the reduction formula (90) is valid only for rapidly decreasing fields $A_\mu^a(x)$. For the nonsingular gauge, the inverse propagator q^2 is replaced by a more complicated expression;

(c) multiplication by the $\langle 0|0_T\rangle_{\text{ins}}$ transition amplitude, which is equal to $\mathrm{d}\rho\rho^{-5}d(\rho)$. Thus, for the matrix element (90) we obtain

$$\langle n \text{ gluons } |\Delta L(x)|0\rangle = \frac{d\rho}{\rho^5} \mathrm{d}(\rho) e^{-ix\Sigma q_k} \prod_{k=1}^{n} \times \left[\int \mathrm{d}x_k e^{-iq_k x_k} (-q_k^2) \varepsilon_{\mu_k}^{a_k} \overline{A_{\mu_k}^{a_k}}(x_k) \right] , \tag{91}$$

where all the quantities on the right-hand side are Euclidean.

The Fourier transform of the instanton solution, which we want in the limit $q\rho \to 0$, is readily found:

$$\int \mathrm{d}x e^{-iqx}(-q^2)\bar{A}^a_\mu(x) = \frac{4\pi i}{g}\bar{\eta}_{a\mu\nu}q_\nu\rho^2 \,. \tag{92}$$

After this, it is easy to recover the complete operator form of ΔL:

$$\Delta L(x) = \frac{d\rho}{\rho^5}d(\rho)\exp\left[-\frac{2\pi^2}{g}\rho^2\bar{\eta}^M_{a\mu\nu}G^a_{\mu\nu}(x)\right],$$
$$\bar{\eta}^M_{a\mu\nu} = \begin{cases} \bar{\eta}_{amn}, & \mu = m,\ \nu = n;\ m,n = 1,2,3\,, \\ i\bar{\eta}_{a4n}, & \mu = 0,\ \nu = n;\ n = 1,2,3\,, \end{cases} \tag{93}$$

where $G^a_{\mu\nu}(x)$ is the operator of the large-scale gluon field. The factorials which occur in the expansion of the exponential cancel against the combinatorial coefficients when the matrix element (91) is taken.

The expression (93) for the interaction of an instanton with an external field was obtained for the first time by Callan, Dashen and Gross [11] by a different and more complicated method. An important point is that we, in contrast to them, have not fixed the external $G^a_{\mu\nu}(x)$ "by hand" but have related it to the field of the large-scale fluctuations.

This is achieved by averaging the Lagrangian (93) over the physical vacuum. The term linear in $G^a_{\mu\nu}$ obviously vanishes as a result of such averaging, and the first nonvanishing correction to the effective density of the instantons is proportional to G^2:

$$\begin{aligned}\langle 0|\Delta L|0\rangle &= \frac{\mathrm{d}\rho}{\rho^5}d_{\mathrm{eff}}(\rho) \\ &= \frac{\mathrm{d}\rho}{\rho^5}d(\rho)\left[1 + \frac{\pi^3\rho^4}{(N^2-1)\alpha_s}\langle 0|G^a_{\mu\nu}G^a_{\mu\nu}|0\rangle + O(\rho^6)\right],\end{aligned} \tag{94}$$

where in the averaging we used the relation

$$\langle 0|G^a_{\mu\nu}G^{a'}_{\mu'\nu'}|0\rangle = \frac{\delta^{aa'}}{N^2-1}\cdot\frac{1}{12}(g_{\mu\mu'}g_{\nu\nu'} - g_{\mu\nu'}g_{\nu\mu'})\langle 0|G^b_{\alpha\beta}G^b_{\alpha\beta}|0\rangle \,. \tag{95}$$

Note that the coupling constant α_s and the operator $(G^a_{\mu\nu})^2$ which occur here are normalized at the point ρ. A quantity that does not depend on the renormalization point (to accuracy $\alpha_s(\rho)$) is the product $\alpha_s G^a_{\mu\nu}G^a_{\mu\nu}$.

To obtain a quantitative estimate of the correction, it is necessary to know the mean square of the intensity of the gluon field in the physical vacuum. If we accept a "canonic" [1] value of the gluon condensate,

$$\langle 0|\frac{\alpha_s}{\pi}G^a_{\mu\nu}G^a_{\mu\nu}|0\rangle \approx 0.012\ \text{GeV}^4\,, \tag{96}$$

then for the group SU(3), the relative correction to $d(\rho)$ can be written in the form

$$\frac{\pi^4\rho^4}{8\alpha_s^2(\rho)}\langle 0|\frac{\alpha_s}{\pi}G^a_{\mu\nu}G^a_{\mu\nu}|0\rangle\,. \tag{97}$$

It reaches unity at a value of ρ equal to

$$\rho_{\text{crit}} \approx \frac{1}{1.15\ \text{GeV}}\,, \tag{98}$$

if for α_s we take $\alpha_s(\rho) = 2\pi/9\ln(1/\Lambda\rho)$ with $\Lambda = 100$ MeV. For $\rho = \rho_{\text{crit}}$, the interaction of the instanton with the vacuum fields of the other fluctuations becomes 100% important. This ρ_{crit} is rather small.

We conclude this subsection by giving a formula that takes into account the higher powers of $G^a_{\mu\nu}$ in the effective instanton density. This formula is based on the hypothesis of dominance of the vacuum intermediate state, which makes it possible to reduce $\langle 0|(G^2)^n|0\rangle$ to $(\langle 0|G^2|0\rangle)^n$. This approximation is analogous to the one used in the many-body theory and for some four-quark operators for which it can be verified to have an accuracy of the order of a few percent.

The factorization leads to the relation

$$\langle 0|\left(\frac{2\pi^2}{g^2}\rho^2\bar\eta^M_{a\mu\nu}G^a_{\mu\nu}\right)^{2k}|0\rangle = (2k-1)!!\left[\frac{4\pi^4}{g^2}\rho^4\langle 0|\,(\bar\eta^M_{a\mu\nu}G^a_{\mu\nu})^2|0\rangle\right]^k\,,$$

by means of which we obtain for the effective instanton density the result

$$d_{\text{eff}}(\rho) = d(\rho)\exp\left[\frac{\pi^4\rho^4}{(N^2-1)\alpha_s^2(\rho)}\langle 0|\frac{\alpha_s}{\pi}G^a_{\mu\nu}G^a_{\mu\nu}|0\rangle\right]\,, \tag{99}$$

which can be represented as the replacement in the expression for $d(\rho)$ of $2\pi/\alpha_s(\rho)$ by

$$\frac{2\pi}{\alpha_s(\rho)} \to \frac{2\pi}{\alpha_s(\rho)}\left[1-\frac{\pi^3\rho^4}{2(N^2-1)\alpha_s^2(\rho)}\langle 0|\frac{\alpha_s}{\pi}(G^a_{\mu\nu})^2|0\rangle\right]\,. \tag{100}$$

Using for $d(\rho)$ the expression (99), we can advance in ρ to $\rho > \rho_{\text{crit}}$. However, when the interaction with the vacuum fields changes the classical action strongly, i.e. when (100) vanishes, the quasiclassical methods cannot be used. This limit under the same assumptions about α_s and $\langle 0|G^2|0\rangle$ is $\rho < 1/500$ MeV.

10.2. *Instanton–Antiinstanton Interactions*

In the first part of the lecture devoted to the quantum-mechanical problem of the double well potential it was explained that the instanton contribution to the vacuum energy is determined by a chain of instantons. The chain was treated as an ensemble of well-separated instantons unaffected by the presence of others. Similar approximation in QCD is called the dilute instanton gas [11]. The instanton gas is not a good approximation in QCD, even at the qualitative level. In the absence of the instanton interactions, there is no way one can make the instantons gas self-consistent. Thus, the situation differs drastically from that we encountered in the quantum-mechanical example discussed in the beginning of this lecture. The basic difference is due to the fact that an intrinsic mass scale controlling the gas "diluteness" appears already at the classical level in the quantum-mechanical problem, while QCD at the classical level has no dimensional parameters. The only mass scale, Λ_{QCD}, emerges as a result of the dimensional transmutation, at the quantum level. It is not surprising then, that the gas model, with no intrinsic mass scale, turns out to be inadequate.

One could try to amend the instanton gas model, by including interactions of instantons between each other. If a classical (or semiclassical) interaction of instantons becomes important in QCD at such values of ρ where $8\pi^2/g^2(\rho)$ is still a large parameter, an instanton-based picture of the QCD vacuum could survive. A crude picture can then be formulated as follows. Instantons and antiinstantons of a relatively small size form an interacting liquid. The "atoms" of this liquid are instantons and antiinstantons. The "atoms" act as potential wells for the light quark propagating in this "medium" [41] (Fig. 11).

This heuristic picture serves as a basis for various versions of the instanton liquid model [41]. We will not go into details of this model for two reasons: (i) it is too closely related to the applications of the instanton calculus while in this lecture we are mostly preoccupied with the basics of the formalism *per se*; (ii) exhaustive reviews exist in the literature (e.g. Ref. [4]). Instead, we will dwell on a general element from which each model of this type begins. If

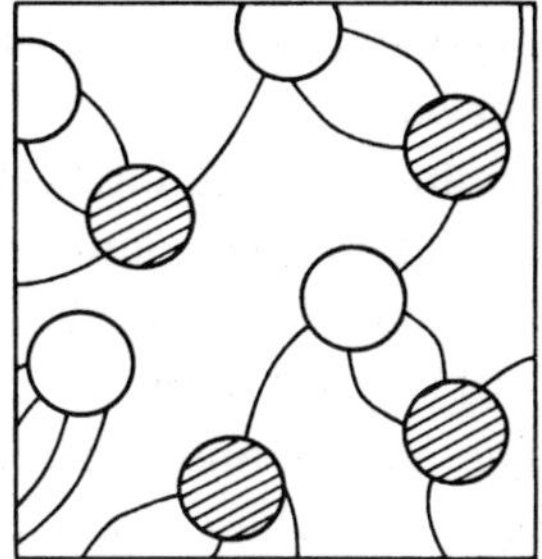

Fig. 11. A schematic picture of the instanton liquid model of the QCD vacuum. The shaded objects are antiinstantons. The lines connecting instantons (antiinstantons) denote interactions and quark exchanges.

one deals with more than one instanton the first issue to be addressed is the question of the instanton–antiinstanton (I–A) interactions at large separations. We speak about the instanton–antiinstanton interaction since two instantons, or two antiinstantons, being the exact solution of the duality equations, do not interact classically — the action of two pseudoparticles is twice the action of one.

The master formula for the instanton of size ρ centered at x "living" in an external background field $G^a_{\mu\nu}(x)$, we have obtained above, will allow us to easily find the leading term in the I–A interaction at large distances, the so-called dipole–dipole interaction.

No assumption is made in Eq. (93) as to the nature of the background field. In particular, this field can be caused by a distant antiinstanton of a larger size. If we substitute in Eq. (93) the value of the gluon field strength tensor induced by the antiinstanton centered at y we will get a formula describing the instanton–antiinstanton (I–A) interaction at a large separation. Since the instantons belong to the Euclidean space, strictly speaking we need a Euclidean analog of Eq. (93). It is quite obvious that the corresponding expression takes the form

$$\mathcal{L}_\rho(x_0) \sim \exp\left(-\frac{2\pi^2\rho^2}{g} O^{ab}\bar{\eta}_{b\mu\nu} G^a_{\mu\nu}(x_0)\right) + \text{h.c.}\,, \tag{101}$$

where O^{ab} is a global color rotation matrix reflecting the relative orientations of the I–A pair in the color space, (it was not explicitly written in Eq. (93) since it was irrelevant for Sec. 10.1), x_0 is the coordinate of the instanton center, the subscript ρ reminds us of its radius, and $\bar{\eta}_{b\mu\nu}$ is the 't Hooft symbol. $G^a_{\mu\nu}(x_0)$ is the operator of the gluon field strength tensor. In principle, beyond the leading approximation, the exponent in Eq. (101) will contain other operators, say with derivatives $D_\alpha G_{\mu\nu}$ or with two or more G's, along with a series in g.

In this way one can determine the I–A interaction as a systematic double expansion, in the ratio ρ/R and in the coupling constant, where R is the distance between instanton and antiinstanton. Thus, the leading dipole–dipole term is obtained from Eq. (101) by substituting the operator $G^a_{\mu\nu}(x_0)$ by the antiinstanton field at x_0. The antiinstanton centered at y_0 should be taken in the singular gauge, see Eq. (65) (where $\bar\eta$ must be substituted by η). It is assumed that $R = |x_0 - y_0| \gg \rho$, of course.

This exercise is pretty straightforward. For pedagogical reasons we find it more instructive to present a somewhat different (and less known) derivation of the I–A interaction that does not use the language of the classical fields at all. It operates with amplitudes of particle emission and allows us to connect the classical problem of the I–A interaction energy with the quantum problem of the instanton-induced cross-sections.

To illustrate how it works we will do the calculation of the I–A interaction in the leading (dipole) approximation. Relevant graphs are depicted in Fig. 12. The instanton with size ρ_1 is placed at x and the antiinstanton with size ρ_2 at the origin; $|x| \gg \rho_{1,2}$ so that the whole approach makes sense.

Figure 12(a) is the basic element of the calculation. We expand the exponent in Eq. (101) and a similar one for the antiinstanton, keep the linear in $G^a_{\mu\nu}$ terms and contract $G(x)$ and $G(0)$ to get

$$\left(\frac{4\pi^2}{g^2}\rho_1^2\rho_2^2\right) O_I^{ab}\bar\eta_{b\mu\nu} O_A^{cd}\eta_{d\alpha\beta}\langle G^a_{\mu\nu}(x) G^c_{\alpha\beta}(0)\rangle\,,$$

where $\langle G^a_{\mu\nu}(x) G^c_{\alpha\beta}(0)\rangle$ is the free Green function of the gauge field. Moreover, in the Green function $\langle A_\mu(x) A_\nu(0)\rangle$ we can only retain the $g_{\mu\nu}$ part, since

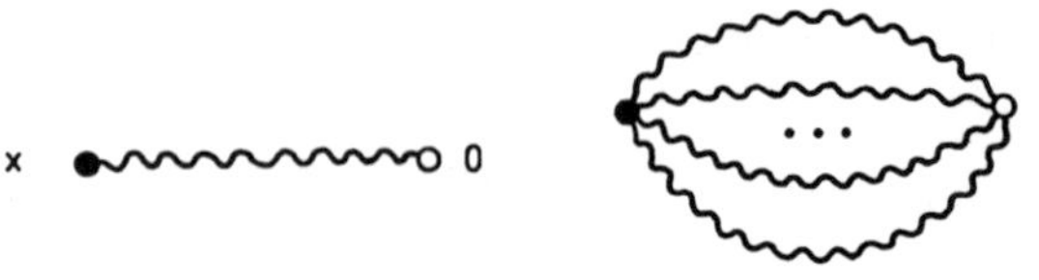

(a) One-gluon exchange (b) Multigluon exchange

Fig. 12. The I–A interaction from the instanton-induced effective Lagrangian (101). The instanton (•) is at the point x while the antiinstanton (∘) is at the origin. The vertices in diagrams (a) and (b) are generated by expanding the exponent in Eq. (101) and keeping only the linear part in each of the $G_{\mu\nu}$ operators appearing in the expansion.

the part $x_\mu x_\nu$ drops out,

$$\langle G^a_{\mu\nu}(x) G^c_{\alpha\beta}(0)\rangle = \frac{2\delta^{ab}}{\pi^2}(g_{\nu\alpha}x_\mu x_\beta + g_{\mu\beta}x_\nu x_\alpha - g_{\nu\beta}x_\mu x_\alpha - g_{\mu\alpha}x_\nu x_\beta)\frac{1}{x^6} + \cdots , \tag{102}$$

where the dots denote the terms which are not contributed due to the fact $\eta_{a\mu\nu}\bar{\eta}_{b\mu\nu} = 0$.

Now, it is not difficult to see that Fig. 12(b) just exponentiate the result [$1/(n!)^2$ from the expansion of the effective Lagrangians is supplemented by $n!$ coming from combinatorics] and we finally get for the I–A amplitude

$$\exp\left[-\frac{8\pi^2}{g^2} - \frac{32\pi^2}{g^2}\rho_1^2\rho_2^2\eta_{a\lambda\mu}\bar{\eta}_{b\lambda\nu}\Omega^{ab}\frac{x_\mu x_\nu}{x^6}\right], \tag{103}$$

where Ω^{ab} is the matrix of the relative orientation of the pseudoparticles, $\Omega^{ab} = O_I^{cb}O_A^{ca}$.

The x^{-4} term above is the dipole–dipole interaction sought for [42]. The procedure can be continued further [43]. The very same term in Eq. (101) generates, through Fig. 13, a part $\propto g^{-2}R^{-6}$ (the graph 13(a) is due to the quadratic term in $G^a_{\mu\nu}$ in one of the vertices). If the Higgs field is included, it generates the terms of order v^2R^{-2}, v^2R^{-4}, etc. through the graphs depicted in Fig. 14. The last term which has been explicitly computed is of order $g^{-2}R^{-8}\ln R$ corresponding to Fig. 15.

Fig. 13. Higher order terms in ρ/R in the I–A interaction due to the quantum corrections in the gluon exchanges. The R^{-6} term is generated by the same effective Lagrangian (101). The vertices attached to the instanton (•) in diagrams (a) and (b) and to the antiinstanton (∘) on diagram (b) are generated by expanding the exponent in Eq. (101) and keeping the quadratic in A_μ part in $G_{\mu\nu}$.

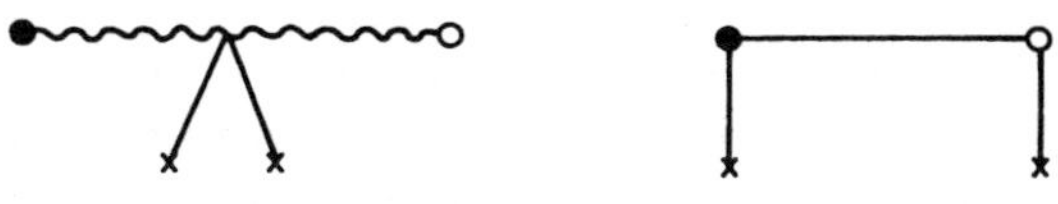

(a) Mass term of the gauge boson (b) Higgs exchange

Fig. 14. Higher order terms in the I–A interaction due to the Higgs field. The crosses denote the vacuum expectation value of the Higgs field.

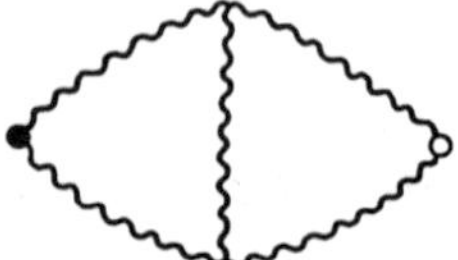

Fig. 15. An example of the two-loop diagram contributing to the I–A interaction at the level $g^{-2}R^{-8}\ln R$.

11. Constrained Instantons and the Valley Method

In the several problems discussed above we dealt with the field configurations that are *not* the exact solution of the classical equations of motion. The instanton in the theories with the spontaneously broken gauge symmetry (the Higgs regime) is an important example. Another example is the ensemble of the I–A "atoms": each given pseudoparticle in the ensemble is affected by the presence of others, and it is quite obvious that the topologically neutral gas cannot, strictly speaking, correspond to a minimum of the action and, hence, is not the exact solution. In other words the action of n pseudoparticles in the I–A ensemble $S_n \neq (8\pi^2 n/g^2)$; instead

$$S_n = \frac{8\pi^2}{g^2}n - S_{\text{int}} \tag{104}$$

where generically the interaction "energy" $S_{\text{int}} > 0$. (From now on we will use the terminology of a static problem in four-dimensional space; the action will be reinterpreted as the energy, and the word "pseudoparticle" becomes appropriate.)

The small-size instanton $\rho \ll m_W^{-1}$, is a legitimate contribution in the theory with the spontaneously broken gauge symmetry. Likewise, in pure gluodynamics well separated I–A pairs should be included in the partition function. It is clear that if the field configuration under consideration is very close to the exact solution (stationary point in the functional integral) it must be taken into account.

Thus, the question arises as to whether one can make quantitative the notion of proximity of the given field configuration to the exact solution and — if yes — what small parameter measures this proximity. A related crucial question is as follows: "If the selection criteria are relaxed and the functional integral is not represented exclusively by the stationary points of the classical action how far can one distance oneself from the exact solutions?"

It is very hard to answer these questions in a fully quantitative manner. The understanding existing in the current literature is heuristic, at best, and since we agreed to avoid controversial issues in this lecture, we will limit ourselves to a discussion of the approximate solutions when they are arbitrarily close to the exact ones; the question of their fate with worsening of the proximity parameter is left aside.

The approach allowing one to deal with the approximate solutions is usually referred to in the literature as *constrained instantons* [28]. Assume we have a field configuration $\phi(\beta)$ continuously depending on some parameter(s) β; ϕ is a generic notation for the set of all relevant fields. Assume that at $\beta \to \beta_0$ the field ϕ tends to become an exact solution. Let us call $\beta = \beta_0$ the limiting point. In the vicinity of the limiting point $\phi(\beta)$ is almost an exact solution. The fact that it is still not the exact solution means that there exist such deformations of ϕ that decrease the action. Typically such "decreasing" deformations are inherent to one — at most, several — directions in the functional space, call them destabilizing directions. Deformations along all other directions increase the action ("energy"). The basic idea of the constrained instantons is as follows. One introduces a constraint in the functional integration measure in such a way as to lock up all destabilizing directions. Then one minimizes the action subject to this constraint. Only those variations of ϕ are allowed which go in the directions perpendicular to the destabilizing ones. In this way one arrives at the constrained instanton. Dynamics in the destabilizing directions is studied separately.

To make a simple physical picture lying behind the program graphic, let us turn to the example of the instanton in the Higgs phase. At $\rho \to 0$ the BPST instanton becomes the exact solution. This is the limiting point which one can approach arbitrarily closely. If $\rho \neq 0$ there exists one direction in the functional space along which the action ("energy") slowly decreases. This direction corresponds to rescaling the instanton solution as a whole to smaller radii. By imposing a constraint we forbid the movement in the functional space along this direction. In the orthogonal subspace any deformation of the field configuration only increases the action, so it is possible to find one which minimizes the action, the constrained instanton. We then calculate the contribution of the constrained instanton in physically observable quantities. At the last stage we eliminate the constraint by integrating the result over all possible values of ρ.

It is worth noting that there is no unambiguous prescription as to how to choose the constraint. Usually, in each particular problem the most convenient and adequate choice is pretty clear from the physical context. For instance in the example above the appropriate constraint must fix the size of the instanton. The classical equations of motion are changed, of course, once the constraint is introduced. New equations do have a solution which at small x behaves like that of Belavin *et al.* while at large distances it decays exponentially, $\mathcal{O}(\exp(-m_W x))$ for the gauge field and $\mathcal{O}(\exp(-m_\chi x))$ for the Higgs field. The corresponding action becomes now ρ dependent and can be readily calculated in the form of expansion in $\rho^2 v^2$, see Sec. 8.6. It is important that the 't Hooft term is $\mathcal{O}(g^2)$ compared to the BPST term, so that the ρ dependence is indeed weak provided that $\rho \ll m_W^{-1}$.

(As a matter of fact, in Sec. 8.6 we managed to obtain the 't Hooft term without explicitly invoking the constrained instanton technique, a fact explainable by specific features of the instanton expression for the scalar field. The heavy artillery of the constrained instantons become relevant at the next-to-leading order.)

A version of the constrained instanton technique most often exploited in practice in connection with the approximate solutions, is the valley or streamline method [44] (for reviews referring to the Yang-Mills theory see e.g. Ref. [45]). The valley method is a variant of the constrained instanton approach, with a specific prescription as to how one should choose the constraints. It is most suitable to the problem of I–A pairs. Let us sketch a physical picture lying behind the valley method in the simplest example, one I–A pair.

We start from a pair of pseudoparticles at very large separations, large compared to their sizes which for simplicity are assumed to coincide for both pseudoparticles. This configuration — our boundary condition — is not an exact solution but is arbitrarily close to that. The field equations will experience a force tending to change the field in the direction of lowering of the total energy of two pseudoparticles. Let us do a *gedankenexperiment* — introduce an auxiliary fifth coordinate, "fifth time", and trace the evolution of the original configuration in the fifth time. In other words, our I and A atoms are set free to move as they wish at $t_5 = 0$.

To visualize the picture further it is instructive to discretize the fifth time. Then at step zero we have $\phi_0 = \phi_I^{R/2} + \phi_A^{-R/2}$ where the superscripts indicate the position of the I (A) centers.

At step one the field is deformed. The variation of ϕ is obtained in the following way. Let us try to change ϕ along different directions in the functional space. Almost every such attempt will lead to a higher energy, with an exception of, say, one direction[15] where the energy of the field configuration $\phi_1 = \phi_0 + \delta\phi_1$ is smaller than that of ϕ_0. At step two we take ϕ_1 as an input and repeat the procedure. In this way we get a chain of field configurations ϕ_i which becomes, in the limit of the continuous fifth time, a one-parametric family $\phi(\beta)$, the bottom of the valley.

In the quantitative terms the bottom of the valley is defined by the following requirement: as we move along the bottom $\phi(\beta)$ at each point the variation of the coordinate of the bottom must be proportional to the force at the given point,

$$\frac{\partial\phi(\beta)}{\partial\beta} \propto \frac{\delta S}{\delta\phi}|_{\phi=\phi(\beta)} \, . \tag{105}$$

The proportionality coefficient is, generally speaking, a function of β sensitive to particular parametrizations of the bottom of the valley. (The coordinates along the bottom can be introduced in different ways.)

To get a clearer picture of the valley method it may be instructive to compare the infinitely-dimensional functional space to an analog mechanical motion of a stream (with a large friction) on two-dimensional surface with a trough (Fig. 16) in the gravitational field. The bottom of the valley in this case is a one-dimensional curve in the three-dimensional space, $\mathbf{x}(\beta)$. The force at

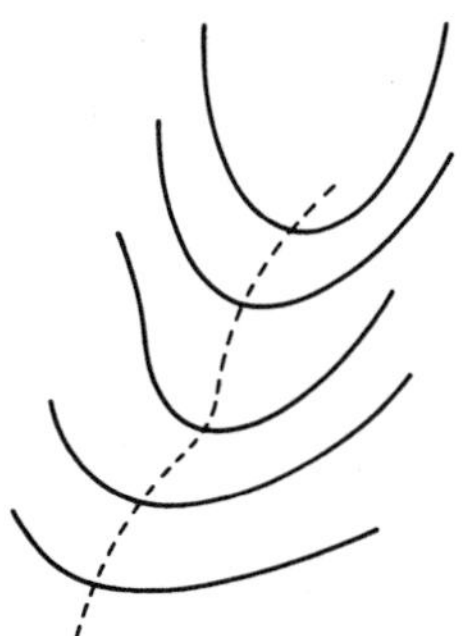

Fig. 16. Mechanical motion of a stream (– – –) on the bottom of a two-dimensional surface, a trough. The surface is steep in the direction perpendicular to the stream trajectory.

[15]It may well happen that the number of such directions is more than one; for clarity we confine ourselves to the simplest case.

the bottom is the gradient of the potential energy V, $F_i = -\partial_i V(\mathbf{x})|_{\mathbf{x}=\mathbf{x}(\beta)}$. The velocity of the stream at the given point at the bottom $\dot{\mathbf{x}}$ is proportional to the force,

$$\dot{\mathbf{x}} = \frac{\partial \mathbf{x}}{\partial \beta}\dot{\beta} \propto -\partial_i V\,. \tag{106}$$

This analogy explains the origin of the name "streamline method".

If the beginning of the valley is a well-defined construction — and one can hardly doubt that — say, a well-separated I–A pair contributes to the observable effects, the question of where and how the valley ends up is rather obscure. Indeed, when the I and A "atoms" approach each other so that the defect of the action becomes comparable to $8\pi^2/g^2$ they tend to annihilate each other. Continuing the journey along the bottom of the valley we smoothly interpolate to a point where the field is weak, the action is $\mathcal{O}(1)$. This point, of course, belongs to ordinary perturbation theory. To avoid double-counting this part of the valley (a flat part) should be definitely excluded from consideration based on instantons. The quasiclassical approximation certainly fails here. So preferably we must stop earlier. Where exactly? Nobody knows. The collapse of the instanton-based approximations might manifest itself as an occurring phase transition when we approach a critical point from the "other" side of the valley (the one corresponding to large separations).

12. Fermions in Instanton Field

12.1. *Very Heavy Quarks*

In this section, we shall briefly discuss how the instanton contribution to the vacuum–vacuum transition amplitude changes when fermions are included in the theory.

It is immediately clear that for a fluctuation with a given scale ρ the influence of "heavy" quarks with mass $m \gg \rho^{-1}$ is small; for in this case the quarks appear at times and distances $\sim 1/m \ll \rho$, at which perturbation theory can be used to calculate the quark loops of the form shown in Fig. 17. The first few terms of the effective Lagrangian that takes into account the fermion

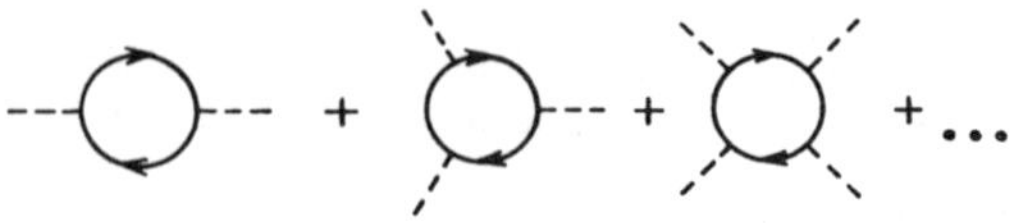

$+ \quad + \quad + \ldots$

Fig. 17. A set of diagrams that one has to calculate to obtain the effective Lagrangian (107).

loops [46] are given below:

$$\Delta L_F = 2\,\mathrm{Tr}\left\{-\frac{1}{4}G_{\mu\nu}^2 \times \frac{g^2}{24\pi^2}\ln\frac{M^2}{m^2} + \frac{1}{16\pi^2}\left(\frac{ig^3}{90m^2}G_{\mu\nu}G_{\nu\gamma}G_{\gamma\mu}\right.\right.$$
$$+\frac{g^4}{72m^4}\left[-(G_{\mu\nu}G_{\mu\nu})^2 + \frac{7}{10}\{G_{\mu\gamma},G_{\gamma\nu}\}_+^2 + \frac{29}{70}[G_{\mu\gamma},G_{\gamma\nu}]_-^2\right.$$
$$\left.\left.\left.-\frac{8}{35}[G_{\mu\nu},G_{\gamma\delta}]_-^2\right]\right)\right\}, \tag{107}$$

$$G_{\mu\nu} = G_{\mu\nu}^a T^a, \qquad \mathrm{Tr}\,T^aT^b = \frac{1}{2}\delta^{ab}.$$

The first term in this expression contains the cut-off parameter M and, obviously, describes the contribution of the quark under consideration to the change in the coupling constant g. Therefore, it is automatically taken into account when the result is expressed in terms of the coupling constant at distances greater than $1/m$.

The second and higher terms in (107) give a series in powers of $1/m^2\rho^2$ on the transition to the Euclidean space and the substitution of the instanton field for $G_{\mu\nu}$. Explicit expressions for all gluon operators in Eq. (107) in terms of the instanton size ρ can be found in Ref. [46].

12.2. *Light (Massless) Quarks*

We now turn to the limiting case of "light" quarks, $m\rho \ll 1$, whose impact on instantons is more radical. We note that for sufficiently small instantons all quarks are light. We calculate the integral over the Fermi fields in the path integral that determines the vacuum–vacuum transition: $\langle 0|0_T\rangle$. In the Euclidean action, a fermion with mass m adds a term of the form [see (45)]

$$S_F^{(E)} = \int d^4x\bar{\psi}(-i\gamma_\mu D_\mu - im)\psi\,, \tag{108}$$

and integration of this with respect to the anticommuting fields leads to

$$\mathrm{Det}(-i\gamma_\mu D_\mu - im)\,. \tag{109}$$

The determinant can be understood as a product of the eigenvalues of the corresponding operator,

$$\mathrm{Det}(-i\gamma_\mu D_\mu - im) = \prod_n(\lambda_n - im)\,, \tag{110}$$

where the real numbers λ_n are the eigenvalues of the Hermitian operator $-i\gamma_\mu D_\mu$:

$$-i\gamma_\mu D_\mu u_n(x) = \lambda_n u_n(x)\,. \tag{111}$$

Of fundamental importance in the study of the limit $m = 0$ is the question of whether certain λ_n vanish, i.e. the question of zero modes of the fermion field. We shall show that the interaction with the instanton field leads to the appearance of one such mode u_0,

$$-i\gamma_\mu D_\mu u_0 = 0\,. \tag{112}$$

We pass to two-component spinors $\chi_{\mathrm{L,R}}$ (we use the standard representation for the γ matrices):

$$u_0 = \begin{pmatrix} 1 \\ -1 \end{pmatrix} \chi_{\mathrm{L}} + \begin{pmatrix} 1 \\ 1 \end{pmatrix} \chi_{\mathrm{R}}\,, \qquad \sigma_\mu^+ D_\mu \chi_{\mathrm{L}} = 0, \quad \sigma_\mu^- D_\mu \chi_{\mathrm{R}} = 0\,, \tag{113}$$

where $\sigma_\mu^\pm = (\boldsymbol{\sigma}, \mp i)$. To the equations for $\chi_{\mathrm{L}}, \chi_{\mathrm{R}}$ we apply the operators $\sigma_\mu^- D_\mu, \sigma_\mu^+ D_\mu$, respectively. Using the relations (57′), the commutator $[D_\mu D_\nu] = -(ig/2)\tau^a G_{\mu\nu}^a$, and the explicit form of $G_{\mu\nu}^a$ [see (63)], we obtain

$$-D_\mu^2 \chi_{\mathrm{L}} = 0, \qquad -D_\mu^2 \chi_{\mathrm{R}} = -4\boldsymbol{\sigma\tau} \frac{\rho^2}{[(x-x_0)^2 + \rho^2]^2} \chi_{\mathrm{R}}\,. \tag{114}$$

The operator $-D_\mu^2$ is a sum of the squares of Hermitian operators: $-D^2 = (-iD_\mu)^2$, i.e. it is positive definite. Therefore, it does not have vanishing eigenvalues (the boundary conditions are imposed at a large but finite distance R) and, therefore, $\chi_{\mathrm{L}} = 0$.

In the equation for χ_{R}, we use a basis in the space of spinor and color indices that diagonalizes the matrix $\boldsymbol{\sigma\tau}$. We recall that $\boldsymbol{\sigma}$ acts on the spinor indices, and $\boldsymbol{\tau}$ on the color indices. This basis corresponds to the addition of the ordinary spin and the color spin to a total angular momentum equal to zero (when $\boldsymbol{\sigma\tau} = -3$) or unity ($\boldsymbol{\sigma\tau} = +1$). It again follows from the positive definiteness of $-D_\mu^2$ that the only suitable case for us is when the total spin is equal to zero, which completely determines the dependence of χ_{R} on the indices:

$$(\boldsymbol{\sigma} + \boldsymbol{\tau})\chi_{\mathrm{R}} = 0, \qquad \chi_{\mathrm{R}}^{\alpha m} \sim \varepsilon^{\alpha m}\,, \tag{115}$$

where $\alpha = 1, 2$ and $m = 1, 2$ are the spin and color indices, respectively.

The dependence on the coordinates can be readily found from the explicit form of D_μ^2, and the final result for the zero mode $u_0(x-x_0)$ (normalized by the condition $\int u^+ u dx = 1$) has the form

$$u_0(x) = \frac{1}{\pi}\frac{\rho}{(x^2+\rho^2)^{3/2}}\begin{pmatrix}1\\1\end{pmatrix}\varphi, \qquad \varphi^{\alpha m} = \frac{1}{\sqrt{2}}\varepsilon^{\alpha m}. \tag{116}$$

We also write down the expression for the zero mode in the singular gauge, $u_0^{\rm sing}(x-x_0)$ (which we shall need),

$$u_0^{\rm sing}(x) = \frac{1}{\pi}\frac{\rho}{(x^2+\rho^2)^{3/2}}\frac{x_\mu\gamma_\mu}{\sqrt{x^2}}\begin{pmatrix}1\\-1\end{pmatrix}\varphi, \tag{117}$$

which is obtained by the multiplication of (116) by the gauge transformation matrix (64′).

12.3. *Tunneling Interpretation in the Presence of Massless Fermions. The Index Theorem*

Since the instanton contribution is proportional to $\det(-i\gamma_\mu D_\mu - im)$, and the operator $i\gamma_\mu D_\mu$ has a zero mode in the instanton field, it is tempting to conclude that in the massless limit the instanton contribution vanishes. How can one reconcile this result with the tunneling interpretation?

Introduction of fermions certainly does not affect the nontrivial topology in the space of the gauge fields. The existence of a noncontractable loop remains intact, and with this loop comes the necessity of considering the wave function of the Bloch type (Sec. 5). The instanton trajectory connects Ψ_n and Ψ_{n+1} under the barrier, and is related to the probability of tunneling. If this probability were to vanish, what could have gone wrong with this picture?

To answer this question we must expand the picture of "tunneling in the $\mathcal{K}$ direction" by coupling the variable $\mathcal{K}$ to (an infinite number of) the fermion degrees of freedom. In order to make the situation more transparent we will slightly distort some details. We will assume that the motion of the system in the $\mathcal{K}$ direction is slow, while the fermion degrees of freedom are fast, so that the approximation of the Born-Oppenheimer type is applicable. In this approximation the motion in the $\mathcal{K}$ direction is treated adiabatically. We first freeze $\mathcal{K}$, then consider the dynamics of the fermion degrees of freedom, integrate them out, and at the last stage return to the evolution of the variable

$\mathcal{K}$. Certainly, in QCD all degrees of freedom are equally fast and no Born-Oppenheimer approximation can be developed. The general feature of the underlying dynamics that we are interested in does not depend, however, on this approximation.

Thus for each given value of $\mathcal{K}$ we must determine the fermion component of the wave function. This is done by building the Dirac sea in the fermion sector, with $\mathcal{K}$ frozen. The structure of the Dirac sea depends on the value of $\mathcal{K}$.

When $\mathcal{K}$ varies adiabatically, the energy of the fermion levels continuously evolves. The points $\mathcal{K} = n$ and $\mathcal{K} = n + 1$, being the gauge copies of each other, are physically identical. This means that the set of the energy levels of the Dirac sea at $\mathcal{K} = n$ is identical to the set at $\mathcal{K} = n + 1$.

This does not mean, however, that the individual levels do not move. When $\mathcal{K} =$ changes by one unit, some fermion levels with positive energy can dive into the negative-energy sea, while those from the sea, with the negative energies, can appear at levels with positive energies. As a whole the set will be intact, but, some levels interchange their positions (Fig. 18).

For each value of $\mathcal{K}$ we build the Dirac sea by filling in all negative-energy states. Let us say at $\mathcal{K} = n$ we built it properly. If in the process of motion

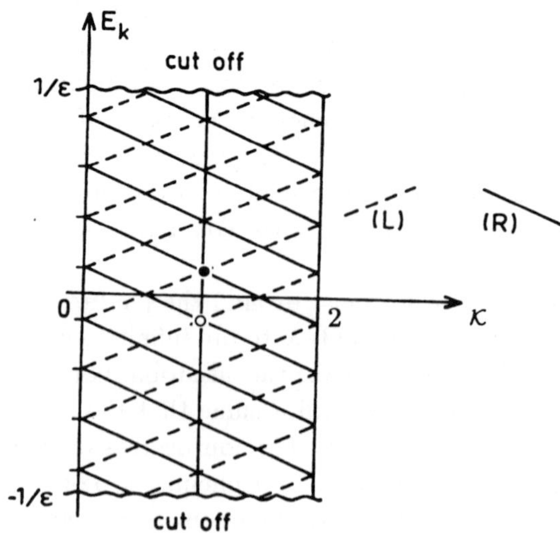

Fig. 18. The fermion energy levels *versus* $\mathcal{K}$.

in the $\mathcal{K}$ direction, at $\mathcal{K} = n + 1/2$, say, one level dives into the sea and one jumps out, this must be interpreted as the fermion production, since the state we end up with at $\mathcal{K} = n + 1$ is an excited state with respect to the proper Dirac sea at $\mathcal{K} = n + 1$. Thus, the tunneling trajectory connects the states $\Psi_n \Phi_n(\text{ferm})$ with $\Psi_{n+1}\Phi_{n+1}(\text{ferm})$ where the fermion components $\Phi_n(\text{ferm})$ and $\Phi_{n+1}(\text{ferm})$ differ by the quantum numbers of the fermion sector. We tried to calculate the probability of the tunneling transition with no change of the fermion state, and got zero. We now understand that we should not be discouraged. This zero could have been expected. The tunnelings occur in such a way that the fermion quantum numbers are forced to change in the process of the tunneling.

The consideration presented above is exact in the two-dimensional Schwinger model (spinor electrodynamics), see the review paper [47] for a pedagogical discussion. As was noted, in QCD a qualitatively similar picture is believed to take place. The argument demonstrating the validity of this picture in QCD is based on the so-called triangle anomaly. Assume for simplicity that we have one massless quark, q. At the classical level both the vector and axial currents

$$V_\mu = \bar{q}\gamma_\mu q, \qquad A_\mu = \bar{q}\gamma_\mu\gamma_5 q$$

are conserved,

$$\partial_\mu V_\mu = 0, \qquad \partial_\mu A_\mu = 0\,.$$

At the quantum level the axial current is anomalous,

$$\partial_\mu A_\mu = \frac{g^2}{16\pi^2} G^a_{\mu\nu}\tilde{G}^a_{\mu\nu}\,.$$

Let us now integrate over x and evaluate both sides of this equation in the instanton field. On the left-hand side we first integrate over the spatial variables $\mathbf{x}$. Then, the left-hand side reduces to

$$\int_{-\infty}^{\infty} dt\partial_0 \int A_0 d^3x = Q_5(t=\infty) - Q_5(t=-\infty)\,. \tag{118}$$

The right-hand side is

$$\frac{g^2}{16\pi^2}\int d^4x (G^a_{\mu\nu}\tilde{G}^a_{\mu\nu})_{\text{inst.}} = 2Q = 2(\mathcal{K}(t=\infty) - \mathcal{K}(t=-\infty))\,.$$

We see that in theory with one massless quark in the instanton transition the chiral charge is forced to change by two units — say, a left-handed quark is converted into a right-handed quark with the unit probability. If we want to get a nonvanishing tunneling probability we have to incorporate this feature. This will be done in the next section.

The variation of the chiral charge, $\Delta Q_5 \neq 0$, in the tunneling transition is in one-to-one correspondence with the occurrence of the zero modes in the Dirac equation in the self-dual fields. The number of the fermion zero modes is related to the topological charge of the gauge field by the so-called Atiyah-Singer (or the index) theorem [48], which has been derived in the instanton context in Ref. [49] (see also Refs. [8, 50, 51]). Specifically, if the number of the normalizable zero modes of positive (negative) chirality is n_+ (n_-) then

$$n_+ - n_- = Q \tag{119}$$

for each Dirac fermion field Ψ in the fundamental representation (since the operator $i\not{D}$ is Hermitian, the equation $\not{D}\Psi = 0$ implies that the equation on $\bar{\Psi}$ has a zero-eigenvalue solution as well). A brief but illuminating discussion of the derivation of Eq. (119) can be found in the review paper of Coleman [8]. As a matter of fact, this theorem is equivalent to the consideration of the triangle anomaly in the axial-vector current presented above.

Let us note in passing that the presence of massless fermions, combined with the triangle anomaly in $\partial_\mu A_\mu$, results in another drastic consequence: the θ term becomes unobservable even if $\theta \neq 0$. Indeed, one can rewrite L_θ as

$$L_\theta = \frac{\theta}{2} \partial_\mu A_\mu \,,$$

i.e. a full derivative of the gauge invariant quantity. Such full derivatives drop out from the action. This is in sharp distinction with the full derivative of the Chern-Simons current, which, as we know, gives a nonvanishing contribution in the action once we switch on the instanton field. The Chern-Simons current is not gauge-invariant.

This argument implies that in the theory with light quarks all θ dependent effects must be proportional to the quark mass.

12.4. *Instanton Density in the Theory with Light Quarks*

To calculate the instanton density we must consider the vacuum-to-vacuum transition amplitude in the theory with light quarks. In addition to the factors we have obtained in pure gluodynamics this amplitude acquires the following factor due to the fermion field (we introduce one such field for the time being)

$$F = \frac{m}{M} \frac{\mathrm{Det}'(-i\gamma_\mu D_\mu)}{\mathrm{Det}'(-i\gamma_\mu D_\mu - iM)} \frac{\mathrm{Det}(-i\gamma_\mu \partial_\mu - iM)}{\mathrm{Det}(-i\gamma_\mu \partial_\mu)},$$

where Det′ denotes the determinant without the zero mode and we take into account the regularization and also the normalization by perturbation theory. In all the nonzero modes, m is taken as equal to zero, so that after the separation in F of the factor m/M the remaining part depends only on the dimensionless parameter $M\rho$. As in pure gluodynamics (see Sec. 9), this dependence must be such that the cut-off parameter M is removed by a renormalization of the coupling constant, i.e. the dependence of F on $M\rho$ must give the renormalization of the coupling constant due to the fermions in the factor $e^{-8\pi^2/g_0^2}$,

$$\Delta_F \frac{8\pi^2}{g^2} = -\ln \frac{F}{m\rho \cdot \mathrm{const}} = \ln M\rho - \frac{1}{3} \ln M\rho \,. \tag{120}$$

The first logarithm derives from the zero mode, the second from the nonzero modes. Comparing the result with formula (81) for gluons, we see that the situation has been changed because of the anticommutativity of the fermion fields: the zero modes of the light quarks lead to the screening of the charge, and the nonzero modes to antiscreening.

In the ordinary perturbation theory, the decomposition in (120) can be associated with the spin-dependent part of the interaction (the first logarithm) and the "charge" part, which is not associated with the spin (the second logarithm). Indeed, the imaginary part of the gluon polarization operator, which derives from the intermediate $q\bar{q}$ state, can be represented in the form

$$\begin{aligned} \mathrm{Im}\, \Pi_{\mu\nu}^{F_{ab}} &= \delta^{ab} \frac{g^2}{2} \int \frac{do}{32\pi^2} [q_\mu q_\nu - g_{\mu\nu} q^2 - (p_1 - p_2)_\mu (p_1 - p_2)_\nu] \\ &= \delta^{ab} \frac{g^2}{16\pi} (q_\mu q_\nu - q_{\mu\nu} q^2) \left(1 - \frac{1}{3}\right). \end{aligned} \tag{121}$$

In this formula, p_1 and p_2 are the particle and antiparticle momenta, $q = p_1 + p_2$, and the integration is over the directions of $\mathbf{p}_1 = -\mathbf{p}_2$ in the center-of-mass system. The second term in (121) differs by only the factor -2 from the contribution of a spinless color doublet. The factor 2 corresponds to the two polarization states, and the minus to the anticommutativity.

We note that for the gluon vacuum polarization there is also an analogous relation between the spin part of the polarization and the zero modes. This is readily seen in the "background" gauge obtained by adding the term (71) to the action. In perturbation theory, one can take as the "external" field, for example, a potential that has only a third color component, and in the loop only "charged" components will propagate. The three-gluon vertex in this gauge has the form of a sum of a color part and a magnetic part, which do not interfere in the polarization operator. The spin (magnetic) part gives the "antiscreening" logarithm, and the charge (color) part (together with the Higgs particles) the "screening" part.

As we already know, in the limit $m \to 0$ the vacuum-to-vacuum transition amplitude tends to zero, and the instanton fluctuation couples the "vacuum" with the state with the one left-handed quark and the one right-handed anti-quark. Alternatively, one can say that the instanton fluctuation couples the state with the left-handed quark with that with the right-handed quark. (Note that if $m = 0$ the left-handed quark does not become the right-handed quark in any order of the perturbation theory.) Let us calculate this coupling. To this end we consider the transition from a single-quark state to a single-quark state; we shall assume that the quark momenta p and p' are small compared with $1/\rho$. Proceeding as in Sec. 10, we use the reduction formula

$$\langle p'|p_T\rangle = -\int dx dx' e^{ip'x'-ipx}\bar{v}^m_\alpha(\not{p}\,')_{\alpha\gamma}\langle 0|T\{q^m_\gamma(x')\bar{q}^k_\beta(x)\}|0\rangle_{\text{ins}}(\not{p}\,)_{\beta\delta}v^k_\delta\,, \tag{122}$$

where $\bar{v}^m_\alpha$ and v^k_δ are the spinors that describe the final and the initial quark (the superscript is the color index, the subscript the spinor index).

We find the instanton contribution to the fermion Green's function by using the relation

$$\langle 0|T\{q^m_\gamma(x')q^k_\beta(x)\}|0\rangle_{\text{ins}} \underset{x_0\to -ix_4}{\longrightarrow} \sum_n \frac{u^m_{(n)\gamma}(x')u^{+k}_{(n)\beta}(x)}{m+i\lambda_n}\langle 0|0_T\rangle_{\text{ins}}\,. \tag{123}$$

In the limit $m \to 0$, the zero mode makes the main contribution, and (123) is finite at $m = 0$ (since $\langle 0|0_T\rangle_{\text{ins}}$ contains $F \propto m$).

Using the explicit form (117) of the zero mode in the singular gauge, we can now readily obtain the result. We formulate it in the form of the expression for the effective Lagrangian that describes all transitions which arise from the instanton fluctuation with scale ρ:

$$\Delta L(x) = \prod_q \left[m_q\rho - 2\pi^2\rho^3\bar{q}_R\left(1+\frac{i}{4}\tau^a\bar{\eta}_{a\mu\nu}\sigma_{\mu\nu}\right)q_L\right] \\ \times \exp\left(-\frac{2\pi^2}{g}\rho^2\bar{\eta}_{b\gamma\delta}G^b_{\gamma\delta}\right)d_0(\rho)\frac{d\rho}{\rho^5}d\hat{o}\,. \qquad (124)$$

This contains a product over all species of light ($m_q\rho \ll 1$) quarks, and $\sigma_{\mu\nu} = (\gamma_\mu\gamma_\nu - \gamma_\nu\gamma_\mu)/2$. In Minkowski space, the symbols $\eta_{a\mu\nu}$ differ from the Euclidean symbols only when μ or $\nu = 0$, and then by a factor i. By $d\hat{o}$ we denote the differential corresponding to the color orientation of the instanton, and it is normalized to unity, $\int d\hat{o} = 1$. A dependence on the orientation enters through the substitution $\bar{\eta}_{a\mu\nu} \to O_{aa'}\bar{\eta}_{a'\mu\nu}$ (O is the matrix of rotations in the color space), which must be made in (124). Further details and explicit color factors for the gauge group SU(3) can be found in Ref. [52]. The quantity $d_0(\rho)$ differs from $d(\rho)$ in Eq. (85) in pure gluodynamics by the multiplication by the factor

$$\exp F\left[-\frac{1}{3}\ln 2 - \frac{17}{36} + \frac{1}{3}(\ln 2\pi + \gamma) + \frac{2}{\pi^2}\sum_{s=1}^{\infty}\frac{\ln s}{s^2}\right] = e^{0.292F}\,, \qquad (125)$$

where F is the number of light fermions. This is for the Pauli-Villars regularization; for the MS scheme, the 0.292 is replaced by -0.495, and by 0.153 for the $\overline{\text{MS}}$ scheme. In addition, in the expression (82) for $8\pi^2/g^2(\rho)$ it is necessary to include the fermion contribution.

For the antiinstanton, ΔL is obtained from (124) by the substitution $\bar{\eta}_{a\mu\nu} \to \eta_{a\mu\nu}$, $q_{\text{L,R}} \to q_{\text{R,L}}$. Note also that all the operators, the constant g, and the masses m_q in ΔL are normalized at the point ρ, so that besides the dependence given explicitly there is a logarithmic dependence on ρ, which is determined by the anomalous dimension of the operator term in ΔL under consideration.

Of particular interest are the instanton-generated fermion vertices; this interaction is frequently called the 't Hooft determinant interaction. The point is that it explicitly demonstrates the breaking of the $U(1)$ symmetry associated with transformations of the form $q' = e^{i\alpha\gamma_5}q$. Naively, such a symmetry holds in a theory with massless quarks. The nontrivial nature of the breaking of this symmetry can be seen from the fact that, for example, in a theory with one quark ΔL describes the transition of a "left-handed" quark into a "right-handed" one, which is impossible in any finite order of perturbation theory for $m = 0$.

12.5. *Global Anomaly*

The effective Lagrangian in Eq. (124) presents a concise formula summarizing the effects due to light quarks in QCD. A thorough inspection of this Lagrangian leads one to a perplexing question. Indeed, let us assume that, instead of QCD, we consider an SU(2) theory with one massless left-handed Weyl fermion transforming as a doublet with respect to SU(2). Usually we deal with the Dirac fermions; one Dirac fermion is equivalent to two Weyl fermions. Now we want to consider a chiral theory. Before the advent of the instantons this theory was believed to be perfectly well-defined. It has no internal anomalies; moreover, in perturbation theory, order by order, one encounters no reasons to make the theory sick. And yet, this theory is pathological. The 't Hooft interaction helps us reveal the pathology.

Indeed, if we start building an effective Lagrangian analogous to Eq. (124) in the SU(2) theory with one massless left-handed Weyl fermion we will immediately discover that this Lagrangian must be *linear* in the fermion field. In the instanton transition for one Dirac fermion $\Delta Q_5 = 2$, but the Weyl fermion $= 1/2$ of the Dirac fermion, and hence $\Delta Q_5 = 1$!

It was obvious to many that something was unusual in this theory. The intuitive feeling of pathology was formalized by Witten who showed [53] that this theory is ill-defined because of the *global anomaly*. Such theory is mathematically inconsistent. It simply does not exist.

One of the possible proofs of the global anomaly is based on the fermion level restructuring in the instanton transition. The key elements are the following: (i) the vacuum-to-vacuum amplitude in the theory with one Weyl fermion is proportional to $\sqrt{\det(i\not{D})}$; (ii) Only one pair of the fermion levels exchange their positions when $\mathcal{K} = n$ goes in $\mathcal{K} = n+1$. For further details see Ref. [53].

13. Continuation in Euclidean Space in the Theories with Chiral Fermions and Supersymmetric Theories

Usually, the first step of every instanton practitioner is to rewrite the theory under consideration as a Euclidean theory. We have done this too in Sec. 6.

This is possible and convenient if the theory one deals with contains only Dirac fermions. For chiral fermions this is not possible. The reason lies in the fact that one cannot define for the chiral Fermi fields the operation of involution (complex conjugation) in the Euclidean space. For this reason one cannot formulate, in particular (unextended) supersymmetric theories (which always involve chiral fermions) in the Euclidean space. In slightly different terms, one cannot introduce the notion of the Majorana spinor in the Euclidean space. Therefore, the problem of the continuation is common to both chiral and supersymmetric theories.

If we want to consider the instanton effects in these theories we have to indicate how the Euclidean continuation is to be done. Here we would like to show that Euclidean continuation is not at all equivalent to constructing the Euclidean version of the theory. One can keep the theory in the Minkowski space-time, and instead of continuing the theory, analytically continue just the solutions of the equations of motion. The fields, the Lagrangian and the action remain Minkowskean. What is really needed for the instanton calculations is the saddle point method of computation of path integrals based on analytic continuation in t for the tunneling processes.

Our method of Euclidean continuation of functional integrals follows the works of Berezin [54]. Let us recall the basic procedure for constructing the functional integrals. If we have a *quantum* Hamiltonian given as a function of the coordinate operator $\hat{q}_i$ and momentum operator $\hat{p}_i$, $\hat{H}(\hat{q}_i, \hat{p}_i)$, then we can unambiguously build a representation for the evolution operator $\exp(-i\hat{H}t)$ in terms of the functional integral

$$\langle q_i^{(+)}|e^{-i\hat{H}T}|q_i^{(-)}\rangle = \int \prod_i Dp_i Dq_i \exp\left\{ i\int_{-T/2}^{T/2} dt \right.$$

$$\left. \times \left[\sum p_i\dot{q}_i - H(p_i(t), q_i(t))\right]\right\}, \qquad (126)$$

where the integration runs over all trajectories satisfying the following conditions

$$q_i\left(t = \pm\frac{1}{2}T\right) = q_i^{(\pm)}, \qquad p_i\left(t = \frac{1}{2}T\right) = p_i\left(t = -\frac{1}{2}T\right).$$

Quite an analogous representation is valid for Fermi systems with the only difference that the integration runs over anticommuting Grassmann parameters.

Now, continuation to the Euclidean space reduces to a single step, transition to an imaginary value of the parameter T, $T = -i\tau$. In other words, instead of $\exp(-iHT)$ we consider $\exp(-H\tau)$. The functional integral representing this operator is built in the same way as above,

$$\langle q_i^{(+)}|e^{-H\tau}|q_i^{(-)}\rangle = \int \prod_i Dp_i Dq_i \exp\left\{ - \int_{-\tau/2}^{+\tau/2} dt \right.$$
$$\left. \times \left[-i\sum_i p_i\dot{q}_i + H(p_i(t), q_i(t)) \right] \right\}. \qquad (127)$$

Performing integration over $Dp_i(t)$ we arrive at the ordinary expression for the Euclidean action. This is for the Bose fields. For the Fermi fields one should calculate the integral over $D\bar{\psi}$, the step usually avoided.

On the other hand, arranging the expression (127) in the Euclidean form is not necessary at all. Indeed, quantization of the theory in the Minkowski space determines the operator $\hat{H}(\hat{p}_i, \hat{q}_i)$ as well as all the conserved charges (commuting operators). The action of these charges on $\hat{p}_i$ and $\hat{q}_i$ fixes their transformation law, which, in turn, generates the corresponding symmetry transformations for integration variables in the expressions like (127), in particular, SUSY transformations (for more details see Chapter VI).

Algebraically, all expressions for the transformations differ from the original ones (defined in the Minkowski space) only by the substitution $x_0 = -ix_4$.

To elucidate our procedure [55] and notations let us consider the vector potential for the instanton solution (more exactly, antiinstanton)

$$A^{\alpha\gamma}_{\beta\dot{\beta}} = -\frac{i}{g}\frac{1}{x^2+\rho^2}(\delta^\alpha_\beta x^\gamma_{\dot{\beta}} + \delta^\gamma_\beta x^\alpha_{\dot{\beta}}). \qquad (128)$$

Here spinor indices are introduced running over 1, 2 both in the color and coordinate spaces. The connection with the ordinary color triplet index is given by the following relation

$$A^{\alpha\gamma} = A^a \left(\frac{\tau^a}{2}\right)^\alpha_\rho \varepsilon^{\gamma\rho}, \qquad \begin{pmatrix} \alpha,\gamma,\rho = 1,2 \\ a = 1,2,3 \end{pmatrix}, \qquad (129)$$

where $(\tau^a)^\alpha_\rho$ stand for the Pauli matrices, $\varepsilon^{\gamma\rho}$ is the antisymmetric Levi-Civita symbol, which plays the role of metric:

$$F^\alpha = \varepsilon^{\alpha\beta}F_\beta, \qquad F_\alpha = \varepsilon_{\alpha\beta}F^\beta, \qquad \varepsilon_{\alpha\beta} = -\varepsilon^{\alpha\beta}, \qquad (130)$$
$$\varepsilon^{12} = -\varepsilon^{21} = 1.$$

The relation between the vector and spinor indices in the coordinate space is as follows

$$A_{\beta\dot\beta} = (\sigma_\mu)_{\beta\dot\beta} A_\nu g^{\mu\nu}, \qquad g_{\mu\nu} = \mathrm{diag}(1,-1,-1,-1)\,, \\ (\sigma^\mu)_{\beta\dot\beta} = (\delta_{\beta\dot\beta}, \boldsymbol{\sigma}_{\beta\dot\beta})\,. \tag{131}$$

(It is instructive to compare the expression for the σ^μ matrices in Eq. (131) with their Euclidean analogs which are defined after Eq. (113).) Furthermore, in particular, this relation refers also to x_μ; explicitly

$$\begin{aligned} x_{1\dot 1} &= x^{2\dot 2} = x_0 - x_3\,, \\ x_{1\dot 2} &= -x^{2\dot 1} = -x_1 + ix_2\,, \\ x_{2\dot 2} &= x^{1\dot 1} = x_0 + x_3\,, \\ x_{2\dot 1} &= -x^{1\dot 2} = -x_1 - ix_2\,. \end{aligned} \tag{132}$$

All the above notations are obviously taken from the Minkowski space. Their Euclidean nature is revealed only in the fact that x_0 is imaginary, $x_0 = -ix_4$. Another compromise to the Euclidean notation, which is rather unnecessary though, is in the definition of x^2,

$$x^2 = -x_\mu x_\nu g^{\mu\nu} = -\frac{1}{2} x^{\alpha\dot\alpha} x_{\alpha\dot\alpha} = -x_0^2 + \boldsymbol{x}^2 = x_4^2 + \boldsymbol{x}^2\,. \tag{133}$$

One can readily convince oneself that the expression (128) coincides with the standard form for the BPST instanton. To this end one should take into account, apart from the expressions given above, the relation between the Minkowskian and Euclidean fields, A_μ and $\hat A_\mu$

$$A_0 = i\hat A_4, \qquad A^m = -\hat A_m \qquad (m = 1,2,3)\,, \tag{134}$$

(the latter are used in the standard approach, see Sec. 6).

The solution of the Dirac equation $\not{D}\Psi = 0$ (the 't Hooft zero mode) has the following form:

$$\Psi^\alpha_\gamma \propto \delta^\alpha_\gamma \frac{\rho}{(x^2+\rho^2)^{3/2}} \tag{135}$$

where the lower (undotted) subscript refers to the left-handed SU(2) subgroup of the Lorentz $O(4)$ while the upper one is the spinorial index of the color SU(2).

References

[1] M. Shifman, A. Vainshtein and V. Zakharov, *Nucl. Phys.* **B147** (1979) 385; 448 [Reprinted in *Vacuum Structure and QCD Sum Rules*, Ed. M. Shifman (North-Holland, 1992) p. 24].

[2] A. Polyakov, *Phys. Lett.* **59B** (1975) 82 [Reprinted in *Instantons in Gauge Theories*, Ed. M. Shifman (World Scientific, Singapore, 1994), p. 19].

[3] A. Belavin, A. Polyakov, A. Schwarz and Yu. Tyupkin, *Phys. Lett.* **59B** (1975) 85 [Reprinted in *Instantons in Gauge Theories*, Ed. M. Shifman (World Scientific, Singapore, 1994), p. 22].

[4] T. Schäfer and E. V. Shuryak, *Instantons in QCD*, *Rev. Mod. Phys.* **70** (1998) 323.

[5] A. Polyakov, *Nucl. Phys.* **B120** (1977) 429.

[6] R. P. Feynman and A. R. Hibbs, *Quantum Mechanics and Path Integrals* (McGraw Hill, New York, 1965).

[7] I. Gradshteyn and I. Ryzhik, *Table of Integrals, Series, and Products* (Academic Press, New York, 1980).

[8] S. Coleman, in *Aspects of Symmetry* (Cambridge University Press, 1985), p. 265.

[9] G. 't Hooft, *Phys. Rev.* **D14** (1976) 3432; (E) **D18** (1978) 2199.

[10] L. D. Landau and E. M. Lifshitz, *Quantum Mechanics*, Third Edition (Pergamon Press, New York, 1989).

[11] C. Callan, R. Dashen and D. Gross, *Phys. Rev.* **D17** (1978) 2717; *Phys. Rev.* **D19** (1979) 1826.

[12] J. F. Willemsen, *Phys. Rev.* **D20** (1979) 3292.

[13] A. S. Schwarz, *Topology for Physicists* (Springer, Berlin, 1994).

[14] A. P. Balachandran, *Skyrmions*, in *High Energy Physics 1985*, Proc. Yale Theoretical Advanced Study Institute on High Energy Physics, New Haven, Conn., June-July, 1985, Eds. M. J. Bowick and F. Gursey (World Scientific, Singapore, 1985), Vol. 1, p. 1.

[15] S. Flüge, *Practical Quantum Mechanics* (Springer, Berlin, 1971), Vol. 1, Problem 28; C. Kittel, *Quantum Theory of Solids* (Wiley & Sons, New York, 1963), Chap. 9; N. Ashcroft and N. Mermin, *Solid State Physics* (Saunders College, Philadelphia, 1976), Chap. 8.

[16] R. Jackiw and C. Rebbi, *Phys. Rev. Lett.* **37** (1976) 172 [Reprinted in *Instantons in Gauge Theories*, Ed. M. Shifman (World Scientific, Singapore, 1994), p. 25].

[17] C. Callan, R. Dashen and D. Gross, *Phys. Lett.* **B63** (1976) 334 [Reprinted in *Instantons in Gauge Theories*, Ed. M. Shifman (World Scientific, Singapore, 1994), p. 29].

[18] K. M. Bitar and S.-J. Chang, *Phys. Rev.* **D17** (1978) 486.

[19] R. J. Crewther, P. Di Vecchia, G. Veneziano and E. Witten, *Phys. Lett.* **B88** (1979) 123; (E)**B91** (1980) 487.

[20] S. Weinberg, *Phys. Rev. Lett.* **40** (1978) 223; F. Wilczek, *Phys. Rev. Lett.* **40** (1978) 279.

[21] R. D. Peccei, *QCD, Strong CP and Axions*, hep-ph/9606475; M. Davier, *Axions: a Review*, in *Hadrons, Quarks and Gluons*, Proc. XXII Rencontre de Moriond, Ed. J. Tran Thanh Van (Editions Frontieres, 1987).

[22] M. Atiyah, N. Hitchin, V. Drinfeld and Yu. Manin, *Phys. Lett.* **65A** (1978) 185 [Reprinted in *Instantons in Gauge Theories*, Ed. M. Shifman (World Scientific, Singapore, 1994), p. 133; V. G. Drinfeld and Yu. I. Manin, *Yad. Fiz.* **29** (1979) 1646 [*Sov. J. Nucl. Phys.* **29** (1979) 845].

[23] R. Jackiw, *Field Theoretic Investigations in Current Algebra*, Sec. 7, in *Current Algebra and Anomalies*, Eds. S. B. Treiman, R. Jackiw, B. Zumino and E. Witten (Princeton University Press, 1985), p. 81; P. Fayet and S. Ferrara, *Phys. Rep.* **32** (1977) 249.

[24] M. Shifman and A. Vainshtein, *Nucl. Phys.* **B362** (1991) 21 [Reprinted in *Instantons in Gauge Theories*, Ed. M. Shifman (World Scientific, Singapore, 1994), p. 97].

[25] R. Jackiw and C. Rebbi, *Phys. Lett.* **B67** (1977) 189; C. Bernard, N. Christ, A. Guth and E. Weinberg, *Phys. Rev.* **D16** (1977) 2967 [Reprinted in *Instantons in Gauge Theories*, Ed. M. Shifman (World Scientific, Singapore, 1994), p. 149; 153].

[26] M. P. Mattis, *Phys. Rep.* **214** (1992) 159; V. A. Rubakov and M. E. Shaposhnikov, *Usp. Fiz. Nauk* **166** (1996) 493 [*Phys. Usp.* **39** (1996) 461]; hep-ph/9603208.

[27] T. Banks and E. Rabinovici, *Nucl. Phys.* **B160** (1979) 349; E. Fradkin and S. Shenker, *Phys. Rev.* **D19** (1979) 3682.

[28] The idea of the constrained instanton was first put forward in Y. Frishman and S. Yankielowicz, *Phys. Rev.* **D19** (1979) 540. The method was later developed by I. Affleck, *Nucl. Phys.* **B191** (1981) 429 [Reprinted in *Instantons in Gauge Theories*, Ed. M. Shifman (World Scientific, Singapore, 1994), p. 247].

[29] F. R. Klinkhamer and N. S. Manton, *Phys. Rev.* **D30** (1984) 2212.

[30] R. Dashen, B. Hasslacher and A. Neveu, *Phys. Rev.* **D10** (1974) 4138.

[31] L. G. Yaffe, *Phys. Rev.* **D40** (1989) 3463; F. Klinkhamer, *Sphalerons and Energy Barriers in the Weinberg–Salam Model*, in Proc. XXV Int. Conf. on High Energy Physics, Singapore, 1990, Ed. K. K. Phua and Y. Yamaguchi (World Scientific, Singapore, 1991), p. 913.

[32] A. Smilga, *Nucl. Phys.* **B459** (1996) 263.

[33] I. Khriplovich, *Yad. Fiz.* **10** (1969) 409 [*Sov. J. Nucl. Phys.* **10** (1970) 235].

[34] T. Appelquist, M. Dine and I. Muzinich, *Phys. Lett.* **B69** (1977) 231.

[35] V. Novikov *et al.*, *Phys. Rep.* **41** (1978) 1, Sec. 1.3.

[36] D. R. T. Jones, *Nucl. Phys.* **B75** (1974) 531; for a mini-review see I. Hinchliffe, *Phys. Rev.* **D54** (1996) 77.

[37] C. Bernard, *Phys. Rev.* **D19** (1979) 3013 [Reprinted in *Instantons in Gauge Theories*, Ed. M. Shifman (World Scientific, Singapore, 1994), p. 109].

[38] G. 't Hooft and M. Veltman, *Nucl. Phys.* **B44** (1972) 189; G. 't Hooft, *Nucl. Phys.* **B62** (1973) 444.

[39] W. A. Bardeen, A. J. Buras, D. W. Duke and T. Muta, *Phys. Rev.* **D18** (1978) 3998.

[40] M. A. Shifman, A. I. Vainshtein and V. I. Zakharov, *Nucl. Phys.* **B165** (1980) 45.

[41] The instanton liquid model gradually developed in the 1980s. The key works are E. V. Shuryak, *Nucl. Phys.* **B198** (1982) 83; D. Diakonov and V. Petrov, *Nucl. Phys.* **B245** (1984) 259; **B272** (1986) 457; E. V. Shuryak, *Nucl. Phys.* **B302** (1988) 559; 574; 599. For references to further works and an exhaustive review see Ref. [4].

[42] D. Förster, *Phys. Lett.* **B66** (1977) 279; C. Callan, R. Dashen and D. Gross, *Phys. Rev.* **D17** (1978) 2717.

[43] A. Yung, *Instanton-Induced Effective Lagrangian in the Gauge-Higgs Theory*, Preprint SISSA 181/90/EP, 1990, unpublished; D. Diakonov and M. Polyakov, *Nucl. Phys.* **B389** (1993) 109; I. Balitsky and A. Schäfer, *Nucl. Phys.* **B404** (1993) 639.

[44] I. Balitsky and A. Yung, *Phys. Lett.* **B168** (1986) 113; I. Balitsky and A. Yung, *Nucl. Phys.* **B274** (1986) 475.

[45] A. Yung, *Valley Method for Instanton Induced Effects in Quantum Field Theory*, in *1991 Summer School in High Energy Physics and Cosmology*, Proc. ICTP 1991 Summer School on High Energy Physics and Cosmology, Ed. E. Gava, K. Narain, S. Randjbar-Daemi, E. Sezgin and Q. Shafi (World Scientific, Singapore, 1992), p. 580; I. Balitsky and A. Schäfer, *Nucl. Phys.* **B404** (1993) 639.

[46] V. A. Novikov, M. A. Shifman, A. I. Vainshtein and V. I. Zakharov, *Fortsch. Phys.* **32** (1985) 585.

[47] M. Shifman, *Phys. Rep.* **209** (1991) 341.

[48] M. Atiyah and I. Singer, *Ann. Math.* **87** (1968) 484; **93** (1971) 119.

[49] A. Schwarz, *Phys. Lett.* **B67** (1977) 172.

[50] L. Brown, R. Carlitz and C. Lee, *Phys. Rev.* **D16** (1977) 417.

[51] D. Friedan and P. Windey, *Nucl. Phys.* **B235** (1984) 395 [Reprinted in Reprint Volume "Supersymmetry", Ed. S. Ferrara (North-Holland/World Scientific, 1987), p. 572.

[52] M. A. Shifman, A. I. Vainshtein and V. I. Zakharov, *Nucl. Phys.* **B163** (1980) 46.

[53] E. Witten, *Phys. Lett.* **B117** (1982) 324 [Reprinted in *Current Algebra and Anomalies*, Ed. S. Treiman *et al.* (Princeton Univ. Press, 1985), p. 429].

[54] F. A. Berezin, *The Method of Second Quantization* (Academic Press, New York, 1966). V. A. Novikov, M. A. Shifman, A. I. Vainshtein and V. I. Zakharov, *Nucl. Phys.* **B260** (1985) 157.

Recommended Literature

Reviews

S. Coleman, *Aspects of Symmetry* (Cambridge University Press, 1985), p. 265.

T. Schäfer and E. V. Shuryak, *Instantons in QCD*, *Rev. Mod. Phys.* **70** (1998) 323.

Books

Instantons in Gauge Theories, Ed. M. Shifman (World Scientific, Singapore, 1994). I bring my apology to E. F. Corrigan, D. B. Fairlie, S. Templeton and P. Goddard whose work (*Nucl. Phys.* **B140** (1987) 31) was inadvertently omitted from Sec. IV of the book. M.S.

R. Rajaraman, *Solitons and Instantons* (North-Holland, Amsterdam, 1982).

Chapter IV

Beginning Supersymmetry (Supersymmetry in Quantum Mechanics)

M. A. Shifman

Theoretical Physics Institute, University of Minnesota, Minneapolis, MN 55455, USA

This lecture was first given at the XXI Winter School of Physics of the Leningrad Institute for Nuclear Physics in 1986 and was published in Russian in the Proceedings of the School. Revised in February 1995.

Abstract

The notion of supersymmetry (SUSY) is introduced in the simplest possible setting — quantum mechanics. One-dimensional motion of the particle with spin in the potential of a special type is supersymmetric, i.e. the Hamiltonian is representable as the square of the supercharges. I discuss simplified analogs of the superspace and superfield formalism. This formalism allows one to construct supersymmetric systems with desired properties in an "industrial way". I explain how it works in quantum mechanics. Then the spontaneous breaking of supersymmetry is considered. General aspects of the phenomenon as well as specific examples are discussed. Much attention is paid to Witten's index, a powerful criterion playing a distinguished role in the studies of the spontaneous SUSY breaking. The concluding part presents a brief encounter with basic supersymmetric field theories. The lecture is intended to be an introduction to general supersymmetric field theory.

Contents

Introduction

If there existed a list of ideas whose impact on high energy physics of the last two decades was the strongest, supersymmetry would probably occupy one of the first lines, perhaps, even the first. It seems most probable that the model that will come after the Standard Model, *the next model of nature*, will be based on supersymmetry. Even if we leave this pragmatic aspect aside, it is easy to understand why supersymmetric theories are so cherished in modern physics. It is impossible not to appreciate the novelty of the construction which connects boson and fermion degrees of freedom by a symmetry transformation. A closer look at the subject opens to newcomers absolutely fascinating and miraculous properties of the supersymmetric dynamics which, I am sure, will amuse people for years to come. In this lecture you will get acquainted with the basic notions and properties of supersymmetric theories in the simplest possible example, supersymmetric quantum mechanics.

Supersymmetry (or SUSY, as it is tenderly called by its practitioners) was born in the beginning of the seventies [1, 2] and has grown now into an industry with many branches. The number of works devoted to SUSY is counted in the thousands, and continues to grow steadily. Now, even experts in the field have trouble orienting themselves in the fast flow of literature, sometimes treating only very specific details. Despite all this, the basic concepts of SUSY are very simple, and can be easily grasped by students at the level of the quantum mechanics class. In other words, all main features of supersymmetry can be demonstrated in quantum mechanics problems. Although this approach is pedagogical and does not lead to the very front line of modern research in SUSY field theory, these lectures will provide a solid introduction and should spur those who are interested to find out more about SUSY. Then, after these lectures, you can turn to textbooks and reviews devoted to four-dimensional supersymmetric field theories in both formalism and application.

This subject, supersymmetric quantum mechanics (invented by Witten [3]), is interesting not only as a pedagogical introduction to SUSY field theory, but also by itself. Among other findings, let me mention an interesting suggestion [4] that SUSY may be realized in atomic systems as a valid symmetry for classifying the spectrum of the alkali–metal atoms. Moreover, it plays a role [5] in expanding the class of the so-called quasi-exactly solvable quantal systems. Such old and famous problems as, say, the hydrogen atom, can be reformulated in the language of SUSY [6], which may provide new insight.

Various aspects of quantum mechanics which are naturally interpreted within SUSY are discussed in Refs. [7, 8, 9]. My task is quite different, however. I am less concerned with the specific quantum-mechanical applications and try to emphasize those features of SUSY which are of the general nature and are equally valid in quantum mechanics *and* field theory. Thus, you may view these lectures as a preparation to the course of supersymmetric field theory. To facilitate the transition at the end I will briefly discuss some of the simplest field-theoretic models.

The original motivation for introducing SUSY in particle physics was the hope of solving two extremely important and acute problems: the problem of the mass hierarchy and the problem of the cosmological term. Both issues seem to be related to the structure of nature at very short distances and are still open despite a large number of SUSY models suggested to explain them. Of course, the realistic formulation of these problems is far beyond the scope of these lectures; the essence, however, can be explained in the toy example of quantum mechanics. Quantum mechanics provides an excellent environment for discussing the vanishing of the ground state energy, exact degeneracy of the excited states, and other general features of SUSY which are not shadowed here by such unpleasant complications as renormalizations and multiple Lorentz indices — indispensable elements of any field theory. Quantum mechanics can be viewed as a limit of the field theory with the number of spatial directions tending to zero. Using this observation we will introduce the notions of the "superfield" and "superspace" in the context of quantum mechanics, the basis of the mathematical apparatus of SUSY.

It is curious that the vanishing of the vacuum energy in the free field theory with the equal number of the boson and fermion degrees of freedom with degenerate masses was noted by Pauli back in 1950 [10]. Later on, in the sixties, Lipkin made an attempt [11] to introduce a symmetry connecting the octet of mesons with that of baryons, a germ of supersymmetry. The idea was abandoned, however, with no consequences.

Our discussion opens with the description of the most primitive SUSY system one can construct — one-dimensional supersymmetric quantum mechanics. Then we will study the general consequences of SUSY and eventually will come to the famous Witten index theorem. At the end of the quantum-mechanical part a dimensionally reduced Wess-Zumino model will be considered as an exercise for training. In the concluding part some simple facts about supersymmetric field theories will be presented.

1. SUSY Quantum Mechanics

The simplest system admitting the notion of SUSY, and all its attributes, was suggested by Witten [3]. An extremely elegant result on the impossibility of spontaneous SUSY breaking under certain conditions, the so-called index theorem, also belongs to him [12]. SUSY quantum mechanics was studied later from various points of view in Refs. [13, 14] (see also Refs. [4–9]). It turns out that a quantum-mechanical problem known to everybody — the one-dimensional motion of an electron in a magnetic field of a special type — possesses SUSY!

In the above I have mentioned that supersymmetry connects bosons and fermions, and you might wonder how this statement corresponds to the one-dimensional problem just mentioned. In the real world of relativistic particles living in four space-time dimensions the bosons and fermions form two distinct and easily recognizable classes of objects: those in the first class have integer spins, Bose statistics, and their canonic quantization procedure involves field commutators. In the second class we deal with half-integer spins, Fermi statistics and anticommutators. In the toy one-dimensional examples to be considered below there is no place, of course, to the rotation transformations, and the issue of integer spins versus half-integer plays no role. It is quite clear that the "boson" and "fermion" degrees of freedom, in the sense we ascribe to these words in the four-dimensional field theory, do not exist in one dimension. We can introduce, however, an operator which divides all quantum-mechanical states into two classes. For the electron in the magnetic field these classes consist of spin-up and spin-down states of the electron, respectively. Somewhat conditionally, I will continue to call these classes "bosonic" and "fermionic" keeping in mind the terminology accepted in field theory. To avoid confusion where necessary I will write the corresponding terms in the quotation marks.

Consider the following Hamiltonian:

$$H = \frac{1}{2}\left(\frac{p^2}{m} + W^2(x) + \sigma_3 \frac{1}{\sqrt{m}}\frac{dW}{dx}\right) \tag{1}$$

where $\hbar$ is set equal to one, p is the momentum operator, σ_3 is the third Pauli matrix, and $W(x)$ is an arbitrary function. Physically, this Hamiltonian describes the motion of the electron along the x-axis in the magnetic field which has only the z component and depends only on x. Indeed, let us compare Eq. (1) with the Schrödinger equation for the electron in the external magnetic field

$$H = \frac{\vec{p}^2}{2m} + \frac{ie}{2m}\text{div}\vec{A} - \frac{e}{m}\vec{A}\vec{p} + \frac{e^2}{2m}\vec{A}^2 + \frac{|e|}{2m}\vec{\sigma}\vec{B}\,. \tag{2}$$

If the first and third component of the vector potential are zero, and the second depends only on x,

$$A_1 = A_3 = 0, \;\; A_2 = \frac{\sqrt{m}}{|e|}W(x)\,,$$

then the general expression (2) immediately reduces to (1) (see e.g. Ref. [15]). It is absolutely essential that the magnetic moment appearing in the last term in (2) is the normal electron magnetic moment. If this is not the case, (2) would not reduce to (1).

The state of the system with the Hamiltonian (1) is described by the two-component spinor,

$$\Psi(x) = \psi(x)\begin{bmatrix}\alpha(x)\\ \beta(x)\end{bmatrix} \tag{3}$$

where $\psi(x)$ is the coordinate wave function. If $\alpha = 1,\ \beta = 0$ the electron spin is directed up, $|\uparrow\rangle$, and just the opposite for $\alpha = 0,\ \beta = 1$. Notice that σ_3 commutes with the Hamiltonian; in other words, the spin is strictly conserved. If, at the initial moment of time, the system is in a pure spin state, say $|\downarrow\rangle$, we may be sure that the evolution of the system leaves it in the same spin state at any time.

The mathematical formulation of SUSY — and this is the central point directly generalizable to field theory — is as follows. The Hamiltonian (1) is representable in the form of the square of an operator called supercharge. In the case at hand there exist two supercharges built from the momentum operator and the Pauli matrices,

$$Q_1 = \frac{1}{2}\left(\sigma_1\frac{p}{\sqrt{m}} + \sigma_2 W\right)\,,$$

$$Q_2 = \frac{1}{2}\left(\sigma_2\frac{p}{\sqrt{m}} - \sigma_1 W\right)\,. \tag{4}$$

It is not difficult to check that they satisfy the following algebra:

$$\{Q_i Q_j\} = \delta_{ij}H, \;\; [Q_i H] = 0, \;\; i,j = 1,2\,. \tag{5}$$

The fact that the algebra of the supercharges involves the anticommutator, along with the normal commutator for Q_i and H, indicates the fermion nature of these operators. Clearly, this is an unusual algebra. Such algebras, which include commutators *and* anticommutators, are called *graded* algebras by mathematicians. Q_i's are called odd elements of the graded algebra (they require anticommutators); H is an even element.

An explanatory remark is in order here. A consistent theory of fermions can only be derived from field theory. In field theory we learn that bosonic operators are characterized by commutators, while fermionic operators are characterized by anticommutators. Still, even in quantum mechanics there exists a hint of this sign distinction. Indeed, we know that the wave function of identical bosons must be symmetrized while that of fermions antisymmetrized. If you continue with quantum mechanics into field theory, you will see that the operators of the supercharge belong to the *spinor* representation of the Lorentz group.

Both supercharges are strictly conserved,

$$\frac{d}{dt}Q_{1,2} = -i[Q_{1,2}, H] = 0\,. \tag{6}$$

Of course, the vanishing of the above commutators is obvious since

$$H = 2Q_1^2 = 2Q_2^2\,. \tag{7}$$

2. Properties of the System

Let us assume, for simplicity, that the function $W(x)$ grows at infinity,

$$|W(x)| \to \infty \text{ as } x \to \infty\,.$$

This condition excludes from consideration continuous spectrum, leaving only discrete levels.

First of all, let us demonstrate the vanishing of the ground state energy — one of the most basic consequences of the unbroken SUSY. When we say that SUSY is unbroken, we do not mean here the possibility of the explicit breaking, since the anticommutation relation (5) is assumed to be valid in the operator sense. However, the symmetry of the Hamiltonian does not guarantee, generally speaking, that the spectrum will have the same symmetry since the spontaneous symmetry breaking can occur. (A well-known example is spontaneous magnetization: spin interactions in metals are rotationally invariant and still, in some metals, the ground state corresponds to all spins aligned along one and the same direction, which obviously breaks the rotational invariance.) I hasten to add that a detailed discussion of this issue is deferred until Sec. 4; we will just assume for now that the spontaneous breaking of SUSY is absent. Mathematically this fact is equivalent to the following requirement: the supercharge operators must annihilate the ground state which I will sometimes call "vacuum" in analogy with field theory. Notationally,

$$Q_1|0\rangle = Q_2|0\rangle = 0\,. \tag{8}$$

To see that this is indeed the case, we need to combine two facts: first, Eq. (7), second, the fact that both supercharges are Hermitean. Then, one can write that

$$H = 2Q_1^+Q_1\,,$$

and it becomes obvious that the energy of any normalizable state is positive. If an arbitrary state is denoted by $|a\rangle$, its energy

$$E_a = \langle a|H|a\rangle = 2\langle a|Q_1^+Q_1|a\rangle = 2\langle b|b\rangle \geq 0\,,$$

where

$$|b\rangle = Q_1|a\rangle\,.$$

The absolute minimum, the zero energy, is achieved only in the case $|b\rangle \equiv 0$, which is equivalent to Eq. (8) and corresponds to supersymmetric "vacuum". If $|b\rangle \neq 0$ there exist two degenerate "vacuum" states, $|a\rangle$ and $|b\rangle$ (see below) connected through supercharge, a situation typical of the spontaneous symmetry breaking. (For those familiar with conventional "boson" symmetries, I add another remark illustrating that the equation

$$Q_{1,2}|0\rangle = 0$$

is standard for an unbroken symmetry. Let us turn, say, to the rotational invariance and recall that the rotational invariance of the ground state means that the rotated state coincides with the original one,

$$\exp(i\omega^a L^a)|0\rangle = |0\rangle\,,$$

where L^a are the generators of the rotations (i.e. the angular momentum operators), $n^a = \omega^a/|\omega|$ is the axis around which we rotate and $|\omega|$ is the rotation angle. Assuming that $|\omega|$ is small and expanding the exponent in a power series, we see that

$$L^a|0\rangle = 0\,. \tag{9}$$

If we now take into account the fact that Q_i are the generators of the supertransformations, the parallel between (8) and (9) becomes clear.)

Using the explicit expression for the supercharges (4) let us find the wave functions satisfying the conditions (8). As a matter of fact, since $Q_2 = -i\sigma_3 Q_1$, it is quite sufficient to solve only the first equation, $Q_1|0\rangle = 0$, which, as it is easy to see, reduces to

$$\frac{d\Psi}{dx} = \sqrt{m}W(x)\sigma_3\Psi(x)\,. \tag{10}$$

This first-order differential equation is trivially solvable,

$$\Psi(x) = C\exp\left(\int_0^x \sqrt{m}W(y)\sigma_3 dy\right) \times (|\uparrow\rangle \text{ or } |\downarrow\rangle)\,, \tag{11}$$

where C is a normalization constant. For the normalizability of the wave function, it is necessary and sufficient that $W(y)$ as $y \to \infty$ has opposite sign than the limit at $y = -\infty$. (An immediate consequence is the fact that $W(y)$ should have an odd number of zeroes — we will elaborate on this later). Now, say

$$W(y) = \text{const} \times y \;\text{ or }\; \text{const} \times y(y^2 - a^2)$$

are suitable functions from the point of view of normalizability. The sign of the constant plays no role. Indeed, if $W(y \to \infty) > 0$, we choose the spin-down state in Eq. (11); then σ_3 in the exponent reduces to -1, and in the opposite case of the spin-up state, of course, $\sigma_3 \to +1$.

Since $Q_i\Psi = 0$ and the Hamiltonian $H = 2Q_1^2 = 2Q_2^2$, the state we have just built is obviously the eigenstate of the Hamiltonian with zero energy. Moreover, this guarantees that — since the spectrum is nonnegative, see above — the state (11) is the ground state of the system, the vacuum.

The assertion of the vanishing ground state energy in the case of the unbroken SUSY is general — it is valid not only in quantum mechanics, but also in field theory. The desire to get the vanishing vacuum energy in the field theory was one of the strongest motives for introducing SUSY models in the phenomenology of quarks and leptons.

Let us trace another general aspect now, a mutual cancellation of corrections to the energy of the "boson" and "fermion" nature in perturbation theory.

The problem is formulated as follows. Assume that the potential of the system at hand contains a small parameter, and instead of calculating the ground state energy exactly, we will try to find it as an expansion in this small parameter, order by order. (In field theory, perturbative corrections are associated with loop Feynman graphs and are referred to as loops. Accordingly, I will sometimes speak of boson and fermion loops). We will discuss in detail how the zero vacuum energy emerges in perturbation theory using as an example the following $W(x)$:

$$W = \sqrt{m}\omega x(1 - \lambda x^2)$$

where ω is a normalization constant (it will become clear shortly why it is denoted ω), and λ is a very small coupling constant. The plot of $W^2(x)$ is given in Fig. 1.

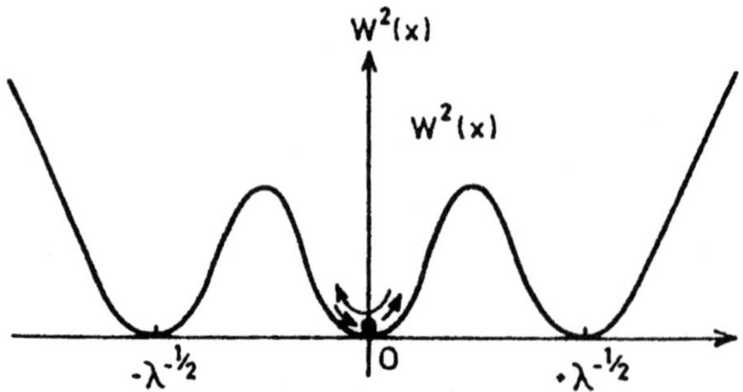

Fig. 1. The profile of the potential energy in Hamiltonian (1).

At the classical level the system placed at one of the minima $-\lambda^{-1/2}$, 0 or $+\lambda^{-1/2}$ (for definiteness, we have chosen the central minimum in Fig. 1), will stay at rest, and its energy will be evidently equal to zero. In quantum mechanics one should take into account the zero-point oscillations contributing to the ground state energy even in the limit of no anharmonicity, $\lambda = 0$. In this limit the "boson" term in the Hamiltonian (1) reduces to $(1/2)(m\omega^2 x^2)$ (the factor ω is introduced in such a way as to yield a standard expression for the Hamiltonian of the harmonic oscillator). The energy of the zero-point oscillations is clearly $\omega/2$. The "fermion" term in the Hamiltonian

$$\frac{\sigma_3}{2}\frac{1}{\sqrt{m}}\frac{dW}{dx} = \frac{\sigma_3}{2}\omega(1 - 3\lambda x^2) \to \frac{\sigma_3}{2}\omega + \mathcal{O}(\lambda). \qquad (12)$$

Now, choosing the negative eigenvalue of σ_3 (i.e. $\Psi(x) \sim |\downarrow\rangle$) one readily convinces oneself that the energy of the ground state

$$E_0 = \frac{\omega}{2} - \frac{\omega}{2} + \mathcal{O}(\lambda). \tag{13}$$

It is instructive to continue the exercise further to check that a similar cancellation takes place at the level $\mathcal{O}(\lambda)$ as well. Indeed, the first-order correction in λ has the form

$$\Delta E = \lambda \int_{-\infty}^{+\infty} dx \left(\frac{m\omega}{\pi}\right)^{1/2} \mathrm{e}^{-m\omega x^2} \left\{-m\omega^2 x^4 + \frac{3}{2}\omega x^2\right\}, \tag{14}$$

where the exponent represents the square of the wave function for the harmonic oscillator in the ground state, while the first and second terms in the braces are due to the anharmonicity of W^2 and dW/dx, respectively. Performing the integral in Eq. (14) we get zero. One can continue this exercise and calculate ΔE in higher orders of perturbation theory. The result is going to be zero in any finite order.

Finally, the last point to be discussed in this section is the degeneracy of all levels with $E \neq 0$. The pairings of levels is one of the manifestations of SUSY. Each "boson" state in the spectrum (with the possible exception of the ground state) is accompanied by a "fermion" partner with the same energy, and *vice versa.* (I remind that the "boson" and "fermion" states in one-dimensional quantum mechanics are just symbolic names; let us call the states of the type $|\downarrow\rangle$ boson while those of the type $|\uparrow\rangle$ fermion). Schematically the energy spectrum is depicted in Fig. 2.

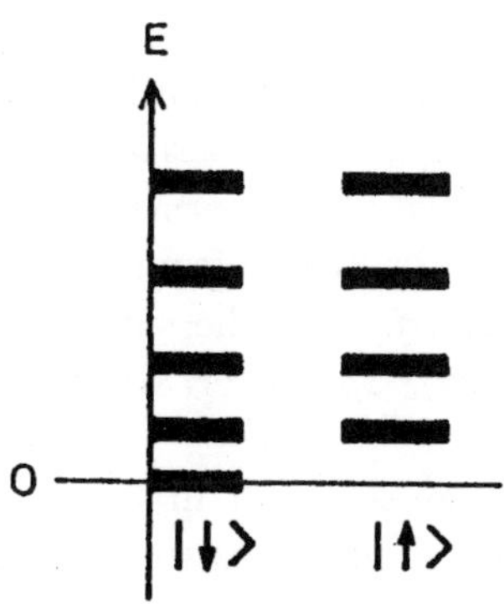

Fig. 2. The energy eigenvalues of the quantum-mechanical system with Hamiltonian (1).

Let $\Psi_{\rm bos}$ be an eigenfunction of the Hamiltonian (1) corresponding to the eigenvalue E. Then $(E/2)^{-1/2}Q_1\Psi_{\rm bos}$ is also the eigenfunction of the Hamiltonian with the same eigenvalue, but of the fermion type,

$$\Psi_{\rm ferm} = \sqrt{\frac{2}{E}}Q_1\Psi_{\rm bos}\,.$$

The factor $(E/2)^{-1/2}$ is needed for the proper normalization. Let us start from the second assertion. Since the supercharge, Q_1, contains only antidiagonal Pauli matrices, $\sigma_{1,2}$, it is quite obvious that Q_1 transforms the spinor of the type $|\downarrow\rangle$ into $|\uparrow\rangle$ and *vice versa*, i.e. boson into fermion, using the terminology accepted above. The first assertion is even more trivial since the supercharges $Q_{1,2}$ commute with the Hamiltonian.

In summary, the simplest quantum-mechanical system considered allows one to demonstrate the basic implications of SUSY:

(i) the vanishing of the ground state energy;

(ii) the exact cancellation of the corrections to the energy of the boson and fermion type in perturbation theory;

(iii) the exact degeneracy of the levels of the boson and fermion type (except possibly the ground state).

3. "Superspace"

At this point you might possibly ask the following questions: — "How could one guess that supersymmetric Hamiltonians exist in nature?" — "Isn't the Hamiltonian (1) unique?" — "If not, how one could construct Hamiltonians like (1)?"

Indeed, in the first works, the search for supersymmetric models was based on trial and error. Later, however, a systematic approach has been developed [17] leading to an industrial method of constructing SUSY Hamiltonians. The method is based on an extension of the normal space due to the introduction of additional fermion coordinates which form, together with the normal coordinates, the so-called *superspace.* In the superspace, the superalgebra can be realized linearly (see below). Moreover, the boson and fermion fields are combined in a single superfield.

As usual, the new language seems at first just a formal technical device facilitating the process of solving problems which could, in principle, be solved by traditional methods. Actually, the transition to the language of superspace and superfields means a new stage considerably expanding our possibilities.

In analogy, the invention of the Feynman graphs paved the way to the theory of renormalizations and was the beginning of mass calculations of radiative corrections in electrodynamics. The Feynman graphs became an indispensable part of modern high energy physics.

The superfield formalism in four-dimensional field theory will not be discussed here; those who are interested are directed to textbooks. Instead, we will treat quantum mechanics as a limit of field theory with zero number of spatial dimensions. Then space-time degenerates into time; correspondingly, superspace will degenerate into supertime. A generic field $\phi(t, \vec{x})$ becomes variable $\phi(t)$. Nevertheless, in our adaptation of the four-dimensional formalism to quantum mechanics, we will still frivolously use the terms "superspace" and "superfield", instead of the proper supertime and supervariable.

The simplest nonsupersymmetric quantum-mechanical problem is that of one degree of freedom $\phi(t)$; the corresponding action has the form

$$S = \int dt\, L(t), \quad L = \frac{1}{2}\left(\frac{d\phi}{dt}\right)^2 - V(\phi) \tag{15}$$

where $V(\phi)$ is the potential energy. Here, and below, the mass of the particle is set equal to 1. The supersymmetric generalization of time t is obtained by adding two Grassmann coordinates, θ and $\bar{\theta}$ ($\bar{\theta}$ can be considered as complex conjugate to θ).

Let me remind some basic facts about Grassmann numbers. (For deeper study I recommend the famous Berezin's book [16].) First, all formal arithmetical properties of these numbers are the same as those of ordinary numbers, with the only exception: the Grassmann numbers anticommute,

$$\theta_1\theta_2 = -\theta_2\theta_1\,. \tag{16}$$

This property, in particular, implies that the square of any Grassmann number vanishes, $\theta_1^2 = \theta_2^2 = 0$.

The differentiation is defined in an intuitively obvious way,

$$\frac{d}{d\theta_i}1 = 0, \quad \frac{d}{d\theta_i}\theta_j = \delta_{ij}, \quad \frac{d}{d\theta_i}(\theta_k\theta_l) = \delta_{ik}\theta_l - \delta_{il}\theta_k\,. \tag{17}$$

The last relation is a combination of the chain rule and the anticommutation property. One can also introduce the notion of the Grassmann integration. In order to preserve all standard properties of integrals, it is necessary and sufficient to assume that

$$\int d\theta_i = 0, \quad \int d\theta_i \theta_j = \delta_{ij}\,. \tag{18}$$

The two-fold integral is to be understood as the product of integrals, etc. Moreover, if C is a number, then

$$d(C\theta_i) = C^{-1} d\theta_i\,. \tag{19}$$

This property stems from the second equation (18). The delta function is also generalizable to the Grassmann variables. One can readily check that

$$\delta(\theta_1 - \theta_2)\delta(\bar{\theta}_1 - \bar{\theta}_2) = (\theta_1 - \theta_2)(\bar{\theta}_1 - \bar{\theta}_2)\,. \tag{20}$$

In spite of its polynomial form, Eq. (20) has all the properties of the conventional Dirac delta function. Indeed, for an arbitrary function, one has

$$\int d\bar{\theta}_1 d\theta_1 \delta(\theta_1 - \theta_2)\delta(\bar{\theta}_1 - \bar{\theta}_2) f(\theta_1, \bar{\theta}_1) = f(\theta_2, \bar{\theta}_2)\,. \tag{21}$$

Notice here that the ordering of the different factors in Eq. (21) is essential. Besides that, unlike the Dirac delta function,

$$\delta(0) = 0, \quad [\delta(\theta_1 - \theta_2)\delta(\bar{\theta}_1 - \bar{\theta}_2)]^2 = 0\,.$$

The Grassmann numbers θ and $\bar{\theta}$ have dimension $length^{1/2}$ while differentials $d\theta$ and $d\bar{\theta}$ are of dimension $length^{-1/2}$.

This introductory information about the Grassmann numbers is sufficient for our purposes. Now, instead of the variable $\phi(t)$, we introduce a supervariable $\Phi(t, \theta, \bar{\theta})$. (An analogous construction in field theory is called superfield, and below I will often use the same name).

The expansion of $\Phi(t, \theta, \bar{\theta})$ in $\theta, \bar{\theta}$ contains a finite number of terms since, as was already mentioned, $\theta^2 = \bar{\theta}^2 = 0$, a specific property of the Grassmann numbers. If one additionally imposes the requirement that Φ is real, $\Phi^* = \Phi$, then

$$\Phi(t, \theta, \bar{\theta}) = \phi(t) + \theta\bar{\psi}(t) + \psi(t)\bar{\theta} + D\theta\bar{\theta} \tag{22}$$

where ϕ and D are real variables of the boson type while ψ and $\bar{\psi}$ are of the fermion type. The coefficient in front of the highest possible power of θ — $\theta\bar{\theta}$ in the case at hand — is traditionally denoted by the letter D and is called the

D term. Analogous to the fact that the space-time (time in our case) linearly realizes the action of the translation generator,

$$t \to t + \tau, \tag{23}$$

the superspace allows one to realize linearly all SUSY generators. Under the supersymmetry transformations,

$$\theta \to \theta + \zeta, \;\; \bar{\theta} \to \bar{\theta} + \bar{\zeta}, \;\; t \to t + i\theta\bar{\zeta} - i\zeta\bar{\theta}, \tag{24}$$

where ζ and $\bar{\zeta}$ are the (Grassmann) parameters of the supertranslations.

Two consecutive transformations with parameters ζ_1 and ζ_2, as one can readily check, reduce to the transformation of the same type plus a time translation (23) with $\tau = i\zeta_2\bar{\zeta}_1 - i\zeta_1\bar{\zeta}_2$. Hence, as a matter of fact, we are dealing here with the realization of the supergroup. The realization (24) is not unique, though; we will return to this issue below.

Let us examine now how different components of the "superfield" $\Phi(t, \theta, \bar{\theta})$ change under supertransformations (24). Substituting Eq. (24) in the expansion (22) and limiting ourselves to the terms linear in ζ, $\bar{\zeta}$ we obtain:

$$\begin{aligned} \phi(t) &\to \phi(t) + \zeta\bar{\psi}(t) + \psi(t)\bar{\zeta}\,, \\ \psi(t) &\to \psi(t) - i\zeta\frac{d\phi(t)}{dt} + \zeta D, \\ \bar{\psi}(t) &\to \bar{\psi}(t) + i\bar{\zeta}\frac{d\phi(t)}{dt} + \bar{\zeta} D, \\ D(t) &\to D(t) - i\frac{d\psi(t)}{dt}\bar{\zeta} + i\zeta\frac{d\bar{\psi}(t)}{dt}. \end{aligned} \tag{25}$$

The terms quadratic in ζ, $\bar{\zeta}$ correspond to conventional time translations. Let us draw attention to the fact that the change of the D term is a full derivative. This property is not specific to the example considered. After some reflection, one can convince oneself that the last component of any superfield, real or chiral (see below), always changes by a full derivative under the supertransformations. Integrating the D term over space (in quantum mechanics, over time), we thus get a superinvariant expression. As a result, we arrive at a simple and efficient prescription for building superinvariant Lagrangians.

Two comments are in order here. First, any linear combination of superfields is again, a superfield. Second, the product of two superfields is also a superfield. If the first assertion is absolutely obvious, the second, perhaps, requires some explanation. It is instructive to check it explicitly, component by component. Let Φ_1 and Φ_2 be two real superfields (see (22)) and $\Phi = \Phi_1\Phi_2$. Then

$$\begin{aligned} \phi &= \phi_1\phi_2\,, \\ \bar\psi &= \bar\psi_1\phi_2 + \bar\psi_2\phi_1\,, \\ \psi &= \psi_1\phi_2 + \psi_2\phi_1\,, \\ D &= D_1\phi_2 + D_2\phi_1 + \bar\psi_1\psi_2 + \bar\psi_2\psi_1\,. \end{aligned} \tag{26}$$

From Eq. (25), we get, for instance, that

$$\begin{aligned} \delta\phi &= \phi_1\delta\phi_2 + \phi_2\delta\phi_1 = \phi_1(\zeta\bar\psi_2 + \psi_2\bar\zeta) + \phi_2(\zeta\bar\psi_1 + \psi_1\bar\zeta) \\ &= \zeta(\phi_1\bar\psi_2 + \phi_2\bar\psi_1) + \text{h.c.} = \zeta\bar\psi + \psi\zeta\,, \end{aligned}$$

i.e. just the transformation law for the lowest component of the superfield we need. Mathematically, this result — the product of two superfields is a new superfield — is explained by the fact that the SUSY transformations are linearly realized on superfields. It is also obvious that differentiating superfields over time, we again get superfields, $\partial\Phi/\partial t$. Differentiating over θ, however, leads us out of the class of superfields. Thus, for instance,

$$\frac{\partial\Phi(t,\theta,\bar\theta)}{\partial\theta}$$

is an object which by no means can be claimed to have components linearly transforming through each other under transformation (24). One can, however, correct the spinor derivative $\partial/\partial\theta$, amending it by an extra term and making it covariant. The covariant spinor derivative is denoted by $\mathcal{D}$ and plays an important role in the superfield formalism,

$$\mathcal{D} = \frac{\partial}{\partial\theta} + i\bar\theta\frac{\partial}{\partial t}\,. \tag{27}$$

Now, $\mathcal{D}$ is constructed in such a way that $\mathcal{D}\Phi$ is also a superfield,

$$\mathcal{D}\Phi = \bar\psi + (i\frac{d\phi}{dt} + D)\bar\theta\,, \tag{28}$$

i.e. its components linearly transform through each other under (24). The argument of $\bar\psi$ and all functions on the right-hand side of Eq. (28) is

$$t - i\theta\bar\theta\,. \tag{29}$$

Here, we encounter a new type of superfield — chiral — and a new realization of SUSY in the superspace. The chiral superfields depend explicitly only on θ or only on $\bar{\theta}$. An implicit dependence on $\theta\bar{\theta}$ enters through the chiral argument (29).

Let us see how the components of (28) transform under the action of the supergenerators. First of all, substituting (24), we convince ourselves that

$$(t - i\theta\bar{\theta}) \to (t - i\theta\bar{\theta}) - 2i\zeta\bar{\theta} - i\zeta\bar{\zeta}\,. \tag{30}$$

In other words, the variation of the combination (29) contains only $\bar{\theta}$ not θ. It is for this reason that $\mathcal{D}\Phi$ is a superfield while $\partial\Phi/\partial\theta$ is not. Using (24) and limiting ourselves to the terms linear in ζ, $\bar{\zeta}$ we find the transformation laws,

$$\delta\left(i\frac{d\phi}{dt} + D\right) = \frac{d\bar{\psi}}{dt}(-2i\zeta),\;\; \delta\bar{\psi} = \left(i\frac{d\phi}{dt} + D\right)\bar{\zeta}\,, \tag{31}$$

which absolutely correspond to Eqs. (25).

If the superfield at hand contains the explicit dependence only on θ,

$$A = \phi(t_{\text{ch}}) + \theta\bar{\psi}(t_{\text{ch}}),\;\; t_{\text{ch}} = t + i\theta\bar{\theta}\,, \tag{32}$$

then it is called chiral. In the opposite case (explicit $\bar{\theta}$ dependence,)

$$\bar{A} = \phi(t_{\text{ach}}) + \psi(t_{\text{ach}})\bar{\theta},\;\; t_{\text{ach}} = t - i\theta\bar{\theta}\,, \tag{33}$$

the field is referred to as antichiral. Notice that the components of the chiral and antichiral fields depend on different arguments, t_{ch} and t_{ach}, respectively. As was mentioned above, under the SUSY transformations,

$$\theta \to \theta + \zeta,\;\; \bar{\theta} \to \bar{\theta} + \bar{\zeta}\,,$$

$$t_{\text{ch}} \to t_{\text{ch}} + 2i\theta\bar{\zeta} + i\zeta\bar{\zeta},\;\; t_{\text{ach}} \to t_{\text{ch}} - 2i\zeta\bar{\theta} - i\zeta\bar{\zeta}\,. \tag{34}$$

It is worth emphasizing that the covariant derivative $\bar{\mathcal{D}}$ should be defined as

$$\bar{\mathcal{D}} = -\frac{\partial}{\partial\bar{\theta}} - i\theta\frac{\partial}{\partial t}\,. \tag{35}$$

At the same time, the expressions $\mathcal{D}\Phi$ and $\bar{\mathcal{D}}\Phi$ are complex conjugate to each other. As a home exercise, I ask you to check that the anticommutator of two covariant derivatives reduces to the ordinary derivative,

$$\{\mathcal{D}\bar{\mathcal{D}}\} = -2i\frac{\partial}{\partial t}\,. \tag{36}$$

Now, the question is: why, by applying the covariant derivative to the real superfield, do we end up with a chiral superfield? It is convenient to formulate the answer in two stages. First, let us give an alternative definition: the superfield A is called (anti)chiral if it satisfies the condition

$$\mathcal{D}\bar{A} = 0\,. \tag{37}$$

Actually, this is just a mathematical transcription of the fact that there is no explicit dependence on θ, i.e. this variable can enter only through the chiral argument for which

$$\mathcal{D}t_{\text{ach}} = 0\,. \tag{38}$$

(For the chiral field, one substitutes $\mathcal{D} \to \bar{\mathcal{D}}$ and $t_{\text{ach}} \to t_{\text{ch}}$). The condition (37) is obviously valid for $\bar{A} = \mathcal{D}\Phi$. Indeed, in this case,

$$\mathcal{D}\bar{A} = \mathcal{D}^2\Phi = 0\,,$$

since the square of the covariant derivative is automatically zero due to its Grassmann nature. The construction described above for quantum mechanics is also applicable, in its basic features, to field theory. The only distinction is technical: time is substituted by the four-vector x_μ, and the variables θ will carry spinor indices.

4. Lagrangian of the SUSY Quantum Mechanics

Now all of our preparatory work has been completed, and we can finally turn to a discussion of a regular method of building supersymmetric theories. The prescription is very simple. First, we start from the expression (15) for the action in ordinary quantum mechanics. Substitute the "field" ϕ by the "superfield" (22), time derivatives by covariant derivatives, and the integral over time by that over supertime,

$$S_{\text{SUSY}} = \int dt d\bar{\theta} d\theta \left[\frac{1}{2}\bar{\mathcal{D}}\Phi\mathcal{D}\Phi - F(\Phi)\right], \tag{39}$$

where $F(\Phi)$ is an arbitrary function of Φ, the so-called superpotential. Now, let us demonstrate that the action obtained in this way is superinvariant. Indeed, the expression in the brackets is a real superfield. Invoking the rules of integration over the Grassmann numbers (see (18)) we readily convince ourselves that the integral $\int d\bar{\theta}d\theta$ singles out the D component of the superfield — all others give no contribution in Eq. (39). Moreover, the variation of the D component under the SUSY transformation reduces to a full derivative over time — a fact, which has been emphasized previously (see (25)). The full derivative over time drops out upon the dt integration. Thus, the Lagrangian stemming from (39) is automatically invariant under the supersymmetry transformations, up to a full derivative.

Perhaps, it may not be immediately evident that the action (39) describes the same quantum-mechanical system which was discussed in detail in Sec. 1 (see Eq. (1)). In particular, one might ask:

— Where is the kinetic term in Eq. (39) and is it properly normalized? — What is the relation between the functions F and W?

To answer these question, it is instructive to rewrite Eq. (39) in the component form. The exercise will be helpful also in mastering the superfield formalism which might seem superfluous in quantum mechanics, but becomes a necessity in SUSY field theories. (Likewise, in conventional field theory, both classic and quantum, one could hardly expect considerable progress if one had to write down all tensor equalities with explicit Lorentz indices, in the component form). Using Eq. (28) for $\mathcal{D}\Phi$ (and complex conjugate for $\bar{\mathcal{D}}\Phi$) and passing from the dt integration to that over dt_{ach} (a linear change of variables), we obtain

$$\frac{1}{2}\int dt d\bar{\theta} d\theta \bar{\mathcal{D}}\Phi\mathcal{D}\Phi$$

$$= \frac{1}{2}\int dt d\bar{\theta} d\theta \left[\psi + \left(-i\frac{d\phi}{dt} + D\right)\theta\right]_{t+2i\theta\bar{\theta}} \left[\bar{\psi} + \left(i\frac{d\phi}{dt} + D\right)\bar{\theta}\right]_t$$

$$= \int dt \left[\frac{1}{2}\left(\frac{d\phi(t)}{dt}\right)^2 + \frac{1}{2}(D(t))^2 + i\frac{d\psi(t)}{dt}\bar{\psi}(t)\right] . \tag{40}$$

Furthermore,

$$\int dt d\bar{\theta} d\theta F(\Phi) = \int dt [F'(\phi)D + F''\bar{\psi}\psi]_t \,, \tag{41}$$

where F' and F'' are the derivatives of the function F over ϕ. Combining Eqs. (41) and (40) we arrive at the following Lagrangian:

$$L_{\mathrm{SUSY}} = \frac{1}{2}\left(\frac{d\phi(t)}{dt}\right)^2 + \frac{1}{2}D^2 - i\bar{\psi}\frac{d\psi}{dt} - F'D - F''\bar{\psi}\psi \,. \tag{42}$$

Notice that the D term enters without derivatives; in other words, it can be eliminated via equations of motion. The fact that the last component of the superfield will enter the Lagrangian without derivatives was obvious from the very beginning. This fact is universal. In field theory the last components of the superfield (D for real superfield and F for chiral) also appear without derivatives and are, thus, auxiliary nondynamical variables to be excluded through the equations of motion. Eliminating D, we get

$$L_{\rm SUSY} = \frac{1}{2}\left(\frac{d\phi(t)}{dt}\right)^2 - \frac{1}{2}(F'(\phi))^2 - i\bar{\psi}\frac{d\psi}{dt} - F''(\phi)\bar{\psi}\psi\,. \tag{43}$$

In principle, this expression for the Lagrangian of SUSY quantum mechanics is suitable for work (see e.g. Ref. [13]). We will make one step further, however, which will make comparison with the Hamiltonian (1) completely trivial. Thus, our task is to proceed to the matrix representation for the fermion variable ψ. Fortunately, this procedure is well-studied since the inverse transition — from the n-particle Schrödinger equation to the field theory prescription — lies in the basis of second quantization and is discussed in detail in classic textbooks (see e.g. Ref. [18]). In the case at hand, the solution looks especially simple, since we deal with only one fermion variable ψ and one complex conjugate $\bar{\psi}$. All one has to do is to explicitly realize anticommutation relations:

$$\{\psi(t)\psi(t)\} = \{\bar{\psi}(t)\bar{\psi}(t)\} = 0,\ \ \{\psi(t)\bar{\psi}(t)\} = 1\,.$$

To this end, one can choose $\bar{\psi}$ and ψ as follows:

$$\bar{\psi} = \begin{pmatrix} 0 & 1 \\ 0 & 0 \end{pmatrix} = \sigma_+,\ \ \psi = \begin{pmatrix} 0 & 0 \\ 1 & 0 \end{pmatrix} = \sigma_-\,. \tag{44}$$

To take into account the Grassmann nature of the variable ψ let us rewrite the last term in (43) as $\frac{1}{2}[\bar{\psi}\psi](-F'')$ and proceed from Lagrangian to Hamiltonian:

$$H_{\rm SUSY} = \frac{1}{2}p^2 + \frac{1}{2}(F')^2 + \frac{1}{2}\sigma_3 F''\,. \tag{45}$$

Now, if one identifies F' with W (and $m = 1$), the expressions (45) and (1) coincide. Concluding the section, let us notice that the Lagrangian (45) possesses a conserved charge

$$q = \bar{\psi}\psi\,,$$

the fermion number of the state considered. Conservation of q stems immediately from the equations of motion

$$\frac{d\psi}{dt} = iF''\psi,\ \ \frac{d\bar{\psi}}{dt} = -iF''\bar{\psi}\,.$$

In the matrix representation, the charge q reduces to

$$q \to \frac{1}{2}(1+\sigma_3)\,.$$

Thus, the convention accepted in Sec. 1 according to which the states of the type $|\downarrow\rangle$ are considered bosonic, and $|\uparrow\rangle$ as fermionic, finds its explanation.

5. The Spontaneous Breaking of SUSY

In Sec. 1 we found the solution of the equation

$$Q_{1,2}\Psi = 0\,; \tag{46}$$

the corresponding wave function describes the eigenstate of the Hamiltonian with zero energy, the vacuum. The solution was normalizable under certain conditions on the function $W(x)$. Namely, the normalizability requires the signs of $W(x)$ at plus and minus infinity to be opposite, i.e. $W(x)$ should have an odd number of zeros. The question is what happens if the condition is not satisfied. An example of such a situation is given in Fig. 3.

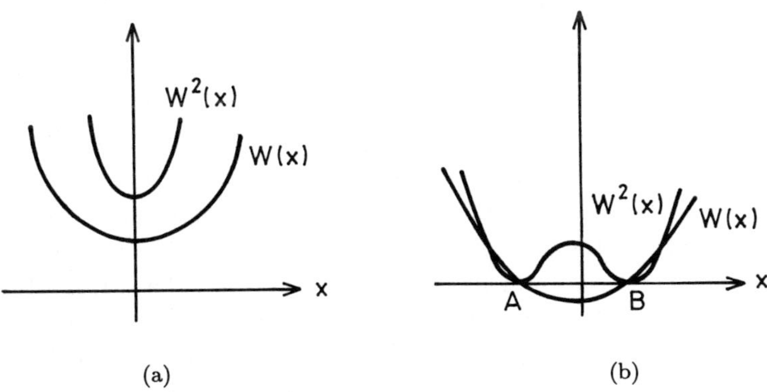

Fig. 3. An example of the function $W(x)$ corresponding to the spontaneous breaking of SUSY.

The formal solution of Eq. (46) given in (11) is now nonnormalizable, since, for any choice of spin state, $|\downarrow\rangle$ or $|\uparrow\rangle$, the exponent in

$$\exp\left(\sqrt{m}\sigma_3\int_0^x W(y)dy\right)$$

is positive either at plus or minus infinity, and the function grows. In other words, there is no normalizable state invariant under the SUSY transformations and, hence, we deal here with the spontaneous symmetry breaking. We know already that for the unbroken supersymmetry the ground state energy vanishes. On the contrary, if SUSY is spontaneously broken, i.e. $Q_{1,2}|0\rangle \neq 0$, then

$$\langle 0|H|0\rangle = 2\langle 0|Q_1 Q_1|0\rangle = 2\langle 0'|0'\rangle > 0, \tag{47}$$

where $|0'\rangle = Q_1|0\rangle$. Hence, the positivity of the vacuum energy is the *necessary and sufficient condition* for the spontaneous breaking of SUSY. Let us return now to the potential depicted by Figs. 3(a) and 3(b). The lowest state in these potentials corresponds to a positive energy — not surprising for the profile of the potential energy given in Fig. 3(a). Indeed, here, even at the classical level, the lowest possible energy is positive,

$$E_{\text{vac}} = \frac{1}{2} W^2(x=0)\,.$$

If the anharmonicity is small, the quantum correction certainly cannot shift the value of E_{vac} down to zero.

As for the potential depicted in Fig. 3(b), in this case, the situation is not so trivial. At the classical level, two states of the system possess zero energy (they are marked by letters A, B in Fig. 3(b)). Moreover, we have demonstrated above that starting from either of these two states and calculating the energy as a series in the coupling constant we get zero order by order in perturbation theory. Indeed, the analysis of Sec. 1 is based only on the fact that $W(x)$ has a zero at $x = x_0$ and that $W(x)$ is expandable in the Taylor series in $(x - x_0)$. If the zero of W at $x = x_0$ is the only one, then SUSY is unbroken. Furthermore, to any finite order, only the behavior of $W(x)$ in a small neighborhood of x_0 is essential, and the existence (or nonexistence) of other zeros, crucial for the occurrence of the normalizable zero energy level, in no way manifests itself in perturbation theory. It seems that one may rejoice: the ground state energy is exactly at zero in perturbation theory and, hence, there is no spontaneous breaking of supersymmetry. Let us not make hasty conclusions, however. The argument presented in the beginning of this section proves that the exact spectrum of the quantum problem corresponding to Fig. 3(b), contains no supersymmetric ground state. Logically, we then have to conclude that nonperturbative effects give a nonvanishing contribution shifting the vacuum energy from zero to a positive value.

What do we learn from the above examples? First of all, in the weak coupling regime, there exist two possible scenarios for the spontaneous SUSY breaking:

(i) the classical potential energy never vanishes, and then $E_{\text{vac}} > 0$ in the classical approximation;

(ii) if at the classical level $E_{\text{vac}} = 0$, then SUSY is preserved in perturbation theory, and the spontaneous breaking of supersymmetry can take place only due to nonperturbative effects (for instance, instantons, see e.g. Refs. [19, 20]).

Second, the fact of breaking (or conservation) of supersymmetry depends only on the global features of the potential energy, such as the sign of asymptotics of $W(x)$, or the overall number of zeroes. Continuous deformations of the function $W(x)$ (localized in x) do not change the situation in this respect. Thus, all continuous potentials can be divided in classes, and if one is interested only in the spontaneous breaking of SUSY, one is free to vary $W(x)$ arbitrarily inside the given class. Just this observation, perhaps, was a hint which motivated Witten to introduce a topological characteristic (later called Witten's index) independent of particular values of parameters like masses, coupling constants, normalization volumes in four-dimensional theories, etc., serving as an indicator for the spontaneous breaking of SUSY. (This characteristic will be discussed in detail in Sec. 6.) The more one thinks the idea over, the more elegant the idea seems. Indeed, one can change the theory under consideration, distorting and simplifying it to the extent that it does not resemble the original theory at all (provided, of course, certain rules of distortion are not violated). One may be sure, however, that if Witten's index I_W of the simplified version coincides with that of the original theory, and if $I_W \neq 0$, then the spontaneous SUSY breaking is absent.

We will defer this issue until the next section; meanwhile, a few remarks elucidating the difference between the ordinary internal symmetries and SUSY are in order. In the conceptual aspect, the main distinction is as follows. The transformations corresponding to internal symmetries, e.g. isotopic invariance, do not affect the space-time variables. The geometric properties of the states before and after transformations are the same. On the other hand, under the SUSY transformations, the bosons pass into fermions, i.e. the spin is changed. Spin is a geometric characteristic. This can be seen from the fact that two consecutive supertransformations are equivalent to a translation (the anticommutator of two supercharges reduces to Hamiltonian, see (5), which, in turn, generates a time translation). Roughly speaking, the SUSY transformation is a square root from translation. Correspondingly, energy plays the role of the order parameter; a nonvanishing vacuum average of the Hamiltonian signals the spontaneous breaking of supersymmetry.

Proceeding from generalities to technical consequences, it is not difficult to give a few examples emphasizing distinctions between the internal symmetries and SUSY. The most clear-cut example seems to be the following. If we deal with a finite number of degrees of freedom (e.g. field theory in a finite volume) internal symmetries cannot be spontaneously broken at all, strictly speaking. Say, in the scalar theory with the action

$$\int_V d^4x[(\partial_\mu\phi)^+(\partial_\mu\phi) - \lambda(|\phi|^2 - \eta^2)^2] \tag{48}$$

(the corresponding potential is depicted in Fig. 4), the ground state wave function is homogeneously smeared over the "bottom of the bottle", and the U(1) invariance of the action (48) is not broken at any finite volume V. As far as SUSY is concerned, in this case the spontaneous breaking can well take place in the finite volume field theory, and even in the limiting case of one degree of freedom, the quantum-mechanical system considered above.

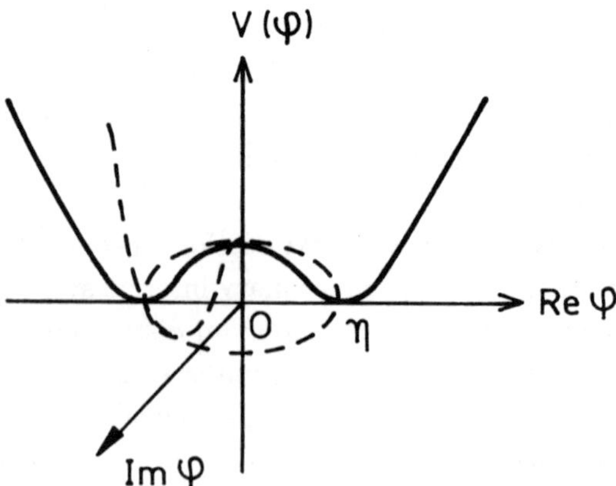

Fig. 4. A characteristic profile of the type 'bottom of a bottle' of the Higgs potential.

There are other essential differences as well. In particular, in the weak coupling regime (now I mean field theory in the infinite volume) the question of spontaneous breaking for the internal symmetry as a rule is decided at the classical level. Indeed, if in the classical approximation, the potential has the form presented in Fig. 4, small quantum corrections cannot smear the minimum corresponding to nonvanishing (large) value of $|\phi|$. The spontaneous breaking of U(1) established at the classical level persists in the quantum theory. On the contrary, for the potential of Fig. 5, with the minimum at the symmetric point $\phi = 0$, the U(1) symmetry is present both in the classical and in the exact theory, since the quantum corrections, being small in the weak coupling regime, cannot convert the potential of the type in Fig. 5 to that in Fig. 4. (Reservation: if, in the classical Lagrangian, there is an exact or approximate degeneracy, then even relatively weak quantum effects lifting the degeneracy can lead to spontaneous symmetry breaking).

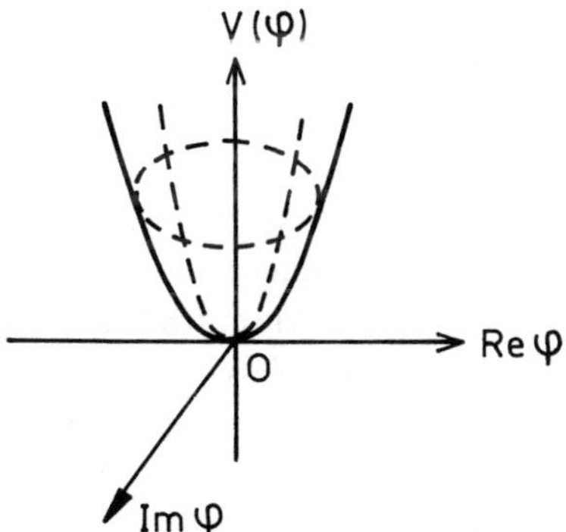

Fig. 5. The potential energy with one minimum.

Now, if SUSY is spontaneously broken at the classical level, it is broken in the exact theory as well, at least in the weak coupling regime. In this respect, the situation is just the same as with the conventional internal symmetries. On the other hand, if it is present at the classical level (then it exists to any finite order of perturbation theory), this does not mean anything since even exponentially small effects can lift the vacuum energy, thus resulting in the spontaneous SUSY breaking. This is certainly rather unusual. The quantum-mechanical example under consideration shows that this, indeed, happens.

6. Witten's Index

Considerations presented in this section are of general character and are equally applicable to quantum mechanics and field theory. An indirect method is described here, allowing one to show, in a very wide class of theories, that the ground state has exactly the zero energy, and hence, SUSY is unbroken. Thus, the range of theories in which the spontaneous SUSY breaking is possible is narrowed down.

In order to show that there is no spontaneous breaking of SUSY, it is sufficient to demonstrate that in the given theory there exists at least one state with the vanishing energy which will then be the supersymmetric vacuum. Following Witten, let us try to find a topological invariant of the theory — a number which would not depend on the volume of the system, coupling constants, and other similar parameters, but would be determined only by the fact of presence (absence) of the zero energy states.

First of all, let us notice that if a state with $E = 0$ exists for any finite volume, by continuity, this state will stay at zero in the limit $V \to \infty$. Therefore in any field theory it is sufficient to consider the spectrum in the finite volume, find a level with $E = 0$, and then one can be sure that SUSY in the original theory is unbroken. However, the finite volume field theory is equivalent to quantum mechanics of many degrees of freedom. Thus, in the most generic formulation, the problem reduces to that from quantum mechanics. Furthermore, let us assume that the Hamiltonian depends on parameters — coupling constants λ , masses m, and so on — and then examine the change of the spectrum under variations of these parameters. (They can vary arbitrarily provided that algebra (5) holds). If the parameters are changing, the levels of the system start moving, they "breathe". In particular, nonzero levels can come down to zero, and the zero energy levels can go up. It is important, however, that both descent and ascent of the levels takes place in pairs, boson–fermion, since, as we already know, all levels with $E > 0$ are paired. This means that the following quantity is invariant under the change of the parameters

$$I_W = \mathrm{Tr}(-1)^F = n_B^{E=0} - n_F^{E=0}\,, \tag{49}$$

where $n_B^{E=0}$ and $n_F^{E=0}$ are the numbers of the 'boson' and 'fermion' zero energy levels, respectively, and F is the fermion number. If $I_W \neq 0$, there exists a state (or states) with $E = 0$, and SUSY is unbroken. The basic idea is that the original Hamiltonian, often very complicated, can be transformed by a deformation of parameters to a simple form suitable for proof that $I_W \neq 0$.

Now, let us illustrate the general strategy in our quantum-mechanical problem. Any function $W(x)$ can be continuously deformed in the following way. The smooth curve is substituted by a few straight lines crossing the zeroes of $W(x)$ and tangent to $W(x)$ in these points. In other words, in the vicinity of each zero, we neglect nonlinear terms. Then, the straight lines are matched to each other by high barriers. For instance, from Fig. 3(b) we get in this way, a funnel shown in Fig. 6 by a solid line.

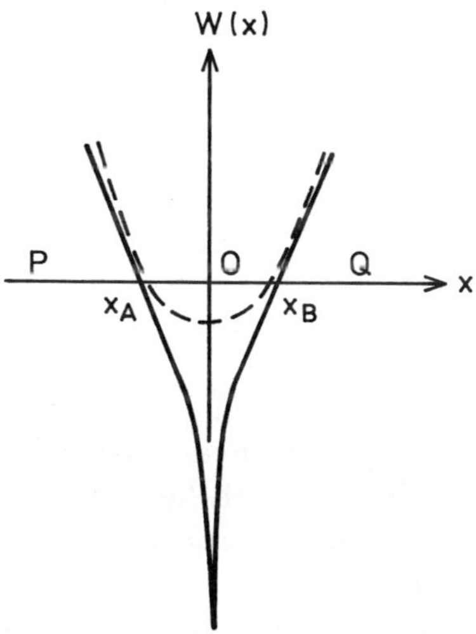

Fig. 6. A continuous deformation of the function $W(x)$. The original function (dashed line) is substituted by straight lines plus a delta-function like matching.

For positive values of x

$$W(x) = \sqrt{m}\omega(x - x_B)\,,$$

while for x negative

$$W(x) = \sqrt{m}\tilde{\omega}(x - x_A)\,,$$

where $\omega(\tilde{\omega})$ is a positive (negative) constant. The presence of a high barrier at $x = 0$ ensures that the wave function vanishes at the origin. Then the equation $Q_{1,2}\Psi = 0$ determining the supersymmetric vacuum has two solutions: one at

$x > 0$ and another at $x < 0$. Looking at Hamiltonian (1) we easily convince ourselves that in the first case (positive ω) the solution with the zero energy requires the spin wave function of the type $|\downarrow\rangle$, while in the second case (negative $\tilde{\omega}$) we have $|\uparrow\rangle$. Let me remind that according to our convention (see Sec. 2) the spinor $|\downarrow\rangle$ is called boson ($F = 0$) and the spinor $|\uparrow\rangle$ is called fermion ($F = 1$). Consequently, the funnel of Fig. 6 implies $n_B^{E=0} = 1$, $n_F^{E=0} = 1$ and hence, $I_W = 0$.

Needless to say, the analysis sketched above for a particular choice of $W(x)$ is immediately generalizable to arbitrary $W(x)$. Within our deformation prescription each zero of the function $W(x)$ entails a zero energy level, boson or fermion, in accordance with the sign of ω. In the general case,

$$
\begin{aligned}
I_W &= n_B^{E=0} - n_F^{E=0} \\
&= (\#\text{ of zeroes of } W \text{ with } \omega > 0) - (\#\text{ of zeroes of } W \text{ with } \omega < 0) \\
&= \begin{cases} 1 \text{ if } W(x \to +\infty) > 0,\ W(x \to -\infty) < 0; \\ 0 \text{ if } W(x \to +\infty)/W(x \to -\infty) > 0; \\ -1 \text{ if } W(x \to +\infty) < 0,\ W(x \to -\infty) > 0. \end{cases}
\end{aligned}
\tag{50}
$$

Thus, the condition ensuring that $I_W \neq 0$ is exactly the condition guaranteeing the existence of the supersymmetric ground state in the exact solution as it was found in Sec. 1. The case with $I_W = 0$ requires a special treatment. The problem at hand corresponds to the spontaneous SUSY breaking, but this is not necessarily the general situation.

The example considered gives an idea of what deformations of the potential constitute the class of the allowed continuous deformations. Say, if $W = Cx(x^2 - a^2)$, we can change C and a and/or add a quadratic term in x without changing the number of zeroes of $W(x)$. More exactly, what is fixed is the difference between the number of zeroes with positive ω's and negative ω's, i.e. I_W. If, however, one allows to change the asymptotics of $W(x)$ by introducing, say, a term like bx^4, then a new zero appears coming from infinity; Witten's index experiences a jump. In other words, such a distortion of the Hamiltonian is to be considered singular and inadmissible.

I hope that now you can fully appreciate the elegance of the construction, the elegance obvious even in quantum mechanics, to say nothing about field theory applications. Practically, it is so important that it is worth reiterating the main points:

(i) Witten's index I_W is calculable in a relatively easy and absolutely reliable way for a wide class of supersymmetric theories;
(ii) if $I_W \neq 0$, there is no spontaneous breaking of SUSY;
(iii) if $I_W = 0$, two scenarios are logically possible: (A) $n_B^{E=0} = n_F^{E=0} = 0$ (spontaneous symmetry breaking); (B) $n_B^{E=0} = n_F^{E=0} \neq 0$ (SUSY is unbroken). The question of what pattern actually takes place requires an additional investigation;
(iv) if SUSY is spontaneously broken at the classical level, then in this approximation, $n_B^{E=0} = n_F^{E=0} = 0$ and thus, Witten's index in such a theory vanishes.

7. Exercise on Witten's Index

Find Witten's index for the system described by the Lagrangian

$$L = \frac{d\phi^\dagger}{dt}\frac{d\phi}{dt} + i\psi^\dagger\frac{d\psi}{dt} + F^\dagger F + \left[m\left(\phi F - \frac{1}{2}(\psi)^2\right) + g(\phi^2 F - \phi(\psi)^2) + \text{h.c.}\right], \quad (51)$$

where ϕ and F are complex variables, ψ is a *two-component* Grassmann variable, $\psi = (\psi^1, \psi^2)$ and $(\psi)^2 \equiv \psi^2\psi^1 - \psi^1\psi^2$. (This Lagrangian occurs under the reduction of the so-called Wess-Zumino model from 4 dimensions to 1. The Wess-Zumino model is the simplest supersymmetric field theory. Notice that F is an auxiliary variable entering without the kinetic term; thus, it can be eliminated via equations of motion).

8. Solution of the Exercise

First of all it is instructive to check that the model (51) is indeed supersymmetric, and write down the corresponding supercharges. Unlike the simplest example considered above, now we have four supercharges (and they are not Hermitean):

$$Q^1 = \sqrt{2}\left(\frac{d\phi^\dagger}{dt}\psi^1 - i\psi^{2\dagger}F\right), \quad Q^2 = \sqrt{2}\left(\frac{d\phi^\dagger}{dt}\psi^2 + i\psi^{1\dagger}F\right), \quad (52)$$

plus Hermitean conjugate. Next, using the equations of motion

$$\frac{d^2}{dt^2}\phi^\dagger = (mF + 2g\phi F + 2g\psi^1\psi^2),$$
$$i\frac{d}{dt}\psi^{1\dagger} = -m\psi^2 - 2g\phi\psi^2, \quad i\frac{d}{dt}\psi^{2\dagger} = m\psi^1 + 2g\phi\psi^1,$$
$$F^\dagger = -(m\phi + g\phi^2),$$

it is not difficult to check that

$$\frac{d}{dt}Q^1 = \frac{d}{dt}Q^2 = 0\,. \tag{53}$$

Clearly, the complex-conjugate charges are also conserved.

The algebra of the supercharges now takes the form

$$\{Q^1 Q^{1\dagger}\} = \{Q^2 Q^{2\dagger}\} = 2H \tag{54}$$

(all other commutators vanish). Here H is the Hamiltonian of the system:

$$H = \pi_\phi \pi_{\phi^\dagger} + |F|^2 + \left[\frac{1}{2}(m + 2g\phi)(\psi)^2 + \text{h.c.}\right],$$

where

$$\pi_\phi = -i\frac{\partial}{\partial\phi},\quad \pi_{\phi^\dagger} = -i\frac{\partial}{\partial\phi^\dagger},\quad (\psi)^2 \equiv \psi^2\psi^1 - \psi^1\psi^2\,.$$

At the next stage we must realize the fermion variables ψ^α in the matrix representation. In the problem at hand the procedure is somewhat more complicated than what we discussed in Sec. 3, since there are two fermion variables now (plus two complex conjugate). The prescription ensuring the canonic commutation relations

$$\{\psi^\alpha \psi^\beta\} = \{\psi^{\alpha\dagger}\psi^{\beta\dagger}\} = 0,\quad \{\psi^\alpha \psi^{\beta\dagger}\} = \delta^{\alpha\beta},$$

is well-known, though, see e.g. Ref. [18].

Let us build ψ^α in the form of the direct product of two 2×2 matrices,

$$\begin{aligned} \psi^1 &= \sigma^- \otimes 1,\quad \psi^2 = \sigma^3 \otimes \sigma^-\,, \\ \psi^{1\dagger} &= \sigma^+ \otimes 1,\quad \psi^{2\dagger} = \sigma^3 \otimes \sigma^+, \end{aligned} \tag{55}$$

where $\sigma^\pm = \frac{1}{2}(\sigma^1 \pm i\sigma^2)$. Then, in the matrix representation, the expression for the Hamiltonian reduces to the following:

$$H = \pi_\phi \pi_{\phi^\dagger} + |F|^2 - [(m + 2g\phi)\sigma^- \otimes \sigma^- + \text{h.c.}]\,. \tag{56}$$

The Hamiltonian acts on the wave functions with 'two spins'.

In the classical approximation the spin interaction can be neglected while the potential of the system

$$V_{\text{pot}} = |F|^2 = |m\phi + g\phi^2|^2$$

has two minima, at $\phi = 0$ and at $\phi = -m/g$, corresponding to zero energy. Furthermore, due to supersymmetry, the energy is at zero at any finite order of perturbation theory. As will be seen below, both the ground states are of the "boson" type, and they have no "fermion" partners. Since the levels can ascend/descend only in pairs we conclude that in the exact solution the Hamiltonian (54) has two exactly zero energy levels, and Witten's index of the system is

$$I_W = 2\,. \tag{57}$$

It is instructive to examine how the ground state of the quantum problem $H\Psi = E\Psi$ turns out to be double-degenerate (at $E = 0$). Notice that because of SUSY, continuous variations of m do not affect I_W. Therefore, surprising though it is, the two-fold degeneracy should take place even in the case when $m = 0$ and $V_{\text{pot}} = g\phi^4$, so that classically there exists only one state with the zero energy. Below we will consider just this massless case.

Let us choose for a start the spin state in the form $|\uparrow\downarrow\rangle$ or $|\downarrow\uparrow\rangle$. Then the spin part of the Hamiltonian (54) acting on these states vanishes. It is obvious that the wave functions corresponding to these spin states are characterized by $E > 0$. Indeed, if the coordinate part of the wave function is denoted by Φ, then with the spin term switched off, we have

$$\langle \Phi \mid (\pi_\phi \pi_{\phi^\dagger} + |F|^2) \mid \Phi \rangle > 0\,.$$

Thus, the wave function of the ground state should have the form

$$\Psi = \Phi_1 |\uparrow\uparrow\rangle + \Phi_2 |\downarrow\downarrow\rangle\,. \tag{58}$$

Now, we take this ansatz, act by supercharges and require zero, $Q\Psi = 0$.

The Lagrangian (56) is invariant under the following transformations:

$$\phi \leftrightarrow \phi^\dagger, \;\; F \leftrightarrow F^\dagger, \;\; \psi^1 \leftrightarrow \psi^{2\dagger}, \;\; \psi^2 \leftrightarrow \psi^{1\dagger}\,. \tag{59}$$

(In field theory these transformations would correspond to C-parity). Under these transformations

$$Q^1 \leftrightarrow Q^{2\dagger}, \;\; Q^2 \leftrightarrow Q^{1\dagger}\,.$$

Therefore, instead of considering four supercharges, Q^α and $Q^{\alpha\dagger}$, which must annihilate the vacuum state, it is quite sufficient to keep two, namely

$$\frac{1}{\sqrt{2}} Q^1 \Psi = \pi_\phi \Phi_1 |\downarrow\uparrow\rangle + iF\Phi_2 |\downarrow\uparrow\rangle = 0,$$

$$\frac{1}{\sqrt{2}}Q^{1\dagger}\Psi = \pi_{\phi^\dagger}\Phi_2|\uparrow\downarrow\rangle + iF^\dagger\Phi_1|\uparrow\downarrow\rangle = 0.$$

These equations, being written down in the explicit form, imply

$$\frac{\partial\Phi_1(\phi,\phi^\dagger)}{\partial\phi} = -g(\phi^\dagger)^2\Phi_2(\phi,\phi^\dagger), \quad \frac{\partial\Phi_2(\phi,\phi^\dagger)}{\partial\phi^\dagger} = -g\phi^2\Phi_1(\phi,\phi^\dagger). \tag{60}$$

After some reflection it is not difficult to arrive at the conclusion that the solutions of the system (58) are as follows:

$$\Phi_1 = X(r), \quad \Phi_2 = Y(r)\mathrm{e}^{i\alpha} \tag{61}$$

or

$$\Phi_1 = Y(r)\mathrm{e}^{-i\alpha}, \quad \Phi_2 = X(r)\,, \tag{62}$$

where $r \equiv |\phi|$ and $\alpha \equiv \arg\phi$, and the functions X, Y satisfy the following system of linear differential equations of the first order:

$$X' = -2gr^2Y, \quad Y' - \frac{Y}{r} = -2gr^2X\,. \tag{63}$$

The solution is expressible in terms of the McDonald functions,

$$X = -r^2K_{2/3}\left(\frac{2gr^3}{3}\right), \quad Y = r^2K_{1/3}\left(\frac{2gr^3}{3}\right), \tag{64}$$

which fall off exponentially at large r.

Thus, we see that there are indeed two ground states,

$$\Psi_{(1)} = -r^2K_{2/3}|\uparrow\uparrow\rangle + r^2K_{1/3}\mathrm{e}^{i\alpha}|\downarrow\downarrow\rangle,$$
$$\Psi_{(2)} = r^2K_{1/3}\mathrm{e}^{-i\alpha}|\uparrow\uparrow\rangle - r^2K_{2/3}|\downarrow\downarrow\rangle, \tag{65}$$

where the argument of the McDonald function is $2gr^3/3$. The orthogonality of $\Psi_{(1)}$ and $\Psi_{(2)}$ is trivially ensured by the angular factor $\exp(i\alpha)$.

Finally, it is worth explaining why both states (63) are of the boson type. The states of the fermion type are obtained from those if one acts by the supercharge operators, and they, obviously, have the structure $|\uparrow\downarrow\rangle$ or $|\downarrow\uparrow\rangle$. In this aspect the problem is drastically different from the one-dimensional system considered in Sec. 1. There we had only two types of states, $|\uparrow\rangle$ and $|\downarrow\rangle$, and under the action of supercharge $|\uparrow\rangle \leftrightarrow |\downarrow\rangle$. This fact explains why in the example of Sec. 1 with multiwell potentials we got alternating $E = 0$ states of the boson and fermion types. Doing the one-dimensional reduction of the four-dimensional field theory we obtain Hamiltonians necessarily containing an even number of the fermion variables. At the same time the supercharge is always linear in ψ. Therefore, the zero energy states are always of one and the same type.

9. First Encounter with Supersymmetric Field Theory

To get acquainted with the supersymmetric field theories first we will make a brief digression in the theory of four-dimensional spinors.

In four dimensions there are two different types of spinors, two-component objects transforming according to the spinor representation of the Lorentz group. Let us denote them by χ_α and $\psi^{\dot\alpha}$ where $\alpha, \dot\alpha = 1, 2$; χ_α is called the left-handed spinor and $\psi^{\dot\alpha}$ the right-handed. The indices of the right-handed spinors are supplied by dots to emphasize the fact that their transformation law does not coincide with that of the left-handed spinors, rather,

$$\eta^\alpha \sim \psi^{\dot\alpha\dagger}$$

where $\sim$ means "transforms as" and † denotes complex conjugation of the spinor. Thus, for instance, if under the proper Lorentz transformations ψ transforms as

$$\psi \to \exp\left(-\frac{1}{2}\vec\omega\vec\sigma\right)\psi\,,$$

χ^α will transform with the complex conjugated matrix. Here $\vec\omega$ are three transformation parameters; notice that the matrix $\exp(-\frac{1}{2}\vec\omega\vec\sigma)$ is not unitary. (Warning: in some popular textbooks the left-handed spinors carry dotted indices and right-handed undotted. This opposite convention is adopted, for instance, in the Landau-Lifshitz course [21] which gives a very nice introduction in the theory of four-dimensional spinors. Another good source which I may recommend for initial reading is Ref. [22]. In this book the undotted spinors are called left-handed, but if you inspect the book more carefully and compare the definitions accepted there with, say, the standard definition of the left-handed neutrino, you will see that actually the undotted spinors of Ref. [22] should have been called right-handed).

Lorentz scalars can be formed as a convolution of two dotted or two undotted spinors, $\chi^\alpha\xi_\alpha$ or $\psi_{\dot\alpha}\eta^{\dot\alpha}$, with one lower and one upper index. The raising and lowering of indices is realized by virtue of the antisymmetric (Levi-Civita) symbol,

$$\psi_{\dot\alpha} = \epsilon_{\dot\alpha\dot\beta}\psi^{\dot\beta}\,, \quad \chi_\alpha = \epsilon_{\alpha\beta}\chi^\beta\,;$$

$$\psi^{\dot\alpha} = \epsilon^{\dot\alpha\dot\beta}\psi_{\dot\beta}\,, \quad \chi^\alpha = \epsilon^{\alpha\beta}\chi_\beta\,,$$

where

$$\epsilon^{12} = \epsilon^{\dot1\dot2} = 1\,; \quad \epsilon_{12} = \epsilon_{\dot1\dot2} = -1$$

so that $\epsilon^{\alpha\beta}\epsilon_{\beta\gamma} = \delta^\alpha_\gamma$. When one raises or lowers the index of ψ the ϵ symbol must be placed to the left of ψ.

Sometimes we will use a shorthand notation when the indices of the spinors are implicit. Then, to distinguish the right-handed spinors from the left-handed ones, the former will be marked by bars, for instance,

$$\chi\xi \equiv \chi^\alpha\xi_\alpha$$

and

$$\bar\psi\bar\eta \equiv \psi_{\dot\alpha}\eta^{\dot\alpha}\,.$$

Notice that in convoluting the undotted indices we first write the spinor with the upper index while for the dotted indices the first spinor has the lower index. This convention is very convenient. The ordering is important since the elements of the spinors are anticommuting Grassmann numbers.

It remains to be added that the vector quantities can be obtained from two spinors — one dotted and one undotted. To this end one introduces four matrices $(\sigma^\mu)_{\alpha\dot\beta}$,

$$(\sigma^\mu)_{\alpha\dot\beta} = \{1, \vec\sigma\}_{\alpha\dot\beta}\,. \tag{66}$$

It is not difficult to check that bilinears

$$\chi^\alpha(\sigma^\mu)_{\alpha\dot\beta}\psi^{\dot\beta} \quad \text{or} \quad \psi^{\dot\alpha\dagger}(\sigma^\mu)_{\alpha\dot\beta}\eta^{\dot\beta}$$

transform as Lorentz vectors (one should not forget that $\psi^{\dot\alpha\dagger} \sim \psi^\alpha$). From two spinors with the lower indices the Lorentz vector is formed by virtue of the matrix

$$(\bar\sigma^\mu)^{\dot\beta\alpha} = \{1, -\vec\sigma\}_{\dot\beta\alpha}\,, \tag{67}$$

namely, $\psi_{\dot\beta}(\bar\sigma^\mu)^{\dot\beta\alpha}\chi_\alpha$.

The two-component spinors discussed above, left-handed and right-handed, are called the Weyl spinors. Each of them describes two degrees of freedom.

The more familiar Dirac spinor (bispinor) is a combination of two Weyl spinors, one left-handed and one right-handed,

$$\Psi = \begin{pmatrix} \psi^{\dot{\alpha}} \\ \chi_\alpha \end{pmatrix} . \tag{68}$$

The Dirac spinor (68) is written in the so-called spinor representation and describes four degrees of freedom. The charge-conjugated bispinor is obtained from Eq. (68) in the following way:

$$\Psi^C = \begin{pmatrix} \epsilon^{\dot{\alpha}\dot{\beta}} \chi^\dagger_\beta \\ \epsilon_{\alpha\beta} \psi^{\dot{\beta}\dagger} \end{pmatrix} .$$

Imposing the condition that the Dirac bispinor is equal to its charge-conjugated we get the Majorana bispinor which, then, obviously, has the form

$$\lambda = \begin{pmatrix} \psi^{\dot{\alpha}} \\ \epsilon_{\alpha\beta} \psi^{\dot{\beta}\dagger} \end{pmatrix} . \tag{69}$$

We see that the Majorana bispinor describes two degrees of freedom and is equivalent to the Weyl two-component spinor. If the Weyl spinor corresponds to the right-handed particle and left-handed antiparticle, the Majorana bispinor describes a "neutral" particle of both polarizations, coinciding with its antiparticle. Both formalisms, Weyl and Majorana, are used in SUSY field theory. In the Majorana notation the bilinears one deals with most commonly take the form

$$\bar{\lambda}\lambda = (\psi_{\dot{\beta}} \psi^{\dot{\beta}} + \text{ h.c.}) , \quad \bar{\lambda}\gamma_5\lambda = -(\psi_{\dot{\beta}} \psi^{\dot{\beta}} - \text{ h.c.})$$

and

$$\bar{\lambda}\gamma^\mu\lambda = 0 , \quad -\frac{1}{2}\bar{\lambda}\gamma^\mu\gamma^5\lambda = \psi^{\dot{\alpha}\dagger} (\sigma^\mu)_{\alpha\dot{\beta}} \psi^{\dot{\beta}} .$$

The definition of the spinor representation gamma matrices used above follows the textbook [21]. Sometimes, instead of the spinor representation, the so-called Majorana representation of gamma matrices is more convenient. In this representation

$$\lambda = \frac{1}{2} \begin{pmatrix} -i(\bar{\psi} - \bar{\psi}^\dagger) \\ (\bar{\psi} + \bar{\psi}^\dagger) \end{pmatrix} , \tag{70}$$

all gamma matrices are purely imaginary, and the operation of charge conjugation reduces to complex conjugation. In the Majorana representation one can say that the Majorana bispinor is merely real.

(Exercise: find the explicit form of gamma matrices corresponding to Eq. (70)).

The excursion we have just undertaken was necessary to elucidate the structure of minimal particle supermultiplets in four-dimensional field theory. As we know, in supersymmetric theories the number of the boson and fermion degrees of freedom must match. The "minimal" fermion is described by the Weyl or Majorana spinor and has two degrees of freedom. Hence, its boson counterpartner must also have two degrees of freedom. This is exactly the number of degrees of freedom of the complex scalar field. Another choice is a massless vector (gauge) field which has exactly two transverse polarizations. Both versions are very important and are at the heart of the modern SUSY field theory.

The first SUSY field-theoretical model, supersymmetric quantum electrodynamics, was constructed by Golfand and Likhtman in 1971 [1]. It took three years to realize that there exists a simpler model which was suggested in 1974 by Wess and Zumino [23] and is now named after them. The dimensionally reduced version of the model was given as a quantum-mechanical exercise in Sec. 7 where we learned that its Witten index is equal to 2. Now let us have a closer look at the four-dimensional model *per se*.

The Lagrangian of the model has the form

$$L = \partial^\mu \phi^\dagger \partial_\mu \phi + i\psi^\dagger{}_{\dot\alpha} \partial_\mu (\bar\sigma^\mu)^{\dot\alpha\alpha} \psi_\alpha + F^\dagger F$$
$$+ \left[m \left(\phi F - \frac{1}{2}(\psi)^2 \right) + g\left(\phi^2 F - \phi(\psi)^2\right) + \text{h.c.} \right], \tag{71}$$

where $\phi(\vec{x},t)$ and $F(\vec{x},t)$ are complex boson fields of spin zero and $\psi(\vec{x},t)$ is a two-component complex (Weyl) spinor field,

$$\psi \to \psi^\alpha , \quad \alpha = 1,2.$$

We use the conventions explained above; the mass parameter m and the coupling constant g are complex numbers.

The field F is auxiliary: it enters the Lagrangian with no derivatives and, hence, can be eliminated from the Lagrangian by virtue of the classical equations of motion,

$$F^\dagger = -m\phi - g\phi^2 \tag{72}$$

plus the complex conjugate equation for F. Thus, our model presents the theory of a self-interacting complex scalar field ϕ, with the Yukawa coupling to the Weyl spinor ψ. The Yukawa coupling constant and the constant of self-interaction are the same, a characteristic feature of supersymmetry.

To check that the model above does have supersymmetry it is sufficient to present a conserved supercurrent which carries a spinor index, in addition to the vector one, of course. In the case at hand the supercurrent has the form (μ is the vector index and β is the spinor one),

$$J^\mu_\beta = \sqrt{2}\left[(\partial_\rho\phi^\dagger)(\sigma^\rho)_{\beta\dot\alpha}(\bar\sigma^\mu)^{\dot\alpha\alpha}\psi_\alpha - i(\sigma^\mu)_{\beta\dot\beta}\psi^{\dagger\dot\beta}F\right] \tag{73}$$

where F is given by (the complex conjugate of) Eq. (72). The complex conjugate of $J_{\mu\beta}$ is also a conserved current, so that in the problem at hand we have a conserved supercharge which is a complex two-component spinor,

$$Q_\beta = \int d^3x J_{0\beta}\,, \quad Q^\dagger_{\dot\beta} = \int d^3x (J_{0\beta})^\dagger\,. \tag{74}$$

The analogs of the quantum-mechanical equations (53) and (54) take the form

$$\dot Q - \dot Q^\dagger = 0$$

and

$$\{Q_\alpha Q^\dagger_{\dot\alpha}\} = 2P_{\alpha\dot\alpha}\,. \tag{75}$$

Here P_μ is the energy-momentum operator (in particular, P_0 is the Hamiltonian of the model); and, as usual, the braces denote anticommutator.

It is instructive to see how the conservation of the supercurrent (73) follows from the equations of motion for ϕ and ψ which we derive from the Lagrangian (71),

$$\Box\phi^\dagger = mF + 2g\phi F - g\psi\psi\,;$$
$$i\partial_\mu(\bar\sigma^\mu)^{\dot\alpha\alpha}\psi_\alpha = m^*\psi^{\dagger\dot\alpha} + 2g^*\phi^\dagger\psi^{\dagger\dot\alpha}\,. \tag{76}$$

The divergence $\partial_\mu J^\mu_\beta$ contains four distinct terms: ∂_μ can act on $\phi^\dagger$ or ψ in the first term in Eq. (73) or on $\psi^\dagger$ or F in the second term. Let us consider them in turn. When ∂_μ acts on $\phi^\dagger$ we get $\partial_\mu\partial_\rho\phi^\dagger$, a combination symmetric in (μ,ρ). Hence, only the symmetric part of the product $(\sigma^\rho)_{\beta\dot\alpha}(\bar\sigma^\mu)^{\dot\alpha\alpha}$ is relevant,

$$(\sigma^\rho)_{\beta\dot\alpha}(\bar\sigma^\mu)^{\dot\alpha\alpha} \to g^{\rho\mu}\delta^\alpha_\beta + \ldots\,.$$

This means that $\partial_\mu\partial_\rho\phi^\dagger$ reduces to $\Box\phi^\dagger$, which is nothing else than the first equation of motion (76). Notice that the terms $g\psi^2$ in the equation of motion will drop out automatically since $\Box\phi^\dagger$ is multiplied by ψ, and $\psi^3 = 0$. The rest of this term is canceled by $-i(\sigma^\mu)_{\beta\dot\beta}(\partial_\mu\psi^{\dagger\dot\beta})F$. When ∂_μ acts on ψ in the first term in Eq. (73) we get the second equation of motion (76); this is canceled by $-i(\sigma^\mu)_{\beta\dot\beta}\psi^{\dagger\dot\beta}\partial_\mu F$.

The last exercise we will do with the Wess-Zumino model is rewriting the Lagrangian (71) in the Majorana notation. Introducing

$$\lambda = \begin{pmatrix} \epsilon^{\dot\alpha\dot\beta}\psi^\dagger_{\dot\beta} \\ \psi_\alpha \end{pmatrix}$$

we get

$$L = \partial^\mu\phi^\dagger\partial_\mu\phi + \frac{i}{2}\bar\lambda\gamma^\mu\partial_\mu\lambda + F^\dagger F$$
$$+ \left[m\left(\phi F - \frac{1}{4}\bar\lambda(1+\gamma^5)\lambda\right) + g\left(\phi^2 F - \frac{1}{2}\phi\bar\lambda(1+\gamma^5)\lambda\right) + \text{h.c.}\right]. \tag{77}$$

Another model we will briefly discuss here is the supersymmetric generalization of the Yang-Mills theory singled out because of its phenomenological importance. Pure Yang-Mills theory is the theory of the interacting vector gauge bosons. In QCD these are gluons, in the electroweak theory, the W bosons. The gauge bosons belong to the adjoint representation of the gauge group which we will assume to be SU(N). Then the number of the gauge bosons is $N^2 - 1$. Each of them has two transverse polarizations, i.e. two degrees of freedom.

Constructing the supersymmetric Yang-Mills theory one adds N^2-1 gauginos, fermion fields belonging to the same adjoint representation of SU(N). To keep the balance between the number of the boson and fermion degrees of freedom the gaugino fields are to be taken to be the Majorana (or Weyl) fermions. Then the Lagrangian of the SUSY Yang-Mills theory can be written as

$$L = -\frac{1}{4}G^a_{\mu\nu}G^{\mu\nu a} + \frac{i}{2}\bar\lambda\gamma^\mu D_\mu\lambda \tag{78}$$

where the Majorana notation is used, $G^a_{\mu\nu}$ is the gauge field strength tensor,

$$G^a_{\mu\nu} = \partial_\mu A^a_\nu - \partial_\nu A^a_\mu + g f^{abc} A^b_\mu A^c_\nu\,, \tag{79}$$
$$a, b, c = 1, 2, \ldots, N^2 - 1\,,$$

g is the gauge coupling constant, f^{abc} stand for the structure constants of the SU(N) group. Moreover, the action of the covariant derivative D_μ on the

Majorana field λ is defined as

$$D_\mu \lambda^a = \partial_\mu \lambda^a + g f^{abc} A_\mu^b \lambda^c \,. \tag{80}$$

Equations (79) and (80) represent the standard gauge interaction. The theory described by the Lagrangian (78) is sometimes called *supersymmetric gluodynamics.*

Since the number of the degrees of freedom match and the interaction is the standard gauge interaction one may suspect that the Lagrangian (78) describes a supersymmetric theory. This is indeed the case. The conserved supercurrent of the SUSY Yang-Mills theory is

$$J^\mu = G^a_{\rho\phi} \sigma^{\rho\phi} \gamma^\mu \lambda^a \tag{81}$$

where the spinorial index is implicit.

It is not difficult to see that the current (81) is conserved. Indeed,

$$\partial_\mu J^\mu = G_{\rho\phi} \sigma^{\rho\phi} \gamma^\mu \overrightarrow{D}_\mu \lambda + G_{\rho\phi} \sigma^{\rho\phi} \gamma^\mu \overleftarrow{D}_\mu \lambda \,.$$

The first term on the right hand side vanishes by virtue of the equation of motion for the gaugino field. In the second term we reduce the product of three gamma matrices to a linear combination of one gamma matrix, using the well-known identities [21]. Then $G_{\rho\phi} \sigma^{\rho\phi} \gamma^\mu \overleftarrow{D}_\mu \lambda$ reduces to the equation of motion for the gauge field,

$$\partial_\mu J^\mu \sim (D^\rho G_{\rho\phi}) \gamma^\phi \lambda \sim (\bar\lambda^a \gamma_\phi \lambda^b) \gamma^\phi \lambda^c f^{abc} \,.$$

Furthermore, the product of three Majorana fields on the right-hand side vanishes. Perhaps, the simplest way to see the vanishing is rearranging two λ's *a lá* Fierz,

$$\begin{aligned}(\bar\lambda^a \gamma_\phi \lambda^b) \gamma^\phi \lambda^c &= \frac{1}{2} (\bar\lambda^a \gamma_\phi \lambda^c) \gamma^\phi \lambda^b + \frac{1}{2} (\bar\lambda^a \gamma_\phi \gamma^5 \lambda^c) \gamma^\phi \gamma^5 \lambda^b \\ &\quad - (\bar\lambda^a \lambda^c) \lambda^b + (\bar\lambda^a \gamma^5 \lambda^c) \gamma^5 \lambda^b \,. \end{aligned} \tag{82}$$

We then observe that for the Majorana field

$$\bar\lambda^a \gamma_\phi \gamma^5 \lambda^b f^{abc} = \bar\lambda^a \lambda^b f^{abc} = \bar\lambda^a \gamma^5 \lambda^b f^{abc} = 0 \,. \tag{83}$$

Equations (82) and (83) imply that

$$(\bar\lambda^a \gamma_\phi \lambda^b) \gamma^\phi \lambda^c f^{abc} = 0$$

and hence the supercurrent (81) is conserved.

Technical derivations given above should not overshadow the striking fact — the occurrence of new conserved quantities of the geometric nature, the supercharges. Each new conservation law for the quantity carrying the Lorentz indices is a very strong property of the theory. As a matter of fact, before the discovery of supersymmetry it was believed that in any dynamically nontrivial four-dimensional field theory the only conserved quantities with the Lorentz indices could be P_μ and $M_{\mu\nu}$, the four-momentum and the angular momentum operators. Every other conserved quantity was assumed to be Lorentz scalar, for instance, the electric charge or the baryon number. This belief was based on the Coleman-Mandula theorem [24]. The idea of this theorem is quite straightforward. If one has too many conserved quantities with the Lorentz indices the S matrix must necessarily be equal to unity. On the contrary, if one deals with interacting four-dimensional theories with the nonzero scattering amplitudes, the conserved quantities transforming as Lorentz tensors that are allowed are P_μ and $M_{\mu\nu}$. Say, if one considers a two-to-two scattering amplitude the conservation of P_μ and $M_{\mu\nu}$ leaves the scattering angles unspecified. Additional conservation laws for Lorentz tensors could be satisfied only for a discrete set of possible scattering angles. Since the scattering amplitude must be a smooth function of the scattering angle the only possible solution then is the vanishing of the two-to-two scattering amplitude for all angles.

The proof of the Coleman-Mandula theorem was based on the assumption that the extra conserved quantities carry vectorial Lorentz indices. The authors did not even think about the possible existence of new conserved quantities of the fermion nature, with the spinorial indices (supercharges). This, thus, was the only loophole in the Coleman-Mandula theorem, the only direction where new conservation laws of the geometric nature could be sought for. The search, with this specific purpose in mind, was carried out by Golfand and Likhtman [1] and was successful; the impact of the discovery is tremendous.

Do further generalizations exist? Can one build nontrivial dynamical theories with more conserved supercharges? Are there interacting theories in four dimensions with the conserved supercharges that, additional to the spinorial index, would carry vector Lorentz indices?

The answer to the second question is negative. The conserved supercharges with additional vector Lorentz indices are actually forbidden by the Coleman-Mandula theorem.

It is easy to understand why they are forbidden. Indeed, the anticommutator of two conserved supercharges is also a conserved quantity. From two ordinary supercharges, of spin 1/2, one can form a conserved vector operator, P_μ, which is allowed by the Coleman-Mandula theorem. However, a would-be supercharge with the vector index is a spin 3/2 object, and the anticommutator of two such objects would contain operators up to Lorentz spin 3. The Coleman-Mandula theorem forbids interacting theories with the conserved Lorentz spin 3 operators.

At the same time, nontrivial theories with a larger number of spin 1/2 conserved supercharges can be built. They are called extended supersymmetric theories. The presence of the extra supercharges extends the algebra (see Eq. (75)). The most general possible superalgebra of the conserved generators generalizing the Poincaré algebra was obtained in Ref. [25].

10. Concluding Remarks and a Little Bit of History

Our brief excursion in supersymmetry is almost over. This is obviously only the beginning of a large journey. Of most practical interest are supersymmetric gauge theories in four dimensions, of which very little was said in this lecture, where we limited ourselves to the simplest notions and concepts. Those of you who want to learn more should turn to the books (the list of the recommended literature is given at the end) and to Chaps. 5 and 6 presenting the state of the art issues.

I would like to conclude the lecture with a few remarks of historic nature. As I have already mentioned, it happened so that supersymmetry was discovered twice. One line of research which culminated in the famous Wess-Zumino paper [2] ascends to a two-dimensional symmetry on the world sheet introduced in the context of the string theory [26]. Wess and Zumino's starting point was a "supergauge transformation" of the fields. The parameters of the supergauge transformations were Grassmann numbers. This approach, as well as its founding fathers, is very well-known. In fact, it was the work of Wess and Zumino that triggered the subsequent explosive development of the subject.

Much less known is the work of Golfand and Likhtman [1] done a few years earlier. Their motivation was algebraic; the idea was to extend the Poincaré algebra to include anticommuting generators. Yuri Golfand died in 1994, in oblivion; I think, a few words about him would be in order here.

The last time I saw him was in Technion, Haifa. A small man, with warm eyes and kind smile. He looked as usual, a little bit out of touch with reality, decoupled from the surrounding world, with thoughts directed inside rather than outside. Occasionallty, I used to bump onto him in the corridors of ITEP, or the Lebedev Physical Institute in Moscow, now this happened in Israel. He told me that he had a dream — to travel in the West, to the United States or France, to attend a physics conference or just "to see the world"... Well, this dream never came true.

For about 40 years Golfand was a member of the Theory Department of the Lebedev Physical Institute (with a long break, see below). The idea of extending the Poincaré algebra by adding the supergenerators preoccupied him for many years. In general, he had very high standards in physics, both with respect to himself and to others, and published rarely. In 1970 he and his student, E. Likhtman, succeeded in building the superalgebra which included the Poincaré algebra plus four spinorial generators and was closed (for the spinorial generators one has to check the anticommutators rather than commutators; let me note in passing that Wess and Zumino in their first work [2] built the superconformal algebra rather then Poincaré). Golfand and Likhtman worked very hard on the field-theoretic realization of their superalgebra. They ended up with a model which represents a massive version of supersymmetric QED. The results were submitted for publication to JETP Letters, while a detailed report was written during the following year and published in the Tamm Memorial Volume [1]. The authors were upset by the fact that their model was chiral and violated parity.

Meanwhile, the sky was becoming cloudy over Golfand. In 1973 his fellow colleagues in the Theory Department fired Golfand as "unproductive". Golfand never tried to conceal his unapologetic attitude to the communist ideas, to put it mildly. Under the circumstances, he applied for permission to emigrate to Israel, a high political crime by Soviet standards of 1970's. The permission was not granted; as a result Golfand was blacklisted and, for the next seven years, was unemployed. Only in 1980, under strong pressure from the world physics community, Golfand was reinstated in his position. In 1990, not long before the demise of the Soviet Union, he was allowed to leave for Israel. I do not think that he had time there to adjust and start feeling himself comfortable...

Acknowledgments

I would like to thank F. Lenz, H. Lipkin and M. Marinov for kindly sharing with me their historical reminiscences.

References

[1] Yu. Golfand and E. Likhtman, *Pisma ZhETF*, **13** (1971) 452 [*JETP Lett.* **13** (1971) 323, reprinted in *Supersymmetry*, Ed. S. Ferrara (North-Holland/World Scientific, 1987), Vol. 1, p. 7]; *On the Extension of the Algebra of Generators of the Poincaré Group by Bispinor Generators*, I. E. Tamm Memorial Volume *Problems of Theoretical Physics*, (Nauka, Moscow 1972), p. 37; D. V. Volkov and V. P. Akulov, *Phys. Lett.* **46B** (1973) 109 [Reprinted in *Supersymmetry*, Ed. S. Ferrara (North-Holland/World Scientific, 1987), Vol. 1, p. 11].

[2] J. Wess and B. Zumino, *Nucl. Phys.* **B70** (1974) 39 [Reprinted in *Supersymmetry*, Ed. S. Ferrara (North-Holland/World Scientific, 1987), Vol. 1, p. 13].

[3] E. Witten, *Nucl. Phys.* **B188** (1981) 513.

[4] V. A. Kostelecky and M. M. Nieto, *Phys. Rev. Lett.* **53** (1984) 2285; *Phys. Rev.* **A32** (1985) 1293.

[5] M. Shifman, *Int. J. Mod. Phys.* **A4** (1989) 3305.

[6] V. A. Kostelecky, M. M. Nieto and D. R. Truax, *Phys. Rev.* **D32** (1985) 2627.

[7] L. E. Gendenshtein and I. V. Krive, *Usp. Fiz. Nauk* **146** (1985) 553 [*Sov. Fiz.- Uspekhi* **28** (1985) 645].

[8] R. Haymaker and A. Rau, *Am. J. Phys.* **54** (1986) 928.

[9] A. Lahiri, P. K. Roy and B. Baghi, *Int. J. Mod. Phys.* **A5** (1990) 1383.

[10] W. Pauli, *Pauli Lectures on Physics*, Vol. 6, *Selected Topics in Field Quantization* (MIT Press, Cambridge, 1973), p. 33.

[11] H. Lipkin, *Phys. Lett.* **9** (1964) 203.

[12] E. Witten, *Nucl. Phys.* **B202** (1982) 253.

[13] P. Salomonson and J. Van Holten, *Nucl. Phys.* **B196** (1982) 509; M. Clandson and M. Halpern, *Nucl. Phys.* **B250** (1985) 689.

[14] A. Smilga, *Nucl. Phys.* **B249** (1985) 413; M. Shifman, A. Smilga and A. Vainshtein, *Nucl. Phys.* **B299** (1988) 79.

[15] M. Crombrugghe and V. Rittenberg, *Ann. Phys.* (*N.Y.*) **151** (1983) 99.

[16] F. Berezin, *The Method of Second Quantization* (Academic Press, New York, 1966).

[17] A. Salam and J. Strathdee, *Nucl. Phys.* **B76** (1974) 477; *Phys. Rev.* **D11** (1975) 1521.

[18] J. D. Bjorken and S. D. Drell, *Relativistic Quantum Fields* (McGraw-Hill, 1965).

[19] S. Coleman, "The Uses of Instantons", in *Aspects of Symmetry* (Cambridge University Press, 1985).

[20] A. Vainshtein *et al.*, "ABC of Instantons", *Usp. Fiz. Nauk* **136** (1982) 553 [*Sov. Fiz.- Uspekhi* **24** (1985) 195].

[21] V. B. Berestetskii, E. M. Lifshitz and L. P. Pitaevskii, *Quantum Electrodynamics*, 2nd Edition (Pergamon Press, Oxford, 1982).

[22] D. Bailin and A. Love, *Supersymmetric Gauge Field Theory and String Theory* (IOP Publishing, Bristol, 1994).

[23] J. Wess and B. Zumino, *Phys. Lett.* **49B** (1974) 52 [Reprinted in *Supersymmetry*, Ed. S. Ferrara (North-Holland/World Scientific, 1987), Vol. 1, p. 77].

[24] S. Coleman and J. Mandula, *Phys. Rev.* **159** (1967) 1251.
[25] R. Haag, J. Lopuszanski and M. Sohnius, *Nucl. Phys.* **B88** (1975) 257 [Reprinted in *Supersymmetry*, Ed. S. Ferrara (North-Holland/World Scientific, 1987), Vol. 1, p. 51].
[26] P. Ramond, *Phys. Rev.* **D3** (1971) 2415; A. Neveu and J. Schwarz, *Nucl. Phys.* **B31** (1971) 86; *Phys. Rev.* **D4** (1971) 1109; J.-L. Gervais and B. Sakita, *Nucl. Phys.* **B34** (1971) 477; 632.

Recommended Literature

Some of the papers quoted above and many other classical papers and reviews on supersymmetry are collected in *Supersymmetry*, Vols. 1 and 2, Ed. S. Ferrara (North-Holland/World Scientific, 1987).

Reviews and Books on Supersymmetric Quantum Mechanics

L. E. Gendenshtein and I. V. Krive, *Usp. Fiz. Nauk* **146** (1985) 553 [*Sov. Fiz.- Uspekhi* **28** (1985) 645].

R. Haymaker and A. Rau, *Am. J. Phys.* **54** (1986) 928.

A. Lahiri, P.K. Roy and B. Baghi, *Int. J. Mod. Phys.* **A5** (1990) 1383.

F. Schwabl, Quantum Mechanics (Springer-Verlag, Berlin, 1988), Chapter 19.

General Reviews

V. I. Ogievetsky and L. Mezincescu, *Usp. Fiz. Nauk*, **117** (1975) 637 [*Sov. Phys. — Uspekhi*, **18** (1975) 960]. Reprint Volume of Physics Reports *Supersymmetry and Supergravity*, Ed. M. Jacob (North-Holland/World Scientific, 1986).

Books

J. Bagger and J. Wess, *Supersymmetry and Supergravity* (Princeton University Press, 1983).

P. West, *Introduction to Supersymmetry and Supergravity* (World Scientific, Singapore, 1986).

S. J. Gates, M. T. Grisaru, M. Rocek and W. Siegel, *Superspace* (The Benjamin/Cummings, 1983).

D. Bailin and A. Love, *Supersymmetric Gauge Field Theory and String Theory* (IOP Publishing, Bristol, 1994).

Chapter V

Nonperturbative Dynamics in Supersymmetric Gauge Theories

M. A. Shifman

Theoretical Physics Institute, University of Minnesota, Minneapolis, MN 55455, USA

An extended version of lectures given at International School of Physics "Enrico Fermi", Varenna, Italy, July 3–6, 1995, Institute of Nuclear Science, UNAM, Mexico, April 11–17, 1996, and Summer School in High-Energy Physics and Cosmology, 10–26 July, 1996, ICTP, Trieste, Italy.

Abstract

I give an introductory review of recent, fascinating developments in supersymmetric gauge theories. I explain pedagogically the miraculous properties of supersymmetric gauge dynamics allowing one to obtain exact solutions in many instances. Various dynamical regimes emerging in supersymmetric quantum chromodynamics and its generalizations are discussed. I emphasize those features that have a chance of survival in QCD and those which are drastically different in supersymmetric and nonsupersymmetric gauge theories.

Unlike most of the recent reviews focusing almost entirely on the progress in extended supersymmetries (the Seiberg-Witten solution of $N = 2$ models), these lectures are mainly devoted to $N = 1$ theories. The primary task is extracting lessons for nonsupersymmetric theories.

Contents

Section 1

Basic Aspects of Nonperturbative Gauge Dynamics

1.1. *Introduction*

All fundamental interactions established in nature are described by non-Abelian gauge theories. The standard model of the electroweak interactions belongs to this class. In this model, the coupling constant is weak, and its dynamics is fully controlled (with the possible exception of a few, rather exotic problems, like the baryon number violation at high energies).

Another important example of the non-Abelian gauge theories is quantum chromodynamics (QCD). This theory has been under intense scrutiny for over two decades, yet remains mysterious. Interaction in QCD becomes strong at large distances. What is even worse, the degrees of freedom appearing in the Lagrangian (microscopic variables — colored quarks and gluons, in the case at hand) are not those degrees of freedom that show up as physical asymptotic states (macroscopic degrees of freedom — colorless hadrons). Color is permanently confined. What are the dynamical reasons for this phenomenon?

Color confinement is believed to take place even in pure gluodynamics, i.e. with no dynamical quarks. Adding massless quarks produces another surprise. The chiral symmetry of the quark sector, present at the Lagrangian level, is spontaneously broken (realized nonlinearly) in the physical amplitudes. Massless pions are the remnants of the spontaneously broken chiral symmetry. What can be said, theoretically, about the pattern of the spontaneous breaking of the chiral symmetry?

Color confinement and the spontaneous breaking of the chiral symmetry are the two most sacred questions of strong non-Abelian dynamics; and the progress in understanding them is painfully slow. In the late 1970s Polyakov showed that in three-dimensional compact electrodynamics (the so-called Georgi-Glashow model, a primitive relative of QCD) color confinement does indeed take place [1]. At approximately the same time a qualitative picture of how this phenomenon could actually happen in four-dimensional QCD was suggested by Mandelstam [2] and 't Hooft [3]. Some insights, though quite limited, were provided by the models of various degrees of fundamentality, and by numerical studies on lattices. This is, basically, all we had before 1994, when a significant breakthrough was achieved in understanding both issues in supersymmetric (SUSY) gauge theories.

Unlike the Georgi-Glashow model in three dimensions mentioned above, which is quite a distant relative of QCD, the four-dimensional supersymmetric gluodynamics and supersymmetric gauge theories with matter come much closer to genuine QCD. Moreover, the dynamics of these theories is rich and interesting by itself, which accounts for the attention they have attracted in the last two or three years. Although the development is not yet complete, the lessons are promising, and definitely deserve thorough studies. Several topics which I consider to be most interesting are discussed below in this lecture course. Before submerging into supersymmetry proper, however, it is worth reiterating the main general ideas which are the key players in this range of questions: the Meissner and the dual Meissner effects, monopoles, Abelian projection of QCD, and so on. The first part is a brief review of these issues intended mostly to refresh the memory and to provide a representative list of pedagogical literature. We will start an excursion into supersymmetric gauge theories in Sec. 2, and gradually proceed from simpler topics to the more complicated ones. The simplest supersymmetric non-Abelian model is SUSY gluodynamics. Simultaneously, it happens to be the closest approximation to QCD (without light quarks). Although there was essentially no progress towards the solution of this theory the seeds of the miraculous properties of the supersymmetric gauge dynamics are clearly visible. I will explain how some exact results (the first example ever in four-dimensional strongly coupled field theory!) can be derived. These results will become a part of our tool kit used in revealing various dynamical scenarios in SUSY gauge theories with matter.

In Sec. 3 we will open a fascinating world of supersymmetric SU(N_c) QCD, a world populated by a variety of unusual regimes governed by nonperturbative supersymmetric dynamics. Here, among other rarities, we will find confinement without spontaneous chiral symmetry breaking, with spontaneous breaking of the baryon number; the so-called *s*-confinement, with additional composite massless fields not related to Goldstone modes of the spontaneously broken global symmetries. We will discover a conformal window — a set of pairs of theories with different gauge groups but identical global symmetries that are dual (equivalent) to each other as far as infrared behavior is concerned. The infrared asymptotics of these theories is (super)conformal. Outside the conformal window we will encounter dual pairs, with one theory coupled superstrongly and another free in the infrared domain. The gauge bosons of the latter can be considered as composite superstrongly bound states in the former theory.

Section 4 is a very brief travel guide to the supersymmetric gauge theories with other gauge groups. The new phenomena encountered are the so-called oblique confinement and triality — the infrared equivalence of three distinct theories. One of them is in the Higgs phase, another in the confinement phase, and the third one is in the oblique confinement phase.

Finally, in Sec. 5, supersymmetry is explicitly (softly) broken by the gluino and squark masses. Ideally, we would like to send these masses to infinity, evolving towards nonsupersymmetric theories, without losing our calculational abilities. Unfortunately, once the gluino and squark masses become large enough, the calculational abilities are lost. We have to settle for small perturbations of the supersymmetric solution. By exploring the dynamical properties of the theory obtained in this way one hopes to get qualitative insights about what happens in the limit of infinitely heavy squarks and gluino.

This review is an extended version of my lecture notes. The pedagogical style of presentation is preserved, where possible. Simple and general issues are first discussed, providing a necessary background for the more advanced theoretical constructions and conclusions. Occasional remarks intended for expert readers will slip, though; they may be ignored in the first reading.

1.2. *Phases of Gauge Theories (Abelian Version)*

Quantum electrodynamics (QED) was historically the first gauge theory studied in detail. Although from the modern perspective it seems to be a very simple model, with no mysteries, it can exhibit at least three different types of behavior. Let us consider supersymmetric version, SQED. The Lagrangian of the model is [4]

$$
\begin{aligned}
\mathcal{L} = & -\frac{1}{4}F_{\mu\nu}^2 + \bar{\psi} i \not{D}\psi - m\bar{\psi}\psi \\
& + (D_\mu\phi)^\dagger(D_\mu\phi) - m^2\phi^\dagger\phi + (D_\mu\chi)^\dagger(D_\mu\chi) - m^2\chi^\dagger\chi \\
& - \frac{e^2}{2}(\phi^\dagger\phi - \chi^\dagger\chi)^2 \\
& + \frac{i}{2}\bar{\lambda}\,\not{\partial}\lambda + e\left\{\bar{\lambda}\frac{1+\gamma^5}{2}\psi\phi^\dagger + \bar{\lambda}\frac{1-\gamma^5}{2}\psi\chi + \text{h.c.}\right\},
\end{aligned}
\tag{1.1}
$$

where ϕ and χ are complex scalar fields with charges $+e$ and $-e$, respectively, (*selectrons*), and λ is the *photino* (Majorana) field. The first line in Eq. (1.1) is simply the conventional quantum electrodynamics of photons and electrons, the second line gives the kinetic and the mass terms of the electron's superpartners, selectrons, the third line is the selectron self-interaction, and, finally, the fourth line presents the photino kinetic term and photino's interactions. As we will see later, supersymmetry guarantees that the overall form of the Lagrangian is preserved under quantum corrections — no new counterterms appear.

Let us first assume that the electron and the selectron masses (which are equal) do not vanish, and $\alpha \equiv e^2/4\pi \ll 1$. For $m \neq 0$ the vacuum state of the theory is unique; it corresponds to the vanishing expectation values of the scalar fields, $\langle\phi\rangle = \langle\chi\rangle = 0$. Apart from the fact that some of the charged particles have spin zero, the theory is very much like QED. If heavy (static) probe charges are introduced, their interaction is just Coulomb, with the potential proportional to

$$V(R) \sim \frac{\alpha(R)}{R}$$

where R is the distance between the probe charges. Classically α is constant, of course, but the quantum renormalization makes α run. The behavior of α is determined by the well-known Landau formula. At large distances α decreases logarithmically; if m is finite, however, the logarithmic fall off is frozen at $R \sim m^{-1}$; the corresponding critical value of α is $\alpha_* = \alpha(R = m^{-1})$. The potential between the distant static charges is

$$V(R) \sim \frac{\alpha_*}{R}, \qquad R \to \infty . \tag{1.2}$$

The dynamical regime with this type of long-distance behavior is referred to as the *Coulomb phase.* In the case at hand we deal with the Abelian Coulomb phase. Similar behavior, Eq. (1.2), can take place in the non-Abelian gauge theories as well. Non-Abelian gauge theories with the long-range potential (1.2) are said to be in the non-Abelian Coulomb phase.

What happens if the mass parameter m is set equal to zero? Note, that in SUSY theories, if the bare value of m is fine-tuned to vanish, it will remain at zero with all quantum corrections.

The most drastic change is evident at first glance, after examining the third line in Eq. (1.1). The minimum of the potential energy is now achieved not only at the vanishing values of the scalar fields but, rather, on a one-dimensional complex manifold. Indeed, ϕ can be an arbitrary complex number; if $\phi = \chi$

the potential energy obviously vanishes. The continuous degeneracy of the classical minimal-energy state (the so-called *vacuum valley*) is a rather typical feature of supersymmetric theories with matter. We will return to the in-depth discussion of this aspect later.

Quantum-mechanically one can say that the expectation value $\langle\phi\chi\rangle \neq 0$ may take an arbitrarily complex value; here $\phi\chi$ is a convenient gauge invariant product parametrizing the vacuum valley. If the scalar fields ϕ and χ take constant nonvanishing values in the vacuum, the Higgs phenomenon takes place [5]: the gauge symmetry (U(1) in the case at hand) is spontaneously broken.

What does one mean by saying that the gauge symmetry is spontaneously broken? The gauge symmetry, in a sense, is not a symmetry at all — rather, it is a description of x physical degrees of freedom in terms of $x+y$ variables; y variables are redundant; the corresponding degrees of freedom are physically unobservable. In other words, only a subspace of all field space ($\psi, \bar{\psi}, \phi, \phi^\dagger, \chi, \chi^\dagger, A_\mu$ in the model under consideration), corresponding to gauge nonequivalent points, describe physically observable degrees of freedom.

Let us first switch off the electric charge, $e = 0$. Then the Lagrangian (1.1) is invariant under the global phase rotations, $\phi \to e^{i\alpha}\phi$, $\chi \to e^{-i\alpha}\chi$, $\psi \to e^{i\alpha}\psi$. The condensation of the scalar fields breaks this invariance. But the invariance of the model is not lost. Under the phase transformation one vacuum goes into another, physically equivalent. Say, if we start from the vacuum characterized by a real value of the order parameter ϕ and χ, in the "rotated" one the order parameter is complex. The spontaneous breaking of any global symmetry leads to a set of degenerate (and physically equivalent) vacua.

If we now switch on the photons, ($e \neq 0$), the degeneracy associated with the spontaneous breaking of the global symmetry is gone. All states related by the phase rotation are gauge-equivalent, and only one of them should be left in the Hilbert space of the theory. In other words, one can always *choose* the vacuum values of ϕ and χ to be real. This is nothing but the gauge condition. Thus, spontaneous breaking of the gauge symmetry does not imply, generally speaking, the existence of a degenerate set of vacua, as is the case with global symmetries. Then, what does it mean, after all?

By inspecting Lagrangian (1.1) it is not difficult to see that if ϕ and χ have nonvanishing (and constant) values in the vacuum, the spectrum of the theory does not contain massless vector particles at all. The photon acquires mass, $M_V^2 = 4e^2|v|^2$, where $v^2 = \langle\phi\chi\rangle$, through the mixing with the "phases" of the fields ϕ and χ. In the supersymmetric model considered, we "cook" in this way a massive vector field and a massive (real) scalar field, both with masses M_V, and a massless complex scalar field, out of the massless photon and the two massless complex scalar fields. (All these boson fields are accompanied by their fermion superpartners, of course).

This regime is referred to as the *Higgs phase.* One massless scalar field is eaten up by the photon field in the process of the transition to the Higgs phase. In the Higgs phase the electric charge is screened by the vacuum condensates. If we put a probe (static) electric charge in the vacuum, the Coulomb potential $\sim 1/R$ it induces at short distances (i.e. distances less than M_V^{-1}) gives place to the Yukawa potential $\sim \exp(-M_V R)/R$ at distances larger than M_V^{-1}. The gauge coupling runs, according to the standard Landau formula, only at distances shorter than M_V^{-1}, and is frozen at M_V^{-1}.

There is one single point in the vacuum valley, the origin, (i.e. $\langle\phi\chi\rangle = 0$) where the gauge symmetry is unbroken. The long-range force due to the massless photons is not screened by the vacuum condensates of the scalar fields. A different type of screening does occur, however, due to the quantum effects. Indeed, the photon propagator is dressed by the virtual pairs of electrons and selectrons. This dressing results in the running of the effective charge $\alpha(R)$,

$$\alpha(R) \sim \frac{1}{\ln R}\,. \tag{1.3}$$

Unlike the massive case, where this running is frozen at $R = m^{-1}$, in the theory with $m = 0$ (and $\langle\phi\chi\rangle = 0$) the logarithmic fall off (1.3) continues indefinitely: at asymptotically large R the effective coupling becomes asymptotically small.

Thus, the asymptotic limit of massless QED is a free photon (and photino) plus massless matter fields whose charge is completely screened. The theory does not have localized asymptotic states and no mass shell at all, no S matrix in the usual sense of this word. Yet, it is well defined in finite volume.

This phase of the theory is referred to as a *free phase.* Sometimes it is also called *the Landau zero-charge phase.* Strictly speaking, the model is ill-defined at short distances where the effective coupling grows and finally hits the Landau pole. To make it self-consistent, at short distances it must be embedded into an asymptotically free theory. This is not difficult to achieve. The Georgi-Glashow model gives an example of such an embedding.

Summarizing, even in the simplest Abelian example we encounter three different phases, or dynamical regimes: the Coulomb phase, the Higgs phase and the free (Landau) phase, depending on the values of the parameters of the model and the choice of the vacuum state (in the case of the vanishing mass parameter, $m = 0$). All these regimes are attainable in non-Abelian models too. The non-Abelian gauge theories are richer, however, since they admit one more dynamical regime, *confinement of color*, a famous property of QCD which attracted so much attention but still defies analytic solution.

1.3. *Non-Abelian Higgs Model; Monopoles*

Interactions of quarks and leptons at the fundamental level are described by non-Abelian gauge theories. The standard model of electroweak interactions is, probably, the most well-studied non-Abelian gauge theory in the Higgs phase. Apart from a few exotic phenomena (e.g. the baryon number violation at high energies, for reviews see Ref. [6]) all processes in this model occur in the weak coupling regime, and are well-understood. The dynamical content is almost exhausted by the perturbation theory; very small nonperturbative corrections are due to instantons.

Since this model is so well-studied and familiar to everybody, it makes more sense to discuss another example of a non-Abelian Higgs phenomenon, the Georgi-Glashow model [7]. This example is instructive and more relevant to the discussion below since this model exhibits magnetic monopoles [8]. A nice review of the Georgi-Glashow model, with special emphasis on this particular aspect, magnetic monopoles and dyons, can be found in Ref. [9].

The gauge group of the model is SU(2), so it has three gauge bosons. The matter sector includes one real scalar field ϕ^a in the adjoint representation of SU(2) (i.e. $a = 1, 2, 3$). The Lagrangian has the form

$$\mathcal{L} = -\frac{1}{4} G^a_{\mu\nu} G^a_{\mu\nu} + \frac{1}{2}(D_\mu \phi^a)(D_\mu \phi^a) - \frac{1}{4}\lambda(\phi^a\phi^a - v^2)^2 \tag{1.4}$$

where λ is a scalar coupling constant, and v is a constant of the dimension of mass determining the vacuum expectation value of the ϕ field. The so-called Bogomol'nyi-Prasad-Sommerfield (BPS) limit [10, 11]

$$\lambda \to 0\,, \qquad v \text{ fixed}\,, \tag{1.5}$$

is most relevant for our purposes. In this limit the scalar self-interaction disappears from the Lagrangian and the equations of motion. The only remnant of the scalar self-interaction term is the boundary condition for the field ϕ at spatial infinity. Indeed, requiring the energy of field configurations to be finite, we single out only those for which

$$\phi^a\phi^a \to v^2 \text{ at } |\mathbf{x}| \to \infty\,.$$

The BPS limit naturally emerges in many supersymmetric theories. If $v \neq 0$ the minimum of the classical energy is achieved for $\phi^a\phi^a = v^2$. In the weak coupling regime, when the gauge coupling constant g is small, the quantum vacuum of the model is characterized by a nonvanishing expectation value of ϕ^2. The vacuum field can always be chosen as follows

$$\phi^3 = v\,, \qquad \phi^{1,2} = 0\,. \tag{1.6}$$

It is not difficult to check that the gauge fields with the color indices 1, 2 propagating in the condensate (1.6) acquire masses $M_V = gv$, and become W bosons, while the gauge field with the color index 3 remains massless. The gauge transformations corresponding to rotations around the third axis leave the condensate (1.6) intact. Correspondingly, the gauge group SU(2) is spontaneously broken down to U(1); A^3 plays the role of the photon of the U(1) gauge theory. The particle spectrum of the theory, apart from this "photon", consists of one neutral massless scalar ϕ^3 (neutral with respect to the U(1) group), and two massive vector particles, $W^\pm = (2)^{-1/2}(A^1 \pm iA^2)$, with the charges ± 1, a rather typical pattern of the non-Abelian Higgs phenomenon. The 1, 2 components of the ϕ field are eaten up: they become the longitudinal components of W's.

Since the residual gauge symmetry is U(1), while at high energies (when the spontaneous symmetry breaking is inessential) we deal with the full original SU(2), which is a compact group, the low-energy electrodynamics obtained after the spontaneous breaking of SU(2) down to U(1) is actually compact. Topological arguments then prompt us [8, 9] that this model has topologically stable localized finite-energy configurations with a nonvanishing magnetic charge, magnetic monopoles. I cannot go into details regarding these objects, referring the interested reader to a vast literature devoted to the subject of the 't Hooft-Polyakov monopoles (recent review papers [12] contain a representative list of references). Here I will only sketch how the monopole mass can be calculated using a limited information coded in the asymptotics of the corresponding fields.

The Lagrangian (1.4) in the BPS limit implies that the energy E of any static field configuration (in the $A_0 = 0$ gauge) can be written as

$$E = \int d^3x \left(\frac{1}{4} G^a_{ij} G^a_{ij} + \frac{1}{2} D_k\phi^a D_k\phi^a \right)$$
$$= \int d^3x \frac{1}{4} (G^a_{ij} - \varepsilon_{ijk} D_k\phi^a)^2 + \int d^3x \frac{1}{2} \varepsilon_{ijk} G^a_{ij} D_k\phi^a \,. \qquad (1.7)$$

Now, the second term is actually an integral over a full derivative; it reduces identically to a two-dimensional integral over the large sphere,

$$\int d^3x \frac{1}{2} \varepsilon_{ijk} G^a_{ij} D_k\phi^a = \int d^3x \partial_k \left(\frac{1}{2} \varepsilon_{ijk} G^a_{ij} \phi^a \right) = \int_{S_2} d\sigma_k \left(\frac{1}{2} \varepsilon_{ijk} G^a_{ij} \phi^a \right), \qquad (1.8)$$

where $d\sigma_k$ is the area element. In deriving Eq. (1.8) it was taken into account that

$$D_\mu \tilde{G}_{\mu\nu} = 0 \,, \qquad \tilde{G}_{\mu\nu} = (1/2) \varepsilon_{\mu\nu\alpha\beta} G_{\alpha\beta} \,.$$

It is not difficult to show that the surface integral in Eq. (1.8) is nothing but the flux of the magnetic field through the large sphere, proportional to the topological charge. In this way we arrive at the following expression for the energy

$$E = \frac{4\pi\kappa v}{g} + \int d^3x \frac{1}{4} (G^a_{ij} - \varepsilon_{ijk} D_k\phi^a)^2 \qquad (1.9)$$

where κ is the topological charge of the configuration considered (see below), related to the magnetic charge m,

$$m = \frac{4\pi\kappa}{g} \,.$$

The second term is obviously positive-definite. Thus, in the sector with the given κ

$$E \geq \frac{4\pi\kappa v}{g} \,; \qquad (1.10)$$

the equality is achieved if and only if

$$G^a_{ij} = \varepsilon_{ijk} D_k\phi^a \,. \qquad (1.11)$$

Equation (1.11) is called the Bogomol'nyi condition. All states satisfying this condition are called the BPS-saturated states; the 't Hooft-Polyakov monopole belongs to this class. The mass of the monopole is, thus, unambiguously related to its magnetic charge, $M_{\rm mon} = |m|v$.

It remains to be added that a topologically nontrivial solution of the Bogomol'nyi condition corresponding to $\kappa = 1$ (the one-monopole solution) has a "hedgehog" form,

$$\phi^a(\mathbf{x}) = \frac{x^a}{r} F(r) , \qquad A_i^a(\mathbf{x}) = \varepsilon_{aij} \frac{x^j}{r} W(r) , \qquad A_0^a(\mathbf{x}) = 0 , \tag{1.12}$$

where

$$r = |\mathbf{x}|$$

and the asymptotics of the functions F and W are as follows:

$$F \to v , \qquad W \to \frac{1}{gr} \text{ at } r \to \infty , \qquad F, W \to 0 \text{ at } \mathbf{x} \to 0 . \tag{1.13}$$

Substituting the *ansatz* (1.12) in the Bogomol'nyi condition we check that it goes through, and get two coupled differential equations for the invariant functions F, W. The solutions of these equations satisfying the boundary conditions (1.13) are [8]

$$F = \frac{1}{gr}\left(\frac{grv}{\tanh(grv)} - 1\right) , \qquad W = -\frac{1}{gr}\left(\frac{grv}{\sinh(grv)} - 1\right) . \tag{1.14}$$

The gauge-invariant definition of the electromagnetic field tensor is

$$F_{\mu\nu} = \hat{\phi}^a G^a_{\mu\nu} - \frac{1}{g}\varepsilon^{abc}\hat{\phi}^a D_\mu \hat{\phi}^b D_\nu \hat{\phi}^c , \qquad \hat{\phi}^a = \phi^a / \sqrt{\phi^b\phi^b} . \tag{1.15}$$

The topological current, whose conservation is obvious, has the form

$$K_\mu = \frac{1}{8\pi}\varepsilon^{\mu\nu\alpha\beta}\varepsilon^{abc}\partial_\nu\hat{\phi}^a\partial_\alpha\hat{\phi}^b\partial_\beta\hat{\phi}^c . \tag{1.16}$$

The corresponding charge, $\kappa = \int d^3x K_0$, counts the windings of the mapping of the two-dimensional large sphere in the configurational space onto S_2, the group space of SU(2)/U(1). The reader is invited to check this statement, using the definition of the topological current.

Using Eq. (1.15) and the solution (1.14) it is not difficult to see that

$$\frac{1}{2}\varepsilon_{\mu\nu\alpha\beta}\partial^\nu F^{\alpha\beta} = \frac{4\pi}{g} K_\mu , \tag{1.17}$$

and, hence, the magnetic charge of the 't Hooft-Polyakov monopole

$$m \equiv \int_{S_2} d\sigma_i B_i$$

is indeed equal to $4\pi/g$. The magnetic charge quantization condition is, thus,

$$gm = 4\pi\,. \tag{1.18}$$

The magnetic charge quantum is seemingly twice larger than for the Dirac monopole [13]. (Note, however, that Eq. (1.18) coincides with the Schwinger quantization condition [14]). This is due to the fact that the electric charge of the W bosons, g, is not the minimal one, in principle. It is conceivable that the probe matter fields in the fundamental (doublet) representation are added in the Georgi-Glashow model. Then, their charge with respect to the U(1) is $g/2$, and the product of the minimal electric charge and the magnetic charge of the monopole is 2π, as required by Dirac's argument.

Shortly after the discovery of the monopoles in the Georgi-Glashow model it was pointed out [15] that the same model also has *dyon* solutions — localized field configurations carrying both, the magnetic and electric charges. The monopole solution (1.12), (1.14) carries no electric charge, since the fields are time independent and $A_0 = 0$; hence, $E_i = -F_{0i} \equiv 0$. One can modify the hedgehog *ansatz* (1.12), keeping its static nature but allowing for $A_0 \neq 0$. In this way one can obtain [15] in the BPS limit an analytic solution in which both integrals

$$\frac{1}{4\pi}\int_{S_2} d\sigma_i B_i \qquad \text{and} \qquad \frac{1}{4\pi}\int_{S_2} d\sigma_i E_i$$

are nonvanishing. These objects, dyons, also supposedly play a role in some mechanisms ensuring color confinement.

1.4. *Phases of Gauge Theories (Non-Abelian Version)*

The Georgi-Glashow model discussed above teaches us that in the non-Abelian case the microscopic variables in the Lagrangian (gauge bosons, adjoint matter fields) do not necessarily coincide with those quanta that can be observed (massive W bosons, magnetic monopoles). In the weak coupling regime the relation between the microscopic and macroscopic degrees of freedom is pretty transparent, though. The theoretical situation in QCD is far from being so cloudless. The QCD Lagrangian is well established. Thus, we know that at short distances the microscopic variables are colored quarks and gluons. The macroscopic degrees of freedom are strongly bound states whose analysis cannot be carried out perturbatively or in the semiclassical approximation. Empirically we know that asymptotic states are colorless hadrons. Thus, if the quarks have fractional charges (2/3 and −1/3), all observable asymptotic states have integer charges and are built from the quark–antiquark pairs (mesons) or three quarks (baryons). Here we encounter for the first time in our brief excursion a new phase of the gauge theory, the *confining phase*.

Consider pure gluodynamics, i.e. the theory of gluons, with no dynamical quarks. The Lagrangian has the familiar form

$$\mathcal{L} = -\frac{1}{4g_0^2} G^a_{\mu\nu} G^a_{\mu\nu} \,, \tag{1.19}$$

where g_0^2 is the gauge coupling at the ultraviolet cut-off. Although this coupling is dimensionless, actually the true parameter characterizing the interactions in the theory is the scale Λ related to g_0^2 as follows

$$\Lambda = M_0 \exp\left[\left(-\frac{8\pi^2}{\beta_0 g_0^2}\right)\right]\left(\frac{8\pi^2}{g_0^2}\right)^{\beta_1/\beta_0^2} \tag{1.20}$$

where β_0 and β_1 are the first and the second coefficients in the Gell-Mann-Low function ($\beta_0 = 11$ in QCD). Unlike the Higgs-phase standard model, QCD is strongly coupled at momenta of order of Λ; all interesting dynamical features of this theory reflect the strong-coupling dynamics. The intricacies of this dynamics are such that the microscopic degrees of freedom — gluons — must disappear at distances larger than Λ^{-1}, giving place to macroscopic degrees of freedom, hadrons. A qualitative picture of how and why this process might take place is believed to be known.

If we place two heavy (static) color charges at a large distance from each other they create a chromoelectric field which is believed to form a flux tube between the charges. This flux tube of the confining non-Abelian theory substitutes the dispersed Coulomb field that one observes between the static charges in electrodynamics (in the Coulomb phase). The flux tube is a string-like one-dimensional object, with the cross-section $\sim \Lambda^{-2}$, and constant string tension $\sigma \sim \Lambda^2$. Then the interaction energy of the two static charges grows linearly with the distance between them, $V(R) = \sigma R$, and they can never be separated asymptotically, since this separation costs infinite energy. The formal signature of this regime is the area law for the Wilson loop.

Do we have any precedents of such a behavior — constant force, linearly rising potential — in the dynamical systems studied previously?

There exists one example known for a long time. Let us return to supersymmetric electrodynamics, Eq. (1.1). The gauge symmetry is U(1), and if $\phi \neq 0$ the theory is in the Higgs phase. A nonrelativistic analog of this theory is nothing but the Ginzburg-Landau model of superconductivity, describing the Bose-condensation of the Cooper electron pairs in the vacuum state. An electrically charged order parameter develops a nonvanishing expectation value. The vector quanta acquire mass, and the electric potential becomes short-range. The magnetic fields are repelled completely from the domain where the condensate develops, the famous Meissner effect. Assume, however, that two static magnetic charges (magnetic monopoles) are placed by hand inside this domain. Since the magnetic flux is conserved, the magnetic field cannot vanish everywhere. It experiences a strong repulsion in the vacuum medium which results in the formation of narrow flux tubes connecting the magnetic charges. The flux tubes are solutions of the classical equations of motion corresponding to the overall 2π change of the phase of the field ϕ when one makes a full rotation around a line connecting the magnetic charges. To avoid singularity the value of the ϕ field in the center of the tube must vanish. These solutions in the Ginzburg-Landau theory were found by Abrikosov (Abrikosov vortices).[1] It is not difficult to calculate the energy of the vortex per unit length — it is a constant at large distances from the sources. In other words, the energy between two magnetic charges in the superconducting medium grows linearly with the separation between the charges. This is exactly what we need for color confinement in QCD.

Turning to QCD we immediately notice two important differences: one conceptual and the other technical. The first difference in QCD is that we want to form chromoelectric, not chromomagnetic flux tubes. The vacuum medium must repel chromoelectric fields. This can only be achieved by the condensation of the magnetic charges, rather than the electric ones, as in the Meissner effect. Thus, if a mechanism of this type ensures color confinement, it must be a *dual Meissner effect.* The problem in QCD is that the classical 't Hooft-Polyakov monopoles do not exist as physical objects (particles).

Second, if in the Ginzburg-Landau model one can adjust parameters to ensure the weak coupling regime, where semiclassical methods are perfectly

[1]A remark for more educated readers: mathematically, the existence of the topologically stable vortices is due to the fact that $\pi_1(\mathrm{U}(1)) = \mathbf{Z}$.

applicable, QCD is a genuinely strong coupled theory, and we do not expect any semiclassical approach to be valid except, perhaps, in qualitative pictures intended for orientation.

A possible solution of the first problem was indicated by 't Hooft [3]. Even though QCD, unlike the Georgi-Glashow model, does not have magnetic monopoles as physical objects, whose existence is a gauge-independent fact, it may still have solutions that *in a certain gauge* look like monopoles. The presence of the appropriate field configurations will, thus, depend on the choice of the gauge. Nevertheless, one may hope, that being found in some gauge, they may turn out to be important for implementing the dual Meissner effect in QCD.

The second problem — the strong coupling regime in QCD — cannot be eliminated in this way, of course. Therefore, to build a fully controllable theoretical description, one must try to implement the 't Hooft-Mandelstam idea, the dual Meissner effect leading to color confinement, beyond the semiclassical approximation. How one could do this in QCD, and whether it is possible at all, is still unclear. At the same time, a remarkable progress was achieved in non-Abelian supersymmetric gauge theories, where in certain instances something similar to the dual Meissner effect could be rigorously proven [16].

We will return to the 't Hooft suggestion of the QCD "monopole" condensation shortly, and now continue our general discussion of the confining phase. It has been already mentioned that in pure gluodynamics (QCD with no quarks) the area law is believed to take place for the Wilson loop. Clearly, we cannot make experiments in pure gluodynamics, but the numerous numerical simulations on the lattices seem to reveal this type of behavior (within usual uncertainties and other natural limitations — finite volume, etc. — inherent to any numerical analysis). There is an invariant clear-cut distinction between the confining and the Higgs phases in the case when all fields appearing in the Lagrangian belong to the adjoint representation. One can consider the Wilson loop

$$W(C) = \mathrm{Tr}\ \exp \oint (iA_\mu dx_\mu)\,, \qquad A_\mu = A_\mu^a T^a$$

in the fundamental representation (i.e. with the generator matrices T^a referring to the fundamental representation; for SU(3), for instance, $T^a = \lambda^a/2$ where λ^a are the Gell-Mann matrices). In the confining phase, for large contours C the Wilson loop $W(C) \sim \exp(-\sigma \cdot (\text{area}))$. The area law reflects the formation of the flux tube of the chromoelectric field attached to the probe fundamental (very) heavy quarks. The color charge cannot be screened, and the flux tube cannot end. It either begins at the color charges or forms closed contours.

In the Higgs phase the color field originating at the point of the color charge is exponentially screened. No long chromoelectric flux tubes exist; the potential between the two separated probe charges saturates at some constant value. Correspondingly, the Wilson loop for large contours behaves as $W(C) \sim \exp(-\Lambda \cdot (\text{perimeter}))$.

If we treat v in Eq. (1.4) as a free parameter, for large values of v we are in the Higgs phase, with the perimeter law for the Wilson loop, while at $v \to 0$ we are presumably in the confining phase, with the area law. At some critical value of v,

$$v_* \sim \Lambda\,,$$

a phase transition from the Higgs to the confinement phase must occur.

Now, if we introduce, additionally, some dynamical fields in the fundamental representation, say, quarks, the Wilson loop no more differentiates between the two regimes. Indeed, the field of the static probe quarks can be screened now by dynamical (anti)quarks. The potential at large distances saturates at a constant value, and the perimeter law always takes place.

As a matter of fact, if the Higgs field itself is in the fundamental representation of the color group, there is no distinction at all between the confinement phase and the Higgs phase. As the vacuum expectation value (VEV) of the Higgs field continuously changes from large values to smaller ones, we continuously flow from the weak coupling regime to the strong coupling one. The spectrum of all physical states, and all other measurable quantities, change smoothly [17]. One can argue that the situation is such in many different ways. Perhaps, the most straightforward line of reasoning is as follows. Using the Higgs field in the fundamental representation one can build gauge invariant interpolating operators for *all* possible physical states. By physical states I mean a part of the Hilbert space, with the gauge equivalent points eliminated. The Källén-Lehmann spectral function corresponding to these operators, which carries complete information on the spectrum, obviously depends smoothly on $\langle\chi^*\chi\rangle$. When the latter parameter is large the Higgs description is convenient, when it is small it is more convenient to think in terms of the bound states. There is no boundary, however. We deal with a single Higgs/confining phase [17].

To elucidate this point in more detail let us consider a specific model. Namely, we will introduce in the Lagrangian (1.4), in addition to the adjoint

Higgs ϕ, a complex doublet field χ^i, $= 1, 2$, with the vacuum expectation value η. We will assume that VEV of the field ϕ vanishes, and will study the η dependence of the physical quantities.

This model has a global SU(2) symmetry, associated with the possibility of rotating the doublet χ^i into the conjugated doublet $\varepsilon^{ij}\chi_j^\dagger$. The SU(2) symmetry of the χ sector becomes explicit if we introduce a matrix field

$$X = \begin{pmatrix} \chi^1 & -\chi^{2\dagger} \\ \chi^2 & \chi^{1\dagger} \end{pmatrix}, \tag{1.21}$$

and rewrite the Lagrangian in terms of this matrix,

$$\Delta\mathcal{L}_\chi = \frac{1}{2}\,\mathrm{Tr}\,D_\mu X^\dagger D_\mu X - \frac{\tilde{\lambda}}{4}\left(\frac{1}{2}\,\mathrm{Tr}\,X^\dagger X - \eta^2\right)^2 . \tag{1.22}$$

All physical states form representations of the global SU(2). Consider, for instance, SU(2) triplets produced from the vacuum by the operators

$$W_\mu^a = -\frac{i}{2}\,\mathrm{Tr}(X^\dagger \overleftrightarrow{D}_\mu X\sigma^a), \qquad a = 1, 2, 3 . \tag{1.23}$$

The lowest-lying states produced by these operators in the weak coupling regime (i.e. when $\langle\chi^\dagger\chi\rangle \gg \Lambda^2$) coincide with the conventional W bosons of the Higgs picture, up to a normalization constant. The mass of the W bosons is $\sim g\eta$. On the other hand, if $\langle\chi^\dagger\chi\rangle \lesssim \Lambda^2$ it is more appropriate to think of the bound states of the χ quanta forming vector mesons, triplet with respect to the global SU(2) ("ρ mesons"). Their mass is $\sim \Lambda$. Continuous evolution of η results in the continuous evolution of the mass of the corresponding states. It is easy to check that the complete set of the gauge invariant operators that can be built in this model spans the whole Hilbert space of the physical states.[2]

I hasten to add that by introducing matter fields in the fundamental representation we do not necessarily kill all phase transitions. For example in the SU(2) model considered above one can put $\eta = 0$ and study the phase transition with respect to the expectation value of the adjoint field ϕ. It is quite obvious that for large v we deal with the Abelian Coulomb phase, while when v is small, confinement presumably takes place. What we do kill is the

[2]The absence of the phase transition and the existence of a unified Higgs/confinement phase in the case when the Higgs field is in the fundamental representation blocks any attempts of modeling "slightly unconfined" quarks by using a mechanism of De Rujula *et al.* [18]. This mechanism simply does not exist.

Wilson loop as the order parameter. However, one can differentiate between the phases by using other criteria.

To this end one assigns some "external" flavor quantum numbers, say, to the quark fields. One of the possibilities is the electric charge.[3] The quarks in QCD are fractionally charged, all up quarks have charges $2/3$ while all down quarks charges $-1/3$. The electromagnetic interaction is external with respect to QCD and has nothing to do with color confinement. It is used just as a marker of particular states. Some other flavor markers will do the job as well, say the Gell-Mann vector SU(3) existing in QCD with the massless u, d, s quarks, or baryon number, and so on.

Assume that *two* scalar fields in the *adjoint* representation are added in QCD, as in Eq. (1.4). This is sufficient to break the QCD gauge group SU(3) completely. When the expectation values of these scalar fields are large, the theory is in the (weakly coupled) Higgs phase. The states with the fractional baryon numbers (and the fractional electric charge; to be referred as *fractionals* below) exist in the spectrum, as asymptotic states. Moreover, since the gauge coupling is small, these states are essentially undressed, and light. The states with the integer baryon numbers composed of the quark–antiquark pairs or triquarks also exist but they are not necessarily bound. Their energy is higher than those with the fractional baryon numbers.

As we move to smaller vacuum expectation values of the scalar fields, the mass of the fractionals grows, since the fields induced become stronger. Bound states of the quark–antiquark pairs or triquarks become energetically advantageous. At a certain point the fractionals become infinitely heavy, so they are not seen in the observable spectrum. Those states which have finite mass and are seen have integer baryon numbers. Presumably, there is a phase transition from the Higgs to the confining phase, although the order parameter is not obvious in the case at hand.

There exists a dynamical scenario in which some fractionals will still be seen in the spectrum, the so-called *oblique confinement* [3, 19]. So far we discussed the dual Meissner effect, with condensed monopoles. The monopole condensation forces the chromoelectric field to form tubes and makes quarks confined. In Sec. 1.3 we got acquainted with the dyon configurations in the

[3] I mean here the genuine electric charge, not to be confused with the "charges" of the QCD monopoles and dyons. The latter are understood as the charges with respect to some U(1) subgroup of the original gauge group, SU(3) in the case of QCD, cf. Sec. 1.3. To consider the true electromagnetic interaction we add an extra U(1). Say, extended QCD including electromagnetism has the gauge group $SU(3)_c \times U(1)_{em}$.

Georgi-Glashow model. They were characterized by nonzero magnetic and electric charges, simultaneously. The existence of dyons is inevitable in any gauge theory with magnetic monopoles. As was shown by Witten [20], introducing a nonvanishing vacuum angle ϑ,

$$\mathcal{L}_\vartheta = \frac{\vartheta g^2}{32\pi^2} G^a_{\mu\nu} G^a_{\mu\nu} \tag{1.24}$$

necessarily generates an electric charge q for a particle with the magnetic charge m,

$$q = \frac{\vartheta g^2}{8\pi^2} m\,. \tag{1.25}$$

It is the condensation of dyons that sets up the phase of the oblique confinement.

Let us discuss the phenomenon in more detail. Figure 1(a) represents the spectrum of possible electric/magnetic charges in the SU(2) theory with $\vartheta = 0$.[4] Elementary states with the electric charge are along the horizontal axis. If the matter fields are in the adjoint representation their quanta have charges $0, \pm 1$ (point A). Two quanta can have charges ± 2, and so on. We may want to introduce quarks in the fundamental (SU(2) doublet) representation; their charges are $\pm 1/2$. Remember, the electric charges are measured in the units of g.

The magnetic monopole lies on the vertical axis (point B). It has $m = 1$, $q = 0$. Other states on the vertical axis are antimonopole, a pair of monopoles and so on. All points which do not belong to the horizontal and vertical axes are bound states of electric and magnetic quanta. Note that the monopole condensation automatically precludes from condensation all states carrying the electric charge, since their mass squared is positive (and infinite). The latter are confined. All states which do not lie on the straight line connecting the origin with the point B are confined by the flux tubes of the chromoelectric field attached to them.

Increasing ϑ we deform the $\{q, m\}$ grid in a continuous way. When ϑ is close to π, but slightly larger than π, we arrive at the grid shown in Fig. 1(b). It is quite obvious that if $\vartheta \neq 0$ or 2π the grid of all possible values of $\{q, m\}$ is oblique (which explains where the name *oblique confinement* comes from). At $\vartheta = \pi$ the objects with the minimal magnetic charge $m = \frac{\vartheta\pi}{g}$ will have

[4] The electric charge here has nothing to do with the conventional electromagnetism. This is the charge with respect to a U(1) singled out by the 't Hooft Abelian projection. See the next section for further details.

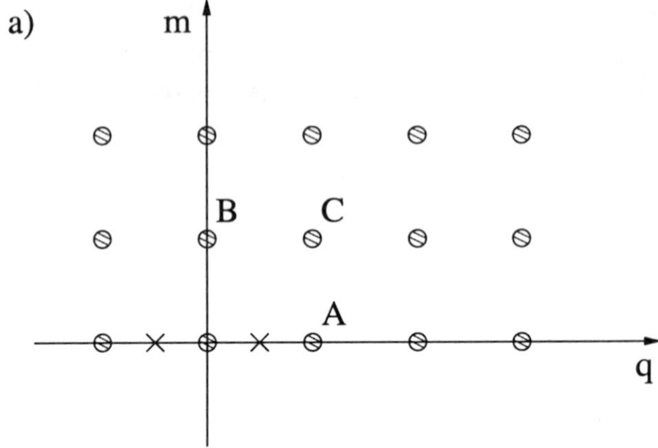

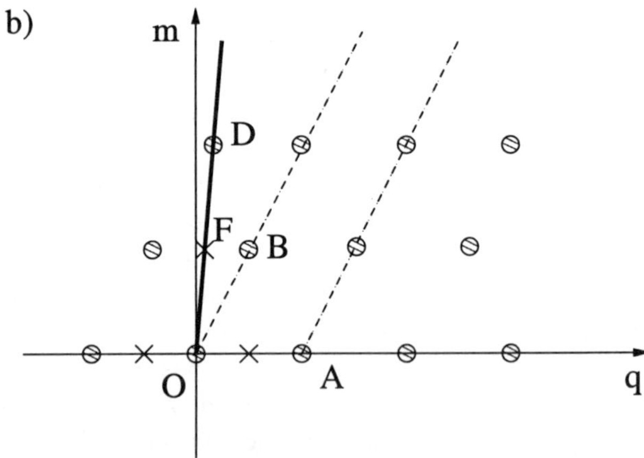

Fig. 1. The grids of the electric-magnetic charges in the SU(2) gauge theory. The charges $\{q, m\}$ are measured with respect to the U(1) subgroup of SU(2), as in Sec. 1.5. The closed circles indicate the charges in the theory with the adjoint (SU(2) triplet) matter. The crosses indicate the electric charge of the matter field in the fundamental (SU(2) doublet) representation. (a) The rectangular grid corresponding to $\vartheta = 0$. The point denoted by C is a bound state of the triplet matter quantum with the monopole. Bound states of the doublet matter quanta with monopoles are possible too (but not indicated). (b) The oblique grid corresponding to $\vartheta = \pi + \varepsilon$, $0 < \varepsilon \ll \pi$. The state denoted by D is a dyon with a small value of the electric charge, which presumably condenses. The point denoted by F is an unconfined bound state of the quark and dyon B. Its external quantum numbers are those of the quark.

$q = \pm 1/2, \pm 3/2$ and so on. It is highly likely that such objects will not condense, since apart from the magnetic charge they have large (and opposite) electric charges. But among the $m = 8\pi/g$ monopoles there is one with the vanishing electric charge, and it is quite conceivable that the condensation of these objects is more energy-expedient. If ϑ is slightly larger than π the state suspect of condensing is that marked by D in Fig. 1(b). All states that do not lie on the straight line connecting the origin with D will be confined. This is a very peculiar confinement, however. Some states, carrying the external quantum numbers of the fundamental quarks, exist in the observable spectrum. The quarks themselves (crosses on the horizontal axis) do not lie on the OD line and, hence, are confined. The bound state of the quark and a dyon (the cross marked by the letter F) does belong to this line, and is *not* confined. Since the dyon has no baryon charge, or any other external quantum number, the state F has exactly the same baryon charge as the fundamental quark, i.e. it is a fractional.

A price has to be paid for having fractionals in the observable spectrum. As explained in Ref. [3], even though the quarks are described by the Fermi fields at the Lagrangian level, the states with the fractional (and nonvanishing) baryon charge to be observed will all be bosons!

1.5. *QCD Monopoles and Abelian Projection*

Let us now discuss how monopole-like configurations could emerge in QCD. Following 't Hooft we will impose an *incomplete* gauge fixing condition in a special way. In the Georgi-Glashow model the Higgs mechanism breaks the original SU(2) down to U(1), which paves the way to the emergence of the monopoles as physical objects. In QCD the gauge group remains unbroken, of course. Our task is to single out an U(1) subgroup by imposing an appropriate gauge condition. The monopole-like solution, obtained in this way certainly cannot be interpreted as a physical particle; it should be viewed rather as a first stage in a construction which, eventually, may answer the question whether the dual Meissner effect takes place in QCD.

That the monopole-like solutions of the classical equations of motion are present in QCD can be seen in many different ways. For instance, assume that (an incomplete) gauge condition is imposed in such a way that the time-dependent gauge transformations are forbidden. Whatever this gauge condition might be, it still does not forbid the gauge transformations which depend on the spatial coordinates, $\mathbf{x}$. For simplicity we will additionally assume that the gauge group is SU(2), and not SU(3) as in genuine QCD. Under this narrow class of gauge transformations $A_0^a A_0^a$ is obviously gauge invariant, since with respect to the spatial-dependent gauge transformations the time component of A transforms homogeneously. If so, the model becomes identical to the BPS limit of the Georgi-Glashow model in the $A_0 = 0$ gauge, which is known to possess the monopole solutions.

Let us elucidate the latter assertion in more detail. The Lagrangian of the SU(2) Yang-Mills theory in the above "gauge", for the static field configurations, takes the form

$$\mathcal{L} = -\frac{1}{4} G_{ij}^a(\mathbf{x}) G_{ij}^a(\mathbf{x}) + \frac{1}{2} D_i A_0^a(\mathbf{x}) D_i A_0^a(\mathbf{x}) , \tag{1.26}$$

where

$$D_i A_0^a = \partial_i A_0 + g\varepsilon^{abc} A_i^b A_0^c$$

is the covariant derivative. If we now rename $A_0^a \to i\phi^a$, the Lagrangian (1.26) will coincide with that of the Georgi-Glashow model in the $A_0 = 0$ gauge (in the BPS limit).[5] The classical equations of motion are identical and should then have identical solutions; in particular, the monopole solution (1.14) goes through. In the case at hand v^2 is the value of $A_0^a A_0^a$ at the spatial infinity. Since the time-dependent gauge transformation is forbidden, this value is well defined.

I pause here to make a few explanatory remarks. The 't Hooft-Polyakov monopole in the BPS limit satisfies the condition (1.11). One immediately recognizes in this condition, in our particular case, the self-duality equation $G_{\mu\nu} = (1/2)\varepsilon_{\mu\nu\alpha\beta} G_{\alpha\beta}$ defining the (multi)instanton solutions. Thus, heuristically it is clear that the "monopole" solution found must be equivalent (up to a gauge transformation) to a chain of instantons. Since the "monopole" mass is finite, the action is infinite; therefore, we deal here with an infinite chain of instantons. The fact that the equivalence does indeed take place was demonstrated in Ref. [21] where it was established that the monopole can be identified with a sequence of equally spaced instantons located at the time axis. The spacing between the instanton centers is $8\pi^2 M_{\rm mon}^{-1} g^{-2}$, and their radii tend to infinity. Some other chains were shown to be gauge-equivalent to dyons [22]. The inverse transition, from monopoles to instantons, was also studied. It was

[5] Since $A_0^a \to i\phi^a$ and is purely imaginary we deal here with an analytic continuation to the Euclidean space.

found that the instanton considered in the Abelian projection (see below) coincides with a closed monopole loop centered at the instanton center [23]. For a discussion of the transition from the "monopoles" to the instanton chains and back see Ref. [24].

Those readers who feel uneasy with the notion of monopoles in QCD can think of the solution above as of a sequence of instantons. The instantons are, of course, more familiar to the QCD practitioners.

A related but somewhat more general line of reasoning leading to the same monopole-like configurations was suggested by 't Hooft [3]. His *Abelian projection* of QCD can be elucidated as follows. Let us consider some gauge noninvariant operator in the adjoint representation of the gauge group transforming homogeneously. In pure gluodynamics this may be, say, G^a_{12}. If quarks are present one might consider $\bar\psi_i\psi^j$ where i and j are color indices. It may be more convenient, however, to introduce (auxiliary) scalar fields in the adjoint representation, ϕ^a, making them sufficiently heavy, so they do not affect the low-energy physics. Generically, this operator will be denoted as $X_i^j \equiv X^a(T^a)_i^j$ where the matrices T^a are the generators of the gauge group. For simplicity we will assume that the gauge group is SU(N); then, $(T^a)_i^j$ are N by N matrices generalizing the Gell-Mann matrices of QCD. The gauge transformation acts on X as $X \to UXU^{-1}$ where U is an arbitrary x dependent matrix from SU(N). It is quite obvious that by choosing $U(x)$ in an appropriate way it is always possible to diagonalize X. In other words, the only components of X^a surviving after the gauge transformation are those corresponding to the Cartan subalgebra of SU(N): $a = 3, 8, 15$ and so on. This "Abelization" of the operator X obviously explains why the gauge is referred to as the Abelian projection.

The gauge condition above is an incomplete gauge, since one can additionally perform gauge transformations corresponding to arbitrary rotations around the third, eighth and so on axes without destroying the Abelian nature of X. The diagonal form of the matrix X is preserved under these rotations: all generators T^3, T^8, T^{15}, ... are diagonal. Thus, in the gauge at hand, $\mathrm{U}(1)^{N-1}$ subgroup is singled out, SU(N) gluodynamics looks similar to QED, with $N-1$ different photons (A^3, A^8, A^{15}, etc.) and $(1/2)N(N-1)$ "matter fields" (all off-diagonal A's) charged with respect to the photon fields. The values of the charges are, generally speaking, different for different photons. They depend on the group constants.

Intuitively it is clear that we are going to have the "monopole" solutions. Formally, this can be observed as follows. In the Abelian projection the operator X is a diagonal matrix, $X = \text{diag}\{\lambda_1, \lambda_2, \ldots, \lambda_N\}$ where the eigenvalues λ may be ordered, $\lambda_1 > \lambda_2 > \cdots > \lambda_N$. In some exceptional points in space, however, two out of N eigenvalues can coincide, say λ_1 and λ_2. This singles out the upper left SU(2) corner of SU(N) with a monopole sitting at the point where $\lambda_1 = \lambda_2$.

Indeed, close to this point we can focus on the upper left two-by-two corner of the matrix X, assuming that the remaining part of the matrix is already diagonal, with noncoinciding eigenvalues. We have to diagonalize only the upper left-hand corner. Near the point $\mathbf{x}_0$ where $\lambda_1 = \lambda_2$ the two-by-two submatrix has the form

$$\lambda \mathbf{1} + \varepsilon^a(\mathbf{x})\sigma^a$$

where $\mathbf{1}$ is the two-by-two unit matrix and σ^a are the Pauli matrices. The condition $\lambda_1 = \lambda_2$ at $\mathbf{x}_0$ means that $\varepsilon^a = 0$ at $\mathbf{x} = \mathbf{x}_0$ ($a = 1, 2, 3$). Since we have to ensure that the three (real) functions vanish, generically this can happen only on manifold of dimension zero, i.e. in isolated points in space, as was mentioned. Moreover, near these points $\varepsilon^a(\mathbf{x})$ has a hedgehog configuration, $\varepsilon^a(\mathbf{x}) \sim (x - x_0)^a$, a characteristic feature of the 't Hooft-Polyakov monopole.

In the absence of a genuinely small parameter in QCD, the semiclassical description sketched above cannot lead us too far beyond a qualitative picture. Definitely, there is no way one can study the monopole condensation, a crucial element of the confinement mechanism-to-be, in this approximation. One can try, however, to apply these ideas in the context of the lattice simulations, hoping to get precious insights from numerical studies. Although work in this direction is far from completion, and many aspects remain unclear (see e.g. Ref. [25]), some initial results are, perhaps, encouraging. In particular, the abundance of the "monopoles" in the vacuum ensemble of the lattice QCD was observed, and a connection with the chiral symmetry breaking conjectured [26]. Approaches combining analytical methods with numerical analysis are under investigation (see e.g. Ref. [27]).

Summarizing, we have learned that the gauge theories can be in the following phases:

(1) Coulomb (Abelian and non-Abelian; in the latter case the infrared limit is conformal, as will be discussed in detail in Sec. 3);
(2) free (Landau);

(3) Higgs (a possible version is a unified Higgs/confining phase);
(4) confining (a possible version is oblique confinement).

Now that we are familiar with the various dynamical scenarios one expects to observe in different gauge theories, we are ready to proceed to supersymmetric theories where, in some cases, it is possible to go far beyond qualitative ideas, towards exact solution, using miracles of supersymmetry.

Section 2
Basics of Supersymmetric Gauge Theories

The dream of every QCD practitioner is to find analytical solutions for two most salient properties of QCD: color confinement and spontaneous breaking of the chiral symmetry. In spite of two decades of vigorous efforts very little progress has been made in this direction. At the same time, exciting developments took place, mostly in the last few years, in supersymmetric gauge theories — close relatives of QCD. These developments can, eventually, lead to a breakthrough in QCD. Even if this does not happen, they are very interesting on their own. It turns out that supersymmetry helps reveal several intriguing and extremely elegant properties which shed light on subtle aspects of the gauge theories in general. In this section we will start our excursion in supersymmetric gauge theories. There is a long way to go, however, before we can discuss a variety of fascinating results recently obtained in this field. As a first step let me briefly review some basic elements of the formalism that will be needed.

2.1. *Introducing Supersymmetry*

Supersymmetry relates bosonic and fermionic degrees of freedom. A necessary condition for any theory to be supersymmetric is the balance between the number of the bosonic and fermionic degrees of freedom, having the same mass and the same "external" quantum numbers, e.g. color. Let us consider several simplest examples of practical importance.

A scalar complex field ϕ has two degrees of freedom (a particle plus antiparticle). Correspondingly, its spinor superpartner is the Weyl (two-component) spinor, which also has two degrees of freedom — say, the left-handed particle and the right-handed antiparticle. Alternatively, instead of working with the complex fields, one can introduce real fields, with the same physical content: two real scalar fields ϕ_1 and ϕ_2 describing two "neutral" spin 0 particles, plus the Majorana (real four-component) spinor describing a "neutral" spin 1/2 particle with two polarizations. (By neutral I mean that the corresponding antiparticles are identical to their particles). This family has a balanced number of the degrees of freedom both in the massless and massive cases. We will see below that in the superfield formalism it is described, in a concise form, by one *chiral superfield.*

When we speak of the quark flavors in QCD we count the Dirac spinors. Each Dirac spinor is equivalent to two Weyl spinors. Therefore, in SQCD each flavor requires two chiral superfields. Sometimes, the superfields from this chiral pair are referred to as subflavors. Two subflavors comprise one flavor.

Another important example is vector particles, gauge bosons (gluons in QCD, W bosons in the Higgs phase). Each gauge boson carries two physical degrees of freedom (two transverse polarizations). The appropriate superpartner is the Majorana spinor. Unlike the previous example the balance is achieved only for massless particles, since the massive vector boson has three, not two, physical degrees of freedom. The superpartner to the massless gauge boson is called gaugino. Notice that the mass can still be introduced through the (super) Higgs mechanism. We will discuss the Higgs mechanism in supersymmetric gauge theories later on.

In counting the degrees of freedom above, the external quantum numbers were left aside. Certainly, they should be the same for each member of the superfamily. For instance, if the gauge group is SU(2), the gauge bosons are "color" triplets, and so are gauginos. In other words, the Majorana fields describing gauginos are provided by the "color" index a taking three different values, $a = 1, 2, 3$.

If we consider the free field theory with the balanced number of degrees of freedom, the vacuum energy vanishes. Indeed, the vacuum energy is the sum of the zero-point oscillation frequencies for each mode of the theory,

$$E_{\rm vac} = \sum_{\mathbf{k}} \{\omega_{\mathbf{k}}^{\rm boson} - \omega_{\mathbf{k}}^{\rm ferm}\}\,. \tag{2.1}$$

I remind that the modes are labeled by the three-momentum $\mathbf{k}$; say, for massive particles

$$\omega_{\mathbf{k}} = \sqrt{m^2 + \mathbf{k}^2}\,.$$

It is important that the boson and fermion terms enter with opposite signs and cancel each other, term by term. This observation, which can be considered as a precursor to supersymmetry, was made by Pauli in 1950 [28]! If interactions are

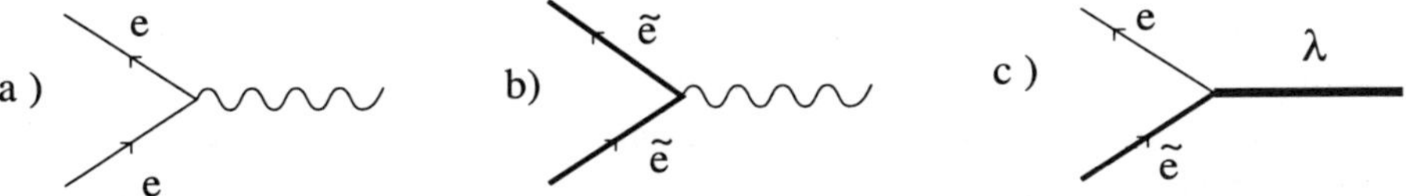

Fig. 2. Interaction vertices in QED and its supergeneralization, SQED. (a) $\bar{e}e\gamma$ vertex; (b) selectron coupling to photon; (c) electron–selectron–photino vertex. All vertices have the same coupling constant. The quartic self-interaction of selectrons is also present, but not shown.

introduced in such a way that supersymmetry remains unbroken, the vanishing of the vacuum energy is preserved in dynamically nontrivial theories.

Balancing the number of degrees of freedom is the necessary but not sufficient condition for supersymmetry in dynamically nontrivial theories, of course. All vertices must be supersymmetric too. This means that each line can be substituted by that of a superpartner. Let us consider, for instance, QED, the simplest gauge theory. We start from the electron–electron–photon coupling (Fig. 2(a)). Now, as we already know, in SQED the electron is accompanied by two selectrons (two, because the electron is described by the four-component Dirac spinor rather than the Weyl spinor). Thus, supersymmetry requires the selectron–selectron–photon vertices, (Fig. 2(b)), with the same coupling constant. Moreover, the photon can be substituted by its superpartner, photino, which generates the electron–selectron–photino vertex (Fig. 2(c)), with the same coupling. In the old-fashioned language of the pre-SUSY era we would call this vertex the Yukawa coupling. In the supersymmetric language this is the gauge interaction since it generalizes the gauge interaction coupling of the photon to the electron.

With the above set of vertices one can show that the theory is supersymmetric at the level of trilinear interactions, provided that the electrons and the selectrons are degenerate in mass, while the photon and photino fields are both massless. To make it fully supersymmetric one should also add some quartic terms, describing self-interactions of the selectron fields, as we will see shortly.

Now, the theory is dynamically nontrivial, the particles — bosons and fermions — are not free and still $E_{\rm vac} = 0$. This is the *first miracle* of supersymmetry.

The above pedestrian (or step-by-step) approach to supersymmetrizing the gauge theories is quite possible, in principle. Moreover, historically the first supersymmetric model derived by Golfand and Likhtman, SQED, was obtained in this way [4]. This is a painfully slow method, however, which is totally out of use at the present stage of the theoretical development. The modern efficient approach is based on the superfield formalism, introduced in 1974 by Salam and Strathdee [29] who replaced the conventional four-dimensional space by the superspace.

2.2. *Superfield Formalism: Bird's Eye View*

I will be unable to explain this formalism, even briefly. The reader is referred to the textbooks and numerous excellent reviews, the list of recommended literature is provided at the end of the lecture. Below some elements are listed mostly with the purpose of introducing relevant notations, to be used throughout the entire lecture course. (Our notation, conventions and useful formulae are collected in Appendix).

If the conventional space-time is parametrized by the coordinate four-vector x_μ, the superspace is parametrized by x_μ and two Grassmann variables, θ and $\bar\theta$. The Grassmann numbers obey all standard rules of arithmetic except that they anticommute rather than commute with each other. In particular, the product of a Grassmann number with itself is zero, for this reason.

With respect to the Lorentz properties, θ and $\bar\theta$ are spinors. As widely known, the four-dimensional Lorenz group is equivalent to $\mathrm{SU}(2)\times\mathrm{SU}(2)$ and, therefore, there exist two types of spinors, left-handed and right-handed, denoted by undotted and dotted indices, respectively; θ_α is the left-handed spinor while $\bar\theta_{\dot\alpha}$ is the right-handed one ($\alpha, \dot\alpha = 1, 2$). The indices of the right-handed spinors are supplied by dots to emphasize the fact that their transformation law does not coincide with that of the left-handed spinors.

The Lorentz scalars can be formed as a convolution of two dotted or two undotted spinors, $\theta^\alpha\theta_\alpha$ or $\bar\theta_{\dot\alpha}\bar\theta^{\dot\alpha}$, with one lower and one upper index. The raising and lowering of indices are realized by virtue of the antisymmetric (Levi-Civita) symbol,

$$\bar\theta_{\dot\alpha} = \varepsilon_{\dot\alpha\dot\beta}\bar\theta^{\dot\beta}\,, \qquad \theta_\alpha = \varepsilon_{\alpha\beta}\theta^\beta\,,$$

where

$$\varepsilon^{12} = \varepsilon^{\dot1\dot2} = 1\,; \qquad \varepsilon_{12} = \varepsilon_{\dot1\dot2} = -1$$

so that $\varepsilon^{\alpha\beta}\varepsilon_{\beta\gamma} = \delta^\alpha_\gamma$. When one raises or lowers the index of θ the ε symbol must be placed to the left of θ.

A shorthand notation when the indices of the spinors are implicit is widely used, for instance,

$$\theta^2 \equiv \theta^\alpha \theta_\alpha$$

and

$$\bar\theta^2 \equiv \bar\theta_{\dot\alpha}\theta^{\dot\alpha}\,.$$

Notice that in convoluting the undotted indices one writes first the spinor with the upper index while for the dotted indices the first spinor has the lower index. The ordering is important since the elements of the spinors are anticommuting Grassmann numbers.

It remains to be added that the vector quantities can be obtained from two spinors — one dotted and the other undotted. Thus, $\theta_\alpha\bar\theta_{\dot\alpha}$ transforms as a Lorentz vector.

Now, we can introduce the notion of supertranslations in the superspace $\{x, \theta, \bar\theta\}$. The generic supertransformation has the form

$$\theta \to \theta + \varepsilon\,, \qquad \bar\theta \to \bar\theta + \bar\varepsilon\,,$$

$$x_{\alpha\dot\beta} \to x_{\alpha\dot\beta} + 2i\varepsilon_\alpha\bar\theta_{\dot\beta} - 2i\theta_\alpha\bar\varepsilon_{\dot\beta}\,. \tag{2.2}$$

The supertranslations generalize conventional translations in the ordinary space.

One can also consider the so-called chiral and antichiral superspaces (chiral realizations of the supergroup); the first one does not explicitly contain $\bar\theta$ while the second does not contain θ. It is not difficult to see that a point from the chiral superspace is parametrized by $\{x_L, \theta\}$, and that from the antichiral superspace is parametrized by $\{x_R, \bar\theta\}$. Here

$$(x_L)_{\alpha\dot\alpha} = x_{\alpha\dot\alpha} - 2i\theta_\alpha\bar\theta_{\dot\alpha}\,, \qquad (x_R)_{\alpha\dot\alpha} = x_{\alpha\dot\alpha} + 2i\theta_\alpha\bar\theta_{\dot\alpha}\,. \tag{2.3}$$

Under this definition the supertransformations corresponding to the shifts in θ and $\bar\theta$, respectively, leave us inside the corresponding superspace. Indeed, if $\theta \to \theta + \varepsilon$ and $\bar\theta \to \bar\theta + \bar\varepsilon$, then

$$(x_L)_{\alpha\dot\beta} \to (x_L)_{\alpha\dot\beta} - 4i\theta_\alpha\bar\varepsilon_{\dot\beta} \quad \text{and} \quad (x_R)_{\alpha\dot\beta} \to (x_R)_{\alpha\dot\beta} + 4i\varepsilon_\alpha\bar\theta_{\dot\beta}\,. \tag{2.4}$$

Superfields provide a very concise description of supersymmetry representations. They are very natural generalizations of the conventional fields. Say, the scalar field $\phi(x)$ in the $\lambda\phi^4$ theory is a function of x. Correspondingly, superfields are functions of x and θ's. For instance, the chiral superfield $\Phi(x_L, \theta)$

depends on θ and x_L (and has no explicit $\bar{\theta}$ dependence). If we Taylor-expand it in the powers of θ we get the following formula:

$$\Phi(x_L, \theta) = \phi(x_L) + \sqrt{2}\psi(x_L)\theta + \theta^2 F(x)\,. \tag{2.5}$$

There are no higher-order terms in the expansion since higher powers of θ vanish due to the Grassmannian nature of this parameter. For the same reason the argument of the last component of the chiral superfield, F, is set equal to x. The distinction between x and x_L is not important in this term. The last component of the chiral superfield is always called F. F terms of the chiral superfields are nondynamical, they appear in the Lagrangian without derivatives. Later, we will see that F terms play a distinguished role.

The lowest component of the chiral superfield is a complex scalar field ϕ, and the middle component is a Weyl spinor ψ. Each field describes two degrees of freedom, so the appropriate balance is achieved automatically. Thus, we see that superfield is a concise form of representing a set of components. The transformation law of the components follows immediately from Eq. (2.4), for instance, $\delta\phi(x) = \sqrt{2}\psi(x)\varepsilon$, and so on.

The antichiral superfields depend on x_R and $\bar{\theta}$. The chiral and antichiral superfields describe the matter sectors of the theories to be studied below. The gauge field appears from the so-called vector superfield V which depends on both, θ and $\bar{\theta}$ and satisfies the condition $V = V^\dagger$. The component expansion of the vector superfield has the form

$$\begin{aligned} V(x, \theta, \bar{\theta}) = {} & C(x) + i\theta\chi(x) - i\bar{\theta}\bar{\chi}(x) + \frac{i}{\sqrt{2}}\theta^2[M(x) + iN(x)] \\ & - \frac{i}{\sqrt{2}}\bar{\theta}^2[M(x) - iN(x)] - 2\theta_\alpha\bar{\theta}_{\dot{\beta}}v^{\alpha\dot{\beta}}(x) \\ & + 2\left\{i\theta^2\bar{\theta}_{\dot{\beta}}\left[\bar{\lambda}^{\dot{\beta}} - \frac{i}{4}\partial^{\alpha\dot{\beta}}\chi_\alpha\right] + \text{h.c.}\right\} + \theta^2\bar{\theta}^2\left[D(x) - \frac{1}{4}\Box C(x)\right]. \end{aligned} \tag{2.6}$$

The components C, D, M, N and $v^{\alpha\dot{\beta}}$ must be real to satisfy the condition $V = V^\dagger$. The vector field $v^{\alpha\dot{\beta}}$ gives its name to the entire superfield.

The last component of the vector superfield, apart from a full derivative, is called the "D term". D terms also play a special role.

Let me say a few words about the gauge transformations. For simplicity I will consider the case of the Abelian (U(1)) gauge group. In the non-Abelian case the corresponding formulae become more bulky, but the essence stays the same.

As well-known, in nonsupersymmetric gauge theories the matter fields transform under the gauge transformations as

$$\phi(x) \to e^{i\alpha(x)}\phi(x)\,, \qquad \phi(x)^\dagger \to e^{-i\alpha(x)}\phi(x)^\dagger\,, \tag{2.7}$$

while the gauge field

$$v_\mu(x) \to v_\mu(x) + \partial_\mu \alpha(x)\,, \tag{2.8}$$

where $\alpha(x)$ is an arbitrary function of x. Equations (2.7) and (2.8) prompt the supersymmetric version of the gauge transformations,

$$\Phi(x_L, \theta) \to e^{i\Lambda}\Phi(x_L, \theta)\,, \qquad \bar{\Phi}(x_R, \bar{\theta}) \to e^{-i\bar{\Lambda}}\Phi(x_R, \bar{\theta}) \tag{2.9}$$

and

$$V \to V - i(\Lambda - \bar{\Lambda}) \tag{2.10}$$

where Λ is an arbitrary chiral superfield, $\bar{\Lambda}$ is its antichiral partner. $\bar{\Phi}e^V\Phi$ is then a gauge invariant combination playing the same role as $\mathcal{D}_\mu\phi^\dagger\mathcal{D}_\mu\phi$ in nonsupersymmetric theories. Let me parenthetically note that supersymmetrization of the gauge transformations, Eqs. (2.9) and (2.10), was the path which led Wess and Zumino [30] to the discovery of the supersymmetric theories (independently of Golfand and Likhtman).

In components

$$C \to C - i(\phi - \phi^\dagger)\,, \qquad \chi \to \chi - \sqrt{2}\psi\,, \qquad M + iN \to M + iN + 2F$$

$$v_{\alpha\dot{\beta}} \to v_{\alpha\dot{\beta}} + \frac{1}{2}\partial_{\alpha\dot{\beta}}(\phi + \phi^\dagger)\,, \qquad \lambda \to \lambda\,, \qquad D \to D\,. \tag{2.11}$$

We see that the C, χ, M and N components of the vector superfield can be gauged away. This is what is routinely done when the component formalism is used. This gauge bears the name of its inventors — it is called the *Wess-Zumino* gauge. Imposing the Wess-Zumino gauge condition in supersymmetric theory one actually does not fix the gauge completely. The component Lagrangian that is arrived at in the Wess-Zumino gauge still possesses the gauge freedom with respect to the nonsupersymmetric (old-fashioned) gauge transformations.

It remains to introduce spinorial derivatives. They will be denoted by capital D and $\bar{D}$,

$$D_\alpha = \frac{\partial}{\partial\theta^\alpha} - i\partial_{\alpha\dot{\beta}}\bar{\theta}^{\dot{\beta}}\,, \qquad \bar{D}_{\dot{\alpha}} = -\frac{\partial}{\partial\bar{\theta}^{\dot{\alpha}}} + i\partial_{\beta\dot{\alpha}}\theta^\beta\,. \tag{2.12}$$

The relative signs in Eq. (2.12) are fixed by the requirements $D_\alpha(x_R)_{\beta\dot\gamma} = 0$ and $\bar D_{\dot\alpha}(x_L)_{\beta\dot\gamma} = 0$.

To make the spinorial derivatives distinct from the regular covariant derivative the latter will be denoted by the script $\mathcal{D}$. The supergeneralization of the field strength tensor of the gauge field has the form

$$W_\alpha(x_L, \theta) = \frac{1}{8}\bar D^2(e^{-V} D_\alpha e^V)$$
$$= i\lambda_\alpha(x_L) - \theta_\alpha D(x_L) - i\theta^\beta G_{\alpha\beta}(x_L) + \theta^2 \mathcal{D}_{\alpha\dot\alpha}\bar\lambda^{\dot\alpha}(x_L)\,, \quad (2.13)$$

where $G_{\alpha\beta}$ is the gauge field strength tensor in the spinorial form.

This brief excursion in the formalism, however boring it might seem, is necessary for understanding physical results to be discussed below. I will try to limit such excursions to the absolute minimum, but we will not be able to avoid them completely. Now, the stage is set, and we are ready to submerge in the intricacies of the supersymmetric gauge dynamics.

2.3. *Simplest Supersymmetric Models*

In this section we will discuss some simple models. Our basic task is to reveal general features playing the key role in various unusual dynamical scenarios realized in supersymmetric gauge theories. One should keep in mind that all theories with matter can be divided in two distinct classes: chiral and nonchiral matter. The second class includes supersymmetric generalization of QCD, and all other models where each matter multiplet is accompanied by the corresponding conjugate representation. In other words, mass term is possible for all matter fields. Even if the massless limit is considered, the very possibility of adding the mass term is very important for dynamics. In particular, dynamical SUSY breaking cannot happen in the nonchiral models.

Models with chiral matter are those where the mass term is impossible. The matter sector in such models is severely constrained by the absence of the internal anomalies in the theory. The most well-known example of this type is the SU(5) model with equal number of chiral quintets and (anti)decuplets. Each quintet and antidecuplet, together, are called *generation*; when the number of generations is three this is nothing but the most popular grand unified theory of electroweak interactions. The chiral models are singled out by the fact that dynamical SUSY breaking is possible, in principle, only in this class. In the present lecture course dynamical SUSY breaking is not our prime concern. Rather, we will focus on various nontrivial dynamical regimes. Most regimes to be discussed below manifest themselves in the nonchiral models, which are simpler. Therefore, the emphasis will be put on the nonchiral models, digression to the chiral models will be made occasionally.

2.3.1. *Supersymmetric gluodynamics*

To begin with we will consider supersymmetric generalization of pure gluodynamics — i.e. the theory of gluons and gluinos. The Lagrangian has the form [31]

$$\mathcal{L} = -\frac{1}{4g^2} G^a_{\mu\nu} G^a_{\mu\nu} + \frac{\vartheta}{32\pi^2} G^a_{\mu\nu} \tilde{G}^a_{\mu\nu} + \frac{i}{2g^2} \bar{\lambda}^a \mathcal{D}_\mu \gamma^\mu \lambda^a \tag{2.14}$$

where $G^a_{\mu\nu}$ is the gluon field strength tensor, $\tilde{G}^a_{\mu\nu}$ is the dual tensor, g is the gauge coupling constant, ϑ is the vacuum angle, and $\mathcal{D}_\mu$ is the covariant derivative. Moreover, λ^a is the gluino field, which can be described either by a four-component Majorana (real) fields or two-component Weyl (complex) fields. Equation (2.14) implies the Majorana representation.

In terms of superfields

$$\mathcal{L} = \frac{1}{4g_0^2} \operatorname{Tr} \int d^2\theta W^2 + \text{H.c.}\,, \tag{2.15}$$

where the superfield W is a color matrix,

$$W = W^a T^a\,,$$

T^a are the generators of the gauge group (in the fundamental representation), $\operatorname{Tr} T^a T^b = (1/2)\delta^{ab}$. It is very important that the gauge constant $1/g_0^2$ in Eq. (2.15) can be treated as a complex parameter. The subscript 0 emphasizes the fact that the gauge couplings in Eqs. (2.15) and (2.14) are different,

$$\frac{1}{g_0^2} = \frac{1}{g^2} - i\frac{\vartheta}{8\pi^2}\,, \tag{2.16}$$

its real part is the conventional gauge coupling while the imaginary part is proportional to the vacuum angle. Thus, the gauge coupling becomes complexified in SUSY theories. This fact has far-reaching consequences.

Equivalence between Eqs. (2.15) and (2.14) is clear from Eq. (2.13). The F component of W^2 includes the kinetic term of the gaugino field (or gluino, I will use these terms indiscriminately),

$$\bar{\lambda} \mathcal{D}_\mu \gamma^\mu \lambda\,,$$

and that of the gauge field,

$$G^a_{\mu\nu}G^a_{\mu\nu} + iG^a_{\mu\nu}\tilde{G}^a_{\mu\nu}\,.$$

Superficially the model looks very similar to conventional QCD; the only difference is that the quark fields belonging to the fundamental representation of the gauge group in QCD are replaced by the gluino field belonging to the adjoint representation in supersymmetric gluodynamics. Like QCD, supersymmetric gluodynamics is a strong coupling non-Abelian theory. Therefore, it is usually believed that

- only colorless asymptotic states exist;
- the Wilson loop (in the *fundamental* representation) is subject to the area law (confinement);
- a mass gap is dynamically generated; all particles in the spectrum are massive.

I would like to stress the word "believe" since the above features are hypothetical. Although the theory does indeed look pretty similar to QCD, supersymmetry brings in remarkable distinctions — some quantities turn out to be exactly calculable. Namely, we know that the gluino condensate develops,

$$\langle\lambda^{a\alpha}\lambda^a_\alpha\rangle = \text{const.} \times \Lambda^3 e^{2\pi i k/N_c} \tag{2.17}$$

where N_c is the number of colors (SU(N_c) gauge group is assumed and the vacuum angle ϑ is set equal to zero), Λ is the scale parameter of supersymmetric gluodynamics, k is an integer ($k = 0, 1, \ldots, N_c - 1$), and the constant in Eq. (2.17) is *exactly* calculable [32, 33]. A discrete Z_{2N_c} symmetry of the model, a remnant of the anomalous U(1), is spontaneously broken by the gluino condensate[6] down to Z_2. Correspondingly, there are N_c degenerate

[6] I hasten to add that it was argued recently [34] that supersymmetric gluodynamics actually has two phases: one with the spontaneously broken Z_{2N_c} invariance, and another, unconventional, phase where the chiral Z_{2N_c} symmetry is unbroken and the gluino condensate does not develop. Dynamics of the chirally symmetric phase is drastically different from what we got used to in QCD. In particular, although no invariance is spontaneously broken, massless particles appear, and no mass gap is generated. This development is too fresh, however, to be included in this lecture course.

The existence of the gluino condensate was anticipated [35], from the analysis of the so-called Veneziano-Yankielowicz effective Lagrangian, even prior to the first dynamical calculation [32]. The Veneziano-Yankielowicz Lagrangian, very useful for orientation, is not a genuinely Wilsonean construction, and one must deal with it extremely cautiously in extracting consequences. For a recent discussion see Ref. [34].

vacua, counted by the integer parameter k. Supersymmetry is unbroken — all vacua have the vanishing energy density.

Moreover, the Gell-Mann-Low function of the model, governing the running of the gauge coupling constant, is also exactly calculable [36],

$$\beta(\alpha) = -\frac{3N_c\alpha^2}{2\pi}\frac{1}{1 - N_c\alpha/(2\pi)}\,. \tag{2.18}$$

By "exactly" I mean that all orders of the perturbation theory are known, and one can additionally show that in the case at hand there are no nonperturbative contributions.

Equations (2.17) and (2.18) historically were the first examples of nontrivial (i.e. nonvanishing) quantities exactly calculated in four-dimensional field theories in the strong coupling regime. These examples, alone, show that the supersymmetric gauge dynamics is full of hidden miracles. We will encounter many more examples in what follows. Eventually, learning more about supersymmetric theories will lead to a better understanding of how Eqs. (2.17) and (2.18) are derived. But this will take some time. Here I would like only to add an explanatory remark regarding the vacuum degeneracy in supersymmetric gluodynamics. At the classical level Lagrangian (2.14) has a U(1) symmetry corresponding to the phase rotations of the gluino fields,

$$\lambda \to e^{i\alpha}\lambda\,. \tag{2.19}$$

The corresponding current is sometimes called the R_0 current; it is a superpartner of the energy-momentum tensor and the supercurrent. The R_0 current exists in any supersymmetric theory. Moreover, in conformally invariant theories — and supersymmetric gluodynamics is conformally invariant at the classical level — it is conserved [37]. In the spinor notation the R_0 current has the form $J_{\alpha\dot{\alpha}} = g^{-2}\bar{\lambda}_{\dot{\alpha}}\lambda_\alpha$, while in the Majorana notation the very same current takes the form $J_\mu = g^{-2}\bar{\lambda}\gamma_\mu\gamma_5\lambda$. (Let me parenthetically note that the vector current of the Majorana gluino identically vanishes. The proof of this fact is left as an exercise). The conservation of the axial current above is broken by the triangle anomaly,

$$\partial^\mu J_\mu = \frac{N_c}{16\pi^2}G^a_{\mu\nu}\tilde{G}^a_{\mu\nu}\,.$$

So, there is no continuous U(1) symmetry in the model. By the same token, the conformal invariance is ruined by the anomaly in the trace of the energy-momentum tensor. As a matter of fact, the divergence of the R_0 current and the trace of the energy-momentum tensor can be combined in one superfield [38] (see Appendix).

However, a remnant of the would-be symmetry remains, in the form of the discrete phase transformations of the type (2.19) with $\alpha = \pi k/N_c$. The gluino condensate further breaks this symmetry to Z_2 corresponding to $\lambda \to -\lambda$. The number of the degenerate vacuum states, N_c, coincides with Witten's index for the SU(N_c) theory [39], an invariant which counts the number of the boson zero energy states minus the number of the fermion zero energy states. If Witten's index is nonvanishing supersymmetry cannot be spontaneously broken, of course.

An interesting aspect, related to the discrete degeneracy of the vacuum states, is the ϑ dependence. What happens with the vacua if $\vartheta \neq 0$? The question was answered in Ref. [33]. The ϑ dependence of the gluino condensate is

$$\langle\lambda\lambda\rangle_\vartheta = \langle\lambda\lambda\rangle_{\vartheta=0} e^{\frac{i\vartheta}{N_c}} . \tag{2.20}$$

This shows that the N_c vacua are intertwined as far as the ϑ evolution is concerned. When ϑ changes continuously from 0 to 2π the first vacuum becomes second, the second becomes third, and so on, in a cyclic way.

2.3.2. *SU(2) SQCD with one flavor*

As the next step on a long road leading us to the understanding of supersymmetric gauge dynamics we will consider SUSY generalization of SU(2) QCD with the matter sector consisting of one flavor. This model will serve us as a reference point in all further constructions.

Since the gauge group is SU(2) we have three gluons and three superpartners — gluinos.

As far as the matter sector is concerned, let us remember that one quark flavor in QCD is described by a Dirac field, a doublet with respect to the gauge group. One Dirac field is equivalent to two chiral fields: a left-handed and a right-handed, both transforming according to the fundamental representation of SU(2). Moreover, the right-handed doublet is equivalent to the left-handed antidoublet, which in turn is equivalent to a doublet. The latter fact is specific to the SU(2) group, whose all representations are (pseudo)real. Thus, the Dirac quark reduces to the two left-handed Weyl doublet fields.

Correspondingly, in SQCD each of them will acquire a scalar partner. Thus, the matter sector will be built from two superfields, S_1 and S_2. In what follows we will use the notation S^{α}_{f} where $\alpha = 1, 2$ is the color index, and $f = 1, 2$ is a "subflavor" index. Two subflavors comprise one flavor. The chiral superfield has the usual form, see Eq. (2.5).

In the superfield language the Lagrangian of the model can be represented in a very concise form

$$\mathcal{L} = \frac{1}{2g_0^2}\,\mathrm{Tr}\int d^2\theta W^2 + \frac{1}{4}\int d^2\theta d^2\bar{\theta}\bar{S}_f e^V S_f + \left(\frac{m_0}{4}\int d^2\theta S^{\alpha f} S_{\alpha f} + \text{H.c.}\right), \tag{2.21}$$

where the superfields V and W_α are matrices in the color space, for instance, $V \equiv V^a \tau^a/2$, with τ^a denoting the Pauli matrices. The subscript 0 indicates that the mass parameter and the gauge coupling constant are bare parameters, defined at the ultraviolet cut off. In what follows we will omit this subscript to ease the notation in several instances where it is unimportant.

If we take into account the rules of integration over the Grassmann numbers we immediately see that the integral over $d^2\theta$ singles out the θ^2 component of the chiral superfields W^2 and S^2, i.e. the F terms. Moreover, the integral over $d^2\theta d^2\bar{\theta}$ singles out the $\theta^2\bar{\theta}^2$ component of the real superfield $\bar{S}e^V S$, i.e. the D term.

Note that the SU(2) model under consideration, with *one flavor* possesses a global SU(2) ("subflavor") invariance allowing one to freely rotate the superfields $S_1 \leftrightarrow S_2$. This symmetry holds even in the presence of the mass term, see Eq. (2.21), and is specific for SU(2) gauge group, with its pseudoreal representations. All indices corresponding to the SU(2) groups (gauge, Lorentz and subflavor) can be lowered and raised by means of the ε symbol, according to the general rules.

The Lagrangian presented in Eq. (2.21) is not generic. Renormalizable models with a richer matter sector usually allow for one more type of F terms, namely

$$\int d^2\theta S^3\,.$$

These terms are called the Yukawa interactions, since one of the vertices they contain corresponds to a coupling of two spinors to a scalar. Strictly speaking, they should be called the super-Yukawa terms, since spinor-spinor-scalar vertices arise also in the (super)gauge parts of the Lagrangian. This jargon is widely spread, however; eventually you will get used to it and learn how to avoid confusion. The combination of the F terms $S^2 + S^3$ is generically referred to as superpotential. The conventional potential of self-interaction of the scalar fields stemming from the given superpotential is referred to as *scalar potential.*

It is instructive to pass from the superfield notations to components. We will do this exercise now in some detail, putting emphasis on those features which are instrumental in the solutions to be discussed below. Once the experience is accumulated the need in the component notation will subside.

Let us start from W^2. The corresponding F term was already discussed in Sec. 2.3.1. There is one new important point, however. In Sec. 2.3.1 we omitted the square of the D term present in $W^2|_F$, see Eq. (2.13),

$$\Delta\mathcal{L}_D^{(W)} = \frac{1}{2g^2} D^a D^a \,. \tag{2.22}$$

If the matter sector of the theory is empty, this term is unimportant. Indeed, the D field enters with no derivatives, and, hence, can be eliminated from the Lagrangian by virtue of the equations of motion. With no matter fields $D = 0$. In the presence of the matter fields, however, eliminating D we get a nontrivial term constructed from the scalar fields, which is of paramount importance. This point will be discussed later; here let me only note that the sign of D^2 in the Lagrangian, Eq. (2.22), is unusually positive.

The next term to be considered is $\int d^2\theta d^2\bar\theta \bar S e^V S$. Calculation of the D component of $\bar S e^V S$ is a more time-consuming exercise since we must take into account the fact that S depends on x_L while $\bar S$ depends on x_R; both arguments differ from x. Therefore, one has to expand in this difference. The factor e^V sandwiched between $\bar S$ and S covariantizes all derivatives. Needless to say that the field V is treated in the Wess-Zumino gauge. It is not difficult to check that

$$\frac{1}{4}\int d^2\theta d^2\bar\theta \bar S_f e^V S_f = \bar\psi_f i \not{\mathcal{D}} \psi_f - \phi_f^\dagger \mathcal{D}^2 \phi_f$$
$$+ i\sqrt{2}[(\psi_f \lambda)\phi_f^\dagger + \text{H.c.}] + D^a \phi_f^\dagger T^a \phi_f \tag{2.23}$$

where T^a are the matrices of the color generators. In the SU(2) theory $T^a = \tau^a/2$. Now we see why the D^2 term is so important in the presence of matter; D^a does not vanish anymore. Moreover, using the equation of motion we can express D^a in terms of the squark fields, generating in this way a quartic self-interaction of the scalar fields,

$$V_D = \frac{1}{2g^2} D^a D^a \,, \qquad D^a = -\frac{g^2}{2}(\phi_1^\dagger \tau^a \phi_1 + \phi_2^\dagger \tau^a \phi_2) \,. \tag{2.24}$$

In the old-fashioned language of the pre-SUSY era one would call the term $(\psi_f \lambda)\phi_f^\dagger$ from Eq. (2.23) the Yukawa interaction. The SUSY practitioner would refer to this term as to the gauge coupling since it is merely a supersymmetric generalization of the quark–quark–gluon coupling. I mention these terms here because later on their analysis will help us establish the form of the conserved R currents.

2.3.3. *Vacuum valleys*

Let us examine the D potential V_D more carefully, neglecting for the time being F terms altogether. As widely known, the energy of any state in any supersymmetric theory is positive-definite. The minimal energy state, the vacuum, has energy exactly at zero. Thus, in determining the classical vacuum we must find all field configurations corresponding to the vanishing energy. From Eq. (2.24) it is clear that in the Wess-Zumino gauge the *classical space of vacua* (somctimes called the moduli space of vacua) is defined by the *D-flatness condition*

$$D^a = 0 \qquad \text{for all } a\,. \tag{2.25}$$

More exactly, Eq. (2.25) is called the Wess-Zumino gauge D flatness condition. Since this gauge is always implied, if not stated to the contrary, we will omit the reference to the Wess-Zumino gauge.

The D potential V_D represents a quartic self-interaction of the scalar fields, of a very peculiar form. Typically in the ϕ^4 theory the potential has one — at most several — minima. In other words, the space of the vacuum fields corresponding to minimal energy, is a set of isolated points. The only example with a continuous manifold of points of minimal energy which was well studied previously is the spontaneous breaking of a global continuous symmetry, say, U(1) (Sec. 1). In this case all points belonging to this vacuum manifold are physically equivalent. The D potential (2.24) has a specific structure — the minimal (zero) energy is achieved along entire directions corresponding to the solution of Eq. (2.25). It is instructive to think of the potential as of a mountain ridge; the D-flat directions then present the flat bottom of the valleys. Sometimes, for transparency, I will call the D-flat directions *the vacuum valleys.* Their existence was first noted in Ref. [40]. As we will see, different points belonging to the bottom of the valleys are physically inequivalent. This is a remarkable feature of the supersymmetric gauge theories.

In the case of the SU(2) theory with one flavor it is not difficult to find the D-flat direction explicitly. Indeed, consider the scalar fields of the form

$$\phi_1 = v \begin{pmatrix} 1 \\ 0 \end{pmatrix}, \qquad \phi_2 = v \begin{pmatrix} 0 \\ 1 \end{pmatrix}, \tag{2.26}$$

where v is an arbitrary complex constant. It is obvious that for any value of v all D^a's vanish. D^1 and D^2 vanish because $\tau^{1,2}$ are off-diagonal matrices; D^3 vanishes after summation over two subflavors.

It is quite obvious that if $v \neq 0$ the original gauge symmetry SU(2) is spontaneously broken totally. Indeed, under the condition (2.26) all three gauge bosons acquire masses $\sim gv$. Thus, we deal here with the supersymmetric generalization of the Higgs phenomenon. Needless to say that supersymmetry is not broken. It is instructive to trace the reshuffling of the degrees of freedom before and after the Higgs phenomenon. In the unbroken phase, corresponding to $v = 0$, we have three massless gauge bosons (6 degrees of freedom), three massless gaugino (6 degrees of freedom), four matter fermions (the Weyl fermions, 8 degrees of freedom), and four matter scalars (complex scalars, 8 degrees of freedom). In the broken phase three matter fermions combine with the gauginos to form three massive Dirac fermions (12 degrees of freedom). Moreover, three matter scalars combine with the gauge fields to form three *massive* vector fields (9 degrees of freedom) plus three massive (real) scalars. What remains massless? One complex scalar field, corresponding to the motion along the bottom of the valley, v, and its fermion superpartner, one Weyl fermion. The balance between the fermion and boson degrees of freedom is explicit.

A gauge invariant description of the system of the vacuum valleys was suggested in Refs. [41, 42] (see also Ref. [40]). In these works it was noted that the set of proper coordinates parametrizing the space of the classical vacua is nothing else but the set of all independent (local) products of the chiral matter fields existing in the theory. Since this point is very important let me stress once more that the variables to be included in the set are polynomials built from the fields of one and the same chirality *only*. These variables are clearly gauge invariant.

At the intuitive level this assertion is almost obvious. Indeed, if there is a D-flat direction, the motion along the degenerate bottom of the valley must be described by some effective (i.e. composite) chiral superfield, which is gauge invariant and has no superpotential. The opposite is also true. If we are able to build some chiral gauge invariant (i.e. colorless) superfield Φ, as a local product of the chiral matter superfields of the theory at hand, then the energy is guaranteed to vanish, since (in the absence of the F terms) all terms which might appear in the effective Lagrangian for ϕ necessarily contain derivatives. Here ϕ is the lowest component of the above superfield Φ. In other words, then, changing the value of ϕ we will be moving along the bottom of the valley.

A formal proof of the fact that the classical vacua are fully described by the set of local (gauge invariant) products of the chiral fields comprising the matter sector is given in work [43] which combines and extends results scattered in the literature [40, 41, 44, 45].

The approach based on the chiral polynomials is very convenient for establishing the fact of the existence (nonexistence) of the moduli space of the classical vacua, and in counting the dimensionality of this space. For instance, in the SU(2) model with one flavor there exists only one invariant, $S^2 \equiv S_{\alpha f} S_{\beta g} \varepsilon^{\alpha\beta} \varepsilon^{fg} = 2v^2$. Correspondingly, there is only one vacuum valley — one-dimensional complex manifold. The remaining three (out of four) complex scalar fields are eaten up in the super-Higgs phenomenon by the vector fields, which immediately tells us that a generic point from the bottom of the valley corresponds to fully broken gauge symmetry.

In other cases we will have a richer structure of the moduli space of the classical vacua. In some instances no chiral invariants can be built at all. Then the D-flat directions are absent.

If the D-flat directions exist, and the gauge symmetry is spontaneously broken, then the constraints of the type of Eq. (2.26) can be viewed as a gauge fixing condition. This is nothing else but the unitary gauge in SUSY. Those components of the matter superfields which are set equal to zero are actually eaten up by the vector particles which acquire the longitudinal components through the super-Higgs mechanism.

Although constructing the set of the chiral invariants is helpful in the studies of the general properties of the space of the classical vacua, sometimes it is still necessary to explicitly parametrize the vacuum valleys, just in the same way as it is done in Eq. (2.26). As we have seen, this problem is trivially solvable in the SU(2) model with one flavor. For higher groups and representations the general situation is much more complicated, and the generic solution is not found. Many useful tricks for finding explicit parametrization of the vacuum valleys in particular examples were suggested in Refs. [40–42]. A few of the simplest examples are considered below. More complicated instances are considered in the literature. For instance, a parametrization of the valleys in the SU(5) model with two quintets and two antidecuplets was given in Ref. [46] and in the E(6) model with the 27-plet in Ref. [47]. The correspondence between the explicit parametrization of the D-flat directions and the chiral polynomials was discussed recently more than once, see e.g. Refs. [48–52]. I would like to single out Ref. [53] where a catalog of the flat directions in the minimal supersymmetric standard model (MSSM) was obtained by analyzing all possible chiral polynomials and eliminating those which are redundant.

Once the existence of the D-flat directions is established at the classical level one may be sure that a manifold of the degenerate vacua will survive at the quantum level, provided no F terms appear in the action which might lift the degeneracy. Indeed, in this case the only impact of the quantum corrections is to provide an overall Z factor in front of the kinetic term, which certainly does not affect the vanishing of the D terms. The F terms which could lift the degeneracy must be either added in the action by hand (e.g. mass terms), or generated nonperturbatively. A remarkable nonrenormalization theorem [54] guarantees that no F terms can be generated perturbatively. We will return to the discussion of this *second miracle* of SUSY in Sec. 2.4.

In this respect the supersymmetric theories are fundamentally different from the nonsupersymmetric ones. Say, in the good old ϕ^4 theory with the Yukawa interaction

$$\mathcal{L} = |\partial_\mu \phi|^2 + \bar{\psi} i \partial_\mu \gamma^\mu \psi + (g\phi\bar{\psi}\psi + \text{H.c.})$$

we could also assume that the mass and self-interaction of the scalar field vanish at the classical level. Then, classically, we will have a flat direction — any constant value of ϕ corresponds to the vanishing vacuum energy. However, this vacuum valley does not survive inclusion of the quantum corrections. Already at the one-loop level both the mass term of the scalar field, and its self-interaction, will be generated, and the continuous vacuum degeneracy will inevitably disappear.

2.3.4. *In search of the valleys*

Although our excursion in the SU(2) model with one flavor is not yet complete, the issue of the D-flat directions is so important in this range of problems that we pause here to do, with pedagogical purposes, a few simple exercises. If you choose to skip this subsection in the first reading it will be necessary to return to it later.

SU(N_c) *model with* N_f *flavors* ($N_f < N_c$) [41]

The matter sector includes $2N_f$ subflavors — N_f chiral fields in the fundamental representation of SU(N_c), $S^{\alpha f}$, and N_f chiral fields in the antifundamental representation, $\tilde{S}^f_\alpha$, where $\alpha = 1, \ldots, N_c$ and $f = 1, \ldots, N_f$. It is quite obvious that one can form N_f^2 chiral products of the type

$$\tilde{S}^f_\alpha S^{\alpha g}\,, \qquad f, g = 1, \ldots, N_f\,. \tag{2.27}$$

All these chiral invariants are independent. Thus, the moduli space of the classical vacua (the vacuum valley) is a complex manifold of dimensionality N_f^2, parametrized by the coordinates (2.27). A generic point from the vacuum valley corresponds to the spontaneous breaking of $\mathrm{SU}(N_c) \to \mathrm{SU}(N_c - N_f)$ (except for the case when $N_f = N_c - 1$, when the original gauge group is completely broken). The number of broken generators is $2N_cN_f - N_f^2$; hence, the same amount of the complex scalar fields are eaten up in the super-Higgs mechanism. The original number of the complex scalar fields was $2N_cN_f$. The remaining N_f^2 degrees of freedom are the moduli (2.27) corresponding to the motion along the bottom of the valley.

In this particular problem it is not difficult to indicate a concrete parametrization of the vacuum field configurations. Indeed, consider a set

$$S_1 = \tilde{S}_1^\dagger = v_1 \begin{pmatrix} 1 \\ 0 \\ 0 \\ \ldots \\ 0 \end{pmatrix}, S_2 = \tilde{S}_2^\dagger = v_2 \begin{pmatrix} 0 \\ 1 \\ 0 \\ \ldots \\ 0 \end{pmatrix}, \ldots, S_{N_f} = \tilde{S}_{N_f}^\dagger = v_{N_f} \begin{pmatrix} 0 \\ 0 \\ \ldots \\ 1 \\ \ldots \\ 0 \end{pmatrix} \tag{2.28}$$

where the unity in S_{N_f} occupies the N_fth line and $v_{1,2,\ldots,N_f}$ are N_f arbitrary complex numbers. It is rather obvious that for this particular set all D terms vanish. In verifying this assertion it is convenient to consider first those D^a's which lie outside the Cartan subalgebra of $\mathrm{SU}(N_c)$. Since the corresponding T^a matrices are off-diagonal each term in the sum $\sum_f (S^\dagger T^a S + \tilde{S}^\dagger T^a \tilde{S})$ vanishes individually. For the generators from the Cartan subalgebra the fundamentals and antifundamentals cancel each other.

The point (2.28) is *not* a generic point from the bottom of the valley. This is clear from the fact that it is parametrized by only N_f complex numbers. To get a generic solution one observes that the theory is invariant under the global $\mathrm{SU}(N_f) \times \mathrm{SU}(N_f)$ flavor rotations (the fundamentals and antifundamentals can be rotated separately). On the other hand, the solution (2.28) is not invariant. Therefore, we can apply a general $\mathrm{SU}(N_f) \times \mathrm{SU}(N_f)$ rotation to Eq. (2.28) without destroying the condition $D^a = 0$. It is quite obvious that the generators belonging to the Cartan subalgebra of $\mathrm{SU}(N_f) \times \mathrm{SU}(N_f)$ do not introduce new parameters. The remaining rotations introduce $N_f^2 - N_f$ complex parameters, to be added to $v_1, \ldots, v_{N_f}$, altogether N_f^2 parameters, as it was anticipated from counting the number of the chiral invariants.

$\mathrm{SU}(N_c)$ *model with* N_f *flavors* ($N_f = N_c + 1$)

The vacuum valley is parametrized by N_f^2 complex parameters, although the number of the chiral invariants is larger, $N_f^2 + 2N_f$. Not all chiral invariants are independent. For further details see Sec. 3.1.

SU(5) *model with one quintet and one (anti)decuplet*

This gauge model describes Grand Unification, with one generation of quarks and leptons. This is our first example of nonchiral matter; it is singled out historically — the instanton-induced dynamical supersymmetry breaking was first found in this model [55]. The quintet field is V^α, the (anti)decuplet field is antisymmetric $X_{\alpha\beta}$. It is quite obvious that there are no chiral invariants at all. Indeed, the only candidate, VVX, vanishes due to antisymmetricity of $X_{\alpha\beta}$. This means that no D-flat directions exist. The same conclusion can be reached by explicitly parametrizing V and X; inspecting then the D-flatness conditions one can conclude that they have no solutions, see e.g. Appendix A in Ref. [56].

SU(5) *model with two quintets and two (anti)decuplets and no superpotential*

This model (with a small tree-level superpotential term) was the first example of the instanton-induced supersymmetry breaking in the weak coupling regime [57]. It presents another example of the anomaly-free chiral matter sector. Unlike the one-family model (one quintet and one antidecuplet) flat directions do exist (in the absence of superpotential). The system of the vacuum valleys in the two-family SU(5) model was analyzed in Ref. [46]. Generically, the gauge SU(5) symmetry is completely broken, so that 24 out of 30 chiral matter superfields are eaten up in the super-Higgs mechanism. Therefore, the vacuum valley should be parametrized by six complex moduli.

Denote two quintets present in the model as V_f^α ($f = 1, 2$), and two antidecuplets as $(X_{\bar g})_{\alpha\beta}$ where $\bar g = 1, 2$ and the matrices $X_{\bar g}$ are antisymmetric in color indices α, β. Indices f and $\bar g$ reflect the $\mathrm{SU}(2)_X \times \mathrm{SU}(2)_V$ flavor symmetry of the model. Six independent chiral invariants are

$$M_{\bar{g}} = V_k X_{\bar{g}} V_l \varepsilon^{kl}\,,$$

$$B_{\bar{g}f} = X_{\bar{g}} X_{\bar{k}} X_{\bar{l}} V_f \varepsilon^{\bar{k}\bar{l}}\,,$$

where the gauge indices in the first line are convoluted in a straightforward manner $V^\alpha X_{\alpha\beta} V^\beta$, while in the second line one uses the ε symbol,

$$X_{\bar{g}} X_{\bar{k}} X_{\bar{l}} V_f = \varepsilon^{\alpha\beta\gamma\delta\rho} (X_{\bar{g}})_{\alpha\beta} (X_{\bar{k}})_{\gamma\delta} (X_{\bar{l}})_{\rho\kappa} (V_f)^\kappa\,.$$

The choice of invariants above implies that there are no moduli transforming as $\{4,\,2\}$ under the flavor group (such moduli vanish).

In this model the explicit parametrization of the valley is far from being obvious, to put it mildly. The most convenient strategy of the search is analyzing the five-by-five matrix

$$D^\alpha_\beta = V^\alpha_f \bar{V}^f_\beta + 2(\bar{X}^{\bar{g}})^{\alpha\gamma} (X_{\bar{g}})_{\gamma\beta} \tag{2.29}$$

where $\bar{V} = V^\dagger$ and $\bar{X} = X^\dagger$. If this matrix is proportional to the unit one, the vanishing of the D terms is guaranteed. (Similar strategy based on analyzing analogs of Eq. (2.29) is applicable in other cases as well).

A solution of the D-flatness condition which contains seven real parameters and was found long ago by Vainshtein *et al.* [46] looks as follows:

$$V_1 = \begin{pmatrix} a_1 \\ 0 \\ 0 \\ a_4 \\ 0 \end{pmatrix}, \qquad V_2 = \begin{pmatrix} 0 \\ a_2 \\ 0 \\ 0 \\ 0 \end{pmatrix},$$

$$X_1 = \frac{1}{\sqrt{2}} \begin{pmatrix} 0 & 0 & 0 & s & 0 \\ 0 & 0 & b & 0 & 0 \\ 0 & -b & 0 & 0 & f \\ -s & 0 & 0 & 0 & 0 \\ 0 & 0 & -f & 0 & 0 \end{pmatrix}, \qquad X_2 = \frac{1}{\sqrt{2}} \begin{pmatrix} 0 & d & 0 & 0 & g \\ -d & 0 & 0 & 0 & 0 \\ 0 & 0 & 0 & 0 & 0 \\ 0 & 0 & 0 & 0 & h \\ -g & 0 & 0 & -h & 0 \end{pmatrix}, \tag{2.30}$$

where

$$|a_1|^2 = r^2 \cos^2\theta\,, \quad |a_4|^2 = r^2 \cos^2\alpha \tan^2\theta\,, \quad |a_2|^2 = r^2 \tan^2\theta \frac{\cos^2\theta - \cos^2\alpha}{\sin^2\theta + \cos^2\alpha}\,,$$

and

$$|s|^2 = \frac{r^2\cos^2\alpha}{\cos^2\theta(\sin^2\theta+\cos^2\alpha)}\,, \qquad |b|^2 = r^2(\sin^2\theta+\cos^2\alpha)\,,$$

$$|f|^2 = r^2\frac{\cos^2\alpha}{\cos^2\theta}\frac{\cos^2\theta-\cos^2\alpha}{\sin^2\theta+\cos^2\alpha}\,,$$

$$|d|^2 = \frac{r^2}{\cos^2\theta}(\cos^2\theta-\cos^2\alpha)\,, \quad |g|^2 = r^2\cos^2\alpha\,, \quad |h|^2 = r^2\sin^2\theta\,. \tag{2.31}$$

Thus, the absolute values of matrix elements are parametrized by three real parameters r, α and θ. Additional four parameters appear *via* phases δ of 9 elements a_i, s, b, f, d, g, h. Three phases, out of nine, are related to gauge rotations and are not observable in the gauge singlet sector. Additionally, there are two constraints,

$$\delta_1 - \delta_4 = \delta_h - \delta_g\,,$$

$$\delta_f - \delta_b = \delta_g - \delta_d\,,$$

which are readily derived from the vanishing of the off-diagonal terms in D^α_β.

Substituting the above expressions it is easy to check that invariants

$$B_{\bar g_1\bar g_2\bar g_3 f} = X_{\{\bar g_1}X_{\bar g_2}X_{\bar g_3\}}V_f$$

symmetrized over g_1, g_2, g_3 (i.e. the $\{4,\,2\}$ representation of $\mathrm{SU}(2)_X\times\mathrm{SU}(2)_V$) do vanish, indeed.

The most general valley parametrization depends on 12 real parameters, while so far we have only seven. This general solution was constructed by T. ter Veldhuis [46] and is discussed in detail in Chapter VI.

2.3.5. *Back to the* SU(2) *model — dynamics of the flat direction*

After this rather lengthy digression into the general theory of the vacuum valleys we return to our simplest toy model, SU(2) with one flavor. The vacuum valley in this case is parametrized by one complex number, v, which can be chosen at will since for any value of v the vacuum energy vanishes. One can quantize the theory near any value of v. If $v \neq 0$, the theory splits into two sectors — one containing massive particles which form SU(2) triplets, and another sector which includes only one massless Weyl fermion and one massless complex scalar field. These massless particles are singlets with respect to both SU(2) groups — color and subflavor.

So far we totally disregarded the mass term in the action, assuming $m = 0$. If $m \neq 0$ the corresponding term in the superpotential lifts the vacuum degeneracy, making the bottom of the valley nonflat. Indeed,

$$F = mv\,,$$

the corresponding contribution in the scalar potential is

$$\Delta V = |mv|^2\,,$$

which makes the theory "slide down" towards the origin of the valley. Since the perturbative corrections do not renormalize the F terms, this type of behavior — sliding down to the origin of the former valley — is preserved to any finite order in the perturbation theory.

What happens if one switches on the nonperturbative effects?

The nonrenormalization theorem [54], forbidding the occurrence of the F terms, does not apply to nonperturbative effects, which, thus, may or may not generate relevant F terms. The possibility of getting a superpotential can be almost completely investigated by analyzing the general properties of the model at hand, with no explicit calculations. Apart from the overall numerical constant, the functional form of the superpotential, if it is generated, turns out to be fixed.

Let me elucidate this point in more detail. First, on what variables can the superpotential depend? The vector fields are massive and are integrated over. Thus, we are left with the matter fields only, and the only chiral invariant is

$$I = S^{\alpha f} S_{\alpha f}\,. \tag{2.32}$$

The superpotential, if it exists, must have the form

$$\mathcal{L}_F = \int d^2\theta f(I(x_L, \theta)) + \text{H.c.} \tag{2.33}$$

where f is some function. Notice that the mass term has just this structure, with $f(z) = z$. We will discuss the possible impact of the mass term later, assuming at the beginning that $m = 0$.

Now, our task is to find the function f exploiting the symmetry properties of the model. At the classical level there exist two conserved currents. One of them, the R_0 current, is the superpartner of the energy-momentum tensor and the supercurrent [58]. The divergence of the R_0 current and the trace of the energy-momentum tensor can be combined in one superfield. The R_0 current exists in any supersymmetric theory, and, moreover, in conformally invariant theories it is conserved. Indeed, since the trace of the energy-momentum tensor vanishes in conformally invariant theories the divergence of R_0 vanishes as well. In our present model the R_0 current corresponds to the following rotations of the fields

$$\lambda_\alpha \to e^{i\beta}\lambda_\alpha\,, \quad \psi^f_\alpha \to e^{-(i/3)\beta}\psi^f_\alpha\,, \quad \phi^f_\alpha \to e^{(2i/3)\beta}\phi^f_\alpha\,. \tag{2.34}$$

If we denote this current by R^0_μ, then

$$R^0_\mu = \frac{1}{g^2}\,\bar\lambda\gamma_\mu\gamma_5\lambda - \frac{1}{3}\sum_f(\bar\psi^f\gamma_\mu\gamma_5\psi^f - 2\phi^{f\dagger}\overleftrightarrow{D}_\mu\,\phi^f)\,. \tag{2.35}$$

The relative phase between ψ and ϕ is established in the following way. Let us try to add the S^3 term to the superpotential. In the model at hand it is actually forbidden by the color gauge invariance, but we ignore this circumstance, since the form of the R_0 current is general, and in other models the S^3 term is perfectly allowed. It violates neither supersymmetry, nor the conformal invariance (at the classical level), which is obvious from the fact that its dimension is 3. The S^3 term in the superpotential produces $\phi\psi^2$ term in the Lagrangian. In this way we arrive at the relative phase between ψ and ϕ indicated in Eq. (2.34). The relative phase between λ and ψ is fixed by requiring the term $\lambda\psi\phi^\dagger$ in the Lagrangian (the supergeneralization of the gauge coupling) to be invariant under the U(1) transformation at hand. In the superfield language the last two transformations in Eq. (2.34) can be concisely written as

$$S^f \to \exp\left(\frac{2i}{3}\beta\right)S^f\,, \qquad \theta_\alpha \to \exp(i\beta)\theta_\alpha\,. \tag{2.36}$$

The second (classically) conserved current is built from the matter fields,

$$J_\mu = \sum_f(\bar\psi^f\gamma_\mu\gamma_5\psi^f + \phi^{f\dagger}\overleftrightarrow{D}_\mu\,\phi^f)\,. \tag{2.37}$$

The corresponding U(1) transformation is

$$\psi^f_\alpha \to e^{i\gamma}\psi^f_\alpha\,, \qquad \phi^f_\alpha \to e^{i\gamma}\phi^f_\alpha\,, \tag{2.38}$$

or, in the superfield notation, $S^f(x_L,\theta) \to \exp(i\gamma)S^f(x_L,\theta)$.

The conservation of both axial currents is destroyed by the quantum anomalies. At one-loop level

$$\partial_\mu R^0_\mu = \frac{5}{3}\frac{1}{16\pi^2} G^a_{\mu\nu}\tilde{G}^a_{\mu\nu}\,, \qquad \partial_\mu J_\mu = \frac{1}{16\pi^2} G^a_{\mu\nu}\tilde{G}^a_{\mu\nu}\,. \tag{2.39}$$

One can form, however, one linear combination of these two currents,

$$R_\mu = R^0_\mu - \frac{5}{3} J_\mu \tag{2.40}$$

which is anomaly free and is conserved even at the quantum level. The occurrence of a strictly conserved axial current, the so-called R current, is a characteristic feature of many supersymmetric models. In what follows we will have multiple encounters with the R currents in various models. The one presented in Eq. (2.40) was given in Ref. [41].

Combining both transformations, Eqs. (2.34) and (2.38), in the appropriate proportion, see Eq. (2.40), we conclude that the SU(2) model under consideration is strictly invariant under the following transformation

$$S^f \to e^{-i\beta} S^f\,, \qquad \theta_\alpha \to e^{i\beta}\theta_\alpha\,. \tag{2.41}$$

This R invariance leaves us with a unique possible choice for the superpotential

$$\mathcal{L}_F = \text{const.} \times \int d^2\theta \frac{1}{I(x_L, \theta)} \to \text{const.} \times \int d^2\theta \frac{\Lambda^5}{S^{\alpha f} S_{\alpha f}} \tag{2.42}$$

where Λ is a scale parameter of the model, and the factor Λ^5 has been written out on the basis of dimensional arguments.

Whether the superpotential is actually generated, depends on the value of the numerical constant above. In principle, it could have happened that the constant vanished. However, since no general principle forbids the F term (2.42), the vanishing seems highly improbable. And indeed, the direct one-instanton calculation in the weak coupling regime (i.e. $v^2 \gg \Lambda^2$) shows [41, 46] that this term is generated.

The impact of the term (2.42) is obvious. The corresponding extra contribution to the self-interaction energy of the scalar field is

$$\Delta V = |F|^2 = 4\frac{\Lambda^{10}}{|v|^6} \tag{2.43}$$

where the numerical constant is included in the definition of Λ. Thus, we see that the instanton-generated contribution ruins the indefinite equilibrium along the bottom of the valley, pushing the theory away from the origin. As a matter of fact, in the absence of the mass term, $m = 0$, the theory does not have any stable vacuum at all since the minimal (zero) energy is achieved only at $|v| \to \infty$. We encounter here an example of the *run-away vacuum* situation.

Switching on the mass term blocks the exits from the valley. Indeed, now

$$F = mv - 2\frac{\Lambda^5}{v^3}\,, \tag{2.44}$$

and the lowest-energy state shifts to a finite value of v. It is easy to see that now there are two points at the bottom of the former valley where the energy vanishes, namely

$$v^2 = \sqrt{6}\frac{\Lambda^{5/2}}{m^{1/2}} \quad \text{and} \quad v^2 = -\sqrt{6}\frac{\Lambda^{5/2}}{m^{1/2}}\,. \tag{2.45}$$

In other words, the continuous vacuum degeneracy is lifted, and only two-fold degeneracy survives; the theory has two vacuum states. The number of the vacuum states could have been anticipated from a general argument based on Witten's index [39].

I pause here to make a few remarks. First, we observe that the supersymmetric version of QCD dynamically has very little in common with QCD. Indeed, the chiral limit of QCD, when all quark masses are set equal to zero, is nonsingular — nothing spectacular happens in this limit except that the pions become strictly massless. At the same time, in the supersymmetric SU(2) model at hand the limit of the massless matter fields results in the run-away vacuum. This situation is quite general, and takes place in many models, although not in all.

Second, the analysis of the dynamics of the flat directions presented above is somewhat simplified. Two subtle points deserve mentioning. The general form of the superpotential compatible with the symmetry of the model was established in the massless limit. In this way we arrived at Eq. (2.42). The mass term was then introduced to avoid the run-away vacuum. If $m \neq 0$ the R current is not conserved any more, even at the classical level. To keep the invariance (2.41) alive one must simultaneously rotate the mass parameter,

$$m \to e^{4i\beta} m\,. \tag{2.46}$$

Let us call this invariance, supplemented by the phase rotation of the mass parameter, an *extended R symmetry*.

One could think of m as of a vacuum expectation value of some auxiliary chiral field, to be rotated in a concerted way in order to maintain the R invariance. (We will discuss this trick later on in more detail). It is clear then that multiplying Eq. (2.42) by any function of the dimensionless complex parameter $\sigma = mS^4/\Lambda^5$ is not forbidden by the extended symmetry. Extra arguments are needed to convince oneself that this additional function actually does not appear. Let us assume it does. Then it should be expandable in the Laurent series of the type

$$\sum_n \frac{C_n}{\sigma^n} .$$

If negative powers of n were present then the function would grow at large $|\sigma|$, a behavior one can immediately reject on physical grounds. The masses of the heavy particles, which we integrate over to obtain the superpotential, are proportional to gv; large values of $|v|$ imply heavier masses, which implies, in turn, that the impact on the superpotential should be weaker. Thus, all C_n's with negative n must vanish. Positive n are not acceptable as well. If positive powers of n were present then the function would blow off at fixed $|v|$ and $m \to 0$. At fixed $|v|$, however, no dynamically nontrivial singularity develops in the theory. The only mechanism which could provide powers of m in the denominator is a chain of instantons connected by one massless fermion line depicted in Fig. 3. The corresponding contribution, however, is one-particle reducible and should not be included in the effective Lagrangian. This concludes our proof of the fact that Eq. (2.42) is exact.

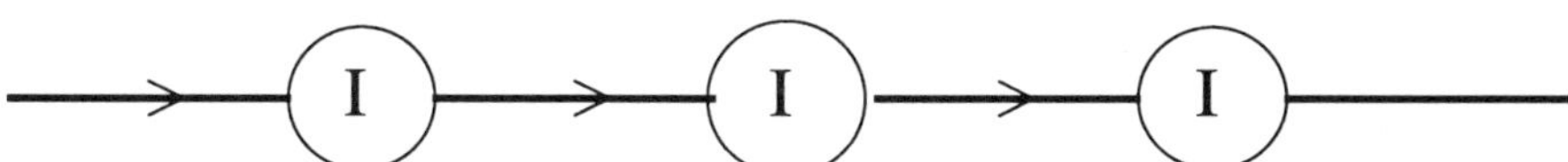

Fig. 3. One-particle reducible graphs which might lead to $1/\sigma$ terms. This mechanism is not (and must not be) included in the superpotential.

The second subtle point is related to the discussion of the anomalies in the R_0 and matter axial currents. The consideration presented above assumes that both anomalies are one-loop. Actually, the anomaly in the R_0 current is multiloop [59]. This fact slightly changes the form of the conserved R current. The very fact of existence of the R current remains intact. All expressions for the currents and charges presented above refer to extreme ultraviolet where the gauge coupling (in asymptotically free theories) tends to zero. The final conclusion that the only superpotential compatible with the symmetry of the model is that of Eq. (2.42) is valid [60].

Thirdly, the consideration above [Eqs. (2.42) and (2.45)] strictly speaking, does not tell us what happens at the origin of the valley, $v^2 = 0$, where all expressions become inapplicable. Logically, it is possible to have an extra vacuum state characterized by $S^2 = 0$. This state would correspond to the strong coupling regime and will not be discussed here. The interested reader is referred to Ref. [34].

One last remark can be made before concluding the section. Eq. (2.43) illustrates why different points from the vacuum valley are physically inequivalent. In the conventional situation of the pre-SUSY era, the spontaneous breaking of a global symmetry, different vacua differ merely by a phase of v. Since physics depends on the ratio $|\Lambda/v|$, this phase is irrelevant. In supersymmetric theories the vacuum valleys are typically noncompact manifolds. Different points are marked not only by the phase of v, but by its absolute value as well. The dimensionless ratio above is different in different vacua. In particular, if $|\Lambda/v| \ll 1$ we are in the weak coupling regime; if $|\Lambda/v| \sim 1$ we are in the strong coupling regime.

2.4. *Miracles of Supersymmetry*

Two of many miraculous dynamical properties of SUSY have been already mentioned — the vanishing of the vacuum energy and the nonrenormalization theorem for F terms. It is instructive to see how these features emerge in perturbation theory.

Let us start from the vacuum energy. Consider a typical two-loop (super) graph shown in Fig. 4.

Each line on the graph represents Green's function of some superfield. We do not even need to know what it is. The crucial point is that (if one works in the coordinate representation) each interaction vertex can be written as an integral over $d^4xd^2\theta d^2\bar{\theta}$. Assume that we substitute explicit expressions for Green's functions and vertices in the integrand, and carry out the integration over the second vertex keeping the first vertex fixed. As a result, we must arrive at an expression of the form

$$\int d^4xd^2\theta d^2\bar{\theta} \times \text{ a function of } x\,,\theta\,,\bar{\theta}\,. \tag{2.47}$$

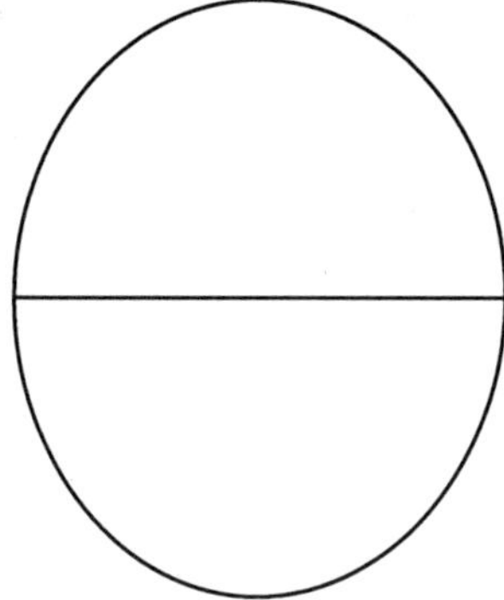

Fig. 4. A typical two-loop supergraph for the vacuum energy.

Since the superspace is homogeneous (there are no points that are singled out, we can freely make translations, any point in the superspace is equivalent to any other point) the function in Eq. (2.47) can be only constant. If so, the result vanishes because of the integration over the Grassmann variables θ and $\bar{\theta}$.

What remains to be demonstrated is that the one-loop vacuum graphs, not representable in the form given in Fig. 4, also vanish. The one-loop (super)graph, however, is the same as for the free particles, and we already know that for free particles $E_{\rm vac} = 0$, see Eq. (2.1), thanks to the balance between the bosonic and fermionic degrees of freedom.

This concludes the proof of the fact that if the vacuum energy is zero at the classical level it remains there to any finite order — there is no renormalization. What changes if, instead of the vacuum energy, we consider renormalizations of the F terms?

The proof presented above can be easily modified to include this case as well. Technically, instead of the vacuum loops, we will now consider loop (super)graphs in a background field.

The basic idea is straightforward. In any supersymmetric theory there are several — at least four — supercharge generators. In a generic background all supersymmetries are broken since the background field is generically not invariant under supertransformations. One can select such a background field, however, that leaves a part of the supertransformations as valid symmetries. For this specific background field some terms in the effective action will vanish, others will not. (Typically, F terms do not vanish while D terms do). The nonrenormalization theorems refer to those terms which do not vanish in the background field chosen.

Consider, for definiteness, the Wess-Zumino model [61],

$$\mathcal{S}^{\rm WZ} = \frac{1}{4}\int d^4x d^2\theta d^2\bar{\theta}\bar{\phi}\phi + \frac{1}{2}\left[\int d^4x d^2\theta(m\phi^2 + g\phi^3) + \text{H.c.}\right]. \qquad (2.48)$$

An appropriate choice of the background field in this case is

$$\bar{\phi}_0 = 0\,, \qquad \phi_0 = C_1 + C_2^\alpha\theta_\alpha + C_3\theta^2\,, \qquad (2.49)$$

where $C_{1,2,3}$ are some constants and the subscript 0 marks the background field. This choice assumes that ϕ and $\bar{\phi}$ are treated as independent variables, not connected by the complex conjugation (i.e. we keep in mind a kind of analytic continuation). The x independent chiral field (2.49) is invariant under the action of $\bar{Q}_{\dot{\alpha}}$, i.e. under the transformations

$$\delta\theta_\alpha = 0\,, \qquad \delta\bar{\theta}_{\dot{\alpha}} = \bar{\varepsilon}_{\dot{\alpha}}\,, \quad \delta x_{\alpha\dot{\alpha}} = -2i\theta_\alpha\bar{\varepsilon}_{\dot{\alpha}}\,.$$

Now, we proceed in the standard way — decompose the superfields

$$\phi = \phi_0 + \phi_{\rm qu}\,, \qquad \bar{\phi} = \bar{\phi}_0 + \bar{\phi}_{\rm qu}\,,$$

where the subscript qu denotes the quantum part of the superfield, expand the action in $\phi_{\rm qu}, \bar{\phi}_{\rm qu}$, drop the linear terms and treat the remainder as the action for the quantum fields. We, then, integrate the quantum fields order by order, keeping the background field fixed. The key element is the fact that in the problem we get for the quantum fields there still exists the exact symmetry under the transformations generated by $\bar{Q}$.

This means that the boson–fermion degeneracy holds, just as in the "empty" vacuum. All lines on the graphs in Fig. 4 have to be treated now as Green's functions in the background field (2.49). After substituting these Green's functions and integrating over all vertices except the first one we come to an expression of the type

$$\int d^4x d^2\bar{\theta} \times \text{a}\,\bar{\theta} \text{ independent function} = 0\,.$$

The $\bar{\theta}$ independence follows from the fact that our superspace is homogeneous in the $\bar{\theta}$ direction even in the presence of the background field (2.49). This completes the proof of the nonrenormalization theorem for the F terms. Note that the kinetic term (D terms) vanishes in the background (2.49), so nothing can be said about its renormalization (and it gets renormalized, of course). The above, somewhat nonstandard, proof of the Grisaru-Roček-Siegel theorem was suggested in Ref. [59].

A word of caution is in order here. Our consideration tacitly assumes that there are no massless fields which can cause infrared singularities. Infrared singular contributions may lead to the so-called *holomorphic anomalies* [62] invalidating the nonrenormalization theorem. We will discuss the property of the holomorphy and the corresponding anomalies later, and now will illustrate how infrared singular D term renormalizations can effectively look as F terms. Consider the D term of the form

$$\int d^2\theta d^2\bar{\theta}\,\frac{D^2}{\Box}F(\phi)\,.$$

It can be rewritten as

$$\int d^2\theta F(\phi)$$

by using the property $[D^2,\bar{D}^2]\propto\Box$ and by integrating by parts in the superspace [44]. It is obvious that the $\Box^{-1}$ singularity can appear only due to massless poles. It was explicitly shown [63] that in the massless Wess-Zumino model such "fake" F terms appear at the two-loop level. The origin of the two-loop and all higher order terms in the Gell-Mann-Low function of supersymmetric gauge theories is the same — they emerge as a "fake" F term which is actually an infrared singular D term [59].

A recent discussion of the "fake" F terms is given in Ref. [64].

2.5. *Holomorphy*

At least some of the miracles of supersymmetry can be traced back to a remarkable property which goes under the name holomorphy. Some parameters in SUSY Lagrangians, usually associated with F terms, are complex rather than real numbers. Mass parameter in the superpotential is an obvious example. Another example mentioned in Sec. 2.3.1 is the inverse gauge coupling, $1/g_0^2$. Now, it is known for a long time, since the mid-eighties, that appropriately chosen quantities depend on these parameters analytically, with possible singularities in certain well defined points. It is obvious that the statement that a function (analytically) depends on a complex variable is infinitely stronger than the statement that a function just depends on two real parameters. The power of holomorphy is such that one can obtain a variety of extremely nontrivial results ranging from nonrenormalization theorems to exact β functions, the first time ever in dynamically nontrivial four-dimensional theories.

In this section we will outline basic steps, keeping in mind that the corresponding technology can be of use more than once in what follows. Let us consider, as an example, SU(2) SQCD with one flavor, Sec. 2.3.5. I have already mentioned that m_0, the complex mass parameter in the action, can be viewed as a vacuum expectation value (of the lowest component) of an auxiliary chiral superfield, let us call it M. It is important that M is singlet with respect to the gauge group, and, thus, say, fermions from M do not contribute to the triangle anomalies. One can think of the corresponding degrees of freedom as of very heavy particles. Then the only role of M is to develop $\langle M \rangle \neq 0$ which obviously does not violate SUSY and provides the mass term.

The theory is strictly invariant under the following phase transformations

$$W_\alpha \to e^{i\beta} W_\alpha\,, \quad S_f \to e^{-i\beta} S_f\,, \quad M \to e^{4i\beta} M\,, \quad \theta_\alpha \to e^{i\beta}\theta_\alpha\,. \qquad (2.50)$$

This is an extended R invariance — extended, because it takes place in the extended theory with the chiral superfield M introduced artificially.

It is rather clear that one chiral superfield can depend only on the expectation value of another superfield of the same chirality — otherwise transformation properties under SUSY would be broken. Thus, the expectation value of W^2 can depend only on that of M; $\bar{M}$ cannot be involved in this relation. Eq. (2.50) then tells us that

$$\langle \mathrm{Tr}\, W^2 \rangle = \mathrm{Const.}(M)^{1/2}\,. \qquad (2.51)$$

In other words, the gluino condensate

$$\langle \lambda\lambda \rangle \propto \sqrt{m_0}\,, \qquad (2.52)$$

and this relation is exact as far as the m_0 dependence is concerned. It holds for small $|m_0|$ when the theory is weakly coupled, as well as for large $|m_0|$, when we are in the strong coupling regime.

A similar assertion is valid regarding the vacuum expectation value of S^2. Now, Eq. (2.50) tells us that

$$\langle S^2 \rangle = \mathrm{Const.}(M)^{-1/2}\,, \qquad (2.53)$$

implying, in turn, that

$$\langle \phi^2 \rangle \propto 1/\sqrt{m_0}\,. \qquad (2.54)$$

It is worth emphasizing that the exact dependence of the condensates on the mass parameter established above refers to the bare mass parameter. If we decided to eliminate the bare mass parameter m_0 in favor of the physical mass of the Higgs field m, we would have to introduce the corresponding Z factor which depends on m in a complicated nonholomorphic way.

Equations (2.52) and (2.54), first derived in Ref. [33] (a similar argument was also given in Ref. [56]), lead to far reaching consequences. Indeed, since the functional dependence of the condensates is fully established, we can calculate the relevant constant at small $|m_0|$, when $\langle\phi^2\rangle$ is large, which ensures the weak coupling regime. I remind that the masses of the gauge bosons in this limit are proportional to

$$M_V^2 \propto |g^2\langle\phi^2\rangle| \, .$$

The result will still be valid for large $|m_0|$, in the strong coupling regime! This line of reasoning [33], based on holomorphy, lies behind many advances achieved recently in SUSY gauge dynamics.

Let me parenthetically note that a nice consistency check to Eqs. (2.52) and (2.54) is provided by the so-called Konishi anomaly [65]. In the model at hand the Konishi relation takes the form

$$\bar{D}^2 \bar{S}^{\alpha f} e^V S^{\alpha f} = 4m_0 S_{\alpha f} S^{\alpha f} + \frac{1}{2\pi^2} \operatorname{Tr} W^2 \, . \tag{2.55}$$

This expression, or more exactly, the second term on the right-hand side, is nothing but a supergeneralization of the triangle anomaly in the divergence of the axial current of the matter fermions, cf. Eq. (2.39). Now, if SUSY is unbroken, the expectation value of the left-hand side must vanish, since the left-hand side is a full superderivative. This fact implies that

$$4m_0 \langle S_{\alpha f} S^{\alpha f}\rangle = -\frac{1}{2\pi^2}\langle \operatorname{Tr} W^2\rangle \tag{2.56}$$

which is consistent with Eqs. (2.52) and (2.54). Another side remark: the exact proportionality of $\langle\phi^2\rangle$ to $1/\sqrt{m_0}$ presents a somewhat different proof of the fact that the instanton-generated superpotential (2.42) is exact even in the presence of the mass term, see Sec. 2.1.5.

One can take advantage of these observations in many ways. One direction is finding the exact β function of the theory. The idea is as follows. First we assume that $|m_0|$ is small and that we are in the weak coupling regime. Then we are able to calculate the gluino condensate — in the weak coupling regime it is saturated by the one-instanton contribution. Since the functional dependence on m_0 is known we can then proceed to the limit of large $|m_0|$, or small ϕ^2. Moreover, the vacuum expectation value of $\lambda\lambda$ is, in principle, a physically measurable quantity. The operator $\lambda\lambda$ has strictly vanishing anomalous dimension since it is the lowest component of the superfield W^2, and the upper component of the same superfield contains the trace of the energy-momentum tensor. This means that if $\langle\lambda\lambda\rangle$ is expressed in terms of the gauge coupling g_0^2 and the ultraviolet cut-off M_0, when one changes the cut off, one should also change g_0^2 in a concerted way, to ensure that $\langle\lambda\lambda\rangle$ stays intact. In this way we obtain a relation between the bare coupling constant and M_0, which is equivalent to the knowledge of the β function.

More concretely, the one-instanton result for the gluino condensate is [66]

$$\langle\lambda\lambda\rangle = \text{Const.} \times \frac{M_0^5 e^{-8\pi^2/g_0^2}}{g_0^4 v_0^2}\,. \tag{2.57}$$

This result is exact; only zero modes in the instanton background contribute in the calculation. It is worth emphasizing that the expectation value of the scalar field appearing in Eq. (2.57) refers to the bare field. The constant on the right-hand side is purely numerical; we will say more about this constant later on, but for the time being its value is inessential.

At the supersymmetric vacuum Eq. (2.56) must hold implying that

$$v_0^2 = \text{Const.}\langle\lambda\lambda\rangle m_0^{-1}\,. \tag{2.58}$$

Combining Eqs. (2.58) and (2.57) we conclude that

$$\langle\lambda\lambda\rangle = \text{Const.}\, M_0^{5/2} m_0^{1/2} e^{-4\pi^2/g_0^2}\frac{1}{g_0^2}\,. \tag{2.59}$$

When analyzing the response of g_0^2 with respect to the variations of M_0 one should keep in mind that m_0 also depends on M_0, implicitly. Indeed, the physical (low-energy) values of the parameters are kept fixed. This means that we fix the renormalized value of the mass, $m = m_0 Z^{-1}$, where Z is the Z factor renormalizing the kinetic term of the matter fields,

$$(\bar{S}^{\alpha f} e^V S^{\alpha f})_0 \to Z(\bar{S}^{\alpha f} e^V S^{\alpha f})_0\,.$$

In this way we arrive at the conclusion that the combination $M_0^{5/2} e^{-4\pi^2/g_0^2}$ $Z^{1/2} g_0^{-2}$ is invariant. Differentiating it with respect to $\ln M_0$ we find the β function,

$$\beta(\alpha) = -\frac{\alpha^2}{2\pi}\frac{5+\gamma}{1-\alpha/\pi}, \tag{2.60}$$

where the β function is defined as

$$\beta(\alpha_0) = d\alpha_0/d\ln M_0, \qquad \alpha_0 = g_0^2/4\pi,$$

and γ is the anomalous dimension of the matter fields,

$$\gamma = d\,\ln Z/d\,\ln M_0\,.$$

Note that due to the SU(2) subflavor symmetry of the model at hand both matter fields, S_1 and S_2, have one and the same anomalous dimension.

Equation (2.60) is a particular case of the general β function, sometimes referred to as the Novikov-Shifman-Vainshtein-Zakharov (NSVZ) β function,

$$\beta(\alpha) = -\frac{\alpha^2}{2\pi}\left(1-\frac{T(G)\alpha}{2\pi}\right)^{-1}\left[3T(G) - \sum_i T(R_i)(1-\gamma_i)\right], \tag{2.61}$$

which can be derived in a similar manner [36]. Here $T(G)$ and $T(R)$ are the so-called Dynkin indices defined as follows. Assume that the gauge group is G, and we have a field belonging to the representation R of the gauge group. If T^a is the generator matrix of the group G in the representation R then

$$\mathrm{Tr}\,(T^aT^b) = T(R)\delta^{ab}\,.$$

More exactly, $T(R)$ is one half of the Dynkin index. Moreover, $T(G)$ is $T(R)$ for the adjoint representation. Note that for the fundamental representation of the unitary groups $T(R) = 1/2$. The sum in Eq. (2.61) runs over all subflavors.

As is clear from its derivation, the NSVZ β function implies the Pauli-Villars regularization. It can also be derived purely perturbatively, with no reference to instantons, using only holomorphy properties of the gauge coupling [67]. The relation of this β function to that defined in other, more conventional regularization schemes is investigated in Ref. [68].

In some theories the NSVZ β function is exact — there are no corrections, either perturbative or nonperturbative, as is the case in the SU(2) model with one flavor. In other models it is exact only perturbatively — nonperturbative corrections do modify it. The most important example [69] of the latter kind is $N = 2$ supersymmetry.

The idea of using the analytic properties of chiral quantities for obtaining exact results was adapted for the case of superpotentials in Ref. [70]. As a matter of fact, I have already discussed some elements of the procedure suggested in Ref. [70], in Sec. 2.3.5, in analyzing possible mass dependence of nonperturbatively generated superpotential in the SU(2) model. Let me summarize here basic stages of the procedure in the general form, and give a few additional examples. Simultaneously, as a byproduct, we will obtain a different proof of some of the nonrenormalization theorems considered in Sec. 2.4. What is remarkable is that, unlike the proof presented in Sec. 2.4, the one given below will be valid both perturbatively and nonperturbatively.

Thus, our task is to establish possible renormalizations of the superpotential $\mathcal{W}$ in a given model. All (complex) coupling constants of the model appearing as coefficients in front of F terms — denote them generically by h_i — are treated as vacuum expectation values of some (auxiliary) chiral superfields. The set of $\{h_i\}$ may include the mass parameters and/or Yukawa constants. Let us assume that if all $h_i = 0$, the model considered possesses a nonanomalous global symmetry group $\mathcal{G}$; however, the couplings $h_i \neq 0$ break this symmetry. Since h_i's are treated now as auxiliary chiral superfields one can always define transformations of these superfields in such a way as to restore the global symmetry $\mathcal{G}$. Then the calculated superpotential depending on the dynamical chiral superfields and on the auxiliary ones, should be invariant under this extended $\mathcal{G}$. This constraint becomes informative if we take into account the fact that the calculated superpotential $\mathcal{W}$ must be a holomorphic function of all chiral superfields. Thus, $\mathcal{W}$ can depend on h_i but cannot depend on h_i^*. A few additional rules apply. The effective superpotential may depend on the dynamically generated scale of the model Λ. It is clear that negative powers of Λ are forbidden since the result should be smooth in the limit when the interaction is switched off. Moreover, if we ensure that the theory is in the weak coupling regime, the possible powers of Λ are only those associated with one, two, three and so on instantons, i.e. $3T(G) - \sum_f T(R_f)$, $2(3T(G) - \sum_f T(R_f))$, etc. since the instantons are the only source of the nonperturbative parameter Λ in the weak coupling regime. Finally, one more condition comes from analyzing the limit of the small bare couplings h_i. This limit can be often treated perturbatively. Sometimes additional massless fields appear in the limit $h_i = 0$, which are absent for $h_i \neq 0$. When these fields are integrated out and are not included in the effective action, $\mathcal{W}$ may develop a singularity at $h_i = 0$.

To illustrate the power and elegance of this approach [70] let us turn again to the Wess-Zumino model, Eq. (2.48). If the bare mass and the coupling constant vanish, $m = g = 0$, then the model has two U(1) global invariances

— one associated with the rotations of the matter field, $\phi \to e^{i\alpha}\phi$, and another one is the R_0 symmetry, $\phi \to e^{2i\beta/3}\phi$, $\theta \to e^{i\beta}\theta$. To maintain both invariances with m and g switched on we demand that

$$m \to me^{-2i\alpha}, \qquad g \to ge^{-3i\alpha}$$

and

$$m \to me^{2i\beta/3}, \qquad g \to g$$

under $\mathrm{U}(1)_\phi$ and $\mathrm{U}(1)_R$, respectively. The most general renormalized superpotential compatible with these symmetries obviously has the form

$$\mathcal{W} = m\phi^2 f\left(\frac{g\phi}{m}\right), \tag{2.62}$$

where f is an arbitrary function. Let us expand it in the power series and consider the coefficient in front of ϕ^n. From Eq. (2.62) it is clear that the corresponding term has the form

$$g^{n-2}m^{3-n}\phi^n .$$

The balance of the powers of the coupling constant and m is such that this contribution could only be associated with the one-particle reducible tree graphs, which should not be included in the effective action. Therefore, we conclude that there is no renormalization of the superpotential, $\mathcal{W} = \mathcal{W}_0$.

An example of a more sophisticated situation is provided by the SU(N) theory with N_f flavors and the tree level superpotential [71]

$$\mathcal{W}_0 = hM\sum_f \tilde{Q}_f Q_f + h'M^3 \tag{2.63}$$

where Q_f is a chiral superfield belonging to the fundamental representation of SU(N) while $\tilde{Q}_f$ is a chiral field in the antifundamental representation. The color indices are summed over; the additional chiral field M is color-singlet.

Now, if h and h' are set equal to zero, M obviously decouples, and the global symmetries of the model are those of the massless SQCD plus one extra global invariance associated with the rotations of the M field. Massless SQCD is invariant under

$$\mathrm{SU}(N_f)_L \times \mathrm{SU}(N_f)_R \times \mathrm{U}(1)_V \times \mathrm{U}(1)_R .$$

The conserved R charge is established from the consideration of the anomaly relations analogous to Eq. (2.39). Indeed, it is not difficult to obtain that

$$\partial_\mu R^0_\mu = \left(N - \frac{N_f}{3}\right)\frac{1}{16\pi^2} G^a_{\mu\nu}\tilde{G}^a_{\mu\nu}\,, \qquad \partial_\mu J_\mu = N_f\,\frac{1}{16\pi^2}\,G^a_{\mu\nu}\tilde{G}^a_{\mu\nu}\,, \tag{2.64}$$

where the R_0 and J currents are defined in parallel to those in the SU(2) model, see Eqs. (2.35) and (2.37). This means that the conserved R current has the form

$$R_\mu = R^0_\mu - \frac{N - (N_f/3)}{N_f} J_\mu\,. \tag{2.65}$$

Using this expression it is not difficult to calculate that the R charge of the matter field is $(N_f - N)/N_f$. The R charge of the M field can be set equal to zero.

Now, the superpotential (2.63) explicitly breaks both, $\mathrm{U}(1)_R$ and $\mathrm{U}(1)_M$. To restore the symmetry we must ascribe to h and h' the following charges

$$h \to \left(-1, 2\frac{N}{N_f}\right), \qquad h' \to (-3, 2)\,,$$

where the first charge is with respect to $\mathrm{U}(1)_M$ while the second charge is with respect to $\mathrm{U}(1)_R$. If $h' = 0$ the model has a rich system of the vacuum valleys. Let us assume that we choose the one for which the expectation value of Q fields vanishes, but the expectation value of $M \neq 0$. Moreover, we will assume that $h\langle M\rangle$ is large. In this "corner" of the valley the matter fields are heavy, and can be integrated over. At very low energies the only surviving (massless) field is M. Our task is to find the effective Lagrangian for the M field. By inspecting the above charge assignments one easily establishes that the most general form of the effective superpotential compatible with all the assignments is

$$\mathcal{W} = h' M^3 f(x)\,, \qquad x = \frac{\Lambda^{3-(N_f/N)}(hM)^{N_f/N}}{h'M^3}\,,$$

where Λ is a dynamically generated scale of the (strongly coupled) SU(N) gauge theory. If $\Lambda \to 0$ there should be no singularities. This implies that $f(x)$ is expandable in positive powers of x. However, the behavior of the effective superpotential at $h' \to 0$ should also be smooth. These two requirements fix the function f up to a constant,

$$f(x) = 1 + \alpha x\,, \qquad \alpha = \mathrm{const.}\,,$$

and

$$\mathcal{W} = h' M^3 + \alpha\, \Lambda^{3-(N_f/N)} (hM)^{N_f/N} \,. \tag{2.66}$$

The first term is the same as in the bare superpotential, the second term is generated nonperturbatively [72]. Note that the nonanalytical behavior at $h = 0$ is due to the fact that at $h = 0$ there are massless matter fields, and we integrated them out assuming that they are massive. The superpotential (2.66) grows with M. This is natural since the interaction becomes stronger as M increases. The superpotential (2.66) leads to a supersymmetric minimum at $M \neq 0$.

Concluding this section, I would like to return to one subtle and very important point in this range of questions, holomorphic anomalies. Consider Eq. (2.59) for the gluino condensate. So far we have studied the analytic dependence of this quantity on m_0. At the same time, however, $1/g_0^2$ is also a coefficient of the F term, which can be viewed as an expectation value of an auxiliary chiral superfield, dilaton/axion. One is tempted to conclude that the dependence of $\langle\lambda\lambda\rangle$ on $1/g_0^2$ must be holomorphic, and its functional form must follow from the consideration of the invariances of the theory. Is this the case?

The answer is yes and no. Let us examine the transformation properties of the SU(2) theory with respect to the matter U(1) rotations, Eq. (2.38), supplemented by the rotation of the mass parameter $m_0 \to m_0 e^{-2i\gamma}$. At the classical level the theory is invariant. The invariance is broken, however, by the triangle anomaly. In order to restore the invariance we must simultaneously shift the vacuum angle ϑ (not to be confused with the supercoordinates θ_α!),

$$\vartheta \to \vartheta - \Delta \mathrm{Arg}(m_0) = \vartheta + 2\gamma \,. \tag{2.67}$$

Taking into account the fact that $\vartheta = \mathrm{Im}\,(-8\pi^2/g_0^2)$ we conclude that if Eq. (2.59) contained no pre-exponential factor g_0^{-2} everything would be perfect — $\lambda\lambda$ would be a holomorphic function of g_0^{-2}, precisely the one needed for invariance of $\lambda\lambda$. The pre-exponential factor g_0^{-2} spoils the perfect picture. As a matter of fact, one can see that in the pre-exponential it is $\mathrm{Re}(g_0^{-2})$ which enters, and the holomorphy in g_0^{-2} is absent. The reason is the holomorphic anomaly associated with infrared effects. In Refs. [59, 62] it was first noted that all formal theorems regarding the holomorphic dependences on the gauge coupling constant are valid provided we define the gauge coupling through the Wilsonian action which, by definition, contains no infrared contributions. What one usually deals with (and refers to as the action) is actually the generator of the one-particle irreducible vertices. In the absence of the infrared singularities these two notions coincide; generally speaking, they are different, however. In particular, the gauge coupling constant in the Wilsonian action, g_W^{-2} is related to g_0^{-2} by the following expression

$$\frac{8\pi^2}{g_W^2} = \frac{8\pi^2}{g_0^2} - T(G)\ln \mathrm{Re} g_0^{-2} \tag{2.68}$$

(in pure gluodynamics, without matter). The gluino condensate is holomorphic with respect to g_W^{-2}.

The holomorphic anomaly in the gauge coupling due to massless matter fields was also observed in Ref. [73] in the stringy context, see also [74].

2.6. *Supersymmetric Instanton Calculus*

As was mentioned more than once, the instanton calculations, combined with specific features of supersymmetry, were instrumental in establishing various precise results in supersymmetric gluodynamics and other theories. We will continue to exploit them in further applications. Needless to say that I will be unable to present supersymmetric instanton calculus to the degree needed for practical uses. The interested reader is referred to Ref. [75] and Chapter VI. Here I will limit myself to a few fragmentary remarks.

Technically, the most remarkable feature making the instanton calculations in the supersymmetric theories by far more manageable than in nonsupersymmetric ones is a residual supersymmetry in the instanton background field. It is clear that picking up a particular external field we typically break (spontaneously) supersymmetry: SUSY generators applied to this field act nontrivially. However, the self-dual (or anti-self-dual) Yang-Mills field, analytically continued to the Euclidean space, to which the instanton belongs, preserves a half of the supersymmetry. Depending on the sign of the duality relation either Q_α or $\bar{Q}_{\dot\beta}$ act trivially, i.e. annihilate the background field [76, 77].

The fact that a part of supersymmetry remains unbroken in the instanton background leads to far-reaching consequences. Indeed, the spectrum of fluctuations around this background remains degenerate for bosons and fermions from one and the same superfield, and the form of the modes is in one-to-one correspondence [77], for all modes except the zero modes. An immediate consequence is the vanishing of the one-loop quantum correction in the instanton background. Unsurprisingly, a more careful study shows [32] that all higher quantum corrections vanish as well.

Thus, the result of any instanton calculation is essentially determined by the zero modes alone. The problem reduces to the quantum mechanics of the zero modes. The structure of the zero modes is governed by a set of relevant symmetries of the theory under consideration [66]. Therefore, all quantities that are saturated by instantons reflect the most general and profound geometrical properties of the theory. One of the examples, the gluino condensate, was already considered above. In Sec. 3 we will discuss another example — the instanton-induced modification of the quantum moduli space in SQCD with $N_f = N_c$.

Historically the first application of instantons in supersymmetric gluodynamics was the calculation of the gluino condensate [32] in the strong coupling regime. I mention this result here because although it is 15 years old, there is an intriguing mystery associated with it.

Let us consider for definiteness the SU(2) gluodynamics. In this case there are four gluino zero modes in the instanton field and hence, there is no direct instanton contribution to the gluino condensate $\langle\lambda\lambda\rangle$. At the same time the instanton does contribute to the correlation function

$$\langle\lambda^a_\alpha(x)\lambda^{a\alpha}(x),\ \lambda^b_\beta(0)\lambda^{b\beta}(0)\rangle\,. \tag{2.69}$$

Here $a, b = 1, 2, 3$ are the color indices and $\alpha, \beta = 1, 2$ are the spinor ones. An explicit instanton calculation shows that the correlation function (2.69) is equal to a nonvanishing constant.

At first sight this result might seem supersymmetry-breaking since the instanton does not generate any boson analog of Eq. (2.69). Surprising though it is, supersymmetry does not forbid (2.69) provided that this two-point function is actually an x independent constant. For purposes which will become clear shortly let us sketch here the proof of the above assertion.

Three elements are of importance: (i) the supercharge $\bar{Q}^{\dot\beta}$ acting on the vacuum state annihilates it; (ii) $\bar{Q}^{\dot\beta}$ commutes with $\lambda\lambda$; (iii) the derivative $\partial_{\alpha\dot\beta}(\lambda\lambda)$ is representable as the anticommutator of $\bar{Q}^{\dot\beta}$ and $\lambda^\beta G_{\beta\alpha}$. (The spinor notation is used.) The second and the third point follow from the fact that $\lambda\lambda$ is the lowest component of the chiral superfield W^2, while $\lambda^\beta G_{\beta\alpha}$ is its middle component.

Now, we differentiate Eq. (2.69), substitute $\partial_{\alpha\dot\beta}(\lambda\lambda)$ by $\{\bar{Q}^{\dot\beta}, \lambda^\beta G_{\beta\alpha}\}$ and obtain zero. Thus, supersymmetry requires the x derivative of (2.69) to vanish [32]. This is exactly what happens if the correlator (2.69) is a constant.

If so, one can compute the result at short distances where it is presumably saturated by small-size instantons, and, then, the very same constant is predicted at large distances, $x \to \infty$. On the other hand, due to the cluster decomposition property which must be valid in any reasonable theory the correlation function (2.69) at $x \to \infty$ reduces to $\langle\lambda\lambda\rangle^2$. Extracting the square root we arrive at a (double-valued) prediction for the gluino condensate.

(The same line of reasoning is applicable in other similar problems, not only for the gluino condensate. The correlation function of the lowest components of any number of superfields of one and the same chirality, if nonvanishing, must be constant. By analyzing the instanton zero modes it is rather easy to catalog all such correlation functions, in which the instanton contribution does not vanish. Thus, for SU(N) gluodynamics one ends up with the N-point function of $\lambda\lambda$. Inclusion of the matter fields, clearly, enriches the list of the instanton-induced "constant" correlators, but not too strongly [78]. The general strategy remains the same as above in all cases).

Many questions immediately come to one's mind in connection with this argument. First, if the gluino condensate is nonvanishing and shows up in a roundabout instanton calculation through (2.69) why it is not seen in the direct instanton calculation of $\langle\lambda\lambda\rangle$? Second, the constancy of the two-point function (2.69) required by SUSY is ensured in the concrete calculation by the fact that the instanton size ρ turns out to be of order of x. The larger the value of x the larger ρ saturates the instanton contribution. For small x this is alright. At the same time at $x \to \infty$ we do not expect any coherent fields with the size of order x to survive in the vacuum; such coherent fields would contradict our current ideas of the infrared-strong confining theories like SUSY gluodynamics. If there are no large-size coherent fields in the vacuum how can one guarantee the x independence of (2.69) at all distances?

A tentative answer to the first question might be found in the hypothesis put forward by Amati *et al.* [56]. It was assumed that, instead of providing us with the expectation value of $\lambda\lambda$ in the given vacuum, instantons in the strong coupling regime yield an average value of $\langle\lambda\lambda\rangle$ in all possible vacuum states. If there exist two vacua, with the opposite signs of $\langle\lambda\lambda\rangle$, the conjecture of Amati *et al.* would explain why instantons in the strong coupling regime do not generate $\lambda\lambda$ directly.

When we do the instanton calculation in the weak coupling regime (the Higgs phase) the averaging over distinct vacua does not take place. In the weak coupling regime, we have a marker: a large classical expectation value of the Higgs field tells us in what particular vacuum we do our instanton calculation. In the strong coupling regime, such a marker is absent, so that the recipe of Amati *et al.* seems plausible.

This is not the end of the story, however. One of the instanton computations which was done in the mid-eighties [66] remained a puzzle defying theoretical understanding for years. The result for $\langle\lambda\lambda\rangle$ obtained in the strong coupling regime (i.e. by following the program outlined after Eq. (2.69)) does *not* match $\langle\lambda\lambda\rangle$ calculated in an indirect way, as we did in Sec. 2.5 — extending the theory by adding one flavor, doing the calculation in the weakly coupled Higgs phase, and then returning back to SUSY gluodynamics by exploiting the holomorphy of the condensate in the mass parameter. In Ref. [66] it was shown that

$$\langle\lambda\lambda\rangle^2_{\rm scr} = \frac{4}{5}\langle\lambda\lambda\rangle^2_{\rm wcr}\,, \tag{2.70}$$

where the subscripts scr and wcr mark the strong and weak coupling regime calculations (for more details see Chapter VI).

The hypothesis of Amati *et al.*, by itself, does not explain the discrepancy (2.70). If there are only two vacua characterized by $\langle\lambda\lambda\rangle = \pm\Lambda^3$, the gluino condensate is not affected by the averaging over these two vacuum states, since the contributions of these two vacua to Eq. (2.69) are equal. If, however, there exist an *extra* zero-energy state with $\langle\lambda\lambda\rangle = 0$ involved in the averaging, the final result in the strong coupling regime is naturally different from that obtained in the weak coupling regime in the *given* vacuum. Moreover, the value of the condensate calculated in the strong coupling approach should be smaller, consistently with Eq. (2.70). At present, there seems to be no other way out of the dilemma [34]. The conclusion of the existence of the extra vacuum with $\langle\lambda\lambda\rangle = 0$ is quite radical, and requires further verification. What is beyond any doubt, however, is that the combination of instanton calculus with holomorphy and other specific features of supersymmetry provides us with the most powerful tool we have ever had in four-dimensional field theories.

It remains to be added that the interest to technical aspects of supersymmetric instanton calculus [66] was revived recently in connection with the Seiberg–Witten solution of the $N = 2$ theory. The solution was obtained [16] from indirect arguments, and it was tempting to verify it by direct instanton calculations [79]. Such calculations require extension of supersymmetric instanton calculus to $N = 2$, which was carried out, in a very elegant way, in Ref. [80], see also Ref. [79].

Concluding this section let me briefly summarize the main lessons.

First, the most remarkable feature of the structure of SUSY gauge theories with matter is the existence of the vacuum valleys — classically flat directions along which the energy vanishes. This degeneracy may or may not be lifted dynamically, at the quantum level. SU(2) model with one flavor is an example of the theory where the continuous degeneracy is lifted, and the quantum vacuum has only discrete (two-fold) degeneracy. If this does not happen, the classically flat directions give rise to *quantum moduli space of supersymmetric vacua.* This feature is the key element of the recent developments pioneered by Seiberg.

Second, holomorphic dependences of various chiral quantities enforced by supersymmetry lie behind numerous miracles occurring in SUSY gauge theories — from specific nonrenormalization theorems to the exact β functions. This is also an important element of dynamical scenarios to be discussed below.

Now the stage is set and we are ready for more adventures and surprises in supersymmetric dynamics.

Section 3

Dynamical Scenarios in SUSY Gauge Theories — Pandora's Box?

In the first two sections I summarized what was known (or assumed) about the intricacies of the gauge dynamics in the eighties. In the following three sections we will discuss the discoveries and exciting results of the recent years. I should say that the current stage of development was opened by Seiberg, and many ideas and insights to be discussed today I learned from him or extracted from the works of his collaborators.

Remarkable facets of the gauge dynamics will open to us. First of all, we will encounter nonconventional patterns of the chiral symmetry breaking. The chiral symmetry breaking is one of the most important phenomena of which very little was known, beyond some empirical facts referring to QCD. In the eighties, when our knowledge of the gauge dynamics was less mature than it is now, it was believed that the massless fermion condensation obeys the so-called maximum attraction channel (MAC) hypothesis [81]. In short, one was supposed to consider the one-gluon exchange between fermions, find a channel with such quantum numbers that the attraction was maximal, and then assume the condensation of the fermion pairs in this particular channel. The concrete quantum numbers of the fermion condensates imply a very specific pattern of the chiral symmetry breaking.

In SQCD we will find patterns *contradicting the MAC hypothesis.* This means that the chiral condensates are not governed by the one-gluon exchange, even qualitatively. The basic tool for exploring the chiral condensates is the 't Hooft matching condition. It was exploited for this purpose several times in the past, in the context most relevant to us in Ref. [82]. Combining supersymmetry (the fact of the existence of the vacuum manifold) with the matching condition drastically enhances the method.

The second remarkable finding is the observation of conformally invariant theories in four dimensions in the strong coupling regime. The crucial instrument in revealing such theories is Seiberg's "electric-magnetic" duality in the infrared domain, connecting with each other two distinct gauge theories — one of them is strongly coupled while the other is weakly coupled. One can view the gluons and quarks of the weakly coupled theory as bound states of the gluons and quarks of its dual partner. If so, composite gauge bosons can exist! The arguments in favor of the "electric-magnetic" duality are again based on the 't Hooft matching condition (combined with supersymmetry) and some additional indirect consistency checks.

3.1. SU(N_c) *QCD with* N_f *Flavors — Preliminaries*

The SU(2) model considered previously is somewhat special since all representations of SU(2) are (pseudo)real. For this reason the flavor sector of this model possesses an enlarged symmetry. Thus, for one flavor we observe the flavor SU(2) symmetry, which is absent if the gauge group is, say, SU(3). Now we will consider a more generic situation. If not stated to the contrary the gauge group is assumed to be SU(N_c) with $N_c > 2$. Peculiarities of the orthogonal groups will be briefly discussed in Sec. 4. In accordance with Witten's index, if the matter sector consists of nonchiral matter allowing (at least, in principle) for a mass term for all matter fields, supersymmetry is unbroken.

To describe N_f flavors one has to introduce $2N_f$ chiral superfields, Q^i in the representation N_c and $\tilde{Q}_j$ in the representation $\bar{N}_c$. To distinguish between the fundamental and antifundamental representations the flavor indices used are superscripts and subscripts, respectively.

The Lagrangian is very similar to that of the SU(2) model,

$$\mathcal{L} = \frac{1}{2g_0^2}\,\mathrm{Tr}\int d^2\theta W^2 + \frac{1}{4}\int d^2\theta d^2\bar{\theta}(Q^\dagger e^V Q + \tilde{Q}^\dagger e^{-V}\tilde{Q}) + \left(\frac{1}{2}\int d^2\theta\, \mathcal{W}(Q,\tilde{Q}) + \text{H.c.}\right), \tag{3.1}$$

where $\mathcal{W}(Q,\tilde{Q})$ is a superpotential which may or may not be present. Then the scalar potential has the form

$$V = \frac{1}{2g^2}D^a D^a + \sum_Q \left|\frac{\partial \mathcal{W}(Q,\tilde{Q})}{\partial Q}\right|^2 + \sum_{\tilde{Q}}\left|\frac{\partial \mathcal{W}(Q,\tilde{Q})}{\partial \tilde{Q}}\right|^2. \tag{3.2}$$

An example of the possible superpotential is a generalized mass term,

$$\mathcal{W} = m_i^j \tilde{Q}_j Q^i$$

where m_i^j is a mass matrix. Most often we will work under the conditions of vanishing superpotential, $\mathcal{W} = 0$.

It is convenient to introduce two $N_f \times N_c$ matrices of the form

$$q = \{Q^1, Q^2, \ldots, Q^{N_f}\}, \qquad \tilde{q} = \{\tilde{Q}_1, \tilde{Q}_2, \ldots, \tilde{Q}_{N_f}\}. \tag{3.3}$$

The rows of these matrices correspond to different values of the color index. Thus, in the first row the color index is 1, in the second row 2, etc., N_c rows altogether. Both matrices can be globally rotated in the color and flavor spaces. Let us assume first that $N_f < N_c$. Then, by applying these rotations one can always reduce the matrix q to the form

$$q = \begin{pmatrix} a_1 & 0 & \cdots & 0 \\ 0 & a_2 & \cdots & 0 \\ \cdots & \cdots & \cdots & \cdots \\ 0 & 0 & \cdots & a_{N_f} \\ 0 & 0 & \cdots & 0 \\ \cdots & \cdots & \cdots & \cdots \end{pmatrix}. \tag{3.4}$$

If $\tilde{q} = q$ we are at the bottom of the vacuum valley — the corresponding energy vanishes. The gauge invariant description is provided by the composite chiral superfield,

$$M_j^i = \tilde{Q}_j Q^i. \tag{3.5}$$

The points belonging to the bottom of the valley are parametrized by the expectation value of M^i_j. Generically, if we are away from the origin, SU(N_c) gauge group is broken down to SU($N_c - N_f$). The first group has $N_c^2 - 1$ generators, the second one has $(N_c - N_f)^2 - 1$ generators. Thus, the number of the chiral fields eaten up in the super-Higgs mechanism is $2N_cN_f - N_f^2$. Originally we started from $2N_cN_f$ chiral superfields; N_f^2 remain massless — exactly the number of degrees of freedom in M^i_j. There are exceptional points. When $\det M = 0$ the unbroken gauge subgroup is larger than SU($N_c - N_f$), and, correspondingly, we have more than N_f^2 massless particles. At the origin of the vacuum valley the original gauge group SU(N_c) remains unbroken.

The situation changes if the number of flavors is equal to or larger than the number of colors. Indeed, if $N_f > N_c$ the generic form of the matrix q, after an appropriate rotation in the flavor and color space, is

$$q = \begin{pmatrix} a_1 & 0 & \cdots & 0 & 0 & \cdots & 0 \\ 0 & a_2 & \cdots & 0 & 0 & \cdots & 0 \\ \cdots & \cdots & \cdots & \cdots & \cdots & \cdots & \cdots \\ 0 & 0 & \cdots & a_{N_c} & 0 & \cdots & 0 \end{pmatrix} . \tag{3.6}$$

The condition defining the bottom of the valley (the vanishing of the energy) is

$$|a_i|^2 - |\tilde{a}_i|^2 = \text{constant independent of } i \tag{3.7}$$

where $i = 1, 2, \ldots, N_c$. At a generic point from the bottom of the valley the gauge group is completely broken. The gauge invariant chiral variables parametrizing the bottom of the valley (the moduli space) now are

$$M^i_j = \tilde{Q}_j Q^i \,, \qquad B = Q^{[i_1} \ldots Q^{i_{N_c}]} \,, \qquad \tilde{B} = \tilde{Q}_{[j_1} \ldots \tilde{Q}_{j_{N_c}]} \,, \tag{3.8}$$

where the color indices in B and $\tilde{B}$ (they are not written out explicitly) are contracted with the help of the ε symbol; the flavor indices $i_1, \ldots, i_{N_c}$ and $j_1, \ldots, j_{N_c}$ then come out automatically antisymmetric, and the square brackets in Eq. (3.8) remind us of this antisymmetrization. *A priori*, the number of the variables B and $\tilde{B}$ is $C^{N_c}_{N_f}$ each, where $C^{N_c}_{N_f}$ are the combinatorial coefficients,

$$C^{N_c}_{N_f} = \frac{N_f!}{N_c!(N_f - N_c)!} \,,$$

since one can pick up N_c flavors out of the total set of N_f in various ways. If we try to calculate now the number of moduli, assuming that all those indicated in Eq. (3.8) are independent, we will see that this number does not match the number of the massless degrees of freedom. Let us consider two examples, $N_f = N_c$ and $N_f = N_c + 1$.

$\boldsymbol{N_f = N_c}$

The original number of the chiral superfields is $2N_f N_c$; since the gauge symmetry is completely broken the number of the "eaten" superfields is $N_c^2 - 1$; the number of the massless degrees of freedom is, thus, $N_f^2 + 1 = N_c^2 + 1$. The number of moduli in Eq. (3.8) is $N_f^2 + 2$. One chiral variable is, thus, redundant.

$\boldsymbol{N_f = N_c + 1}$

The number of the chiral superfields is $2N_f N_c$; since the gauge symmetry is completely broken the number of the "eaten" superfields is $N_c^2 - 1$; the number of the massless degrees of freedom is N_f^2. The number of moduli in Eq. (3.8) is $N_f^2 + 2N_f$, i.e. $2N_f$ chiral variables are, thus, redundant.

In the first case, $N_f = N_c$, the constraint eliminating the redundant chiral variable is

$$\det\{M\} = B\tilde{B}\,, \tag{3.9}$$

while in the second example, $N_f = N_c + 1$, by using merely the definitions of the moduli in Eq. (3.8), it is not difficult to obtain

$$B_i\{M\}^i_j = \{M\}^i_j \tilde{B}^j = 0\,, \tag{3.10}$$

and

$$\text{minor}\,\{M\}^j_i = B_i \tilde{B}^j\,, \tag{3.11}$$

where the left-hand side of the last equation is the minor of the matrix $\{M\}$ (i.e. $(-1)^{i+j}\times$ determinant of the matrix obtained from M by omitting the ith row and the jth column). Note that $\det\{M\}$ vanishes in this case. For brevity we will sometimes write Eq. (3.11) in a somewhat sloppy form

$$\det M (M^{-1})^j_i = B_i \tilde{B}^j\,. \tag{3.12}$$

At the classical level one could, in principle, eliminate the redundant chiral variables using Eqs. (3.9) or (3.12). One should not hurry with this elimination, however, since at the quantum level the classical moduli fields are replaced by the vacuum expectation values of M^j_i and B, $\tilde{B}$, and although generically the total number of the massless degrees of freedom does not change, the quantum version of constraints (3.9) and (3.12) may (and will be) different. Moreover, at some specific points from the valley the number of the massless degrees of freedom may increase, as we will see shortly.

SQCD with N_f flavors and no tree-level superpotential has the following global symmetries free from the internal anomalies:

$$\mathrm{SU}(N_f)_L \times \mathrm{SU}(N_f)_R \times \mathrm{U}(1)_B \times \mathrm{U}(1)_R \tag{3.13}$$

where the conserved R current is introduced in Eq. (2.65), and the quantum numbers of the matter multiplets with respect to these symmetries are collected in Table 1.

Table 1. The quantum numbers of the matter fields with respect to the global symmetries (3.13) in SQCD with N_f flavors. The R charges indicated refer to the lowest components of the superfields.

	$\mathrm{SU}(N_f)_L$	$\mathrm{SU}(N_f)_R$	$\mathrm{U}(1)_B$	$\mathrm{U}(1)_R$
Q	N_f	1	1	$(N_f - N_c)/N_f$
$\tilde{Q}$	1	$\bar{N}_f$	-1	$(N_f - N_c)/N_f$

(For discussion of the subtleties in the R current definition see Ref. [60]. These subtleties, being conceptually important, are irrelevant for our consideration).

The $\mathrm{SU}(N_f)_L \times \mathrm{SU}(N_f)_R \times \mathrm{U}(1)_B$ transformations act only on the matter fields in an obvious way, and do not affect the superspace coordinate θ. As for the extra global symmetry $\mathrm{U}(1)_R$ it is defined in such a way that it acts nontrivially on the supercoordinate θ and, therefore, acts differently on the spinor and the scalar or the vector components of superfields. The R charges in Table 1 are given for the lowest component of the chiral superfields. If the R charge of the boson component of the given chiral superfield is r then the R charge of the fermion component is, obviously, $r-1$. A part of the above global symmetries is spontaneously broken by the vacuum expectation values of M^i_j and/or $B, \tilde{B}$.

Unlike the $N_c = 2$, $N_f = 1$ model discussed in Sec. 2 instantons do not lift the classical degeneracy, and the bottom of the valley remains flat. The easiest way to see this is to consider a generic point from the bottom of the valley, far away from the origin, where the theory is in the weak coupling regime, and try to write the most general superpotential, compatible with all exact symmetries (it must be symmetric even under those symmetries which may turn out to be spontaneously broken) [41, 42]. The symmetry under $\mathrm{SU}(N_f)_L \times \mathrm{SU}(N_f)_R \times \mathrm{U}(1)_B$ is guaranteed if we assume that the superpotential $\mathcal{W}$ depends on $\det M$. What about the R symmetry?

For $N_f = N_c$ the R charge of the matter superfield vanishes, as is clear from Table 1. Since the superpotential must have the R charge 2, it is obvious that it cannot be generated. For $N_f > N_c$ the R charge of the matter fields does not vanish, and, in principle, one could have written

$$\mathcal{W} \propto \left(\frac{\Lambda^{3N_c - N_f}}{\det M} \right)^{\frac{1}{N_c - N_f}} ,$$

an expression which has the right dimension (3) and the correct R charge (2). However, the dimension of Λ does not match the instanton expression which can produce only $\Lambda^{3N_c - N_f}$ (and in the weak coupling regime the instanton is the only relevant nonperturbative contribution). What is even more important, for $N_f > N_c$ the determinant of M vanishes identically. This fact alone shows that no superpotential can be generated, and the flat direction remains flat [83, 41, 42].

The argument above demonstrates again the power of holomorphy. In nonsupersymmetric theories one could build a large number of invariants involving $Q, \tilde{Q}, Q^\dagger$ and $\tilde{Q}^\dagger$. In SUSY theories, as far as the F terms are concerned, one is allowed to use only Q and $\tilde{Q}$ which constraints the possibilities to the extent when nothing is left.

In summary, for $N_f \geq N_c$ the vacuum degeneracy is not lifted. At the origin of the space of moduli, where $B = \tilde{B} = 0$ and M has fewer than $N_c - 1$ nonzero eigenvalues, the gauge symmetry is not fully broken. At these points, the classical moduli space is singular. Far away from the origin, when the expectation values of the squark fields are large, the distinction between the classical and quantum moduli space should be unimportant. In the vicinity of the origin, however, this distinction may be crucial. Our next task is to investigate this distinction. Needless to say that just the vicinity of the origin is the domain of most interesting dynamics. Since the Higgs fields are in the fundamental representation, we are always in the Higgs/confining phase. Far away from the origin the theory is in the weak coupling regime and is fully controllable by the well understood methods of the weak coupling. In the vicinity of the origin the theory is in the strong coupling regime. The issues to be investigated are the patterns of the spontaneous breaking of the global symmetries and the occurrence of the composite massless degrees of freedom at large distances. Here each nontrivial theoretical result or assertion is a precious asset, a miraculous achievement.

3.2. *The Quantum Moduli Space*

Relations (3.9) and (3.12) are constraints on the classical composite fields. Since in the quantum theory the vacuum valley is parametrized by the *expectation values* of the fields, which may get a contribution from quantum fluctuations, these relations may alter. In other words, the quantum moduli space need not exactly coincide with the classical one. Only in the limit when the vacuum expectation values of the fields parametrizing the vacuum valley become large, much larger than the scale parameter of the underlying theory, we must be able to return to the classical description.

To see that the quantum moduli space does indeed differ from the classical one we will consider here, following Ref. [48], the same two examples, $N_f = N_c$ and $N_f = N_c + 1$. The general strategy used in these explorations is the same as the one discussed in detail in Sec. 2, in connection with SU(2) SQCD [33]: in order to analyze the theory along the classically flat directions one adds the appropriately chosen mass terms (sometimes, other superpotential terms as well), solves the theory in the weak coupling regime, and then analytically continues to the limit where the classical superpotential vanishes.

$\boldsymbol{N_f = N_c}$

Introduce a mass term for all quark flavors, or, more generically, the quark mass matrix

$$m_i^j Q^i \tilde{Q}_j\,, \tag{3.14}$$

(it can always be diagonalized, of course). If we additionally assume that the mass terms for $N_f - 1$ flavors are small and the mass term for one flavor is large then we find ourselves in a situation where an effective low-energy theory is that of $N_c - 1$ flavors. From Sec. 2 we already know that in this case the SU(N_c) symmetry is totally broken spontaneously, the theory is in the weak coupling phase, instantons generate a superpotential, and this superpotential, combined with Eq. (3.14), leads to [41, 42, 66, 56]

$$M_j^i = \langle Q^i \tilde{Q}_j \rangle = \Lambda^2 (\det m)^{\frac{1}{N_c}} \left(\frac{1}{m}\right)_j^i\,,$$

$$B = \langle Q^{i_1} \ldots Q^{i_{N_c}} \rangle \varepsilon_{i_1 \ldots i_{N_c}} = 0 \,, \quad \tilde{B} = \langle \tilde{Q}_{j_1} \ldots \tilde{Q}_{j_{N_c}} \rangle \varepsilon^{j_1 \ldots j_{N_c}} = 0 \,. \quad (3.15)$$

Although this result was obtained under a very specific assumption on the values of the mass terms, holomorphy tells us that it is exact. In particular, one can tend $m_i^j \to 0$, thus returning to the original massless theory. Eq. (3.15) obviously implies that

$$\det M - B\tilde{B} = \Lambda^{2N_c} \,. \quad (3.16)$$

It is instructive to check that this relation stays valid even if $B \neq 0$, $\tilde{B} \neq 0$. To this end one must introduce, additionally, a superpotential $\beta B + \tilde{\beta}\tilde{B}$ where β and $\tilde{\beta}$ are some constants, and redo the instanton calculations. If $\beta \neq 0$ and $\tilde{\beta} \neq 0$, the instanton-induced superpotential changes, nonvanishing values of B and $\tilde{B}$ are generated, the vacuum expectation values M_j^i change as well, but the relation (3.16) stays intact.

Far from the origin, where the semiclassical analysis is applicable, the quantum moduli space (3.16) is close to the classical one. A remarkable phenomenon happens near the origin [48]. In the classical theory where the gluons were massless near the origin, the classical moduli space was singular. Quantum effects eliminated the massless modes by creating a mass gap.[7] Correspondingly, the singular points with $B = \tilde{B} = 0$ and the vanishing eigenvalues of M are eliminated from the moduli space.

In the weak coupling regime dynamics is rather trivial and boring. Let us consider the most interesting domain of the vacuum valley, near the origin, in more detail, "in a microscope". There are several points that are special, they are characterized by an enhanced global symmetry. For instance, if

$$B = \tilde{B} = 0 \qquad \text{and} \qquad M_j^i = \Lambda^2 \delta_j^i \,, \quad (3.17)$$

the original global $SU(N_f) \times SU(N_f)$ symmetry is spontaneously broken down to the diagonal $SU(N_f)$, while the $U(1)_R$ remains unbroken (the R charges of Q and $\tilde{Q}$ vanish, see Table 1). We are in the vicinity of the origin, where all moduli are either of order of Λ^2 or vanish. Hence, the fundamental gauge dynamics of the quark (squark) matter is strongly coupled. We are in the strong coupling regime.

The spontaneous breaking of the global symmetry implies the existence of the massless Goldstone mesons which, through supersymmetry, entails, in

[7] The latter statement is not quite correct. Massless moduli fields still persist. What is important, however, is that the gluons acquire a dynamical "mass".

turn, the occurrence of the massless (composite) fermions. These fermions reside in the superfields M, B, and $\tilde{B}$. Their quantum numbers with respect to the unbroken symmetries are indicated in Table 2.

Table 2. The quantum numbers of the composite massless fermions with respect to the unbroken global symmetries in SQCD with $N_f = N_c$.

	$\mathrm{SU}(N_f)$	$\mathrm{U}(1)_B$	$\mathrm{U}(1)_R$
ψ_M	$N_f^2 - 1$	0	-1
ψ_B	1	N_f	-1
$\psi_{\tilde{B}}$	1	$-N_f$	-1
ψ_Q	N_f	1	-1
$\psi_{\tilde{Q}}$	$\bar{N}_f$	-1	-1
λ	1	0	1

For convenience Table 2 summarizes also the quantum numbers of the fundamental fermions — quarks and gluino. A remark is in order concerning the multiplet of the massless fermions ψ_M. Since M is an $N_f \times N_f$ matrix, naively one might think that the number of these fermions is N_f^2. Actually we must not forget that we are interested in small fluctuations of the moduli fields M, B and $\tilde{B}$ around the expectation values (3.17) subject to the constraint (3.16). It is easy to see that this constraint implies that the matrix of fluctuations $M_j^i - \Lambda^2 \delta_j^i$ is traceless, i.e. the fluctuations form the adjoint ($(N_f^2 - 1)$-dimensional) representation of the diagonal $\mathrm{SU}(N_f)$.

Massless composite fermions in the gauge theories are subject to a very powerful constraint known as the *'t Hooft consistency condition* [84]. As was first noted in Ref. [85], the triangle anomalies of the AVV type in the gauge theories with the fermion matter imply the existence of infrared singularities in the matrix elements of the axial currents. (Here A and V stand for the axial and vector currents, respectively). These singularities are unambiguously fixed by the short-distance (fundamental) structure of the theory even if the theory at hand is in the strong coupling regime and cannot be solved in the infrared. The massless composite fermions in the theory, if present, must arrange themselves in such a way as to match these singularities. If they cannot, the corresponding symmetry is spontaneously broken, and the missing infrared singularity is provided by the Goldstone-boson poles coupled to the corresponding broken generators. This device — the 't Hooft consistency condition, or anomaly matching — is widely used in the strongly coupled gauge theories: from QCD to technicolor, to supersymmetric models; it allows one to check various conjectures about the massless composite states. (For a pedagogical review see, for example, Ref. [86]).

In our case we infer the existence of the massless fermions from the fact that a set of moduli exists, plus supersymmetry. Why do we need to check the matching of the AVV triangles? If we know for sure the pattern of the symmetry breaking — which symmetry is spontaneously broken and which is realized linearly — the matching of the AVV triangles for the unbroken currents must be automatic. The condensates indicated in Eq. (3.17) suggest that the axial SU(N_f) is spontaneously broken while the R current and the baryon current are unbroken. Suggest, but do not prove! For in the strong coupling regime other (nonchiral) condensates might develop too. For instance, on general grounds one cannot exclude the condensate of the type $\langle M^i_j B^\dagger \rangle$ which will spontaneously break the baryon charge conservation. Since this superfield is nonchiral the holomorphy consideration is inapplicable. If the anomalous triangles with the baryon current do match, it will be a strong argument showing that no additional condensates develop, and the pattern of the spontaneous symmetry breaking can be read off from Eq. (3.17). Certainly, this is not a complete rigorous proof, but, rather, a very strong indication.

What is extremely unusual in the pattern implied by Eq. (3.17) is the survival of an unbroken axial current (the axial component of the R current). We must verify that this scheme of the symmetry breaking is compatible with the spectrum of the massless composite fermions residing in the superfields M, B, and $\tilde{B}$.

The 't Hooft consistency conditions, to be analyzed in the general case, refer to the so-called *external* anomalies of the AVV type. More exactly, one considers those axial currents, corresponding to global symmetries of the theory at hand, which are nonanomalous inside the theory *per se*, but acquire anomalies in weak external backgrounds. For instance, in QCD with several flavors the singlet axial current is internally anomalous — its divergence is proportional to $G\tilde{G}$ where G is the gluon field strength tensor. Thus, it should not be included in the set of the 't Hooft consistency conditions to be checked. The nonsinglet currents are nonanomalous in QCD itself, but become anomalous if one includes the photon field, external with respect to QCD. These currents must be checked. The anomaly in the singlet current does not lead to the statement of the infrared singularities in the current while the anomaly in the nonsinglet currents does. Those symmetries that are internally anomalous, are nonsymmetries.

In our case we first list all those symmetries which are supposedly realized linearly, i.e. unbroken. After listing all relevant currents we then saturate the corresponding triangles. The diagonal $\mathrm{SU}(N_f)$ symmetry which remains unbroken is induced by the vector current, not axial. The same is true with regards to $\mathrm{U}(1)_B$. The conserved (unbroken) R current has the axial component. Therefore, the list we must consider includes the following triangles

$$\mathrm{U}(1)_R^3, \qquad \mathrm{U}(1)_R\,\mathrm{SU}(N_f)^2, \qquad \mathrm{U}(1)_R\mathrm{U}(1)_B^2.$$

One more triangle is of a special nature. One can consider the gravitational field as external, and study the divergence of the R current in this background. This divergence is also anomalous,

$$\partial^\mu(\sqrt{-g}R_\mu) = \frac{1}{192\pi^2}[\mathrm{const.}]\,\varepsilon_{\mu\nu\lambda\delta}R^{\mu\nu\sigma\rho}R^{\lambda\delta}_{\sigma\rho}\,, \tag{3.18}$$

where g is the metric and $R^{\mu\nu\sigma\rho}$ is the curvature of the gravitational background. The constant in the square brackets depends on the particle content of the theory, and must be matched at the fundamental and composite fermion level. This gravitational anomaly in the R current is routinely referred to as $\mathrm{U}(1)_R$. Thus, altogether we have to analyze four triangles. Let us start, for instance, from $\mathrm{U}(1)_R\mathrm{SU}(N_f)^2$. The relevant quantum numbers of the fundamental-level fermions (ψ_Q, $\psi_{\tilde{Q}}$ and λ), and the composite fermions (ψ_M, ψ_B and $\psi_{\tilde{B}}$) are collected in Table 2. At the fundamental level we have to take into account only ψ_Q and $\psi_{\tilde{Q}}$ since only these fields have both $\mathrm{U}(1)_R$ charge and transform nontrivially with respect to $\mathrm{SU}(N_f)$. The corresponding triangle is proportional to $-N_fT_{\mathrm{fund}} - N_fT_{\mathrm{anti-fund}} \equiv -N_f$. The factor N_f appears since we have N_f fundamentals and N_f antifundamentals. (Remember, $N_c = N_f$ in the model at hand.) Here T is (one half of) the Dynkin index defined as follows. Assume we have the matrices of the generators of the group G in the representation R. Then

$$\mathrm{Tr}\,(T^aT^b)_R = T_R\,\delta^{ab}\,.$$

For the fundamental representation $T = 1/2$ while for the adjoint representation of $\mathrm{SU}(N)$ the index $T = N$. Now, let us calculate the same triangle at the level of the composite fermions. From Table 2 it is obvious that we have to consider only ψ_M, and the corresponding contribution is $-T_{\rm adjoint} = -N_f$. The match is perfect.

The balance in the $\mathrm{U}(1)_R^3$ triangle looks as follows. At the fundamental level we include ψ_Q, $\psi_{\tilde{Q}}$ and λ and get $2N_f^2(-1)^3 + (N_f^2 - 1) = -N_f^2 - 1$. At the composite fermion level we include ψ_M, ψ_B and $\psi_{\tilde{B}}$ and get $(N_f^2-1)(-1)^3-2 = -N_f^2 - 1$.

By the same token one can check that $\mathrm{U}(1)_B^2\mathrm{U}(1)_R$ triangle gives $-2N_f^2$ both at the fundamental and composite levels. The $\mathrm{U}(1)_R$ case requires a special comment. The coupling of all fermions to gravity is universal. Therefore, the coefficient in Eq. (3.18) merely counts the number of the fermion degrees of freedom weighed with their R charges. At the fundamental level we, obviously, have $-N_f^2 - N_f^2 + (N_f^2 - 1) = -N_f^2 - 1$, while at the composite level the coefficient is $-(N_f^2 - 1) - 1 - 1 = -N_f^2 - 1$. Again, the match is perfect.

Thus, the massless fermion content of the theory is consistent with the regime implied by Eq. (3.17) — spontaneous breaking of the chiral $\mathrm{SU}(N_f)_R\times$ $\mathrm{SU}(N_f)_L$ down to vector $\mathrm{SU}(N_f)$. The baryon and the $\mathrm{U}(1)_R$ currents remain unbroken. This regime is rather similar to what we have in ordinary QCD. The unconventional aspect, as was stressed above, is the presence of the conserved unbroken R current which has the axial component.

This does not mean, however, that all points from the vacuum valley are so reminiscent of QCD. Other points are characterized by different dynamical regimes, with drastic distinctions in the most salient features of the emerging picture. To illustrate this statement let us consider, instead of Eq. (3.17), another point

$$B = -\tilde{B} = \Lambda^{N_f}\,, \qquad M_j^i = 0\,. \tag{3.19}$$

This point is characterized by a fully unbroken chiral $\mathrm{SU}(N_f)\times\mathrm{SU}(N_f)$ symmetry, in addition to the unbroken R symmetry. The only broken generator is that of $\mathrm{U}(1)_B$.

This regime is exceptionally unusual from the point of view of the QCD practitioner. As a matter of fact, the emerging picture is directly opposite to what we got used to in QCD: the axial $\mathrm{SU}(N_f)$ generators remain unbroken while the vector baryon charge generator is spontaneously broken.

As is well-known, spontaneous breaking of the vector symmetries is forbidden in QCD [87]. The no-go theorem of Ref. [87] is based only on the very general features of QCD — namely, on the vector nature of the quark–gluon vertex. Where does the no-go theorem fail in SQCD?

The answer is quite obvious. The spontaneous breaking of the baryon charge generator in SQCD, apparently defying the no-go theorem of Ref. [87], is due to the fact that in SQCD we have scalar quarks (and the quark–squark–gluino interaction) which invalidates the starting assumptions of the theorem.

Moreover, in QCD general arguments, based on the 't Hooft consistency condition and N_c counting, strongly disfavor the possibility of the linearly realized axial SU(N_f) [88]. Although I do not say here that the consideration of Ref. [88] proves the axial SU(N_f) to be spontaneously broken in QCD, the living space left for this option is extremely narrow. The linear realization is not ruled out at all only because the argument of Ref. [88] is based on an assumption regarding the N_c dependence (discussed below) which is absolutely natural but still was not derived from the first principles. Certain subtleties which cannot be explained now due to time limitations might, in principle, invalidate this assumption. Leaving aside these — quite unnatural — subtleties one can say that the linear realization of the axial SU(N_f) is impossible in QCD.

At the same time, this is exactly what happens in SQCD in the regime specified by Eq. (3.19). Again, the scalar quarks are to blame for the failure of the argument presented in Ref. [88]. In QCD it is difficult to imagine how massless baryons could saturate anomalous triangles since the baryons are composed of N_c quarks; the corresponding contribution naturally tends to be suppressed as $\exp(-N_c)$ at large N_c. In SQCD there exist fermion states built from one quark and one (anti)squark whose contribution to the triangle is not exponentially suppressed.

After this introductory remark it is time to check that the 't Hooft consistency conditions are indeed saturated. The triangles to be analyzed are

$$\mathrm{SU}(N_f)^3\,,\qquad \mathrm{SU}(N_f)^2\mathrm{U}(1)_R\,,\qquad \mathrm{U}(1)_R^3\,,\quad \text{and } \mathrm{U}(1)_R\,.$$

The SU(N_f) symmetry is either $\mathrm{SU}(N_f)_R$ or $\mathrm{SU}(N_f)_L$, but the triangles are the same for both. It is necessary to take into account the fact that the fluctuations around the expectation values (3.19) subject to the constraint (3.16) are slightly different than those indicated in Table 2. Namely, the matrix of fluctuations M^i_j need not be traceless any longer; correspondingly, there are N_f^2 fermions in this matrix transforming as $(N_f,\, \bar{N}_f)$ representation of $\mathrm{SU}(N_f)_R \times \mathrm{SU}(N_f)_L$. At the same time the fluctuations of B and $\tilde{B}$ are not independent now, so that $B - \Lambda^{N_f} = \tilde{B} + \Lambda^{N_f}$. One should count only one of them. The $\mathrm{U}(1)_R$ quantum numbers remain intact, of course.

With this information in hands, matching of the triangles becomes a straightforward exercise. For instance, the $\mathrm{SU}(N_f)^3$ triangle obviously yields $N_f D_{\mathrm{fund}}$ both at the quark and composite levels. Here D_{fund} is the cubic Casimir operator for the fundamental representation defined as follows

$$\mathrm{Tr}\,(T^a\{T^b, T^c\}) = d^{abc} D\,;$$

the matrices of the generators T^a are taken in the given representation, and the braces denote the anticommutator; d^{abc} stands for the d symbols. The $\mathrm{SU}(N_f)^2\mathrm{U}(1)_R$ triangle yields $-N_f T_{\mathrm{fund}}$ both at the quark and composite levels. Both triangles, $\mathrm{SU}(N_f)^3$ and $\mathrm{SU}(N_f)^2\mathrm{U}(1)_R$, are saturated by ψ_M. Passing to $\mathrm{U}(1)_R^3$ we must add the gluino contribution at the fundamental level and that of ψ_B at the composite level. At both levels the coefficient of the $\mathrm{U}(1)_R^3$ triangle is $-N_f^2 - 1$. Finally, $\mathrm{U}(1)_R$ counts the number of degrees of freedom weighed with the corresponding R charges. The corresponding coefficient again turns out to be the same, $-N_f^2 - 1$.

Summarizing, the massless composite fermions residing in the moduli superfields $M, B, \tilde{B}$ saturate all anomalies induced by the symmetries that are supposed to be realized linearly. The conjecture of the unbroken $\mathrm{SU}(N_f)_R \times \mathrm{SU}(N_f)_L$ and spontaneously broken $\mathrm{U}(1)_B$ at the point (3.19) goes through. As a matter of fact, some of these anomaly matching conditions were observed long ago, in Refs. [41, 42].

Sometimes it is convenient to mimic the constraint (3.16) by introducing a Lagrange multiplier superfield X with the superpotential

$$\mathcal{W} = X(\det M - B\tilde{B} - \Lambda^{2N_f})\,. \tag{3.20}$$

We could treat, in a similar fashion, any point belonging to the quantum moduli space (3.16). For instance, we could travel from (3.17) to (3.19) observing how the regime continuously changes from the broken axial $\mathrm{SU}(N_f)$ to the broken baryon number.

Concluding this part we remind that the case of the gauge group SU(2) is exceptional. Indeed, in this case, the matter sector consisting of two fundamentals and two antifundamentals has SU(4) global flavor symmetry, rather than SU(2) × SU(2). This is because all representations of SU(2) are (pseudo)real, and fundamentals can be transformed into antifundamentals and *vice versa* by applying the $\varepsilon^{\alpha\beta}$ symbol. This peculiarity was already discussed in detail in Sec. 2. Under the circumstances the pattern of the global symmetry breaking is somewhat different, and the saturation of the anomaly triangles must be checked anew. Although this is a relatively simple exercise, we will not do it here. The interested reader is referred to Ref. [48] and Chapter VI.

$\boldsymbol{N_f = N_c + 1}$

The general strategy is the same as in the previous case. We introduce the mass term (3.14) assuming that two eigenvalues of the mass matrix are large while others are small. Then two heavy flavors can be integrated over, leaving us with the theory with $N_f = N_c - 1$, which can be analyzed in the weak coupling regime. A superpotential is generated on the vacuum valley. Using this superpotential it is not difficult to get the vacuum expectation values of the moduli fields $M, B_i, \tilde{B}^j$. They turn out to be constrained by the following relation [48]:

$$\det M \left(\frac{1}{M}\right)_i^j - B_i \tilde{B}^j = \Lambda^{2N_c - 1} m_i^j . \tag{3.21}$$

Note that the vanishing of the determinant, $\det M = 0$, which at the classical level automatically follows from the definition of M_j^i, is gone for the quantum VEV's $\langle Q^i \tilde{Q}_j \rangle$. This is most readily seen if the mass matrix $m_i^j \to m\delta_i^j$. In this case $\langle Q^i \tilde{Q}_j \rangle \sim (\Lambda^{2N_c - 1} m)^{1/N_c} \delta_j^i$, $i, j = 1, \ldots, N_f$.

In the massless limit $m_i^j \to 0$ the quantum constraint (3.21) coincides with the classical one (3.11), or (3.12). Thus, the quantum and classical moduli spaces are identical. Every point from the vacuum valley can be reached by adding appropriate perturbations to the Lagrangian (i.e. mass terms and/or $\beta B + \tilde{\beta}\tilde{B}$).

The only point which deserves special investigation is the origin, which, unlike the situation $N_f = N_c$, remains singular. This is a signature of massless fields. Classically we have massless gluons and massless moduli fields. In the strong coupling regime we expect the gluons to acquire a dynamical mass gap. The classical moduli subject to the constraint (3.12) need not be the only composite massless states, however. Other composite massless states may form too. We will see shortly that they actually appear. At the origin, when $M = B = \tilde{B} = 0$, *all* global symmetries of the Lagrangian are presumably unbroken. In particular, the axial SU(N_f) is realized linearly. Although we have already learned, from the previous example, that such a regime seems to be attainable in SQCD (in sharp contradistinction with QCD), the case $N_f = N_c + 1$ is even more remarkable — we want *all* global symmetries to be realized linearly. (For $N_f = N_c$, in the vacuum where the axial SU(N_f) symmetry was unbroken, the baryon charge generators were spontaneously broken.)

At the origin (and near the origin) the theory is in the strong coupling regime. Let us examine the behavior of the theory in this domain more carefully. When the expectation values of all moduli fields vanish, the global symmetry $\mathrm{SU}(N_f) \times \mathrm{SU}(N_f) \times \mathrm{U}(1)_B \times \mathrm{U}(1)_R$ is unbroken provided no other (nonchiral) condensates develop. Is this solution self-consistent?

To answer this question we will try to match all corresponding anomalous AVV triangles; in this case we have seven triangles,

$$\mathrm{SU}(N_f)^3\,, \qquad \mathrm{SU}(N_f)^2\mathrm{U}(1)_B\,, \qquad \mathrm{SU}(N_f)^2\mathrm{U}(1)_R\,,$$

$$\mathrm{U}(1)_R\mathrm{U}(1)_B^2\,, \qquad \mathrm{U}(1)_R^3\,, \qquad \mathrm{U}(1)_R^2\mathrm{U}(1)_B\,, \qquad \mathrm{U}(1)_R\,. \tag{3.22}$$

They must be matched by the composite massless baryons residing in M, B_i and $\tilde{B}^j$. As we will see shortly, to achieve the matching we will need to consider all components of M, B_i and $\tilde{B}^j$ as independent, ignoring the constraint

$$\det M \left(\frac{1}{M}\right)_i^j - B_i \tilde{B}^j = 0 \tag{3.23}$$

defining the vacuum valley both at the classical and the quantum levels. In other words, we will have to deal with a larger number of massless fields than one could infer from the parametrization (3.23) of the vacuum valley. The constraint (3.23) on the vacuum valley will reappear due to the fact that the expanded set of the massless fields gets a superpotential $\mathcal{W}(M, B, \tilde{B})$. The requirement of the vanishing of the F term will return us Eq. (3.23).

Thus, our first task is to verify the matching. The quantum numbers of the fundamental quarks and the composite massless fermions can be inferred from Table 1. For convenience we collect them in Table 3.

Table 3. The quantum numbers of the massless fermions in SQCD with $N_f = N_c + 1$.

	$\mathrm{SU}(N_f)_L$	$\mathrm{SU}(N_f)_R$	$\mathrm{U}(1)_B$	$\mathrm{U}(1)_R$
ψ_M	N_f	$\bar{N}_f$	0	$\frac{2}{N_f} - 1$
ψ_B	$\bar{N}_f$	1	$N_f - 1$	$-\frac{1}{N_f}$
$\psi_{\tilde{B}}$	1	N_f	$-N_f + 1$	$-\frac{1}{N_f}$
ψ_Q	N_f	1	1	$\frac{1}{N_f} - 1$
$\psi_{\tilde{Q}}$	1	$\bar{N}_f$	-1	$\frac{1}{N_f} - 1$
λ	1	1	0	1

Since we already have a considerable experience in matching the AVV triangles, I will not discuss all triangles from Eq. (3.22). As an exercise let us do just one of them, namely $\mathrm{U}(1)_R^3$. In this case, at the fundamental level we have the $\psi_Q, \psi_{\tilde{Q}}$ and λ triangles which yield

$$2N_cN_f\left(\frac{1}{N_f}-1\right)^3+N_c^2-1\,.$$

At the composite level the $\mathrm{U}(1)_R^3$ anomalous triangle is contributed by ψ_M (N_f^2 degrees of freedom), ψ_B and $\psi_{\bar{B}}$ (each has N_f degrees of freedom). Thus, we get

$$N_f^2\left(\frac{2}{N_f}-1\right)^3-2N_f\left(\frac{1}{N_f}\right)^3.$$

Both expressions reduce to

$$-N_f^2+6N_f-12+\frac{8}{N_f}-\frac{2}{N_f^2}\,.$$

Other triangles match too, in a miraculous way. Namely,

$$\begin{aligned}
\mathrm{SU}(N_f)^3 &\to (N_f-1)D_{\text{fund}}\,,\\
\mathrm{SU}(N_f)^2\mathrm{U}(1)_R &\to -\frac{(N_f-1)^2}{N_f}T_{\text{fund}}\,,\\
\mathrm{U}(1)_B^2\mathrm{U}(1)_R &\to -2(N_f-1)^2\,,\\
\mathrm{SU}(N_f)^2\mathrm{U}(1)_B &\to (N_f-1)\,T_{\text{fund}}\,,\\
\mathrm{U}(1)_R^2\mathrm{U}(1)_B &\to 0\,,\\
\mathrm{U}(1)_R &\to -N_f^2+2N_f-2\,.
\end{aligned}$$

The matching discussed above, was observed many years ago in Ref. [56] where the spectrum of the composite massless particles corresponding to the unconstrained M, B and $\tilde{B}$ was conjectured.

Thus, the above spectrum of the composite massless particles appearing at the origin of the vacuum valley in the $N_f = N_c+1$ theory is self-consistent. We know, however, that the vacuum valley in the model at hand is characterized by Eq. (3.23). The situation seems rather puzzling. How does the constraint (3.23) appear?

The answer to this question was given by Seiberg [48]. If the massless fields, residing in the unconstrained M, B and $\tilde{B}$, acquire a superpotential, then the vacuum values of the moduli fields are obtained through the condition of vanishing F terms. The "right" superpotential will lead to Eq. (3.23) automatically.

So, what is the right superpotential? If it is generated, several requirements are to be met. First, it must be invariant under all global symmetries of the model, including the R symmetry. Second, the vacuum valleys obtained from this superpotential must correspond to Eq. (3.23). Third, away from the origin the only massless fields must be those compatible with the constraint (3.23).

All these requirements are satisfied by the following superpotential [48]:

$$\mathcal{W} = \frac{1}{\Lambda^{2N_f-3}} \left(B_i M^i_j \tilde{B}^j - \det M \right). \tag{3.24}$$

It is obvious that the condition of the vanishing of F terms corresponding to M^i_j identically coincides with Eq. (3.23); moreover, the vanishing of F terms corresponding to B and $\tilde{B}$ yields to remaining constraints, $B_i M^i_j = \tilde{B}^j M^i_j = 0$. Once we move away from the origin, the moduli M^i_j grow, the fields $B_i, \tilde{B}^j$ acquire masses and can be integrated out. This eliminates $2N_f$ degrees of freedom. This is exactly the amount of the redundant degrees of freedom, see Sec. 3.1. The emerging low-energy theory for the remaining degrees of freedom M^i_j has no superpotential. When $M^i_j \gg \Lambda^2$, the fields $B_i, \tilde{B}^j$ are very heavy, and the low-energy description based on Eq. (3.24) is no longer legitimate. It is interesting to trace the fate of the baryons $B_i, \tilde{B}^j$ in the process of this evolution from small to large values of M^i_j. This question has not been addressed in the literature so far.

Let us pause here to summarize the features of the dynamical regime taking place in the $N_f = N_c + 1$ model. The space of vacua is the same at the classical and quantum levels, the origin being singular due to the existence of the massless degrees of freedom. Since we have Higgs fields in the fundamental representation the theory is in the Higgs/confining phase; at the origin and near the origin the theory is strongly coupled and "confines" in the sense that the physics is adequately described in terms of gauge invariant composites and their interactions. We think that at the origin all global symmetries of the Lagrangian are unbroken. The number of the massless degrees of freedom here is larger than the dimensionality of the space of vacua. To get the right description of the space of vacua one needs a superpotential, and such a superpotential is generated dynamically. It is a holomorphic function of the massless composites. The vacuum valley for this superpotential coincides with the quantum moduli space of the original theory. As we move along the vacuum valley away from the origin there is no phase transition — the theory smoothly goes into the weak coupling Higgs phase. The "extra" massless fields become massive, and irrelevant for the description of the vacuum valley.

The dynamical regime with the above properties acquired a special name — now it is referred to as *s-confinement.*

Seiberg's example of the s-confining theory was the first, but not the last. Other theories with similar behavior were found, see e.g. Refs. [89–95]. The set of the s-confining models includes even such an exotic one as the gauge group G_2 (this is an exceptional group), with five fundamentals [94, 95]. As a matter of fact, it is not difficult to work out a general strategy allowing one to carry out a systematic search of all s-confining theories. This was done in Ref. [96]. Without submerging into excessive technical details let me outline just one basic point of the procedure suggested in Ref. [96].

A necessary condition of the s confinement is the generation of a superpotential at the origin of the moduli space, a holomorphic function of relevant moduli fields. Generically, the form of this superpotential, dictated by the R symmetry plus the dimensional arguments is

$$\mathcal{W} \propto \left[\prod_\ell S_\ell^{2T(R_\ell)} \Lambda^{3T(G)-\sum_\ell T(R_\ell)}\right]^{1/(\sum_\ell T(R_\ell)-T(G))} . \tag{3.25}$$

The product (sum) runs over all matter fields present in the theory. For instance, in the case of SQCD for each flavor we have to include two subflavors. I remind that $T(R)$ is (one half of) the Dynkin index. Particular combinations of the superfields in the product are not specified; they depend on the particular representations of the matter fields with respect to the gauge group. What is important is only the fact that they all are homogeneous functions of S_ℓ's, of order $2T(R_\ell)$. Note that the combination appearing in Eq. (3.25) is the only one which has the correct properties under renormalization, i.e. compatible with the NSVZ β function.

Now, if we want the origin to be analytic (and this is a feature of the s-confinement, by definition), we must ensure that

$$\sum_\ell T(R_\ell) - T(G) = 1 \tag{3.26}$$

(more generically, 1/integer). This severely limits the choice of possible representations since the Dynkin indices are integers. For instance, if the matter sector is vector-like, there exist only two options: (i) Seiberg's model, SU(N_c) color with $N_c + 1$ flavors (i.e. $N_c + 1$ fundamentals and $N_c + 1$ antifundamentals; (ii) SU(N_c) color with one antisymmetric tensor plus its adjoint plus three flavors.

Not to make a false impression I hasten to add that some models that satisfy Eq. (3.26), are not s-confining. A few simple requirements to be met, which comprise a sufficient condition for s-confinement, are summarized in Ref. [96], which gives also a full list of the s-confining theories.

3.3. *Conformal Window. Duality*

Our excursion towards larger values of N_f must be temporarily interrupted here — the methods we used so far fail at $N_f = N_c + 2$. One can show that the quantum moduli space coincides with the classical one, just as in the case $N_f = N_c + 1$. However, at the origin of the moduli space, description of the large-distance behavior of the theory in terms of the massless fields residing in M, B and $\tilde{B}$ does not go through. These degrees of freedom are irrelevant for this purpose; the dynamical regime of the theory in the infrared is different. To see that M, B and $\tilde{B}$ do not fit suffice it to try to saturate the 't Hooft triangles corresponding to the unbroken global symmetries, in the same vein as we did previously for $N_f = N_c + 1$. There is no matching! As we will see shortly, the dynamical regime does indeed change in passing from $N_f = N_c + 1$ to $N_f = N_c + 2$. The correlation functions of the $N_f = N_c + 2$ theory at large distances are those of a free theory, like in massless electrodynamics. But the number of free degrees of freedom ("photons" and "photinos") is different from what one might expect naively. Namely, we will have three "photons" and three "photinos" in the case at hand, in addition to $(N_c + 2) \times 2$ free "fermion" fields. These photons and photinos, in a sense, may be considered as the bound states of the original gluons, gluinos, quarks and squarks.

To elucidate this, rather surprising, picture we will have to make a jump in our travel along the N_f axis, leave the domain of N_f close to N_c for a while, and turn to much larger values of N_f. The critical points on the N_f axis are $3N_c$ and $3N_c/2$. That's where a *conformal window* starts and ends. We will return to the $N_f = N_c + 2$ and $N_f = N_c + 1$ theories later on.

At first, let me remind a few well-known facts from ordinary nonsupersymmetric QCD. The Gell-Mann-Low function in QCD has the form [97]

$$\beta_{\rm QCD}(\alpha_s) = -\beta_0 \frac{\alpha_s^2}{2\pi} - \beta_1 \frac{\alpha_s^3}{4\pi^2} - \dots ,$$
$$\beta_0 = 11 - \frac{2}{3} N_f , \qquad \beta_1 = 51 - \frac{19}{3} N_f . \tag{3.27}$$

At small α_s it is negative since the first term always dominates. This is the celebrated asymptotic freedom. With the scale μ decreasing the running gauge coupling constant grows, and the second term becomes important. Generically the second term takes over the first one at $\alpha_s/\pi \sim 1$, when all terms in the α_s expansion are equally important, i.e. in the strong coupling regime. Assume, however, that for some reasons the first coefficient β_0 is abnormally small, and this smallness does not propagate to higher orders. Then the second term catches up with the first one when $\alpha_s/\pi \ll 1$, we are in the weak coupling regime, and higher order terms are inessential. Inspection of Eq. (3.27) shows that this happens when N_f is close to 33/2, say 16 or 15 (N_f has to be less than 33/2 to ensure asymptotic freedom). For these values of N_f the second coefficient β_1 turns out to be negative! This means that the β function develops a zero in the weak coupling regime, at

$$\frac{\alpha_s^*}{2\pi} = \frac{\beta_0}{-\beta_1} \ll 1 . \tag{3.28}$$

(Say, if $N_f = 15$ the critical value is at 1/44). This zero is nothing but the infrared fixed point of the theory. At large distances $\alpha_s \to \alpha_s^*$, and $\beta(\alpha_s^*) = 0$, implying that the trace of the energy-momentum vanishes, and the theory is in the conformal regime. There are no localized particle-like states in the spectrum of the theory; rather we deal with massless unconfined interacting quarks and gluons; all correlation functions at large distances exhibit a power-like behavior. In particular, the potential between two heavy static quarks at large distances R will behave as $\sim \alpha_s^*/R$. The situation is not drastically different from conventional QED. The corresponding dynamical regime is, thus, a non-Abelian Coulomb phase. As long as α_s^* is small, the interaction of the massless quarks and gluons in the theory is weak at all distances, short and large, and is amenable to the standard perturbative treatment (renormalization group, etc.), QCD becomes a fully calculable theory.

There is nothing remarkable in the observation that, for a certain choice of N_f, quantum chromodynamics becomes conformal and weakly coupled in the infrared limit. Belavin and Migdal played with this model over 20 years ago [98]. They were quite excited explaining how great it would be if N_f in our world were close to 16, and the theory would be in the infrared conformal regime, with calculable anomalous dimensions. Later on this idea was discussed also by Banks and Zaks [99]. Alas, we do not live in the world with $N_f \approx 16$.

What is much more remarkable is the existence of the infrared conformal regime in SQCD for large couplings, $\alpha_*/\pi \sim 1$. This fact, as many others in the given range of questions, was established by Seiberg [100]. The discovery of the strong coupling conformal regime [100] is based on the so-called *electric-magnetic duality.* Although the term suggests the presence of electromagnetism and the same kind of duality under the substitution $\mathbf{E} \leftrightarrow \mathbf{B}$ one sees in the Maxwell theory, actually both elements, "electric-magnetic" and "duality" in the given context are nothing but remote analogies, as we will see shortly.

Analysis starts from consideration of SQCD with N_f slightly smaller than $3N_c$. More exactly, if

$$\varepsilon \equiv 1 - \frac{N_f}{3N_c} \tag{3.29}$$

we assume that $N_c \to \infty$, and $0 < \varepsilon \ll 1$. It is assumed also that we are at the origin of the moduli space — no fields develop VEV's. By examining the NSVZ β function, Eq. (2.61), it is easy to see that in this limit the first coefficient of the β function is abnormally small, and the second coefficient is positive and is of a normal order of magnitude,

$$\beta_0 = 3N_c\varepsilon\,, \qquad \beta_1 = -3N_c^2 + \mathcal{O}(\varepsilon)\,. \tag{3.30}$$

To get β_1 I used the fact that

$$\gamma(\alpha) = -\frac{N_c^2-1}{2N_c}\,\frac{\alpha}{\pi} + \mathcal{O}(\alpha^2) \tag{3.31}$$

in the model considered (for a pedagogical review see, for example, the last paper in Ref. [67]). There is a complete parallel with the conformal QCD, with 15 or 16 flavors, discussed above. The numerator of the NSVZ β function vanishes at

$$\gamma(\alpha_*) \equiv \gamma_* = 1 - 3\frac{N_c}{N_f} \to -\varepsilon \quad \text{or} \quad \frac{N_c\alpha_*}{2\pi} = \varepsilon\,. \tag{3.32}$$

The vanishing of the β function marks the onset of the conformal regime in the infrared domain; the fact that α_* is small means that the theory is weakly coupled in the infrared (it is weakly coupled in the ultraviolet too since it is asymptotically free).

Here comes the breakthrough observation of Seiberg. Compelling arguments can be presented indicating that the original theory with the SU(N_c) gauge group (let us call this theory "electric"), and another theory, with the SU($N_f - N_c$) gauge group, the same number of flavors N_f as above, and a specific Yukawa interaction (let us call this theory "magnetic"), flow to one and the same limit in the infrared asymptotics. The corresponding Gell-Mann-Low functions of both theories vanish at their corresponding critical values of the coupling constants. Both theories are in the non-Abelian Coulomb (conformal) phases. By inspecting Eq. (2.61) it is easy to see that, when α_* in the electric theory approaches zero (i.e. $\varepsilon \to 0$), in the magnetic theory $\gamma(\alpha_*)$ approaches -1, i.e. the theory becomes strongly coupled. The opposite is also true. When the magnetic theory becomes weakly coupled, i.e.

$$N_f \to \frac{3}{2} N_c \text{ from above}, \qquad (\alpha_*)_{\text{magn}} \to 0, \tag{3.33}$$

in the electric theory $(\gamma_*)_{\text{electr}} \to -1$, and the electric theory is strongly coupled in the infrared. This reciprocity relation is, probably, the reason why the correspondence between the two theories is referred to as the electric-magnetic duality. It is worth emphasizing that the correspondence takes place only in the infrared limit. By no means the above two theories are totally equivalent to each other; their ultraviolet behavior is completely different. If $N_f < (3N_c)/2$ the magnetic theory loses asymptotic freedom. Thus, the *conformal window*, where both theories are asymptotically free in the ultraviolet and conformally invariant in the infrared extends in the interval $3N_c/2 \leq N_f \leq 3N_c$.

The fact that the conformal window cannot stretch below $3N_c/2$ is seen from the consideration of the electric theory *per se*, with no reference to the magnetic theory. Indeed, the total (normal + anomalous) dimension of the matter field $Q\tilde{Q}$ in the infrared limit is equal to

$$d = 2 + \gamma_* = \frac{3(N_f - N_c)}{N_f} \,.$$

No physical field can have dimension less than unity; this is forbidden by the Källén-Lehmann spectral representation. If $d = 1$ the field is free. The dimension d reaches unity exactly at $3N_c/2$. Decreasing N_f further and assuming that the conformal regime $\beta(\alpha_*) = 0$ is still preserved would violate the requirement $d \geq 1$.

Let us describe the electric and magnetic theories in more detail. I will continue to denote the quark fields of the first theory as Q, while those of the magnetic theory will be denoted by $\mathcal{Q}$. Both have N_f flavors, i.e. $2N_f$ chiral superfields in the matter sector. The same number of flavors is necessary to ensure that the global symmetries of both theories are identical.

The magnetic theory, additionally, has N_f^2 colorless "meson" superfields $\mathcal{M}_j^i$ whose quantum numbers formally are such as if they were built from a quark and an antiquark. The meson superfields are coupled to the quark ones of the magnetic theory through a superpotential

$$\mathcal{W} = f \mathcal{M}_j^i \mathcal{Q}_i \tilde{\mathcal{Q}}^j \,. \tag{3.34}$$

The quantum numbers of the fields belonging to the matter sectors of the magnetic and electric theories are summarized in Tables 4 and 5.

Table 4. The quantum numbers of the massless fields of the "electric" theory from the dual pair.

	$SU(N_f)_L$	$SU(N_f)_R$	$U(1)_B$	$U(1)_R$
ψ_Q	N_f	0	1	$-N_c/N_f$
$\psi_{\tilde{Q}}$	0	$\bar{N}_f$	-1	$-N_c/N_f$

Table 5. The quantum numbers of the massless fields of the "magnetic" theory from the dual pair.

	$SU(N_f)_L$	$SU(N_f)_R$	$U(1)_B$	$U(1)_R$
$\psi_{\mathcal{Q}}$	$\bar{N}_f$	0	$N_c/(N_f - N_c)$	$(N_c - N_f)/N_f$
$\psi_{\tilde{\mathcal{Q}}}$	0	N_f	$-N_c/(N_f - N_c)$	$(N_c - N_f)/N_f$
$\chi = \psi_{\mathcal{M}}$	N_f	$\bar{N}_f$	0	$(N_f - 2N_c)/N_f$

The quantum numbers of the meson superfield are fixed by the superpotential (3.34). Note a very peculiar relation between the baryon charges of the quarks in the electric and magnetic theories. This relation shows that the quarks of the magnetic theory cannot be expressed, in any polynomial way, through the quarks of the electric theory. The connection of one to the other is presumably extremely nonlocal and complicated. The explicit connection between the operators in the dual pairs is known only for a handful of operators which have a symmetry nature [101].

We can now proceed to the arguments that establish the equivalence of these two theories in the infrared limit. The main tool we have at our disposal for establishing the equivalence is again the 't Hooft matching, the same line of reasoning as was used above in verifying various dynamical regimes in $N_f = N_c$ and $N_f = N_c + 1$ models. Since we are at the origin of the moduli space, all global symmetries are unbroken, and one has to check six highly nontrivial matching conditions corresponding to the various triangles with $\mathrm{SU}(N_f)$, $\mathrm{U}(1)_R$ and $\mathrm{U}(1)_B$ currents at their vertices.

The presence of fermions from the meson multiplet $\mathcal{M}$ is absolutely crucial for this matching. Specifically, one finds for the one-loop anomalies in both theories [100]:

$$\begin{aligned}
\mathrm{SU}(N_f)^3 &\to N_c D_{\text{fund}}\,,\\
\mathrm{SU}(N_f)^2\mathrm{U}(1)_R &\to -\frac{N_c^2}{N_f}T_{\text{fund}}\,,\\
\mathrm{SU}(N_f)^2\mathrm{U}(1)_B &\to N_c T_{\text{fund}}\,,\\
\mathrm{U}(1)_B^2\mathrm{U}(1)_R &\to -2N_c^2\,,\\
\mathrm{U}(1)_R^3 &\to N_c^2 - 1 - 2\frac{N_c^4}{N_f^2}\,,\\
\mathrm{U}(1)_R &\to -N_c^2 - 1\,.
\end{aligned} \tag{3.35}$$

For example, in the $\mathrm{U}(1)_R^3$ anomaly in the electric theory the gluino contribution is proportional to $N_c^2 - 1$ and that of the quarks to $-(N_c/N_f)^3 2N_f N_c = -2N_c^4/N_f^2$; altogether $N_c^2 - 1 - 2N_c^4/N_f^2$ as in (3.35). In the dual theory one gets from gluino and quarks another contribution, $(N_f - N_c)^2 - 1 - 2(N_f - N_c)^4/N_f^2$. Then the fermions χ from the meson multiplet $\mathcal{M}$ add extra $[(N_f - 2N_c)/N_f]^3 N_f^2$, which is precisely the difference.

The last line in Eq. (3.35) corresponds to the anomaly of the R current in the background gravitational field. In the electric $\mathrm{SU}(N_c)$ theory the corresponding coefficient is

$$N_c^2 - 1 - 2(N_c/N_f)N_c N_f = -N_c^2 - 1$$

while in the magnetic theory it is $-(N_f - N_c)^2 - 1$ from quarks and gluinos and $[(N_f - 2N_c)/N_f]N_f^2$ from the $\mathcal{M}$ fermions, i.e. in the sum again $-N_c^2 - 1$.

It is not difficult to check the matching of other triangles from Eq. (3.35). The dependence on N_c and N_f is rather sophisticated, and it is hard to imagine that this is an accidental coincidence. The fact that the electric and magnetic theories described above have the same global symmetries is an additional argument in favor of their (infrared) equivalence. Of course, they have different gauge symmetries: SU(N_c) in the first case and SU($N_f - N_c$) in the second. The gauge symmetry, however, is not a regular symmetry; in fact, it is not a symmetry at all. Rather, it is a redundancy in the description of the theory. One introduces first more degrees of freedom than actually exist, and then the redundant variables are killed by the gauge freedom. That is why the gauge symmetry has no reflection in the spectrum of the theory. Therefore, distinct gauge groups do not preclude the theories from being dual, generally speaking. On the contrary, the fact that such dual pairs are found is very intriguing; it allows one to look at the gauge dynamics from a new angle.

I have just said that various dual pairs of supersymmetric gauge theories are found. To avoid misunderstanding I hasten to add that although Seiberg's line of reasoning is very compelling it still falls short of *proving* the infrared equivalence. The theory in the strong coupling regime is not directly solved, and we are hardly any closer now to the solution than we were a decade ago. The infrared equivalence has the status of a good solid conjecture substantiated by a number of various indirect arguments we have at our disposal (Sec. 3.4).

If we accept this conjecture we can make a remarkable step forward compared to the conformal limit of QCD studied in the weak coupling regime in the 70s and 80s. Indeed, if N_f is close to $3N_c$ (but slightly lower), i.e. we are near the right edge of the conformal window, the weakly coupled electric theory is in the conformal regime. Since it is equivalent (in the infrared) to the magnetic theory, which is strongly coupled at these values of N_f we, thus, establish the existence of a *strongly coupled* superconformal gauge theory. Moreover, when N_f is slightly higher than $3N_c/2$, i.e. near the left edge of the conformal window, the magnetic theory is weakly coupled and in the conformal regime. Its dual, the electric theory, which is strongly coupled near the left edge of the conformal window, must then be in the conformal regime too. In the middle of the conformal window, when both theories are strongly coupled, strictly speaking we do not know whether or not they continue to stay superconformal. In principle, it is possible that they both leave the conformal regime. This could happen, for instance, if α_*, the solution of the equation $\gamma(\alpha_*) = 1 - 3N_cN_f^{-1}$, (temporary) becomes larger than $2\pi/T(G)$, the position of the zero of the *denominator* of the NSVZ β function, as we go further away from the point $N_f = 3N_c$ in the direction of $N_f = 3N_c/2$, and then α_* becomes smaller than $2\pi/T(G)$ again, as we approach $N_f = 3N_c/2$. Such a scenario, although not ruled out, does not seem likely, however.

3.4. *Traveling Along the Valleys*

So far, the dual pair of theories was considered at the origin of the vacuum valley. Both theories, electric and magnetic, have the vacuum valleys, and a natural question arises as to what happens if we move away from the origin.[8] As a matter of fact, this question is quite crucial, since if the theories are equivalent in the infrared, a certain correspondence between them should persist not only at the origin, but at any other point belonging to the vacuum valley. If a correspondence can be found, it will only strengthen the conjecture of duality.

Thus, let us start from the electric theory and move away from the origin. Consider for simplicity a particular direction in the moduli space, namely,

$$Q = \tilde{Q} = \begin{pmatrix} a_1 \\ 0 \\ \cdots \\ 0 \end{pmatrix}, \tag{3.36}$$

where $Q, \tilde{Q}$ are the superfields comprising, say, the first flavor. Moving along this direction we break the gauge symmetry SU(N_c) down to SU($N_c - 1$); $2N_c - 1$ chiral superfields are eaten up in the (super)-Higgs mechanism providing masses to $2N_c - 1$ W bosons. Below the mass scale of these W bosons the effective theory is SQCD with SU($N_c - 1$) gauge group and $N_f - 1$ flavors. (Additionally there is one SU($N_c - 1$) singlet, but it plays no role in the gauge dynamics). It is not difficult to see that decreasing both N_f and N_c by one unit in the electric theory we move rightwards along the axis $N_f/3N_c$. In other words, we move towards the right edge of our conformal window, making the electric theory weaker. From what we already know, we should then expect that the magnetic theory becomes stronger.

Let us take a closer look at the magnetic theory. The vacuum expectation value (3.36) is reflected in the magnetic theory as the expectation value of the

[8]This question was suggested to me by C. Wetterich. Note that if in the electric theory the vacuum degeneracy manifests itself in arbitrary vacuum expectations of Q and $\tilde{Q}$, in the magnetic theory the expectation values of $\mathcal{Q}$ and $\tilde{\mathcal{Q}}$ vanish. The flat direction corresponds to arbitrary expectation value of $\mathcal{M}$.

(1,1) component of the meson field $\mathcal{M}$. No Higgs phenomenon takes place, but, rather, $\mathcal{M}_1^1 \neq 0$. Then, thanks to the superpotential (3.34), the magnetic quark $\mathcal{Q}_1, \tilde{\mathcal{Q}}_1$ gets a mass, and becomes irrelevant in the infrared limit. The gauge group remains the same, $\mathrm{SU}(N_f - N_c)$, but the number of active flavors reduces by one unit (we are left with $N_f - 1$ active flavors). This means that the first coefficient of the Gell-Mann-Low function of the magnetic theory becomes more negative and the critical value $\alpha_{*\mathrm{magn}}$ increases. The theory becomes coupled stronger, in full accord with our expectations.

Let us now try the other way around. What happens if we introduce the mass term to one of the quarks in the electric theory, say the first flavor? The gauge group remains, of course, the same, $\mathrm{SU}(N_c)$. However, in the infrared domain the first flavor decouples, and we are left with $N_f - 1$ active flavors. The first coefficient in the β function of the electric theory becomes more negative; hence, the critical value $\alpha_{*\mathrm{electr}}$ increases. We move leftwards, towards the left edge of the conformal window. Correspondingly, the electric theory becomes coupled stronger, and we expect the magnetic one to be coupled weaker.

What is the effect of the mass term in the magnetic theory? It is rather obvious that the corresponding impact reduces to the introduction of a mass term in the superpotential (3.34),

$$\mathcal{W}' = f \mathcal{M}_j^i \mathcal{Q}_i \tilde{\mathcal{Q}}^j + m \mathcal{M}_1^1 . \tag{3.37}$$

Extending the superpotential is equivalent to changing the vacuum valley. Indeed, the expectation values of $\mathcal{Q}_1$ and $\mathcal{Q}_1$ do not vanish anymore. Instead, the condition of the vanishing of the F term implies

$$\frac{\partial \mathcal{W}}{\partial \mathcal{M}_1^1} = \mathcal{Q}_1 \tilde{\mathcal{Q}}^1 + m = 0 . \tag{3.38}$$

If m is large, Eq. (3.38) implies, in turn, that the magnetic squarks of the first flavor develop a vacuum expectation value, the magnetic theory turns out to be in the Higgs phase, the gauge group $\mathrm{SU}(N_f - N_c)$ is spontaneously broken down to $\mathrm{SU}(N_f - N_c - 1)$, and one magnetic flavor is eaten up in the super-Higgs mechanism. We end up with a theory with the gauge group $\mathrm{SU}(N_f - N_c - 1)$ and $N_f - 1$ flavors. The $\mathcal{M}_1^i$ and $\mathcal{M}_i^1$ components of the meson field become sterile in the infrared limit. In this theory the first coefficient of the β function is less negative, $\alpha_{*\mathrm{magn}}$ is smaller, we are closer to the left edge of the conformal window, as was expected.

Summarizing, we see that Seiberg's conjecture of duality is fully consistent with the vacuum structure of both theories. As a matter of fact, this observation may serve as an additional evidence in favor of duality. Simultaneously it makes perfectly clear the fact that, if duality does take place, it can be valid only in the infrared limit; by no means the two theories specified above are fully equivalent.

One may ask what happens if we continue adding mass terms to the electric quarks of the first, second, third, etc. flavors. Adding large mass terms we eliminate the flavors one by one. In other words, we launch a cascade taking us back to smaller values of N_f. The electric theory becomes stronger and stronger coupled. Simultaneously the dual magnetic theory is coupled weaker and weaker. When the number of active flavors reaches $3N_c/2$, $\alpha_{*\rm magn} = 0$. Eliminating the quark flavors further we leave the conformal window — the magnetic theory loses asymptotic freedom and becomes infrared-trivial, with the interaction switching off at large distances. We find ourselves in the free magnetic phase, or the Landau phase. By duality, the correlation functions in the electric theory (which is superstrongly coupled in this domain of N_f) must have the same trivial behavior at large distances. To be perfectly happy we would need to know the relation between all operators of the electric and magnetic theory, so that given a correlation function in the electric theory we could immediately translate it in the language of essentially free magnetic theory. Alas ... as was already mentioned, this relation is basically unknown. It can be explicitly found only for some operators of a geometric nature. Even though the general relation was not found, the achievement is remarkable. For the first time ever the gauge bosons of the weakly coupled theory (magnetic) are shown to be "bound states" of a strongly coupled theory (electric).

The last but one step in the reduction process is when the number of active flavors is $N_c + 2$. The gauge group of the electric theory is $\mathrm{SU}(N_c)$ while that of the magnetic one is SU(2). Under the duality conjecture the large distance behavior of the superstrongly coupled electric theory is determined by the massless modes of the essentially free magnetic theory: three "photons", three "photinos", $2(N_c + 2)$ fields of the type $\mathcal{Q}$ and $2(N_c + 2)$ fields of the type $\tilde{\mathcal{Q}}$. We can return to the remark made in the very beginning of this section — it is explained now. It becomes clear why all attempts to describe the infrared behavior of the theory in terms of the variables $M, B, \tilde{B}$ failed in the $N_f = N_c + 2$ case: these are not proper massless degrees of freedom.

The last step in the cascade that can still be done is reducing one more flavor, by adding a large mass term in the electric theory, or the corresponding entry of the matrix $\mathcal{M}$ in the superpotential of the magnetic theory. At this last stage, the super-Higgs mechanism in the magnetic theory completely breaks the remaining gauge symmetry. We end up with N_c+1 massless flavors interacting with $(N_c+1)\times(N_c+1)$ meson superfield $\mathcal{M}^i_j$ (plus a number of sterile fields inessential for our consideration). This remaining meson superfield $\mathcal{M}^i_j$ is assumed to have no (or small) expectation values, so that we stay near the origin of the vacuum valley. Moreover, it is not difficult to see that the instantons of the broken SU(2) generates the superpotential of the type $\det\mathcal{M}$. This is nothing but a supergeneralization of the 't Hooft interaction [102]. Indeed, the instanton generates fermion zero modes, one mode for each ψ_Q and $\psi_{\tilde{Q}}$. In the absence of the Yukawa coupling $f\mathcal{M}Q\tilde{Q}|_F$, there is no way to contract these zero modes, and their proliferation results in the vanishing of the would-be instanton-induced superpotential. The Yukawa coupling $f\phi_{\mathcal{M}}\psi_Q\psi_{\tilde{Q}}$ lifts the zero modes, much in the same way as the mass term does. (Supersymmetrization of the result is achieved through the vertex $f\psi_{\mathcal{M}}\phi_Q\psi_{\tilde{Q}}$ where the boson field ϕ_Q is induced by the zero mode of ψ_Q and the gluino zero mode). In this way we arrive at the instanton-induced superpotential $\det\mathcal{M}$. The full superpotential of the magnetic theory, describing the interaction of massless (or nearly massless) degrees of freedom, is

$$\mathcal{W} = f\mathcal{M}Q\tilde{Q} + \det\mathcal{M}\,. \tag{3.39}$$

Compare it with Eq. (3.24) which was derived for the interaction of the massless degrees of freedom of the $N_f = N_c+1$ model by exploiting a totally different line of reasoning. Up to a renaming of the fields involved, the coincidence is absolute! Thus, the duality conjecture allows us to rederive the s-confining potential in the "electric" theory, SQCD with the gauge group SU(N_c) and $N_f = N_c+1$. The fact that two different derivations lead to one and the same result further strengthens the duality conjecture, implying that actually it is more than a conjecture: the full infrared equivalence of Seiberg's "electric" and "magnetic" theories does indeed occur.

Patterns of Seiberg's duality in more complicated gauge theories than that discussed above, including nonchiral matter sectors, were studied in a number of publications. The dual pairs proliferate! By now an entire zoo of dual pairs is densely populated. Finding a "magnetic" counterpart to the given "electric" theory remains an art, rather than science — no general algorithm exists which would allow one to generate dual pairs automatically, although there is a collection of some helpful hints and recipes. Systematic searches for dual partners to every given supersymmetric theory is an intriguing and fascinating topic. However at the present stage it is too technical to be included in this lecture course. Even a brief discussion of the corresponding advances would lead us far astray. Interested readers are referred to the original literature, see for example Refs. [103–105].

Section 4

New Phenomena in Other Gauge Groups (Orthogonal Groups)

After 1994 many followers worked on dynamical aspects of non-Abelian SUSY gauge theories. In many instances the development went not in depth but, rather, on the surface. Various "exotic" gauge groups and matter representations were considered, supplementing the list of models considered in the previous section by several new examples with essentially the same dynamical behavior. The corresponding discussion might be interesting to experts but is hardly appropriate here. Along with thorough explorations of the previously known patterns some interesting nuances in dynamical scenarios were revealed *en route*. In this section we will focus on these new dynamical phenomena. We briefly review below some exciting findings "after Seiberg". Preference will be given to the results of general interest — multiple inequivalent branches, oblique confinement and electric-magnetic triality.

New phenomena takes place in SO(N_c) SUSY gauge theories with matter in the vector representation of the gauge group. Unlike the SU(N_c) theories with the fundamental quarks, where there is no invariant distinction between the Higgs and the confinement regimes, in SO(N_c) these phases (as well as the oblique confinement phase) can be distinguished. Indeed, the Wilson loop in the spinor representation of SO(N_c) cannot be screened by the dynamical quarks and, therefore, presents a gauge invariant order parameter for confinement.

As was explained in Sec. 3, technically the most fruitful idea is considering N_f, the number of the quark flavors, as a free parameter. Note, that in the orthogonal groups, we do not need subflavors — one chiral superfield in the *vector* representation presents one flavor. It is useful to note that the adjoint representation in the orthogonal groups has dimension $N_c(N_c-1)/2$ and $T(G) = N_c - 2$. The vector representation has dimension N_c and $T(\text{vect}) = 1$.

The group SO(3) is exceptional: the vector representation is the same as adjoint, and $T(G) = T(\text{vect}) = 2$.

We will consider below SO(N_c) theories with $N_f = N_c - 4$, $N_c - 3$ and $N_c - 2$ which exhibit inequivalent phase branches, massless glueball/exotic states at the origin of the moduli space and oblique confinement [106] (see also Ref. [107]). A brief digression in SO(3) theories will provide us with the first example of the electric-magnetic triality.

All results obtained for the orthogonal groups can be readily adapted for the symplectic groups by formally extrapolating the parameter N in SO(N) to negative values [108], so that the dynamical behavior of the $Sp(N)$ theories [109] merely parallels that of SO(N).

4.1. *Two Branches of $N_f = N_c - 4$ Theory*

Assume that $N_c > 4$ and $N_f = N_c - 4$. The structure of the vacuum valleys is very similar to that discussed in detail in Secs. 2 and 3 for the unitary groups. The moduli matrix is

$$M^{ij} = Q^i Q^j \,, \tag{4.1}$$

where the summation over the color index (running from 1 to N_c) is implicit. A generic point from the vacuum valley $\langle M^{ij}\rangle \neq 0$ corresponds to the spontaneous breaking of the gauge symmetry SO(N_c) down to SO(4). Since SO(4) = SU(2) × SU(2) (we will mark one of these subgroups by subscript L and another by R), at low energies, below the masses of those gauge bosons that become heavy W's, we have two SU(2) theories of gluons/gluinos coupled to massless axion/dilaton fields. These two low-energy theories are in the strong coupling regime; the corresponding scale parameters $\Lambda_{L,R}$ are related to the fundamental parameter Λ of the high energy theory as

$$\Lambda^6_{L,R} = \frac{\Lambda^{2N_c-2}}{\det M} \,, \tag{4.2}$$

where I omitted an irrelevant numerical constant. Equation (4.2) is most easily verified by matching the running α from the NSVZ β functions of the original SO(N_c) theory and the low-energy SO(4) in the limiting case when M^{ij} is proportional to the unit matrix.

The gluino condensate can develop independently in the $\text{SU}(2)_L$ and $\text{SU}(2)_R$ theories,

$$\langle\lambda\lambda\rangle_{L,R} = \pm\Lambda^3_{L,R} \,. \tag{4.3}$$

The existence of the gluino condensate implies the following superpotential

$$\mathcal{W}(M) = \langle\lambda\lambda\rangle_L + \langle\lambda\lambda\rangle_R = (\varepsilon_L + \varepsilon_R)\left(\frac{\Lambda^{2N_c-2}}{\det M}\right)^{1/2}, \qquad (4.4)$$

where $\varepsilon_{L,R}$ are the phase factors, $\varepsilon_{L,R} = \pm 1$, corresponding to the two possible signs of the gluino condensates. Here I used Eq. (4.2) for the low-energy scale parameter. The relation between the gluino condensate in the low-energy theory and the emerging superpotential is perfectly the same as that observed in the SU(N_c) theories long ago [41].

The crucial novel element is the occurrence of physically distinct solutions: (i) $\varepsilon_L = \varepsilon_R = +1$; (ii) $\varepsilon_L = \varepsilon_R = -1$; (iii) $\varepsilon_L = +1$, $\varepsilon_R = -1$; and (iv) $\varepsilon_L = -1$, $\varepsilon_R = +1$. The solutions (i) and (ii) are related by a discrete symmetry of the model and are equivalent. By the same token, (iii) and (iv) are equivalent. It is quite evident, however, that the first pair leads to a familiar picture, with a superpotential destroying the vacuum valley, while the second pair of solutions corresponds to no superpotential — the classical valley persists as the quantum moduli space. In other words, one and the same fundamental theory has two branches, two drastically different realizations.

Let us take a closer look at the second branch. The most interesting point is the origin, $M = 0$. At this point the full SO(N_c) gauge symmetry is unbroken. Classically all $N_c(N_c-1)/2$ gauge bosons are massless — the theory is singular at the origin. The singularity might manifest itself in the form of the kinetic terms of the fields M^{ij}. In the quantum theory, due to confinement, this singularity must be smoothed out (much in the same way as it happens in the SU(N_c) theory with $N_f = N_c$). The only apparent reason why the singularity might still survive is the emergence of some additional massless bound states.

The conclusion that there are no extra massless particles at $\langle M\rangle = 0$, beyond those residing in the moduli fields M^{ij}, is based, as usual, on the 't Hooft matching condition. In the absence of condensates the global (unbroken) symmetry of the model[9] is $\mathrm{SU}(N_f) \times \mathrm{U}(1)_R$. One can check that the anomalous *AVV* triangles calculated at the fundamental level are exactly saturated by the massless fermions from M^{ij} [106]. Thus, following the standard logic, we believe that the only massless particles are represented by the moduli fields, and their kinetic terms are nonsingular. If so, we encounter here

[9] Additional discrete symmetries exist. They are irrelevant in this consideration and will be briefly discussed later.

another example of confinement without the spontaneous breaking of the chiral symmetry.

4.2. $N_f = N_c - 3$ (Massless Glueballs and Exotic States)

Adding one more flavor, i.e. considering $N_f = N_c - 3$, allows one to break the gauge symmetry down to SO(3). It is convenient to consider first a special point from the vacuum valley where the expectation values of $N_c - 4$ flavors are large, and the VEV of the last flavor is small. Then the pattern of the breaking of gauge symmetry is of two stages. At the first stage the symmetry is broken down to $\mathrm{SU}(2)_L \times \mathrm{SU}(2)_R$, as in Sec. 4.1; then it is further broken down to a diagonally embedded SO(3). Omitting the details of the derivation, let me quote the superpotential generated due to the gaugino condensation in the low-energy theory [106],

$$\mathcal{W}(M) = (1+\varepsilon)\,\frac{\Lambda^{2N_c-3}}{\det M}\,, \qquad \varepsilon = \pm 1\,. \tag{4.5}$$

Again, as in Sec. 4.1, we have two physically inequivalent phase branches (corresponding to $\varepsilon = 1$ and $\varepsilon = -1$), one with a dynamically generated superpotential and the other with a moduli space of the quantum vacua. The latter is of most interest. What can be said about the low-energy spectrum on this branch?

If a mass term is given to the last flavor the solution is required to smoothly pass into the $\varepsilon_L \varepsilon_R = -1$ branch of the $N_f = N_c - 4$ theory as the mass parameter becomes large. If the kinetic term of the fields M^{ij} is nonsingular, and there are no other massless states (other than those in M^{ij}), the smooth transition to this branch is impossible. Indeed, adding the tree level superpotential

$$\mathcal{W} = m M^{i_\ell i_\ell} \tag{4.6}$$

(i_ℓ is the last flavor) and integrating out $M^{i_\ell i_\ell}$ under the assumption that the kinetic term is nonsingular we end up with no supersymmetric solution [110].

A possible scenario ensuring the desired smooth transition is the emergence of additional massless fields at the origin. Since additional massless fields are absent at generic points from the vacuum valley, they must have a superpotential giving them a mass at $M \neq 0$. A simplest possibility is a massless N_f-plet q_i with the superpotential

$$\Delta\mathcal{W} = -M^{ij} q_i q_j\,. \tag{4.7}$$

Combining now Eqs. (4.6) and (4.7) and integrating out the heavy superfield $M^{i_\ell i_\ell}$ we do obtain two physically equivalent solutions with no superpotential. The two-fold ambiguity is due to the fact that $\langle q_{i_\ell} \rangle = \pm\sqrt{m}$. This is precisely the $\varepsilon_L \varepsilon_R = -1$ branch of the $N_f = N_c - 4$ theory.

From the superpotential (4.7) we infer that the $\mathrm{U}(1)_R$ charge of the superfield q_i is $1/N_f$. It is not difficult to check that with this particular massless sector (M^{ij} plus q_i) and this charge assignment all anomalous AVV triangles corresponding to the conserved global symmetries $\mathrm{SU}(N_f) \times \mathrm{U}(1)_R$ are matched [106].

One can ask whether it is possible to build, from the fundamental fields of the theory, an interpolating local gauge invariant product with the external quantum numbers coinciding with those of q_i. The answer to this question is positive,

$$q_i \sim (Q)^{N_c - 4} W^2 \,. \tag{4.8}$$

In other words, we are free to interpret the massless states q_i as exotic quark-gluon bound states. The emergence of these states is remarkable by itself. At $N_c = 4$ we deal with massless glueballs (cf. Ref. [34]).

4.3. $N_f = N_c - 2$ *(Massless Monopoles/Dyons; Oblique Confinement)*

Theories with $N_f = N_c - 2$ turn out to have much in common with the extended-SUSY $N = 2$ model whose solution was found by Seiberg and Witten [16] (see also Refs. [49, 111]). In particular, massless monopoles and dyons appear at certain points on the moduli space. Their condensation causes confinement (oblique confinement). To be able to understand the corresponding dynamics, and the details of relevant derivations, in full, one must master the results of Ref. [16]. Certainly, we cannot submerge in this topic now, nor do I see any reasons why we should undertake this endeavor. The $N = 2$ model, specific features of the Seiberg-Witten solution and the specific tools used to reveal them, is a very vast topic extensively covered in numerous dedicated reviews. The reader is referred to the list of recommended literature at the end. We will settle here for a descriptive discussion of the $\mathrm{SO}(N_c)$ model at hand which, hopefully, gives an undistorted general idea of the basic ingredients.

We start from the following obvious observation: the $\mathrm{SO}(N_c)$ model with $N_f = N_c - 2$ flavors has a moduli space parametrized by the matrix of the expectation values $\langle M^{ij} \rangle$; a generic point from the vacuum valley corresponds to the spontaneous breaking of the original SO(N_c) gauge symmetry down to SO(2). Since SO(2) is the same as U(1) the theory has a *massless photon.* Hence, the theory is in the Coulomb phase. The sector of massless states consists of all moduli fields, the massless photon and all its superpartners. In other words, at low energies we deal with QED; the moduli fields are electrically neutral. Those states that are electrically charged (e.g. the gauge bosons from SO(N_c)/SO(2)) are generically massive.

The value of the QED coupling constant depends on where exactly we are on the vacuum valley. Denote the inverse (complexified) coupling constant by T,

$$T \equiv \frac{1}{g^2} + i\,\frac{\vartheta}{8\pi^2}\,. \tag{4.9}$$

The low-energy electromagnetic coupling constant is a holomorphic function of the moduli. Because of the SU(N_f) flavor symmetry of the model it can actually depend only on a specific combination of the moduli, namely,

$$U \equiv \det M\,, \qquad M^{ij} = Q^i Q^j\,. \tag{4.10}$$

If $\det M \gg \Lambda^{2N_c-4}$ the relation between T and the high-energy coupling constant of the original SO(N_c) theory T_0 is given by a familiar one-loop formula,

$$T = T_0 - \frac{1}{8\pi^2}\,\ln\,\frac{M_0^{2N_c-4}}{U}\,. \tag{4.11}$$

This expression is perturbatively exact. It does receive nonperturbative corrections, however, which modify the U dependence of T at small U.

To see how this happens we, first, observe that the matter field R charge vanishes with respect to the anomaly-free U(1)$_R$ symmetry. (The reader is invited to check this statement, as well as all other numerical factors mentioned in this section). In other words, the R charge of M^{ij} is zero. This implies, in particular, that no superpotential can be generated along the valley — the reason is the same as in the SU(N_c) model with $N_f = N_c$. By the same reason the F terms of the form

$$\left[W^2\left(\frac{\Lambda^{2N_c-4}}{U}\right)^k\right]_F,$$

where k is an integer, can be generated by instantons; $k = 1$ corresponds to the one-instanton correction, $k = 2$ to two-instanton and so on. These F terms are equivalent to the instanton corrections in T of the type

$$\left(\frac{\Lambda^{2N_c-4}}{U}\right)^k ; \tag{4.12}$$

and, as a matter of fact, they do appear. Instead of relying on the R-charge arguments one could just count the number of the fermion zero modes in the instanton background, with the same conclusion.

Summation of all instanton terms (4.12), one by one, would be a brute force approach to calculating $T(U)$. Nobody has attempted to follow this road, of course. A roundabout way based on subtle considerations of analytical and other general properties [106] leads to an exact formula for $T(U)$ in terms of elliptic integrals. To write the formula we first introduce the curve

$$y^2 = x^3 + x^2(-U + 8\Lambda^{2N_c-4}) + 16\Lambda^{4N_c-8}x \,, \tag{4.13}$$

which obviously has branch points at $x = 0, \infty$ and

$$x_\pm = \frac{1}{2}\left[U - 8\Lambda^{2N_c-4} \pm \sqrt{U(U - 16\Lambda^{2N_c-4})}\right]. \tag{4.14}$$

The exact dependence $T(U)$ is given by the ratio

$$T(U) = \frac{1}{16\pi i}\frac{\int_{x_-}^{x_+} dx/y(x)}{\int_0^{x_-} dx/y(x)} . \tag{4.15}$$

The branches of the square roots are defined in such a way that for positive (real) a the square root $\sqrt{a}$ is positive. Then at large U Eq. (4.15) reproduces the perturbative logarithmic U dependence (4.11). It allows us to go further, however, and examine the behavior of the complexified coupling constant in the entire complex plane. From Eq. (4.14) it is perfectly clear that at U tending to zero and to $16\Lambda^{2N_c-4}$ the denominator of Eq. (4.15) develops a logarithmic singularity since $x_+ \to x_-$. The singularities in the effective gauge coupling constant can appear only if there are massless particles in the physical spectrum which carry electric and/or magnetic charges and are coupled to our photon. We, thus, conclude that in two points on the moduli space, $U = 0$ and $16\Lambda^{2N_c-4}$ the sector of the massless states is extended: apart from the photon and the moduli fields it includes some massless electrically/magnetically charged particles.

More exactly, we define two submanifolds, $\mathcal{M}_1$ and $\mathcal{M}_2$, of the vacuum manifold (both are noncompact)

$$\langle M^{ij}\rangle = M_*^{ij} \,, \quad \det M_* = 0 \text{ on } \mathcal{M}_1 \text{ or } \det M_* = 16\Lambda^{2N_c-4} \text{ on } \mathcal{M}_2 \,. \tag{4.16}$$

Examining the character of the singularity in T allows one to say which particular massless particles contribute. In this way it was found [106] that on the second manifold $\mathcal{M}_2$ dyons are massless. They have both electric and magnetic charges ± 1. Since outside $\mathcal{M}_2$ they are massive, a superpotential of a special form (vanishing on $\mathcal{M}_2$) must be generated. If the massless dyons are denoted by $E^{\pm}$, the superpotential must obviously have the form

$$\mathcal{W} = (U - 16\Lambda^{2N_c-4})E^+E^-\left[1 + \mathcal{O}\left(\frac{U - 16\Lambda^{2N_c-4}}{\Lambda^{2N_c-4}}\right)\right]. \tag{4.17}$$

As soon as we leave $\mathcal{M}_2$ and pass to $\mathcal{M}_1$, the dyons E become massive. Instead, other particles lose their masses on $\mathcal{M}_1$, so that their contribution to T ensures a proper singularity. Equation (4.7), which takes place in the $N_f = N_c - 3$ theory, gives us a hint that the number of distinct species of massless monopoles on $\mathcal{M}_1$ is larger than one and is related to N_f. The hint is based on the anticipation of a smooth transition from $N_f = N_c - 2$ to the case of $N_f = N_c - 3$, see below.

Let us introduce $2N_f$ chiral superfields $q_i^{\pm}$, $i = 1, 2, \ldots, N_f$. The superfields q_i^+ describe monopoles with the magnetic charge $+1$, q_i^- monopoles with the magnetic charge -1, with a superpotential term roughly speaking of the form

$$\mathcal{W} = M^{ij} q_i^+ q_j^- \,. \tag{4.18}$$

The moduli field M^{ij} is supposed to belong to $\mathcal{M}_1$. If the rank of this matrix is r, where r is obviously less than N_f, then Eq. (4.18) corresponds to $N_f - r$ massless supermultiplets. The electric and magnetic charges of q, E and the original "quarks" Q are related. Namely, the relation is such that one can think of E as of a bound state of Qq,

$$E^{\pm} \sim q_i^{\pm} Q^i \,.$$

At the origin of the vacuum valley, when $M^{ij} = 0$, all N_f species $q_i^{\pm}$ are massless. At this point, the full global symmetry of the model, $\mathrm{SU}(N_f) \times \mathrm{U}(1)_R$, is presumably unbroken. As usual, we check the self-consistency of this hypothesis by matching the 't Hooft triangles. To this end we calculate four triangles, $\mathrm{U}(1)_R$, $\mathrm{U}(1)_R^3$, $\mathrm{SU}(N_f)^3$ and $\mathrm{SU}(N_f)^2\mathrm{U}(1)_R$, at the fundamental level (λ and ψ_Q) and at the composite level. The massless fermions at the composite level are those residing in M^{ij}, $q_i^{\pm}$ and the photon supermultiplet. As was expected, all four AVV triangles do match [106]!

The last question to be addressed in this section is the issue of transition from the $N_f = N_c - 2$ theory to $N_f = N_c - 3$ upon adding the mass term to one of the original quarks. The tree level superpotential is the same as in Eq. (4.6). It is not difficult to see that with the mass term added the whole vacuum valley shrinks to the submanifolds $\mathcal{M}_1$ and $\mathcal{M}_2$. In other words, at $m \neq 0$ (see Eq. (4.6)) the zero vacuum energy is achieved on two solutions: $M^{ij} \in \mathcal{M}_1$ and $M^{ij} \in \mathcal{M}_2$.

The second solution corresponds to the condensation of dyons, $\langle E^+E^-\rangle \sim m$. The nonvanishing expectation value of the dyon field entails confinement of the electric charges, i.e. the original quarks of the theory are confined, in accordance with the arguments presented in Sec. 1. As a matter of fact, since the condensed objects are dyons, we deal here with the *oblique* confinement, see Sec. 1.4. Simultaneously, the former massless photon becomes massive. The would-be moduli M^{ij}, where $i, j = 1, 2, \ldots, N_f - 1$, survive as light fields with the superpotential (4.5) with positive ε. Thus, this solution smoothly passes into the $\varepsilon = 1$ branch of the $N_f = N_c - 3$ theory.

For the first solution, $M^{ij} \in \mathcal{M}_1$, adding the mass term (4.6) for the quarks of the last flavor entails the condensation of monopoles, $\langle q^+_{N_f} q^-_{N_f}\rangle \neq 0$. The fact that they do condense immediately follows from the minimization of the superpotential

$$M^{ij} q_i^+ q_j^- + m M^{N_f N_f} \,.$$

The first term in this superpotential, Eq. (4.18), is actually somewhat simplified; I have omitted details irrelevant for qualitative conclusions but important in quantitative analysis. Since the monopoles condense, all fields carrying the electric charges — in particular, the original quarks — are confined. The massless unconfined fields are the moduli M^{ij} with $i, j = 1, 2, \ldots, N_f - 1$, and the chiral superfields

$$q_i \sim (q_i^+ q^-_{N_f} - q_i^- q^+_{N_f}) \,, \qquad i = 1, 2, \ldots, N_f - 1 \,. \tag{4.19}$$

A superpotential is generated coupling the above moduli to $q_i q_j$, cf. Eq. (4.7). Thus, the solution $M^{ij} \in \mathcal{M}_1$ does indeed pass into the $\varepsilon = -1$ branch of the $N_f = N_c - 3$ theory upon adding the mass term to one of the quarks. Comparing the results described above with the dynamical portrait of the $N_f = N_c - 3$ theory we conclude that $N_c - 3$ monopoles (4.19) that remain massless can actually be interpreted as massless exotics or glueballs in terms of the original degrees of freedom of the fundamental Lagrangian, cf. Eq. (4.8).

Most of the phenomena discussed in this section were originally discovered in another model, the extended $N = 2$ SUSY theory with the SU(2) gauge group [16]. In some of the examples [16], where the matter hypermultiplets were present, there are monopoles and dyons too, whose condensation leads to confinement and oblique confinement of the quarks belonging to the fundamental representation of SU(2). We already know, however, that in the presence of the scalar fields in the fundamental representation there is no invariant distinction between the Higgs phase and the phase of confinement. One can only speak of the strong coupling *versus* weak coupling regimes, which are smoothly connected (Sec. 1). The SO(N_c) example considered here presents a picture of three physically distinct inequivalent phases: Higgs, confinement and oblique confinement.

4.4. *Electric-Magnetic-Dyonic Triality*

Traveling further along the N_f axis, towards higher values of N_f, we find ourselves in a situation where the original SO(N_c) theory is dual in the infrared domain to another theory, with the gauge group SO($N_f - N_c + 4$), the same number of the dual quarks as in the original theory, and an additional gauge singlet field M^{ij}. Leaving aside certain nuances, we find a full parallel to Seiberg's duality for the unitary groups, see Secs. 3.3 and 3.4. In particular, the interval $\frac{3}{2}(N_c - 2) < N_f < 3(N_c - 2)$ is the conformal window of the SO(N_c) theories, where both the electric and magnetic theories flow to the same nontrivial infrared fixed point (superconformal, or non-Abelian Coulomb phase). To the left of the conformal window, at $N_c - 2 < N_f \leq \frac{3}{2}(N_c - 2)$, the magnetic degrees of freedom are free in the infrared (the magnetic Landau phase). They can be regarded as superstrong bound states of the original (electric) gluons and matter, just in the same way as the corresponding domain in the SU(N_c) theories.

Instead of continuing this excursion in the direction of higher N_f for the generic values of N_c presenting a dynamical picture which is already pretty familiar, we turn to an exceptional case of the gauge group SO(3), where a new phenomenon occurs. We will limit ourselves to a specific choice of the matter sector — two triplet "quark" fields, Q_a^1 and Q_a^2 where $a = 1, 2, 3$. This particular problem is nicely reviewed in Refs. [107, 112]. Many other instructive examples are considered in Refs. [106, 113].

The remarkable new phenomenon just mentioned is the occurrence of *two* theories dual to the original one. If the original theory is "electric", one of the dual theories is "magnetic" and the other is "dyonic". Thus, instead of duality, one can speak of *triality.* Moreover, a symmetry which is explicit in the original electric theory is realized in the magnetic and dyonic theories as a *quantum symmetry.* Such (quantum) symmetries are not easily visible in the Lagrangian because they are implemented by the nonlocal transformations of the fields.

Let us take a closer look at the model we will be dealing with. First, at the classical level the vacuum valley is obviously parametrized by the three gauge invariant moduli M^{ij} (note that $M^{12} = M^{21}$). The expectation value of the first triplet Q^1 breaks the gauge symmetry SO(3) down to SO(2), which is then further broken by the expectation value of the second triplet Q^2. Thus generically, if $\det M^{ij} \neq 0$, the gauge group is completely broken and the theory is in the Higgs phase.

If the values of all moduli are large, all gauge bosons are very heavy, and we are in the quasiclassical (weak coupling) regime. Nonperturbative effects can come only from instantons. Do instantons generate a superpotential which lifts the vacuum degeneracy?

The theory at hand has a global flavor symmetry SU(2) associated with the presence of two quark triplets. Moreover, it has a conserved R current. The first fact implies that, if there is a superpotential, it can depend only on $\det M$. Moreover, the $\mathrm{U}(1)_R$ charges of the matter fields are such that $\det M$ could enter the superpotential only linearly. It is clear then that no superpotential is generated by the instantons. The same conclusion follows from a straightforward analysis of the zero modes in the instanton background. Thus, the classical vacuum valley remains intact upon the inclusion of quantum effects.

As long as we stay away from the origin of the three-dimensional complex manifold of the moduli (i.e. $\det M \neq 0$), we are in the Higgs phase. At long distances from the origin the theory is weakly coupled. Needless to say that the emerging dynamical portrait is quite boring.

As usual, surprises await us near the origin of the moduli space where the strong coupling regime is attainable. At $\det M = 0$ we expect to find more varied dynamics. Let us see whether our expectations come true.

At first we consider a part of the vacuum valley where $\det M = 0$ but $M \neq 0$. This part obviously corresponds to the spontaneous breaking of the gauge SO(3) down to SO(2) = U(1). The massless-state sector consists of one photon, a pair of electrically charged fields plus some neutral fields. In other words, in this "corner" we deal with massless SQED. According to Landau, the electric charge is totally screened at large distances, so we find ourselves in the free electric phase. This dynamical regime has been discussed previously; therefore, our finding is not too exciting.

If all moduli are small, $\langle M^{ij}\rangle \to 0$, the SO(3) gauge group is unbroken and the strong coupling regime sets in. It is natural that the most nontrivial situation can and does take place here [106].

The scale of strong interactions is determined by Λ. The first question would be what fields are light in this scale, if at all. In the theory under consideration we do know that the moduli fields M are light. If we approach from the large M side, from the Higgs phase, the moduli fields have no superpotential. However, this does not mean, that no superpotential is generated in the strong coupling regime in *other* phases of the theory. The situation reminds that in supersymmetric gluodynamics where the gluino condensate may or may not develop depending on the phase of the theory (see Sec. 2.6 and Ref. [34]). Another rather close parallel exists with the s-confining theories, where a superpotential for the light degrees of freedom is generated at the origin, while no superpotential is allowed in the Higgs weak coupling regime (Sec. 3.2). If the superpotential is generated by strong interactions then, from the R charge counting and the dimensional arguments, we conclude [106] that it must have the form

$$\mathcal{W} = \frac{\eta}{8\Lambda}\det M + \frac{1}{2}\,\mathrm{Tr}\, mM\,, \tag{4.20}$$

where the second (tree-level) mass term is added by hand, in anticipation of future applications. It presents a mass deformation

$$\Delta\mathcal{L}_m = \frac{1}{2} m_{ji} Q^i Q^j$$

of the original massless theory; m_{ji} is the quark mass matrix. So far we considered $m = 0$. The parameter η in Eq. (4.20) is a numerical constant. In the Higgs phase $\eta = 0$. As we will see shortly, there are two other solutions, $\eta = \pm 1$; the first solution corresponds to oblique confinement, the second to confinement.

The occurrence of the nontrivial solutions with $\eta = \pm 1$ can be detected in the following way. Assume that the mass matrix is diagonal, $\{m_{ij}\} = \mathrm{diag}(m_1, m_2)$, and

$$m_{1,2} \neq 0\,, \qquad m_1/m_2 \to 0\,. \tag{4.21}$$

Since the second quark is much heavier than the first one, we can integrate out Q^2, reducing the theory to that with one triplet quark Q^1 which has a tiny mass m_1. In the limit $m_1 \to 0$ one deals with the $N = 2$ model solved by Seiberg and Witten [16]. As widely known, this model possesses a single modular field M^{11} which is locked by the mass term m_1 at $\pm 4\tilde{\Lambda}^2$ where the monopoles/dyons become massless and condense. Here $\tilde{\Lambda}$ is the scale of the strong interactions in the low-energy SO(3) theory with one triplet quark. Using the NSVZ β function it is easy to obtain an exact relation between both scale parameters, $\tilde{\Lambda}^2 = m_2\Lambda$. The monopole/dyon condensate is proportional to m_1. Thus, in the limit (4.21) we expect $M^{11} = \pm 4m_2\Lambda$ and $M^{22} \to 0$. The positive value of M^{11} corresponds to the monopole condensation (confinement), the negative value to the dyon condensation (oblique confinement).

Now, we observe that exactly this pattern of the M^{ij} condensates emerges from Eq. (4.20). Namely, minimizing the superpotential (4.20) one gets [106]

$$\langle M^{ij} \rangle = -4\eta\Lambda \det m (m^{-1})^{ij} . \tag{4.22}$$

Of special interest is the point $m_2 \to 0$ (along with $m_1/m_2 \to 0$). In this limit, on the one hand, all moduli vanish and, on the other hand, the monopoles and dyons become massless simultaneously. These objects are mutually nonlocal. Their presence at the point $\langle M^{ij} \rangle = 0$ is most naturally interpreted as the onset of the conformal behavior, cf. Sec. 3.3 and Refs. [49, 100].[10] In other words, at $\langle M^{ij} \rangle = 0$ the theory is, presumably, in the non-Abelian Coulomb phase.

The fact that the first term in the superpotential (4.20) is generated by strong interactions of the original "electric" quarks and gluons is established above on the basis of indirect arguments. A question which immediately comes to mind is whether one can get this superpotential in a more direct way. In a sense, the answer is positive.

As we already know, one of the most powerful tools from the magic toolkit of supersymmetry is Seiberg's duality. If we are able to identify a dual partner equivalent to the original "electric" theory in the infrared, which is simpler than the original model (for instance, the dual partner can be weakly coupled while the original theory is strongly coupled), then the problem is solved. Although the procedure for building dual models is not formalized, there is a standard strategy based on the 't Hooft matching and additional

[10]Another example of a similar dynamical scenario was found in the SU(N_c) theory with one adjoint matter field and several fundamentals and antifundamentals [114].

self-consistency checks. Implementing this strategy one finds [106] that the original theory has two duals in the case at hand! Thus, as a matter of fact, we deal with a triplet of theories. For the reasons which will become clear shortly the first dual theory will be called T_1 and the second T_{-1}.

Both dual theories dubbed $T_{\pm 1}$ have the gauge groups SO(3) and the matter sectors including two triplet quarks q_i, $(i = 1, 2)$. Additionally, they include three gauge singlets M^{ij} where $i, j = 1, 2$ and $M^{ij} = M^{ji}$. I stress that the field M^{ij} is to be treated as elementary in the theories $T_{\pm 1}$; it has mass dimension two. The gauge singlets M^{ij} interact with q_i through a superpotential, much in the same way as in the "magnetic" theory of Sec. 3.3. The theories T_1 and T_{-1} differ from one another by the form of the superpotential,

$$\mathcal{W}_\varepsilon = \frac{1}{12\Lambda} M^{ij}(q_i q_j) + \varepsilon \left(\frac{1}{24\Lambda} \det M + \frac{1}{24\Lambda} \det\{q_i q_j\} \right) \tag{4.23}$$

where $\varepsilon = 1$ for T_1 ("magnetic" dual) and $\varepsilon = -1$ for T_{-1} ("dyonic" dual). For simplicity the scale parameters of the original theory and its duals are set equal. This assumption is inessential and can be readily lifted. The superpotential (4.23) is nonrenormalizable. This is unimportant too, since the theories $T_{\pm 1}$ are effective low-energy theories anyway. They are suitable only for describing large distance behavior.

The theories $T_{\pm 1}$ have the D-flat directions, which are just the same as in the original "electric" model. They are parametrized by the three gauge invariant chiral products $N_{ij} = q_i q_j$. The composite fields N_{ij} are would-be moduli in the dual theories. Interaction with the elementary fields M^{ij} makes them massive. The coefficients in Eq. (4.23) are fixed by duality. One can also check that the relative weight of the various terms in Eq. (4.23) is correct by flowing down from other theories [106].

The simplest phase of the original "electric" theory is the Higgs phase. In this phase superpotential is not generated, $\eta = 0$. Our goal is exploring the confinement/oblique confinement phases of the "electric" theory. To this end we start with the simplest (Higgs) phase of the theories $T_{\pm 1}$. In this phase interaction of the dual quarks with the dual gluons generates no superpotential, so that the interaction of the would-be moduli is exhausted by Eq. (4.23). If we integrate out the massive composite fields N by using the equations of motion[11] the superpotential for the gauge singlet elementary field M obviously takes the form

---

[11] The equations of motion imply that $\mathrm{Tr}\, MN = -4\varepsilon \det M$ and $\det N = 4 \det M$.

$$\frac{-\varepsilon}{8\Lambda} \det M\,.$$

This is exactly the first term in Eq. (4.20), with $\eta = -\varepsilon$. Thus, the Higgs phase in the "magnetic" theory ($\varepsilon = 1$) corresponds to the confinement phase of the original "electric" theory. By the same token, the Higgs phase in the "dyonic" theory ($\varepsilon = -1$) corresponds to the oblique confinement phase of the "electric" theory. We see that the first term in the superpotential Eq. (4.20), which in the "electric" theory appears as a result of complicated strong-interaction dynamics, is present essentially at the tree level in the "magnetic" and "dyonic" theories.

Without going into further details let me note that the Higgs phase of the "electric" theory corresponds to oblique confinement in T_1 and to confinement in T_{-1}. Moreover, one can check that dualizing $T_{\pm 1}$ — each of these theories has two dual partners — one obtains permutations of the same three theories.

In summary, the phase structure of the SO(3) theory with $N_f = 2$ is so rich that it exhibits virtually all phases of the gauge theories considered so far. If the quark mass term in the original model is set to zero, in a generic point from the vacuum valley, $\det M \neq 0$, we find ourselves in the Higgs phase. On the noncompact two-dimensional subspace, $\det M = 0$, $M \neq 0$, there exists an unbroken U(1), with the massless photon. Since the electrically charged fields are massless, the theory is Landau-screened in the infrared. We deal here with the free electric phase. Finally, if $M = 0$ we are in the conformal (non-Abelian Coulomb) phase. Adding a small quark mass term we force the monopoles (dyons) to condense pushing the theory in the confinement or oblique confinement phase. The "electric" theory gives a weak coupling description of the Higgs branch of the theory, T_1 gives a weak coupling description of the confining branch of the original theory and T_{-1} gives a weak coupling description of the oblique confinement branch of the original theory.

4.5. *Quantum Symmetry*

The electric theory discussed above has a discrete symmetry Z_8. This symmetry is a remnant of the anomalous chiral rotations of the matter fields. (I remind that the Konishi current is anomalous). The fact that a discrete subgroup Z_8 survives becomes quite evident e.g. from the instantons. The number of the zero modes of the matter fermions is eight.

Although the Z_8 symmetry is obvious in the original "electric" Lagrangian, in the Lagrangian of the dual magnetic theory we only see Z_4. The full Z_8

is not visible at the Lagrangian level; it must (presumably) be realized as a nonlocal symmetry of the quantum states [106]. Such a situation is familiar in the string context where it is referred to as *quantum symmetries.* More sophisticated patterns of the quantum symmetries of the dual pairs are treated in Ref. [104].

The confining and the oblique confinement branches of the SO(3) theory discussed in Sec. 4.4 are related by a spontaneously broken global discrete symmetry. Therefore, the magnetic and the dyonic theories are similar. This particular example of the dyonic theories is not unique. More dyonic theories were revealed in Ref. [106]. In these more complicated examples there is no global symmetry which can make the theories similar. The electric, magnetic and dyonic theories are really distinct.

Section 5

Towards QCD

In spite of discouraging developments in recent years I still believe that high energy physics is an empirical science whose ultimate task is describing and understanding the laws of Nature. For those who share this opinion the major question in studying various dynamical scenarios in supersymmetric gauge theories is

What lessons can be drawn for nonsupersymmetric theories, first and foremost QCD?

From this standpoint one can view the previous sections of this lecture course as an extended introduction.

Although there are few quantitative achievements in the direction of actual QCD, if at all, some qualitative insights were obtained. One useful lesson has been already discussed: we have seen that the MAC approach does not necessarily lead to correct answers for the chiral symmetry breaking condensates developing in the strong coupling regime. This lesson is negative, however. We will consider below several issues where SUSY might provide with positive answers.

QCD is *the* theory of strong interactions in Nature, describing three light (but not massless) quarks coupled to the octet of gluons. The number of colors is three, not two or four, and the light quark masses are such as they are, not smaller or larger. However, from the early days of QCD it became clear that it is extremely advantageous to treat the theory in a more flexible way, as a "laboratory". The "laboratory" aspect is the second face of QCD. For, instance, setting the quark masses to zero allows one to develop the chiral perturbation theory. Considering the multicolor limit, $N_c \to \infty$, gives crucial insights in phenomenologically important problems (e.g. the Zweig rule, or factorization in the weak nonleptonic decays), and serves as a consistent basis for the Skyrme model of baryons and many other approaches.

The recent advances in supersymmetric gauge theories teach us the same story, confirming the old wisdom: whatever parameters we have in complicated and messy theories describing our world, it is worth trying to treat them as free parameters. Changing them may reveal novel features of the model, hidden in more conservative approaches, and provide important insights. The analysis of various gauge groups and matter sectors reveals a rich spectrum of different dynamical scenarios in SUSY gauge theories, some of which, in this or that form, may be relevant to QCD.

The conformal window was discussed in detail in Sec. 3.3. In QCD it exists too. The right edge of the window lies at $N_f = 16$ (if $N_c = 3$). Unlike SUSY QCD, however, we do not know exactly where the left edge lies. One could only dream of analytical methods like those presented in Sec. 3.3. From numerical studies in lattice QCD one may conjecture that the conformal window in QCD extends down to $N_f \approx 7$ [115].

Towards the left side of the conformal window the critical value α_{s*} becomes large, and the theory is strongly coupled.

At $N_f = 3$ experiment tells us that the chiral symmetry breaking takes place and the quarks are confined. Does the chiral symmetry breaking set in simultaneously with confinement?

Generally speaking, these two phenomena are distinct, and it is conceivable that they do not occur simultaneously. For many years it was believed that confinement of color in QCD implies the chiral symmetry breaking [116]. Are we sure if that is the case today? Supersymmetric examples teach us to be ready for surprises. Confinement without the chiral symmetry breaking is still an open possibility in the theory with, say, 5 or 6 massless quarks, although I hasten to add that no indications exist in favor of such a scenario at present. Moreover, the reverse — the chiral symmetry breaking without confinement — seems quite plausible. *A priori* it is not ruled out that the chiral symmetry breaking without confinement occurs at the edge of the conformal window.

Thus, we see that even the most global questions concerning the dynamical behavior in QCD remain unanswered. That is why new understanding of supersymmetric gauge dynamics gave rise to great expectations among QCD practitioners. The following strategy may prove to be fruitful. One starts from a supersymmetric theory where a solution of a particular dynamical aspect of interest is known. One then introduces explicit (soft) supersymmetry breaking by adding mass terms to the gluinos and/or squarks. When these mass terms are sent to infinity we find ourselves in a nonsupersymmetric theory, with decoupled gluinos/squarks. Ideally, we would like to calculate in this limit. Unfortunately, this is still beyond our reach. What can be carried out, however, is the analysis of the theory with small enough SUSY breaking mass terms. One may hope that the trend revealed in this way continues to take place in the domain of larger gluino/squark masses. In this way we get a qualitative picture of what is to be expected in QCD and other nonsupersymmetric theories in the strong coupling regime. In some cases, on the contrary, one can predict phase transitions in the gluino/squark masses. Although such an outcome is less interesting than the possibility of solving issues in QCD, it still provides us with information about gauge dynamics which may turn out to be useful.

Below we will see how this strategy works in some simple examples.

5.1. ϑ Dependence and the Puzzle of "Wrong" Periodicity

Since the mid-seventies it is known [117] that there is a hidden parameter, the vacuum angle ϑ in QCD and pure Yang-Mills theory (no quarks). The vacuum wave function is of the Bloch type. Moreover, in the absence of strictly massless quarks the physical quantities depend on ϑ, and this dependence must be periodic, with the period 2π.

On the other hand, various Ward identities can be used in certain instances to show that the ϑ parameter enters through the ratio ϑ/N where N is some integer, typically, the number of colors or the number of light flavors. For, instance, the Witten-Veneziano formula reads [118]

$$m^2_{\eta'} f^2_{\eta'} = 36 \left. \frac{d^2 \varepsilon_{\rm nlq}}{d\vartheta^2} \right|_{\vartheta=0} , \tag{5.1}$$

where ε is the vacuum energy density, the subscript nlq indicates that it has to be "measured" in pure Yang-Mills theory (no light quarks), $m_{\eta'}$ and $f_{\eta'}$ are the η' meson mass and the coupling constant, and the limit $N_c \to \infty$ is implied. The second derivative on the right-hand side is nothing but the topological succeptability of the vacuum. As well-known [118]

$$m_{\eta'}^2 = \mathcal{O}\left(\frac{1}{N_c}\right), \qquad f_{\eta'}^2 = \mathcal{O}(N_c), \qquad \varepsilon_{\rm nlq} = \mathcal{O}(N_c^2), \tag{5.2}$$

which implies, in turn, that the left-hand side of Eq. (5.1) is $\mathcal{O}(N_c^0)$. To make the right-hand side of the correct order in N_c one is forced to assume that the vacuum energy density in pure Yang-Mills theory depends on ϑ/N_c rather than on ϑ.

A similar conclusion can be reached by analyzing the chiral Ward identities in the theory with light (but not exactly massless) quarks. If the number of the light quarks is N_f, the topological succeptability $d^2\varepsilon_{\rm lq}/d\vartheta^2$ can be demonstrated [119] to be proportional to $1/N_f$. The subscript lq marks the light quarks. Since on general grounds one expects $\varepsilon_{\rm lq}$ to be a linear function of N_f, we conclude that the vacuum energy density in the world with the light quarks switched on must depend on ϑ through ϑ/N_f. In QCD the number of colors and the number of light quarks is three. So, both the Witten-Veneziano argument and that of Crewther point to the expected ϑ dependence of physical quantities of the type $f(\vartheta/3)$, making one suspect that something is wrong with the 2π periodicity.

These observations gave rise to numerous speculations that the standard construction [117] of the vacuum wave function of the Bloch type based on instantons was wrong; significant effort was invested in searches for additional degenerate "pre-vacua" which should have been included in the construction of the "correct" Bloch wave function, but were actually overlooked. If these additional "pre-vacua" existed, a new superselection rule should have been imposed (a concise review of the topic can be found in Ref. [120]). The speculations were seemingly further supported by 't Hooft's torons — field configurations defined in a finite volume for pure Yang-Mills SU(N_c) theories and having fractional topological charges proportional to $1/N_c$ [121].

This situation — an apparently wrong ϑ dependence — is already familiar to us. As we saw in Sec. 2.3.1, the gluino condensate in SUSY gluodynamics depends on ϑ through the factor $\exp(i\vartheta/N_c)$. When ϑ continuously evolves from 0 to 2π we do *not* return to the same value of the gluino condensate. The way out is obvious. One has N_c degenerate and physically equivalent vacua, each of the Bloch type. In order to restore the correct 2π periodicity one has just to rename these vacua after $\vartheta \to \vartheta + 2\pi$. Thus, all physically observable quantities will be actually 2π-periodic, in spite of the fact that the gluino condensate in the *given* vacuum is not.

The set of N_c states intertwined under the ϑ evolution consists of distinct vacua, totally disconnected from each other in infinite volume, rather than merely "pre-vacua" to be included in the Bloch wave function. These distinct vacua reflect the spontaneous breaking of Z_{2N_c} down to Z_2. Thus, the controversy "distinct vacua *versus* superselection rule" is solved in favor of the former option [122]. A clear-cut signal confirming the distinct vacua scenario is the emergence of the domain walls interpolating between the spatial regions of the different vacua, with a finite energy density [122]. The existence of the domain walls rules out the superselection rule scenario.

In short, the problem is fully solved in SUSY gluodynamics, case closed. Now, what can be said about nonsupersymmetric Yang-Mills theory, with no fermion fields?

To get a hint we introduce the gluino mass term, m_g. As long as $m_g \ll \Lambda$, theoretical analysis is reliable — we merely perturb the theory slightly near its supersymmetric limit and calculate the effects due to this perturbation in the first order in m_g.

The most drastic impact of $m_g \neq 0$ is lifting the vacuum degeneracy. If the gauge group is SU(3) the gluino mass term explicitly breaks Z_6 down to Z_2. Correspondingly, we will now have three nondegenerate states: one with a negative energy density $\varepsilon = -m_g\Lambda_0^3$, and two degenerate states, with positive energy densities $\varepsilon = -m_g\Lambda_0^3\cos(2\pi/3) = m_g\Lambda_0^3/2$. The parameter m_g is chosen to be real and positive; then the gluino condensate in the ground state is negative and equal to ($\vartheta = 0$)

$$\langle\lambda\lambda\rangle \equiv -\Lambda_0^3/2\,,$$

where Λ_0 is a positive parameter of the dimension of mass. The sign of the gluino condensate can be established from a general analysis parallelizing that of Ref. [123]. The first state has the lowest-energy and is the genuine vacuum. Two other states are excited quasistable states, with the spontaneously broken CP. For very small m_g these false "vacua" are almost stable. The lifetime decreases, however, as m_g approaches Λ. The family of states entangled under the ϑ evolution now consists of three physically *inequivalent* states.

In the limit of small m_g the expectation value of the operator $G^2 + iG\tilde{G}$ (G is the gluon field strength tensor) is proportional to that of $m_g\lambda\lambda$. So, both can be used as appropriate order parameters for the states discussed above. In particular, the quasistable states are characterized by nonvanishing (and opposite) values of $G\tilde{G}$. When gluino becomes heavy and decouples, it is the gluon operator that survives as the order parameter.

Although at $m_g \neq 0$ the degeneracy of three states inherent to the supersymmetric limit is gone, the ϑ evolution still intertwines these three states together. They interchange their positions when ϑ varies from 0 to 2π. The one with the negative energy density, the true vacuum at $\vartheta = 0$, takes the place of one of the two false vacua, which formerly had positive energy density, and *vice versa.* The definition of the genuine vacuum (the lowest-energy state) has to be changed *en route*, at $\vartheta = \pi$. All physical quantities (which for every given ϑ must be defined in the genuine vacuum) are 2π-periodic, although by naively inspecting the gluon condensate at small ϑ we could easily have drawn a false conclusion of a false ϑ dependence through $\exp(i\vartheta/3)$.

Now, what happens when m_g becomes comparable to Λ, and, eventually, much larger than Λ, and the gluino decouples leaving us with pure Yang-Mills theory? There are two logical possibilities. The extra minima might just disappear. However, to match the Witten-Veneziano formula it is natural to assume that it is not what happens. Two false "vacua" may survive in QCD as quasistable states, separated by a barrier from the genuine vacuum. They still form a triplet of the states entangled with each other under the ϑ evolution. Certainly, there is no small parameter which might ensure a large lifetime; one can only hope for an interplay of numerical factors.

If the above picture is correct, the extra quasistable states may show up, as droplets of the false vacuum, in the nuclear-nuclear collisions, or in any environment where the hot quark-gluon plasma is formed and then cools down, much in the same way as the droplets of the disoriented chiral condensate [124].

The fact that N_c distinct vacua in supersymmetric gluodynamics form a family whose members interchange their positions each time ϑ reaches 2π, 4π and so on is known for over a decade [33]. If $m_g = 0$ the ϑ dependence of the gluino condensate is unobservable. As soon as m_g becomes nonvanishing the vacuum energy exhibits a ϑ dependence through the condensates G^2 and $m_g\lambda^2$. The ϑ dependence is now observable. Needless to say it remains observable in the limit $m_g \to \infty$, i.e. in (nonsupersymmetric) gluodynamics.

Inclusion of the matter fields clearly makes the corresponding analysis more complicated. There exists a limit, however, in which one can reliably deal [125] with supersymmetry breaking terms in the same vein we dealt with in SUSY gluodynamics. Introduce N_f flavors with a supersymmetric mass term, $m_{\rm SUSY} \neq 0$. The vacuum structure remains intact — we still have N_c degenerate vacua at zero, intertwined in one family. The ϑ dependence is unobservable since there exists a classically conserved anomalous current (built from the gluino and squark fields). Assume now that a SUSY breaking term is introduced as an F term of a spurion field μ,

$$\Delta\mathcal{L} = \mu Q\tilde{Q}|_F\,, \qquad F_\mu \neq 0\,, \tag{5.3}$$

so that

$$F_\mu \ll m_{\rm SUSY}\,.$$

It is pretty obvious that the SUSY breaking effects in the D terms will be proportional to $|F_\mu|^2$. The quark–squark mass splitting will be linear in F_μ and is fully controllable. The term (5.3) does not allow the squark field to be rotated. Correspondingly, the classically conserved anomalous current is gone, and gone with it is the ϑ independence of the physical quantities.

The issue of the ϑ dependence in appropriately deformed $N = 2$ theories is discussed in Ref. [126].

5.2. *Questions and Lessons in Supersymmetric QCD*

To get as close as possible to the real world we have to incorporate light quarks. The number of the light quarks is three, and so is the number of colors, so that the case $N_f = N_c$ is of most interest. SQCD with $N_f = N_c$ was thoroughly discussed in Sec. 3.2. If the mass term of the matter superfields is set to zero the vacuum valley of the theory is parametrized by the moduli M^i_j, B and $\tilde{B}$ subject to constraint (3.16). Any point from this manifold is a legitimate vacuum. If all moduli have large values, the gauge symmetry is completely broken, all gauge bosons are very heavy (in the scale Λ), and the theory is obviously in the weak coupling regime. Since the scalar fields (squarks) are in the fundamental representation we deal with a unified Higgs/confinement phase. In Sec. 3.2 we discussed what happens when one enters the part of the valley where the moduli become small and the strong coupling regime sets in. In this part of the valley two points were obviously singled out: (i) the one corresponding to the standard pattern of the chiral symmetry breaking, $B = \tilde{B} = 0$, $M^i_j = \Lambda^2\delta^i_j$; and (ii) the point with the unbroken chiral symmetry and broken baryon charge, $B = -\tilde{B} = \Lambda^{N_f}$, $M^i_j = 0$, see Eqs. (3.17) and (3.19). Here we will briefly consider a SUSY-breaking deformation of this model analyzed in Refs. [127, 128]. To obtain more concise expressions from now on Λ will be put to unity. All dimensionful quantities will be measured in the units of Λ.

Following Ref. [127] we will assume that the squarks and/or gluinos get a mass term, added in the Lagrangian by hand. These mass terms represent a soft supersymmetry breaking,

$$\Delta\mathcal{L} = -m_Q^2(|Q|^2 + |\tilde{Q}|^2) - (m_g\lambda^2 + \text{h.c.}) \tag{5.4}$$

where Q and $\tilde{Q}$ are the lowest components of the corresponding superfields. This particular expression is not very general. Other soft supersymmetry breaking terms are possible. The choice (5.4) is singled out by the fact that at $m_g = 0$ all global symmetries of the original supersymmetric model, see Eq. (3.13), are preserved. A nonvanishing gluino mass, $m_g \neq 0$, explicitly breaks the conservation of the R current, leaving all other global symmetries intact. Correspondingly, the model with $m_g \neq 0$ will be referred to as the $\not{R}$ model, while that with $m_g = 0$ as the R model. Tending $m_Q, m_g \to \infty$ we return back to the chiral limit of QCD. If m_g is kept at zero while $m_Q \to \infty$ the theory we arrive at is not QCD but, rather, the theory of quarks and gluons plus one Majorana fermion in the adjoint representation. It also exhibits the strong coupling behavior and is interesting on its own.

The deformation of SQCD caused by the gluino mass term $m_g \neq 0$ switches on supersymmetry breaking in a smooth way. The point $m_g = 0$ is nonsingular, so that one can study all correlation functions of interest at small m_g as an expansion in m_g. In this way, the vacuum energy density was calculated to order $\mathcal{O}(m_g)$ in the previous section.

Introduction of the squark mass term is more problematic. Indeed, for the negative values of m_Q^2 the theory becomes unstable and tends to develop large VEV's of the scalar fields [128] (see also Ref. [129]). Strictly speaking, the theory with negative m_Q^2 is not defined at all unless we add a quartic term for stabilization. Therefore, at $m_Q^2 = 0$ a phase transition must take place, and the expansion near $m_Q^2 = 0$ is potentially dangerous. We will return to this question later on.

For the time being let us assume that $m_Q^2 > 0$. In the presence of the squark mass ($m_Q^2 > 0$) the vacuum degeneracy is gone, the theory is locked in a concrete lowest-energy state which is, generally speaking, nondegenerate. If $m_Q^2 \ll \Lambda^2$ one may hope that the predictive power we had in the supersymmetric limit is not lost, and everything is calculable.

Saying that everything is calculable is an exaggeration. Although we do know what happens with the F terms under the deformation specified above, there is an ambiguity in the D terms of the relevant fields. These terms were unknown even in the limit of exact supersymmetry. Adding a soft supersymmetry breaking does not help to pin them down, of course. To be able to predict the impact of the squark/gluino masses one has to accept certain assumptions about the kinetic terms. Since we are aiming at a qualitative picture anyway, it is natural to make a simplest possible choice still capturing general features that must be inherent to any sensible kinetic term. As long as our conclusions do not strongly depend on details of the kinetic terms at hand, they seem to be trustworthy.

At first we will focus on the R model. A few comments on the $\not{R}$ will be given at the end of this section.

In the supersymmetric limit the particles residing in $M, B, \tilde{B}$ are massless. When a mass term m_Q^2 is switched on some of the particles still remain massless, while others acquire a mass. If $m_Q^2 \ll \Lambda^2$ the acquired mass is small, and one can limit oneself to the linear approximation. To answer physically interesting questions we must build an effective Lagrangian for the light degrees of freedom, find the location of the vacuum state, and the spectrum.[12]

Intuitively it is clear that the squark mass pushes the vacuum towards small values of M, B and $\tilde{B}$. As a matter of fact, if it were not for the constraint (3.17) the lowest-energy state would be achieved at $M = B = \tilde{B} = 0$. The constraint (3.17) does not allow M, B and $\tilde{B}$ to vanish simultaneously. Upon a brief reflection, two points (3.17) and (3.19), which were singled out even in the absence of supersymmetry breaking, become prime suspects.

The effective low-energy Lagrangian for the light composite fields depends on a number of constants in the D terms and in the SUSY breaking part. The latter constants appear when one rewrites the mass term (5.4), introduced at the fundamental level, in terms of the composite fields M, B and $\tilde{B}$. The values of the constants remain undetermined. Depending on the relative weight of the M and B parts the minima at the points (3.17) and (3.19) "breathe". One of them lies higher than the other. Although both logically possible versions are analyzed in Ref. [127], there are good reasons to believe that the minimum at $B = \tilde{B} = 0$, $M^i_j = \Lambda^2 \delta^i_j$ must be deeper, while the minimum at $B = -\tilde{B} = 1$, $M^i_j = 0$ is irrelevant (i.e. it is local and quantum-mechanically unstable).

Why? Our starting point was supersymmetric QCD with the *massless* matter superfields. We could have started, instead, from the massive SQCD, with a small supersymmetric mass term for all matter superfields. Then, this

[12]Eventually we will proceed to the second stage — what happens when m_Q and m_g become large. It is needless to say that this question can be addressed only at the qualitative level.

new theory could have been deformed by adding the SUSY-breaking term (5.4) implying heavier squarks. Such an approach is even more realistic, since the light quarks in QCD, although light, are not exactly massless. However, if we start, from the *massive* SQCD, there is no vacuum degeneracy from the very beginning; the only solution for the vacuum is given by Eq. (3.17).

I remind that this vacuum solution, $B = \tilde{B} = 0$, $M^i_j = \Lambda^2\delta^i_j$, corresponds to the conventional dynamical pattern with the spontaneously broken axial chiral symmetry and unbroken baryon charge. This is just the pattern we deal with in QCD. The "wrong" vacuum (3.19) corresponds to the unbroken chiral symmetry and spontaneously broken baryon charge.

In the R model the $\mathrm{U}(1)_R$ symmetry is not broken explicitly (since $m_g = 0$); the condensate $\langle M^i_j\rangle = \delta^i_j$ does not break it either. The success of the 't Hooft matching in SQCD (Sec. 3.2) implies that there is no spontaneous breaking of $\mathrm{U}(1)_R$ in the limit of *exact* supersymmetry. When the SUSY-breaking mass term m_Q^2 is introduced, a phase transition may occur, leading to a spontaneous breaking of $\mathrm{U}(1)_R$. As we already know, the point $m_Q^2 = 0$ is singular anyway. What we do not know is whether the $\mathrm{U}(1)_R$ breaking phase transition occurs right at $m_Q^2 = 0$, or it takes place at a finite value of m_Q^2. The former option, disregarded in Ref. [127], is somewhat exotic but seemingly cannot be ruled out on general grounds.

To begin with, we may assume, following [127], that the R symmetry is not spontaneously broken in some finite range of m_Q^2,

$$0 < m_Q^2 \leq (m_Q^2)_* \,.$$

Then the light particle spectrum at m_Q^2 from this interval is almost evident. All fermions, ψ_M, ψ_B and $\psi_{\tilde{B}}$, must remain massless. As for their boson counterparts, some of them ($N_f^2 - 1$ pseudoscalar bosons) have to remain massless since they are the Goldstone bosons of the spontaneously broken global $\mathrm{SU}(N_f)$ symmetry. All others get masses generated by the SUSY-breaking term (5.4). The constraint (3.17) leads to a subtlety which deserves mentioning. The matrix M has N_f^2 complex elements. It can be conveniently parametrized as

$$M = \exp\left(\frac{\pi_V + i\pi_A}{\sqrt{2N_f}}\right)\exp(\pi_V^a t^a)\exp(i\pi_A^a t^a) \tag{5.5}$$

where t^a are the generators of $\mathrm{SU}(N_f)$, all π's are real, and $a = 1, 2, \ldots, N_f^2 - 1$. The fields B and $\tilde{B}$ provide two extra complex parameters, so that altogether we deal with $N_f^2 + 2$ complex fields. The constraint (3.17) eliminates one complex field, namely,

$$\pi_V + i\pi_A = \sqrt{(2/N_f)}B\tilde{B} \tag{5.6}$$

plus cubic and higher-order terms. The fermionic partner of $\pi_V + i\pi_A$ is also removed from the theory. Since the order parameter of the spontaneously broken $\mathrm{SU}(N_f)_L \times \mathrm{SU}(N_f)_R \to \mathrm{SU}(N_f)_V$ is $\langle M_j^i \rangle = \delta_j^i$, it is easy to see that the corresponding Goldstone bosons are π_A^a. The masses of the remaining $N_f^2 + 3$ (real) bosons are proportional to m_Q.

How do we learn that the composite fermions are massless even though $m_Q^2 \neq 0$? The global (unbroken) symmetry of the model at hand is $\mathrm{SU}(N_f)_V \times \mathrm{U}(1)_B \times \mathrm{U}(1)_R$. Since the R current has the axial component, one has to match the 't Hooft triangles. In the limit of the exact supersymmetry they do match (Sec. 3.2) — the quark and gluino contribution at the fundamental level is matched by that of the massless composite fermions ψ_M, ψ_B and $\psi_{\tilde{B}}$. To ensure that the matching persists at $m_Q^2 \neq 0$ (with the unbroken R symmetry), we are forced to conclude that the masses of ψ_M, ψ_B and $\psi_{\tilde{B}}$ remain at zero even though the squarks have a mechanical mass.

The masslessness of ψ_M is surprising, to put it mildly. Indeed, ψ_M is a composite state built of ψ_Q and $\tilde{Q}$. The mass term of $\tilde{Q}$ is a free parameter, and it is hard to imagine that the strong interaction fine-tunes itself in such a way that the mass of the composite state does not depend on this free parameter at all.

A possible way out of this hurdle was indicated in Ref. [127]. It was noted that the squarks Q and $\tilde{Q}$ have exactly the same quantum numbers (color, flavor and Lorentz) as the gluino–quark composites,

$$Q \to \lambda^\alpha (\psi_Q)_\alpha\,, \qquad \tilde{Q} \to \lambda^\alpha (\psi_{\tilde{Q}})_\alpha\,, \tag{5.7}$$

where α is the Lorentz index, and the color and flavor indices are suppressed. Therefore, one may think of ψ_M as of a state built from ψ_Q and $(\lambda\psi_{\tilde{Q}})$. Since neither ψ's nor λ's have mechanical masses in the R model, the masslessness of ψ_M seems less counterintuitive. An extremist evolution of this idea leads to a hypothetical scenario [127], according to which there is no phase transition in m_Q^2 at all, and massless composites $(\psi_Q\lambda\psi_{\tilde{Q}})$ survive in the limit $m_Q^2 \to \infty$, when the theory at hand flows to QCD supplemented by one extra (Majorana) spinor in the adjoint representation.

The possibility which seems preferable to me is the R-breaking phase transition at $m_Q^2 = 0$. If we postulate that at any positive nonvanishing value of m_Q^2 an R violating condensate, say, $\langle\lambda\lambda\rangle$ develops, there will be no need in saturating the 't Hooft triangles by massless fermions. Then they will be saturated by a Goldstone meson, the "ninth" Goldstone, built from a mixture of $\bar{\lambda}\lambda$ and $\bar{\psi}\psi$. The theory emerging in this way is very similar to the conventional QCD.

Thus, we see that the deformation of SQCD by the squark mass term does not take us too far. Too many unknowns remain, precluding us from making definite conclusions about the dynamical behavior of the nonsupersymmetric theory even for small values of m_Q^2. The issue is clearly not ripe enough, further effort is needed before insights in QCD dynamics can be obtained.

As far as the $\not R$ model is concerned, here the route to QCD is much smoother. The R symmetry is explicitly broken by $m_g \neq 0$ from the very beginning. At small m_g the composite fermions can be shown to have masses proportional to m_g. If we take the limit $m_g \to \infty$ first, the gluino field decouples while the mass of the composite fermions presumably is frozen at Λ. $N_f^2 - 1$ bosons survive as massless Goldstones; the remaining $N_f^2 + 3$ are still light. Increasing now the value of m_Q^2 in the positive direction we decouple all squarks, and make $N_f^2 + 3$ light bosons heavy (i.e. of order Λ). The Goldstones that survive in this limit are familiar QCD pions.

Concluding, let me note that the potential of this approach is far from being exhausted. It seems that this range of ideas — SUSY-breaking deformations of supersymmetric models whose solution we know — can bring lavish fruits. Various supersymmetric models can be used as a starting point. In particular, one can start from $N = 2$ supersymmetry [130]. No matter where the starting point is, the destination is the same, QCD.

Will we reach this destination in the foreseeable future?

6. Conclusions

This lecture course summarizes advances in theoretical understanding of nonperturbative phenomena in the strong coupling regime. If before the SUSY era, the number of exact nonperturbative results in four-dimensional field theory could be counted on one hand, with the advent of supersymmetry a wide spectrum of problems relevant to the most intimate aspects of strong gauge dynamics have found exact solutions. Mysteries unravel. Our understanding of gauge theories is dramatically deeper now than it was a decade ago. When preparing these lectures, I intended to share with you, all the excitement and joys associated with the continuous advances in this field spanning over 15 years. Hopefully, the message I tried to convey will be appreciated in full.

Supersymmetric gauge dynamics is very rich, but life is richer, still. The world surrounding us is not supersymmetric. It remains to be seen whether the remarkable discoveries and the elegant, powerful methods developed in supersymmetric gauge theories will prove to be helpful in solving the messy problems of real-life particle physics. So far, not much has been done in this direction. In today's climate it is rare that the question of practical applications is even posed. I hope that we have reached a turning point: high-energy theory will return to its empiric roots. The command we obtained of supersymmetric gauge theories will be a key which will open to us a Pandora's box of problems of Quantum Chromodynamics, *the* theory of our world. Pandora opened the jar that contained all human blessings, and they were gone. Will the achievements obtained in supersymmetric gauge theories be lost in the Planckean nebula?

Acknowledgments

I would like to thank B. Chibisov, I. Kogan, A. Kovner, G. Veneziano and especially A. Vainshtein for numerous discussions and helpful comments. I am grateful to S. Ferrara, F. Fucito, K. Intriligator, V. Miransky, M. Sato and P. West, for pointing out misprints and omissions in the first version, and to B. Chibisov for his kind assistance in the preparation of the manuscript. Parts of these lecture notes were written during my stay at CERN Theory Division in fall 1996 and at Isaac Newton Institute for Mathematical Sciences, Cambridge University, in March 1997. I am grateful to colleagues from the above groups for hospitality.

This work was supported in part by DOE under the grant number DE-FG02-94ER40823.

7. Appendix: Notation, Conventions, Useful Formulae

In this Appendix the key elements of the formalism used in supersymmetric gauge theories are outlined. Basic formulae are collected for convenience.

The notation I follow is close to that of the canonic textbook of Bagger and Wess [44]. There are some distinctions, though. The most important of them is the choice of the metric. Unlike Bagger and Wess, I use the standard metric $g_{\mu\nu} = (+ - --)$. There are also distinctions in normalization, see Eq. (A.17).

The left-handed spinor is denoted by undotted indices, e.g. η_β. The right-handed spinor is denoted by dotted indices, e.g. $\bar\xi^{\dot\beta}$. (This convention is standard in supersymmetry but is opposite to the one accepted in the textbook [131].) The Dirac spinor Ψ then takes the form

$$\Psi = \begin{pmatrix} \bar\xi^{\dot\beta} \\ \eta_\beta \end{pmatrix} . \tag{A.1}$$

Lowering and raising of the spinorial indices is done by applying the Levi-Civita tensor from the left,

$$\chi^\alpha = \varepsilon^{\alpha\beta}\chi_\beta\,, \qquad \chi_\alpha = \varepsilon_{\alpha\beta}\chi^\beta\,, \tag{A.2}$$

and the same for the dotted indices, where

$$\varepsilon^{\alpha\beta} = -\varepsilon^{\beta\alpha}\,, \qquad \varepsilon^{12} = -\varepsilon_{12} = 1\,. \tag{A.3}$$

The products of the undotted and dotted spinors are defined as follows:

$$\eta\chi = \eta^\alpha\chi_\alpha = -\eta_\alpha\chi^\alpha\,, \qquad \bar\eta\bar\chi = \bar\eta_{\dot\alpha}\bar\chi^{\dot\alpha}\,. \tag{A.4}$$

Under this convention $(\eta\chi)^+ = \bar\chi\bar\eta$. Moreover,

$$\theta^\alpha\theta^\beta = -\frac{1}{2}\varepsilon^{\alpha\beta}\theta^2\,, \qquad \theta_\alpha\theta_\beta = \frac{1}{2}\varepsilon_{\alpha\beta}\theta^2\,,$$

$$\bar\theta^{\dot\alpha}\bar\theta^{\dot\beta} = \frac{1}{2}\varepsilon^{\dot\alpha\dot\beta}\bar\theta^2\,, \qquad \bar\theta_{\dot\alpha}\bar\theta_{\dot\beta} = -\frac{1}{2}\varepsilon_{\dot\alpha\dot\beta}\bar\theta^2\,. \tag{A.5}$$

The vector quantities (representation $(\frac{1}{2}\,,\frac{1}{2})$) are obtained in the spinorial formalism by multiplication by

$$(\sigma^\mu)_{\alpha\dot\beta} = \{1, \boldsymbol{\tau}\}_{\alpha\dot\beta} \tag{A.6}$$

where $\boldsymbol{\tau}$ stands for the Pauli matrices, for instance,

$$A_{\alpha\dot\beta} = A_\mu(\sigma^\mu)_{\alpha\dot\beta}\,. \tag{A.7}$$

Note that

$$A_\mu B^\mu = \frac{1}{2}A_{\alpha\dot\beta}B^{\alpha\dot\beta}\,, \qquad A_{\alpha\dot\beta}A^{\gamma\dot\beta} = \delta_\alpha^\gamma A_\mu A^\mu\,. \tag{A.8}$$

The square of the four-vector is understood as

$$A^2 \equiv A_\mu A^\mu = \frac{1}{2}A_{\alpha\dot\beta}A^{\alpha\dot\beta}\,. \tag{A.9}$$

If the matrix $(\sigma^\mu)_{\alpha\dot\beta}$ is "right-handed" it isconvenient to introduce its "left-handed" counterpart,

$$(\bar\sigma^\mu)^{\dot\beta\alpha} = \{1, -\boldsymbol{\tau}\}_{\dot\beta\alpha} \,. \tag{A.10}$$

The matrices that appear in dealing with the representations (1, 0) and (0, 1) are

$$(\boldsymbol{\sigma})^\alpha_\beta = \boldsymbol{\tau}_{\alpha\beta}\,, \qquad (\boldsymbol{\sigma})^{\alpha\beta} = \varepsilon^{\beta\delta}\boldsymbol{\sigma}^\alpha_\delta\,, \tag{A.11}$$

and the same for the dotted indices. The matrices $(\boldsymbol{\sigma})^{\alpha\beta}$ are symmetric, $(\boldsymbol{\sigma})^{\alpha\beta} = (\boldsymbol{\sigma})^{\beta\alpha}$. In the explicit form

$$(\boldsymbol{\sigma})^{\alpha\beta} = \{\tau^3, i\mathbf{1}, -\tau^1\}_{\alpha\beta} \,. \tag{A.12}$$

Note that with our definitions

$$(\boldsymbol{\sigma})_{\alpha\beta} = \{-\tau^3, i\mathbf{1}, \tau^1\}_{\alpha\beta} \,.$$

The left (right) coordinates $x_{L,R}$ and covariant derivatives are

$$\begin{aligned} (x_L)_{\alpha\dot\alpha} &= x_{\alpha\dot\alpha} - 2i\theta_\alpha\bar\theta_{\dot\alpha}\,, \qquad & (x_R)_{\alpha\dot\alpha} &= x_{\alpha\dot\alpha} + 2i\theta_\alpha\bar\theta_{\dot\alpha}\,, \\ D_\alpha &= \frac{\partial}{\partial\theta^\alpha} - i\partial_{\alpha\dot\alpha}\bar\theta^{\dot\alpha}\,, \qquad & \bar D_{\dot\alpha} &= -\frac{\partial}{\partial\bar\theta^{\dot\alpha}} + i\theta^\alpha\partial_{\alpha\dot\alpha}\,, \end{aligned} \tag{A.13}$$

so that

$$\begin{aligned} &\{D_\alpha \bar D_{\dot\alpha}\} = 2i\partial_{\alpha\dot\alpha}\,, \\ &\bar D_{\dot\beta}(x_L)_{\alpha\dot\alpha} = 0\,, \qquad D_\beta(x_R)_{\alpha\dot\alpha} = 0\,, \\ &\bar D_{\dot\beta}(x_R)_{\alpha\dot\alpha} = -4i\theta_\alpha\varepsilon_{\dot\beta\dot\alpha}\,, \qquad D_\beta(x_L)_{\alpha\dot\alpha} = 4i\bar\theta_{\dot\alpha}\varepsilon_{\beta\alpha} \,. \end{aligned} \tag{A.14}$$

The law of the supertranslation is

$$\begin{aligned} \theta \to \theta + \varepsilon\,, &\qquad \bar\theta \to \bar\theta + \bar\varepsilon\,, \\ x_{\alpha\dot\beta} &\to x_{\alpha\dot\beta} - 2i\theta_\alpha\bar\varepsilon_{\dot\beta} + 2i\varepsilon_\alpha\bar\theta_{\dot\beta}\,, \\ (x_L)_{\alpha\dot\beta} &\to (x_L)_{\alpha\dot\beta} - 4i\theta_\alpha\bar\varepsilon_{\dot\beta}\,, \\ (x_R)_{\alpha\dot\beta} &\to (x_R)_{\alpha\dot\beta} + 4i\varepsilon_\alpha\bar\theta_{\dot\beta} \,. \end{aligned} \tag{A.15}$$

The integrals over the Grassmann variable are normalized as follows

$$\int d^2\theta\theta^2 = 2\,, \qquad \int d^4\theta\theta^2\bar\theta^2 = 4\,, \tag{A.16}$$

and we define

$$\{\cdots\}_F = \frac{1}{2}\int d^2\theta\{\cdots\}, \qquad \{\cdots\}_D = \frac{1}{4}\int d^4\theta\{\cdots\}. \tag{A.17}$$

A generic non-Abelian SUSY gauge theory has the Lagrangian

$$\mathcal{L} = \left\{\frac{1}{4g_0^2}\,\mathrm{Tr}\int d^2\theta W^2 + \text{H.c.}\right\} + \frac{1}{4}\sum_i \int d^4\theta \bar{Q}_i e^V Q_i + \left\{\frac{1}{2}\int d^2\theta \mathcal{W}(\{Q_i\}) + \text{H.c.}\right\}, \tag{A.18}$$

where

$$\frac{1}{g_0^2} = \frac{1}{g^2} - \frac{i\vartheta}{8\pi^2},$$

is the (complexified) gauge coupling constant, the sum in Eq. (A.18) runs over all matter superfields Q_i present in the theory, and $\mathcal{W}(\{Q_i\})$ is a generic superpotential. Most commonly one deals with the superpotential corresponding to the mass term of the matter fields. In many models, cubic terms are gauge invariant; then they are allowed too (and do not spoil renormalizability of the theory).

Furthermore, the superfield W_α, which includes the gluon strength tensor, is defined as follows:

$$W_\alpha = \frac{1}{8}\,\bar{D}^2(e^{-V}D_\alpha e^V) \tag{A.19}$$

where V is the vector superfield. In the Wess-Zumino gauge

$$V = -2\theta^\alpha\bar{\theta}^{\dot\alpha}A_{\alpha\dot\alpha} - 2i\bar{\theta}^2(\theta\lambda) + 2i\theta^2(\bar{\theta}\bar{\lambda}) + \theta^2\bar{\theta}^2 D\,, \tag{A.20}$$

$V = V^a T^a$ and T^a stands for the generators of the gauge group G. In the fundamental representation of SU(N), a case of most practical interest,

$$\mathrm{Tr}(T^aT^b) = \frac{1}{2}\delta^{ab}.$$

For a general representation R of any group G we define

$$\mathrm{Tr}(T^aT^b)_R = T(R)\delta^{ab}.$$

If R is the adjoint representation, $T\,(\text{adjoint}) \equiv T(G)$.

The supergauge transformation has the form

$$Q_i \to e^{i\Lambda} Q_i \,, \qquad e^V \to e^{i\bar{\Lambda}} e^V e^{-i\Lambda} \,, \qquad W_\alpha \to e^{i\Lambda} W_\alpha e^{-i\Lambda} \,, \tag{A.21}$$

where Λ is an arbitrary chiral superfield ($\bar{\Lambda}$ is antichiral). In components

$$W_\alpha = i(\lambda_\alpha + i\theta_\alpha D - \theta^\beta G_{\alpha\beta} - i\theta^2 \mathcal{D}_{\alpha\dot{\alpha}} \bar{\lambda}^{\dot{\alpha}}) \tag{A.22}$$

where λ_α is the gluino (Weyl) field, $\mathcal{D}_{\alpha\dot{\alpha}}$ is the covariant derivative, and $G_{\alpha\beta}$ is the gluon field strength tensor in the spinorial notation.

The standard gluon field strength tensor transforms as $(1,0)+(0,1)$ with respect to the Lorentz group. Projecting out pure (1, 0) is achieved by virtue of the $(\sigma)_{\alpha\dot{\beta}}$ matrices,

$$G_{\alpha\beta} = -\frac{1}{2} G_{\mu\nu} (\sigma^\mu)_{\alpha\dot{\beta}} (\sigma^\nu)_{\beta\dot{\delta}} \varepsilon^{\dot{\beta}\dot{\delta}} = (\mathbf{E} - i\mathbf{B})(\boldsymbol{\sigma})_{\alpha\beta} \,. \tag{A.23}$$

Then

$$G^{\alpha\beta} G_{\alpha\beta} = 2(\mathbf{B}^2 - \mathbf{E}^2 + 2i\mathbf{E}\mathbf{B}) = G_{\mu\nu} G_{\mu\nu} - i G_{\mu\nu} \tilde{G}_{\mu\nu}$$

where

$$\tilde{G}_{\mu\nu} = \frac{1}{2} \varepsilon_{\mu\nu\alpha\beta} G^{\alpha\beta} \,, \qquad (\varepsilon_{0123} = -1) \,. \tag{A.24}$$

The supercurrent supermultiplet has the following general form

$$\begin{aligned} J_{\alpha\dot{\alpha}} = R^0_{\alpha\dot{\alpha}} &- \frac{1}{2} \left\{ i\theta^\beta \left(J_{\beta\alpha\dot{\alpha}} - \frac{2}{3} \varepsilon_{\beta\alpha} \varepsilon^{\gamma\delta} J_{\delta\gamma\dot{\alpha}} \right) + \text{H.c.} \right\} \\ &- \theta^\beta \bar{\theta}^{\dot{\beta}} \left(J_{\alpha\dot{\alpha}\beta\dot{\beta}} - \frac{1}{3} \varepsilon_{\alpha\beta} \varepsilon_{\dot{\alpha}\dot{\beta}} \varepsilon^{\gamma\delta} \varepsilon^{\dot{\gamma}\dot{\delta}} J_{\gamma\dot{\gamma}\delta\dot{\delta}} \right) + \ldots \end{aligned} \tag{A.25}$$

where $R^0_{\alpha\dot{\alpha}}$ is the R_0 current, $J_{\beta\alpha\dot{\alpha}}$ is the supercurrent, and $J_{\alpha\dot{\alpha}\beta\dot{\beta}}$ is related to the energy-momentum tensor, $\theta_{\mu\nu}$, in the following way

$$J_{\alpha\dot{\alpha}\beta\dot{\beta}} = -(\sigma^i)_{\alpha\beta} (\sigma^j)_{\dot{\alpha}\dot{\beta}} \{\theta^{ij} + \theta^{00} g^{ij} - \varepsilon^{ijk} \theta^{0k}\} + \frac{1}{2} \varepsilon_{\alpha\beta} \varepsilon_{\dot{\alpha}\dot{\beta}} \theta^\mu_\mu \,; \tag{A.26}$$

here $i,j,k = 1,2,3$, $g_{\mu\nu}$ is the metric tensor and matrices $(\sigma^i)_{\alpha\beta}$ are defined in Eq. (A.11).

The general anomaly relation (three "geometric" anomalies) is

$$\begin{aligned} \bar{D}^{\dot{\alpha}} J_{\alpha\dot{\alpha}} = \frac{1}{3} D_\alpha \Bigg\{ & \left[3\mathcal{W} - \sum_i Q_i \frac{\partial \mathcal{W}}{\partial Q_i} \right] \\ & - \left[\frac{3T(G) - \sum_i T(R_i)}{16\pi^2} \operatorname{Tr} W^2 + \frac{1}{8} \sum_i \gamma_i \bar{D}^2 (\bar{Q}_i e^V Q_i) \right] \Bigg\} \,, \end{aligned} \tag{A.27}$$

where γ_i are the anomalous dimensions of the matter fields Q_i. The general Konishi anomaly has the form

$$\frac{1}{8}\bar{D}^2(\bar{Q}_i^+ e^V Q_i) = \frac{1}{2}Q_i\frac{\partial \mathcal{W}}{\partial Q_i} + \frac{\sum_i T(R_i)}{16\pi^2}\,\mathrm{Tr}\,W^2\,. \tag{A.28}$$

In conclusion I present the full component expression for the simplest SU(2) model with one flavor (two subflavors), assuming that the superpotential in the case at hand reduces to the mass term of the quark (squark) fields. This model was discussed in detail in Secs. 2.3.2 and 2.3.5. If the index f denotes the subflavors, $f = 1, 2$,

$$\begin{aligned}\mathcal{L} = {} & \frac{1}{g^2}\left\{-\frac{1}{4}G^a_{\mu\nu}G^a_{\mu\nu} + \lambda^{\alpha,a}i\mathcal{D}_{\alpha\dot\alpha}\bar\lambda^{\dot\alpha,a} + \frac{1}{2}D^aD^a\right\}\\ & + \psi^{f\alpha}i\mathcal{D}_{\alpha\dot\alpha}\bar\psi^{f\dot\alpha} + (\mathcal{D}_\mu\phi^{+f})(\mathcal{D}_\mu\phi^f) + F^{+f}F^f\\ & + i\sqrt{2}(\phi_1^{\dagger}\lambda\psi_1 + \phi_1\bar\lambda\bar\psi_1 + \phi_2^+\lambda\psi_2 + \phi_2\bar\lambda\bar\psi_2)\\ & + \frac{1}{2}D^a(\phi_1^+T^a\phi_1 + \phi_2^+T^a\phi_2)\\ & + m(\phi_1F_2 + \phi_1^+F_2^+ + \phi_2F_1 + \phi_2^+F_1^+ + \psi_1\psi_2 + \bar\psi_1\bar\psi_2)\,.\end{aligned} \tag{A.29}$$

In this model the component expression for the supercurrent is

$$\begin{aligned}J_{\alpha\beta\dot\beta} = 2\Bigg\{ & \frac{1}{g^2}\left[iG^a_{\beta\alpha}\bar\lambda^a_{\dot\beta} - 3\varepsilon_{\beta\alpha}D^a\bar\lambda^a_{\dot\beta}\right] + \sqrt{2}[(\partial_{\alpha\dot\beta}\phi^+)\psi_\beta - i\varepsilon_{\beta\alpha}F\bar\psi_{\dot\beta}]\\ & - \frac{\sqrt{2}}{6}\left[\partial_{\alpha\dot\beta}(\psi_\beta\phi^+) + \partial_{\beta\dot\beta}(\psi_\alpha\phi^+) - 3\varepsilon_{\beta\alpha}\partial^\gamma_{\dot\beta}(\psi_\gamma\phi^+)\right]\Bigg\}\,.\end{aligned} \tag{A.30}$$

References

[1] A. Polyakov, *Nucl. Phys.* **120** (1977) 429.

[2] S. Mandelstam, *Phys. Reports* **23** (1976) 245.

[3] G. 't Hooft, in *1981 Cargése Summer School Lecture Notes on Fundamental Interactions, NATO Adv. Study Inst. Series B: Phys.*, Vol. 85, Ed. M. Lévy *et al.* (Plenum Press, New York, 1982) [Reprinted in G. 't Hooft, *Under the Spell of the Gauge Principle* (World Scientific, Singapore, 1994), p. 514]; *Nucl. Phys.* **B190** (1981) 455.

[4] Yu. Golfand and E. Likhtman, *Pis'ma ZhETF*, **13** (1971) 452 [*JETP Lett.* **13** (1971) 323]; see also in *Problems of Theoretical Physics*, I. E. Tamm Memorial Volume (Nauka, Moscow, 1972), p. 37; E. P. Likhtman, *Irreducible representations, extensions of the algebra of Poincaré generators by bispinor generators, Kratk. Soob. Fiz. — Short Comm. Phys. FIAN*, 1971, No. 5, p. 197.

[5] P. W. Higgs, *Phys. Lett.* **12** (1964) 132; **13** (1964) 508; *Phys. Rev.* **145** (1966) 1156; F. Englert and R. Brout, *Phys. Rev. Lett.* **13** (1964) 321; G. S. Guralnik, C. R. Hagen and T. W. B. Kibble, *Phys. Rev. Lett.* **13** (1964) 585.

[6] M. Mattis, *Phys. Reports* **214** (1992) 159; V. A. Rubakov and M. E. Shaposhnikov, *Usp. Fiz. Nauk* **166** (1996) 493 [*Phys. Uspekhi* **39** (1996) 461].

[7] H. Georgi and S. Glashow, *Phys. Rev. Lett.* **28** (1972) 1494.

[8] A. Polyakov, *JETP Lett.* **20** (1974) 194; G. 't Hooft, *Nucl. Phys.* **B79** (1974) 276.

[9] R. Rajaraman, *Solitons and Instantons* (North-Holland, Amsterdam, 1987), Chap. 3.

[10] E. Bogomol'nyi, *Sov. J. Nucl. Phys.* **24** (1976) 449.

[11] M. K. Prasad and C. H. Sommerfield, *Phys. Rev. Lett.* **35** (1976) 760.

[12] J. Harvey, in Proc. *1995 Summer School in High-Energy Physics and Cosmology*, Eds. E. Gava, A. Masiero, K. S. Narain, S. Randjbar-Daemi and Q. Shafi (World Scientific, Singapore, 1997) [hep-th/9603086]; L. Alvarez-Gaumé and S. F. Hassan, *Fortsch. Phys.* **45** (1997) 159.

[13] P. A. M. Dirac, *Proc. Roy. Soc.* **A133** (1931) 60. The classical source on the Dirac monopole is T. T. Wu and C. M. Yang, *Phys. Rev.* **D12** (1975) 3845.

[14] J. Schwinger, *Phys. Rev.* **144** (1966) 1087. This paper as well as all classical Dirac's papers on the monopole are reprinted in Russian in *The Dirac Monopole*, Eds. B. Bolotovskii and Yu. Usachev (Mir Publishers, Moscow, 1970).

[15] B. Julia and A. Zee, *Phys. Rev.* **D11** (1975) 2227.

[16] N. Seiberg and E. Witten, *Nucl. Phys.* **B426** (1994) 19; (E) **B430** (1994) 485; *Nucl. Phys.* **B431** (1994) 484.

[17] T. Banks and E. Rabinovici, *Nucl. Phys.* **B160** (1979) 349; E. Fradkin and S. Shenker, *Phys. Rev.* **D19** (1979) 3682.

[18] A. De Rujula, R. C. Giles and R. L. Jaffe, *Phys. Rev.* **D17** (1978) 285.

[19] J. Cardy and E. Rabinovici, *Nucl. Phys.* **B205** (1982) 1; J. Cardy, *Nucl. Phys.* **B205** (1982) 17.

[20] E. Witten, *Phys. Lett.* **B86** (1979) 283.

[21] P. Rossi, *Nucl. Phys.* **B149** (1979) 170.

[22] Yu. Simonov, *Yad. Fiz.* **42** (1985) 557 [*Sov. J. Nucl. Phys.* **42** (1985) 352].

[23] M. Chernodub and F. Gubarev, in *Nonperturbative Approaches to Quantum Chromodynamics*, Proc. Int. Workshop of the European Center for Theoretical Studies in Nuclear Physics and Related Areas, Trento, 1995, Ed. D. Diakonov (Gatchina, 1995), p. 217; H. Suganuma, K. Itakura, H. Toki and O. Miyamura, *ibid.*, p. 224; R. C. Brower, K. N. Orginos and C.-I. Tan, *Phys. Rev.* **D55** (1997) 6313.

[24] Yu. Simonov, in *Selected Topics in Nonperturbative QCD*, Proc. Inter. School of Physics, "Enrico Fermi", Course 80, Varenna, Italy, 1995, Eds. A. Di Giacomo and D. Diakonov (IOS Press, Oxford, 1996), p. 339 [hep-ph/9509403]; M. Polikarpov, *Nucl. Phys. Proc. Suppl.* **53** (1997) 134.

[25] L. Del Debbio, M. Faber, J. Greensite and S. Olejnik, *Nucl. Phys. Proc. Suppl.* **53** (1997) 141.

[26] A. Kronfeld, G. Schierholz and U.-J. Wiese, *Nucl. Phys.* **B293** (1987) 461; S. Hioki, S. Kitahara, S. Kiura, Y. Matsubara, S. Ohno and T. Suzuki, *Phys. Lett.* **B272** (1991) 326; S. Kitahara, Y. Matsubara and T. Suzuki, *Prog. Theoret. Phys.* **93** (1995) 1.

[27] H. Suganuma, S. Umisedo, S. Sasaki, H. Toki and O. Miyamura, *Aust. J. Phys.* **50** (1997) 233; A. Hart and M. Teper, *Nucl. Phys. Proc. Suppl.* **53** (1997) 497; T. Suzuki *et al.*, *Nucl. Phys. Proc. Suppl.* **53** (1997) 531.

[28] W. Pauli, *Pauli Lectures on Physics*, Vol. 6, *Selected Topics in Field Quantization* (MIT Press, Cambridge, 1973), p. 33.

[29] A. Salam and J. Strathdee, *Nucl. Phys.* **B76** (1974) 477; *Phys. Rev.* **D11** (1975) 1521.

[30] J. Wess and B. Zumino, *Nucl. Phys.* **B70** (1974) 39; *Phys. Lett.* **B49** (1974) 52.

[31] S. Ferrara and B. Zumino, *Nucl. Phys.* **B79** (1974) 413.

[32] V. Novikov, M. Shifman, A. Vainshtein and V. Zakharov, *Nucl. Phys.* **B229** (1983) 407.

[33] M. Shifman and A. Vainshtein, *Nucl. Phys.* **B296** (1988) 445.

[34] A. Kovner and M. Shifman, *Phys. Rev.* **D56** (1997) 2396.

[35] G. Veneziano and S. Yankielowicz, *Phys. Lett.* **B113** (1982) 321; T. R. Taylor, G. Veneziano and S. Yankielowicz, *Nucl. Phys.* **B218** (1983) 493.

[36] V. Novikov, M. Shifman, A. Vainshtein and V. Zakharov, *Nucl. Phys.* **B229** (1983) 381; *Phys. Lett.* **B166** (1986) 334.

[37] A. Salam and J. Strathdee, *Nucl. Phys.* **B87** (1975) 85; P. Fayet, *Nucl. Phys.* **B90** (1975) 104.

[38] M. Grisaru, in *Recent Developments in Gravitation* (Cargése Lectures, 1978), Eds. M. Levy and S. Deser (Plenum Press, New York, 1979), p. 577, and references therein.

[39] E. Witten, *Nucl. Phys.* **B202** (1982) 253.

[40] F. Buccella, J.-P. Derendinger, C. Savoy and S. Ferrara, *Phys. Lett.* **B115** (1982) 375, and in Proc. Europhys. Study Conf. *Unification of the Fundamental Particle Interactions. II*, Eds. J. Ellis and S. Ferrara (Plenum Press, New York, 1983), p. 349.

[41] I. Affleck, M. Dine and N. Seiberg, *Nucl. Phys.* **B241** (1984) 493.

[42] I. Affleck, M. Dine and N. Seiberg, *Nucl. Phys.* **B256** (1985) 557.

[43] M. Luty and W. Taylor, *Phys. Rev.* **D53** (1996) 3399.

[44] J. Bagger and J. Wess, *Supersymmetry and Supergravity*, 2nd Edition (Princeton University Press, Princeton, 1992).

[45] E. Witten, *Nucl. Phys.* **B403** (1993) 159.

[46] A. Vainshtein, V. Zakharov and M. Shifman, *Usp. Fiz. Nauk*, **146** (1985) 683 [*Sov. Phys. – Usp.* **28** (1985) 709]; T. ter Veldhuis, hep-th/9811132.
[47] I. Kogan, A. Morozov, M. Olshanetsky and M. Shifman, *Yad. Fiz.* **43** (1986) 1587 [*Sov. J. Nucl. Phys.* **43** (1986) 1022].
[48] N. Seiberg, *Phys. Rev.* **D49** (1994) 6857; *Nucl. Phys.* **B435** (1995) 129.
[49] K. Intriligator and N. Seiberg, *Nucl. Phys.* **B431** (1994) 551; K. Intriligator, R. Leigh and N. Seiberg, *Phys. Rev.* **D50** (1994) 1092.
[50] K. Intriligator, R. Leigh and M. Strassler, *Nucl. Phys.* **B456** (1995) 567.
[51] E. Poppitz and L. Randall, *Phys. Lett.* **B336** (1994) 402; J. Bagger, E. Poppitz and L. Randall, *Nucl. Phys.* **B426** (1994) 3.
[52] S. Giddings and J. M. Pierre, *Phys. Rev.* **D52** (1995) 6065.
[53] T. Gherghetta, C. Kolda and S. Martin, *Nucl. Phys.* **B468** (1996) 37.
[54] J. Wess and B. Zumino, *Phys. Lett.* **B49** (1974) 52; J. Iliopoulos and B. Zumino, *Nucl. Phys.* **B76** (1974) 310; P. West, *Nucl. Phys.* **B106** (1976) 219; M. Grisaru, M. Roček and W. Siegel, *Nucl. Phys.* **B159** (1979) 429.
[55] Y. Meurice and G. Veneziano, *Phys. Lett.* **B141** (1984) 69; I. Affleck, M. Dine and N. Seiberg, *Phys. Lett.* **B137** (1984) 187.
[56] D. Amati, K. Konishi, Y. Meurice, G. Rossi and G. Veneziano, *Phys. Rep.* **162** (1988) 557.
[57] I. Affleck, M. Dine and N. Seiberg, *Phys. Rev. Lett.* **52** (1984) 1677; Y. Meurice and G. Veneziano, *Phys. Lett.* **B141** (1984) 69.
[58] S. Ferrara and B. Zumino, *Nucl. Phys.* **B87** (1975) 207.
[59] M. Shifman and A. Vainshtein, *Nucl. Phys.* **B277** (1986) 456.
[60] I. Kogan, M. Shifman and A. Vainshtein, *Phys. Rev.* **D53** (1996) 4526.
[61] J. Wess and B. Zumino, *Phys. Lett.* **B49** (1974) 52.
[62] M. Shifman and A. Vainshtein, *Nucl. Phys.* **B359** (1991) 571.
[63] I. Jack, D. R. T. Jones and P. West, *Phys. Lett.* **B258** (1991) 382.
[64] E. Poppitz and L. Randall, *Phys. Lett.* **B389** (1996) 280.
[65] T. E. Clark, O. Piguet and K. Sibold, *Nucl. Phys.* **B159** (1979) 1; K. Konishi, *Phys. Lett.* **B135** (1984) 439; K. Konishi and K. Shizuya, *Nuov. Cim.* **A90** (1985) 111.
[66] V. Novikov, M. Shifman, A. Vainshtein and V. Zakharov, *Nucl. Phys.* **B260** (1985) 157.
[67] A. Vainshtein, V. Zakharov and M. Shifman, *Yad. Fiz.* **43** (1986) 1596 [*Sov. J. Nucl. Phys.* **43** (1986) 1028], Sec. 3; see also *Pis'ma ZhETF* **42** (1985) 182 [*JETP Lett.* **42** (1985) 224]; for a comprehensive discussion see M. Shifman, *Int. J. Mod. Phys.* **A11** (1996) 5761.
[68] I. Jack, D. R. T. Jones and C. G. North, *Nucl. Phys.* **B486** (1997) 479.
[69] N. Seiberg, *Phys. Lett.* **206B** (1988) 75.
[70] N. Seiberg, *Phys. Lett.* **B318** (1993) 469.
[71] I. Affleck, M. Dine and N. Seiberg, *Phys. Rev. Lett.* **51** (1983) 1026.
[72] I. Affleck, M. Dine and N. Seiberg, *Phys. Lett.* **B137** (1984) 187; *Phys. Rev. Lett.* **52** (1984) 493; *Phys. Lett.* **B140** (1984) 59.
[73] L. Dixon, V. Kaplunovsky and J. Louis, *Nucl. Phys.* **B355** (1991) 649.

[74] H. Li and K. Mahanthappa, *Phys. Lett.* **B319** (1993) 152; *Phys. Rev.* **D49** (1994) 5532.
[75] M. Shifman (Ed.), *Instantons in Gauge Theories* (World Scientific, Singapore, 1994), Chap. VII.
[76] B. Zumino, *Phys. Lett.* **B69** (1977) 369.
[77] A. D'Adda and P. Di Vecchia, *Phys. Lett.* **B73** (1978) 162.
[78] G. C. Rossi and G. Veneziano, *Phys. Lett.* **B138** (1984) 195.
[79] N. Dorey, V. Khoze and M. Mattis, *Phys. Rev.* **D54** (1996) 2921; *Phys. Rev.* **D54** (1996) 7832; *Phys. Lett.* **B388** (1996) 324; *Phys. Lett.* **B390** (1997) 205; H. Aoyama, T. Harano, M. Sato and S. Wada, *Phys. Lett.* **B388** (1996) 331; K. Ito and N. Sasakura, *Nucl. Phys.* **B484** (1997) 141; *Mod. Phys. Lett.* **A12** (1997) 205; F. Fucito and T. Travaglini, *Phys. Rev.* **D55** (1997) 1099; T. Harano and M. Sato, *Nucl. Phys.* **B484** (1997) 167; Y. Yoshida, hep-th/9610211 (unpublished); M. Slater, *Phys. Lett.* **B403** (1997) 57.
[80] A. Yung, *Nucl. Phys.* **B485** (1997) 38.
[81] S. Dimopoulos, *Nucl. Phys.* **B168** (1980) 69; M. Peskin, *Nucl. Phys.* **B175** (1980) 197.
[82] M. Vysotsky, I. Kogan and M. Shifman, *Yad. Fiz.* **42** (1985) 504 [*Sov. J. Nucl. Phys.* **42** (1985) 318].
[83] A. C. Davis, M. Dine and N. Seiberg, *Phys. Lett.* **B125** (1983) 487.
[84] G. 't Hooft, in *Recent Developments in Gauge Theories*, Eds. G. 't Hooft *et al.*, (Plenum Press, New York, 1980).
[85] A. Dolgov and V. Zakharov, *Nucl. Phys.* **B12** (1971) 68.
[86] M. Shifman, *Phys. Rep.* **209** (1991) 161.
[87] C. Vafa and E. Witten, *Nucl. Phys.* **B234** (1984) 173.
[88] S. Coleman and E. Witten, *Phys. Rev. Lett.* **45** (1980) 100.
[89] E. Poppitz and S. Trivedi, *Phys. Lett.* **B365** (1996) 125; P. Pouliot, *Phys. Lett.* **B367** (1996) 151.
[90] K. Intriligator and P. Pouliot, *Phys. Lett.* **B353** (1996) 471.
[91] P. Cho and P. Kraus, *Phys. Rev.* **D54** (1996) 7640.
[92] C. Csáki, W. Skiba and M. Schmaltz, *Nucl. Phys.* **B487** (1997) 128.
[93] K. Intriligator and N. Seiberg, *Nucl. Phys.* **B444** (1995) 125.
[94] P. Pouliot, *Phys. Lett.* **B359** (1995) 108; P. Pouliot and M. Strassler, *Phys. Lett.* **B370** (1996) 76; *Phys. Lett.* **B375** (1996) 175.
[95] I. Pesando, *Mod. Phys. Lett.* **A10** (1995) 1871; S. Giddings and J. Pierre, *Phys. Rev.* **D52** (1995) 6065.
[96] C. Csáki, M. Schmaltz and W. Skiba, *Phys. Rev. Lett.* **78** (1997) 799; *Phys. Rev.* **D55** (1997) 7840.
[97] For a concise review and relevant references see e.g. I. Hinchliffe, *Phys. Rev.* **D54** (1996) 77.
[98] A. Belavin and A. Migdal, *Pis'ma ZhETF* **19** (1974) 317 [*JETP Lett.* **19** (1974) 181]; *Scale Invariance and Bootstrap in the Non-Abelian Gauge Theories*, Landau Institute Preprint-74-0894, 1974 (unpublished).
[99] T. Banks and A. Zaks, *Nucl. Phys.* **B196** (1982) 189.

[100] N. Seiberg, *Nucl. Phys.* **B435** (1995) 129.

[101] D. Kutasov, A. Schwimmer and N. Seiberg, *Nucl. Phys.* **B459** (1996) 455.

[102] G. 't Hooft, *Phys. Rev.* **D14** (1976) 3432; (E) **D18** (1978) 2199.

[103] P. Pouliot, *Phys. Lett.* **B359** (1995) 108; D. Kutasov, *Phys. Lett.* **B351** (1995) 230; D. Kutasov and A. Schwimmer, *Phys. Lett.* **B354** (1995) 315; P. Pouliot and M. Strassler, *Phys. Lett.* **B370** (1996) 76; **B375** (1996) 175; K. Intriligator, *Nucl. Phys.* **B448** (1995) 187; K. Intriligator, R. Leigh and M. Strassler, *Nucl. Phys.* **B456** (1995) 567.

[104] J. Distler and A. Karch, *Fortsch. Phys.* **45** (1997) 517.

[105] J. Brodie and M. Strassler, hep-th/9611197 (unpublished); P. Cho, *Phys. Rev.* **D56** (1997) 5260; and references therein.

[106] K. Intriligator and N. Seiberg, *Nucl. Phys.* **B444** (1995) 125.

[107] K. Intriligator and N. Seiberg, in Proc. Conf. *Future Perspectives in String Theory (Strings '95)* Eds. I. Bars, P. Bouwknegt, J. Minahan, D. Nemeschansky, K. Pilch, H. Saleur and N. Warner (World Scientific, Singapore, 1996) [hep-th/9506084].

[108] N. Maru and S. Kitakado, *Mod. Phys. Lett.* **A12** (1997) 691.

[109] K. Intriligator and P. Pouliot, *Phys. Lett.* **B353** (1995) 471; K. Intriligator, *Nucl. Phys.* **B448** (1995) 187; R. Leigh and M. Strassler, *Phys. Lett.* **B356** (1995) 492.

[110] K. Intriligator, N. Seiberg and S. Shenker, *Phys. Lett.* **B342** (1995) 152.

[111] A. Klemm, W. Lerche, S. Theisen and S. Yankielowicz, *Phys. Lett.* **B344** (1995) 169; P. Argyres and A. Faraggi, *Phys. Rev. Lett.* **74** (1995) 3931.

[112] K. Intriligator and N. Seiberg, *Nucl. Phys. Proc. Suppl.* **45BC** (1996) 1 [hep-th/9509066].

[113] S. Elitzur, A. Forge, A. Giveon and E. Rabinovici, *Nucl. Phys. Proc. Suppl.* **49** (1996) 174.

[114] A. Kapustin, *Phys. Lett.* **B398** (1997) 104.

[115] Y. Iwasaki *et al.*, *Z. Phys.* **C71** (1996) 343.

[116] A. Casher, *Phys. Lett.* **B83** (1979) 395.

[117] V. Gribov, 1976, unpublished; R. Jackiw and C. Rebbi, *Phys. Rev. Lett.* **37** (1976) 172; C. Callan, R. Dashen and D. Gross, *Phys. Lett.* **B63** (1976) 172.

[118] E. Witten, *Nucl. Phys.* **B156** (1979) 269; G. Veneziano, *Nucl. Phys.* **B159** (1979) 213.

[119] R. Crewther, *Phys. Lett.* **B70** (1977) 349; *Phys. Lett.* **B93** (1980) 75. For a review see G. A. Christos, *Phys. Rept.* **116** (1984) 251.

[120] A. Smilga, *Phys. Rev.* **D54** (1996) 7757.

[121] G. 't Hooft, *Comm. Math. Phys.* **81** (1981) 267.

[122] G. Dvali and M. Shifman, *Phys. Lett.* **B396** (1997) 64; (E) **B407** (1997) 452.

[123] T. Banks and A. Casher, *Nucl. Phys.* **B169** (1980) 103.

[124] A. Anselm, *Phys. Lett.* **B217** (1989) 169; *Phys. Lett.* **B266** (1991) 482; J. D. Bjorken, *Int. J. Mod. Phys.* **A7** (1992) 4189; K. Rajagopal and F. Wilczek, *Nucl. Phys.* **B399** (1993) 395; J.-P. Blaizot and A. Krzywicki, *Phys. Rev.* **D46** (1992) 246.

[125] N. Evans, S. Hsu and M. Schwetz, *Phys. Lett.* **B404** (1997) 77; see also earlier works of these authors cited there.
[126] K. Konishi, *Phys. Lett.* **B392** (1997) 101; N. Evans, S. Hsu and M. Schwetz, *Nucl. Phys.* **B484** (1997) 124.
[127] O. Aharony, J. Sonnenschein, M. Peskin and S. Yankielowicz, *Phys. Rev.* **D52** (1995) 6157.
[128] E. D'Hoker, Y. Mimura and N. Sakai, *Phys. Rev.* **D54** (1996) 7724.
[129] E. D'Hoker, Y. Mimura and N. Sakai, hep-ph/9611458 (unpublished).
[130] L. Alvarez-Gaumé, J. Distler, C. Kounnas and M. Marino, *Int. J. Mod. Phys.* **A11** (1996) 4745; L. Alvarez-Gaumé and M. Marino, *Int. J. Mod. Phys.* **A12** (1997) 975; L. Alvarez-Gaumé, M. Marino and F. Zamora, *Int. J. Mod. Phys.* **A13** (1998) 403.
[131] V. B. Berestetskii, E. M. Lifshits and L. P. Pitaevskii, *Quantum Electrodynamics*, (Pergamon Press, New York, 1982).

Recommended Literature

It is assumed that the reader is familiar with the textbooks on supersymmetry:

J. Bagger and J. Wess, *Supersymmetry and Supergravity* (Princeton University Press, 1983).

P. West, *Introduction to Supersymmetry and Supergravity* (World Scientific, Singapore, 1986).

S. J. Gates, M. T. Grisaru, M. Roček and W. Siegel, *Superspace* (The Benjamin/Cummings, 1983).

D. Bailin and A. Love, *Supersymmetric Gauge Field Theory and String Theory* (IOP Publishing, Bristol, 1994).

A solid introduction to supersymmetric instanton calculus is given in *Instantons in Gauge Theories*, Ed. M. Shifman (World Scientific, Singapore, 1994), Chap. VII.

A brief survey of those aspects of supersymmetry which are most relevant to the recent developments can be found in J. Lykken, *Introduction to Supersymmetry*, hep-th/9612114.

Reviews on Exact Results in SUSY Gauge Theories and Related Issues

N. Seiberg, *The Power of Holomorphy — Exact Results in 4D SUSY Field Theories*, in Proc. *VI International Symposium on Particles, Strings, and Cosmology (PASCOS 94)*, Ed. K. C. Wali (World Scientific, Singapore, 1995) [hep-th/9408013].

K. Intriligator and N. Seiberg, *Lectures on Supersymmetric Gauge Theories and Electric — Magnetic Duality*, *Nucl. Phys. Proc. Suppl.* **45BC** (1996) 1 [hep-th/9509066].

K. Intriligator and N. Seiberg, *Phases of $N = 1$ Supersymmetric Gauge Theories and Electric — Magnetic Triality*, in Proc Conf. *Future Perspectives in String Theory (Strings '95)* Eds. I. Bars, P. Bouwknegt, J. Minahan, D. Nemeschansky, K. Pilch, H. Saleur and N. Warner (World Scientific, Singapore, 1996) [hep-th/9506084].

D. Olive, *Exact Electromagnetic Duality*, *Nucl. Phys. Proc. Suppl.* **45A** (1996) 88 [hep-th/9508089].

P. Di Vecchia, *Duality in Supersymmetric Gauge Theories*, *Surveys High Energ. Phys.* **10** (1997) 119 [hep-th/9608090].

L. Alvarez-Gaumé and S. F. Hassan, *Introduction to S-Duality in $N = 2$ Supersymmetric Gauge Theories*, *Fortsch. Phys.* **45** (1997) 159 [hep-th/9701069].

W. Lerche, *Notes on $N = 2$ Supersymmetric Yang-Mills Theory*, *Nucl. Phys. Proc. Suppl.* **55B** (1997) 83 [hep-th/9611190].

A. Bilal, *Duality in $N = 2$ SUSY Yang-Mills Theory: A Pedagogical Introduction to the Work of Seiberg and Witten*, hep-th/9601007.

S. Ketov, *Solitons, Monopoles, and Duality: From Sine-Gordon to Seiberg-Witten*, *Fortsch. Phys.* **45** (1997) 237 [hep-th/9611209].

M. Peskin, *Duality in Supersymmetric Yang-Mills Theory*, hep-th/9702094.

AUTHOR INDEX

SUBJECT INDEX

PHYSICS LIBRARY
Imperial College, London. SW7 2BZ
0171 594 7871

This book may be recalled after 2 weeks if required by another reader
2/00

28 MAY 2002